ADVANCES IN CHEMICAL PHYSICS

VOLUME XIV

ADVANCES IN CHEMICAL PHYSICS—VOLUME XIV

I. Prigogine and S. Rice—Editors

Correlation Effects in Atoms and Molecules

Edited by R. LEFEBVRE and C. MOSER

Centre de Mécanique Ondulatoire Appliquée, Paris, France

INTERSCIENCE PUBLISHERS

a division of John Wiley & Sons Ltd., London - New York - Sydney - Toronto

FIRST PUBLISHED 1969

LIBRARY OF CONGRESS CATALOG CARD NUMBER 58-9935

SBN 471 52143 4

MADE IN GREAT BRITAIN AT THE PITMAN PRESS, BATH

INTRODUCTION

Since the advent of wave mechanics, now more than forty years ago, the search for accurate representations of the many electron wave function for atoms and molecules has been actively pursued. One should, however, remember that the major activity in the application of wave mechanics to atomic and molecular physics has been the building up to models for the classification and understanding of the large body of experimental data. The powerful electronic computers now available make it possible to imagine that within a few years it will be feasible to calculate accurate Hartree–Fock wave functions even for a polyatomic system. The correlation effects will not for many years to come be amenable to treatment at the same level of accuracy, even considering foreseeable power of computing.

Recently several new approaches have been presented for the many (but not too many) body problems posed by the electrons of an atom or a molecule. They are a reflection, to some extent, of previous advances made in other fields of physics (electron gas, nuclei . . .). We felt it would be timely to organize an Advanced Summer Institute on Correlation Effects in Atoms and Molecules so that students in the field of atomic and molecular physics could be confronted with both the traditional and less traditional approaches. This volume contains most of the lectures given at the Institute which was held in Frascati, Italy, in July, 1967.

We hope that the reading of this book will encourage the young (and still young) to work in this field. The purpose in pursuing the correlation problem is not to put the experimentalist on the bread line. The interest is in the development of new physical concepts and new interpretations of experimental data. The yield in these directions of the past few years is, we feel, very encouraging.

We would like to take this opportunity to express our thanks to the Science Committee of NATO for financial support and also the French Atomic Energy Commission for making it possible to have participants from Eastern European countries. We would also like to express our warm gratitude to Professor Gozzer, Director of the European Center for Education, for allowing us to use the Villa Falconieri to hold the Institute. And finally we want to thank all participants in the Institute for what we believe was a very stimulating happening.

R. LEFEBVRE

C. M. MOSER

Paris, July, 1968

CONTRIBUTORS TO VOLUME XIV

K. A. BRUECKNER, Department of Physics, University of California at San Diego, La Jolla, U.S.A.

JIŘÍ ČÍŽEK, Institute of Physical Chemistry, Czechoslovak Academy of Sciences, Prague, Czechoslovakia

F. GRIMALDI, Commissariat a L'Energie Atomique, 29, Rue de la Fédération, Paris, France

A. P. JUCYS, Institute of Physics and Mathematics of the Academy of Sciences of Lithuanian S.S.R., V. Kapsukas State University Vilnius, Vilnius, Lithuanian S.S.R.

B. R. JUDD, The Johns Hopkins University, Baltimore, Maryland, U.S.A.

V. A. KAMINSKAS, Institute of Physics and Mathematics of the Academy of Sciences of Lithuanian S.S.R., V. Kapsukas State University Vilnius, Vilnius, Lithuanian S.S.R.

HUGH P. KELLY, Physics Department, University of Virginia, Charlottesville, U.S.A.

PER-OLOV LÖWDIN, Quantum Theory Project, University of Florida, Gainesville, Florida, U.S.A., and Department of Quantum Chemistry, University of Uppsala, Uppsala, Sweden

R. K. NESBET, IBM Research Laboratory, Monterey and Cottle Roads, San Jose, California 95114, U.S.A.

P. G. H. SANDARS, The Clarendon Laboratory, Oxford University, England

OKTAY SINANOĞLU, Sterling Chemistry Laboratory, Yale University, New Haven, Connecticut, U.S.A., and Teorik Kimga Bölümü Orta Doğu Teknik Üniversitesi, Ankara, Turkey

V. V. TOLMACHEV, Mathematics-Mechanics Faculty, University of Moscow, Moscow, U.S.S.R., and Theoretical Department, Institute of Electrochemistry, Academy of Sciences, Moscow, U.S.S.R.

CONTENTS

ATOMIC BETHE–GOLDSTONE EQUATIONS

R. K. NESBET, *IBM Research Laboratory, Monterey and Cottle Roads, San Jose, California 95114, U.S.A.*

CONTENTS

1. INTRODUCTION

In any discussion of the electronic structure of atoms and molecules it is important to recognize the central role of the Hartree–Fock approximation. As a computational method, the Hartree–Fock approximation is simply a variational formulation of the intuitive concept of an independent particle model. The Bohr model of the hydrogen atom can be extended to many-electron atoms only on condition that an independent particle model is qualitatively valid. The molecular-orbital theory, with its detailed success in describing the electronic excitation spectra of molecules, depends also on the general qualitative validity of an independent particle model.

In recent years, as a consequence of the evolution of computational technique, Hartree–Fock wave functions have been obtained for atoms and for some molecules. The results, particularly for the electric moments of diatomic molecules, provide a demonstration of the validity of the Hartree–Fock approximation as a description of the gross electronic density distribution in many-electron systems. This work on small molecules has recently been reviewed,[1] and more recent computations continue to show substantial agreement with experiment.

This success of the Hartree–Fock approximation has important implications about the utility of various methods proposed for improving upon it. With regard to properties of the electronic charge distribution, including electric and magnetic polarizabilities, what is required in general is a quantitative refinement of qualitatively useful results. This must put emphasis on systematic methods that are capable of refinement to an ultimately exact result. Many-particle perturbation theory is an example of such a method. The present article will describe an alternative method, with this convergent property, based on a hierarchy of Bethe–Goldstone equations of successively higher order.

Although the Hartree–Fock approximation gives a very accurate value of the total energy of a many-electron system, the total energy is in itself of little direct interest. Experimentally one measures a chemical dissociation energy, or a spectroscopic excitation energy. In either case the energy value of interest is the difference between two much larger quantities. Differences of correlation energy (by definition the difference between exact non-relativistic energy and the Hartree–Fock value) can make a substantial contribution to the dissociation energy of molecules. For example, the net contribution of correlation energy to the ground-state dissociation energy of the very stable molecule N_2 is approximately half of the total dissociation energy.[2]

It is clear from this that correlation energy, in contrast to correlation effects on the gross electronic charge distribution, can be of qualitative as well as quantitative significance in chemistry and spectroscopy.

In a Hartree–Fock wave function there is a one-to-one correspondence between particles and orbitals (one-particle wave functions). These terms are often used interchangeably in discussing the independent particle model. Thus, the fact that all occupied canonical orbitals in a Hartree–Fock determinantal wave function are eigenfunctions of the same integro-differential operator is equivalent to saying that all the electrons move in a common potential field. This is the self-consistent field of Hartree,[3] including the exchange operators introduced by Fock.[4]

Since each electron sees only the average potential field of all electrons, the Hartree–Fock approximation cannot describe the detailed dynamical correlation of particles interacting through the two-particle terms in the many-particle Hamiltonian. A strongly repulsive or attractive two-particle potential modifies the many-particle wave function near the zeroes of the relative coordinates r_{ij} in a way that cannot be described by the single determinant wave function of the Hartree–Fock approximation. The Coulomb interaction between electrons also gives rise to long-range correlation effects of considerable physical significance. These include the London dispersion force or van der Waals potential and the polarization potential that acts at large distances on an electron scattered by a neutral atom or molecule.

Section 2 of this article is a brief review of Hartree–Fock theory.

In Section 3 three proposed methods of improving the Hartree–Fock approximation will be discussed: the method of geminals (paired-electron wave functions); superposition of two-particle configurations; and the use of Bethe–Goldstone equations to describe electron pairs. Formal many-particle perturbation theory will not be used here. A detailed presentation of perturbation theory is given in an accompanying article by Kelly.[5] Reviews of correlation theory, applicable to atoms and molecules, have been given by Löwdin,[6] Sinanoğlu,[7] and by the present author.[8]

Section 4 develops the main argument of this work, essentially the definition of a computing algorithm based on a hierarchy of generalized Bethe–Goldstone equations of nth order. Here n refers to a number of electrons whose mutual interaction is taken into account by solution of an n-particle variational problem equivalent to an n-particle Schrödinger equation.

The theory developed here is expressed in terms of variational calculations. For this reason, actual computations reduce to the evaluation of matrix elements of the many-electron Hamiltonian in a basis system of Slater determinants, followed by successive diagonalization of a hierarchy of submatrices of the Hamiltonian matrix.

Results of computations are summarized in Section 5. Application of this formalism to scattering theory is discussed in Section 6. Some details of the computational organization of this work are given in Section 7.

2. REVIEW OF THE HARTREE–FOCK THEORY[8]

Any Schrödinger wave function for an N-electron system can be represented as a linear combination of a complete orthonormal set of Slater determinants. To construct such a complete set, it is convenient to assume the existence of some arbitrary countable complete set of one-particle functions ϕ_i, referred to as orbitals. Let N of these orbitals be used to define a reference state Slater determinant

$$\Phi_0 = \det \phi_1(1) \ldots \phi_N(N) \tag{1}$$

where det implies antisymmetrization and multiplication by a normalizing constant. The special subset of orbitals occupied in Φ_0 will be distinguished from the remainder of the complete orbital set by using indices $i, j, \ldots \leq N$ for the former and $a, b, \ldots > N$ for the latter, referred to as unoccupied orbitals with respect to reference state Φ_0.

Slater determinants forming a complete orthonormal set can be constructed from Φ_0 by replacing occupied orbitals by unoccupied orbitals. A general member of this set can be denoted by

$$\Phi_{ijk\ldots}^{abc\ldots} = \det \phi_1 \cdots \phi_a \cdots \phi_b \cdots \phi_c \cdots \phi_N \tag{2}$$

where orbitals a, b, c have replaced i, j, k in the reference state,

$$\Phi_0 = \det \phi_1 \cdots \phi_i \cdots \phi_j \cdots \phi_k \cdots \phi_N \tag{3}$$

subject to the ordering convention

$$i < j < k < \ldots \leq N < a < b < c < \ldots \tag{4}$$

Equation (4) determines all phase factors.

Consider an N-particle Schrödinger Hamiltonian operator

$$H(1 \ldots N) = \sum_i K(i) + \sum_{ij} Q(ij) \tag{5}$$

where each index denotes the set of discrete or continuous variables needed to describe a single particle. The single index i is summed from 1 to N and the pair ij is summed over $i < j \leq N$. For electrons, neglecting relativistic corrections, $K(i)$ is the sum of kinetic energy, nuclear Coulomb attraction, and external electromagnetic field, while $Q(ij)$ is the interelectronic Coulomb repulsion. Matrix elements of Eq. (5) are given by Condon and Shortley.[9] Since Φ_0 is arbitrary all matrix elements in a complete set of Slater determinants can be deduced from matrix elements $(\Phi_\mu, H\Phi_0)$, where Φ_μ is any member of the complete set.

Because Eq. (5) contains only one- and two-particle operators, $(\Phi_\mu, H\Phi_0)$ must vanish if Φ_μ and Φ_0 differ by more than two orbitals. The only non-vanishing matrix elements are

$$(\Phi_{ij}^{ab}, H\Phi_0) = (ab|R|ij) \tag{6}$$

$$(\Phi_i^a, H\Phi_0) = (a|\mathscr{H}_0|i) \tag{7}$$

$$\begin{aligned}(\Phi_0, H\Phi_0) &= \sum_i (i|K|i) + \sum_{ij}(ij|R|ij) \\ &= \tfrac{1}{2}\sum_i[(i|K|i) + (i|\mathscr{H}_0|i)]\end{aligned} \tag{8}$$

where

$$R(12) = Q(12)(1 - P_{12}) \tag{9}$$

and

$$\mathscr{H}_0 = K + \sum_{i=1}^{N}(i|R|i) \tag{10}$$

In Eq. (9), P_{12} is a permutation operator that interchanges indices 1 and 2. The operator $\mathscr{H}_0$ will be referred to as the Hartree–Fock single-particle Hamiltonian, for reference state Φ_0. In general it is convenient to define

$$\mathscr{H}_\mu = K + \sum_{i(\mu)}(i|R|i) \tag{11}$$

where $i(\mu)$ denotes the set of orbitals occupied in Φ_μ, any member of the complete set of Slater determinants. Then, in the notation of Eq. (2),

$$\mathscr{H}_{ij\ldots}^{ab\ldots} - \mathscr{H}_0 = \sum_{ab\ldots}(a|R|a) - \sum_{ij\ldots}(i|R|i) \tag{12}$$

From Eqs. (8) and (12) it follows, in the notation

$$D_{ij\ldots}^{ab\ldots} = (\Phi_{ij\ldots}^{ab\ldots}, H\Phi_{ij\ldots}^{ab\ldots}) \tag{13}$$

that the diagonal energy difference is

$$\begin{aligned}D_{ij\ldots}^{ab\ldots} - D_0 = &\sum_a(a|\mathscr{H}_0|a) - \sum_i(i|\mathscr{H}_0|i) \\ &+ \sum_{ab}(ab|R|ab) + \sum_{ij}(ij|R|ij) - \sum_{ia}(ia|R|ia)\end{aligned} \tag{14}$$

where the orbital indices are summed only over values explicitly indicated in the symbol $D_{ij\ldots}^{ab\ldots}$.

In principle an exact stationary state eigenfunction of H can be represented by

$$\Psi = \Phi_0 + \sum_{\mu}{}' \Phi_\mu c_\mu \tag{15}$$

where μ runs over the set of determinants $\Phi_{ij\ldots}^{ab\ldots}$, excluding Φ_0. The normalization implied by Eq. (15) is

$$(\Phi_0, \Psi) = 1 \tag{16}$$

The Schrödinger equation

$$H\Psi = E\Psi \tag{17}$$

can be expressed as a series of linear equations for the coefficients c_μ, $\mu \neq 0$,

$$E - D_0 = \sum_{\mu}{}' (\Phi_0, H\Phi_\mu) c_\mu \tag{18a}$$

$$(E - D_\mu) c_\mu = (\Phi_\mu, H\Phi_0) + \sum_{\substack{\nu \\ \neq 0,\mu}} (\Phi_\mu, H\Phi_\nu) c_\nu \tag{18b}$$

Eqs. (7), (8), and (18a) imply

$$E - D_0 = \sum_i \sum_a (i|\mathscr{H}_0|a) c_i^a + \sum_{ij} \sum_{ab} (ij|R|ab) c_{ij}^{ab} \tag{19}$$

an exact formula for the energy eigenvalue E.

The reference state, Φ_0, has been completely arbitrary up to this point. In the Hartree–Fock theory, a single Slater determinant serves as an approximation to the exact function Ψ. If this is to be physically valid, the coefficients c_μ in Eq. (15) must be small when Φ_0 is taken to be the Hartree–Fock determinant. In this case, Eq. (19) is a formula for the correlation energy, $E - D_0$.

The one-particle terms can be removed from this equation either by requiring all coefficients c_i^a, expressed in general through Eq. (18b), to vanish or by imposing the condition

$$(i|\mathscr{H}_0|a) = 0, \ i \leq N < a \tag{20a}$$

Either of these conditions determines the orbital basis set up to an arbitrary unitary transformation that does not combine occupied and unoccupied orbitals. The condition

$$c_i^a = 0, \ i \leq N < a \tag{20b}$$

is used in Brueckner's theory.[10]

Eq. (20a) is one of a number of statements of Hartree–Fock theory, all equivalent to the requirement that D_0 should be stationary, subject to preservation of the form of Φ_0 as a normalized Slater determinant. These statements, which follow immediately from each other, are

$$\begin{aligned} \delta D_0 &= 0,\ \Phi_0 \text{ normalized} \\ \operatorname{Re}(\Phi_i^a, H\Phi_0)\mathrm{d}\alpha^* &= 0,\ i \le N < a,\ \mathrm{d}\alpha \text{ arbitrary} \\ (\Phi_i^a, H\Phi_0) &= 0,\ i \le N < a \\ (\phi_a, \mathscr{H}_0\phi_i) &= 0,\ i \le N < a \\ \mathscr{H}_0\phi_i &= \sum_j \phi_j \varepsilon_{ji},\ i \le N \end{aligned} \tag{21}$$

Since $\mathscr{H}_0$ is Hermitian, the matrix ε_{ji} is Hermitian and can be diagonalized by a unitary transformation of the occupied orbitals ϕ_i. This leads to a canonical form of the Hartree–Fock equations

$$\mathscr{H}_0\phi_i = \varepsilon_i\phi_i,\ i \le N \tag{22}$$

Under such a unitary transformation of occupied orbitals, a Slater determinant is invariant except for a possible phase factor. The Hermitian quadratic form in Eq. (10) is invariant.

If Φ_0 is a Hartree–Fock function, Eqs. (21) and (8) imply that the only non-vanishing matrix elements $H_{\mu 0}$ are those for doubly excited determinants Φ_{ij}^{ab}. In the Rayleigh–Schrödinger perturbation theory it follows that the first-order wave function is

$$\Psi^{(1)} = \text{linear combination } \{\Phi_{ij}^{ab}\} \tag{23}$$

with no singly excited terms. This result, derived by Brillouin[11] and by Møller and Plesset,[12] implies that all correlation corrections to the mean value of a one-particle operator

$$F = \sum_i F(i) \tag{24}$$

are of second order and higher. This is the reason for the relatively high accuracy of Hartree–Fock values for properties of molecular electronic charge distributions.

If estimates of the coefficients c_μ in Eq. (15) are known, then Eqs. (18) give an improved estimate of any particular coefficient, and form the basis of a rapidly convergent iterative method of solving the eigenvalue problem.[13] Various forms of perturbation theory can be derived by transferring part of the diagonal term in Eq. (18b) from the left- to right-hand member before expanding the iterative process. In the Rayleigh–Schrödinger method, $(E - D_0)c_\mu$ is transferred to the right and expanded iteratively. This still leaves an arbitrary choice of the diagonal matrix elements of the unperturbed and perturbing operators, subject to the requirement that the former must be diagonal in the assumed basis of Slater determinants.

If all diagonal elements are removed from the perturbing operator, the first-order estimate of c_{ij}^{ab} is

$$c_{ij}^{ab} = -(ab|R|ij)/(D_{ij}^{ab} - D_0) \tag{25}$$

In the standard many-particle perturbation theory,[12] the two-electron integrals in Eq. (14) are treated as part of the perturbing operator, leading to

$$c_{ij}^{ab} = -(ab|R|ij)/(\varepsilon_a + \varepsilon_b - \varepsilon_i - \varepsilon_j) \tag{26}$$

where the eigenvalues ε_i from Eq. (22) are generally used instead of the diagonal elements of $\mathscr{H}_0$ indicated by Eq. (14).

Under either the Hartree–Fock (Eq. (20a)) or Brueckner (Eq. (20b)) conditions on the orbital set, Eq. (19) reduces to

$$E - D_0 = \sum_{ij}\sum_{ab}(ij|R|ab)c_{ij}^{ab} \tag{27}$$

This exact formula for the correlation energy can be expressed as a linked cluster expansion when the form of perturbation theory is used that gives Eq. (26) as the first-order estimate of c_{ij}^{ab}.[14]

3. PROPOSALS FOR CALCULATION OF CORRELATION EFFECTS

Perturbation theory indicates that the dominant corrections to the Hartree–Fock approximation come from virtual two-particle excitations of the Hartree–Fock reference state. From formulas such as Eqs. (25), (26), and (27), it is clear that correlation energy can be expressed exactly in terms of the coefficients c_{ij}^{ab}, and that these coefficients, in lowest order, are proportional to the two-particle matrix elements $(ab|R|ij)$. In a formula such as Eq. (27), it should be possible to reduce the number of significant terms by making individual terms as large as possible. Because of the nature of the integrals $(ab|R|ij)$, this suggests that the freedom of representation of the occupied orbitals of a Slater determinant (an arbitrary unitary transformation) should be used to obtain the most highly localized orbitals ϕ_i that can be constructed, and then to localize the unoccupied orbitals ϕ_a in a corresponding manner. This will have the effect in Eq. (25) of maximizing the numerator, which occurs quadratically in Eq. (27), for terms that represent a local virtual excitation of a localized orbital pair, and the magnitude of other terms in Eq. (27) will be reduced.

From these formal considerations, which correspond well with physical intuition about the nature of correlation effects, it is to be expected that correlation effects could most readily be computed by methods that make use of localized orbitals, and that are capable of dealing accurately with the virtual excitations of orbital pairs. Since the electronic shells in an atom are already spatially distinct from each other in the usual Hartree–Fock approximation, the use of localized orbitals has been emphasized in the literature primarily for extended systems, molecules, or solids.

Hurley, Lennard–Jones, and Pople[15] proposed the use of a variational trial function

$$\Psi = \det \psi_{12}(1, 2)\psi_{34}(3, 4) \ldots \psi_{N-1,N}(N - 1, N) \qquad (28)$$

where det symbolizes the total antisymmetrizing operator, multiplied by a normalization constant. The two-particle wave functions ψ_{ij}, called *geminals* in the more recent literature, describe localized pairs of electrons. For singlet pairs, ψ_{ij} is a generalization of the doubly occupied localized spatial orbital that would occur in a Hartree–Fock wave function. However, ψ_{ij} can give an accurate description of the correlation of two electrons. By constraining the geminals ψ_{ij} to be mutually orthogonal, with respect to integration over one common coordinate space, a simple energy formula can be derived for Ψ.[15] The geminal functions themselves correspond to chemical ideas of molecular structure based on inner shells, valence bonds, and lone pairs of electrons.

The paired-electron wave function, Eq. (28), has the advantage of describing what is expected to be the dominant effect of electronic correlation, modification of pairs of doubly occupied localized orbitals. This is accomplished with a wave function adapted to direct variational calculations.

The obvious disadvantage of the paired-electron wave function is that it makes no provision for correlation between different localized electron pairs. From Eqs. (25) or (26), such effects occur in the lowest order of perturbation theory. Physically, they correspond to the London dispersion force. Thus, Eq. (28) could describe a system of helium atoms very accurately with regard to the intra-atomic correlation energy, but it would neglect the van der Waals interaction that in practice dominates the interatomic potential at nuclear separations greater than the collision diameter.

Since second-order perturbation theory (for the energy) explicitly includes all of the terms in Eq. (23), but with coefficients fixed by the first-order perturbation formula, it has been proposed that a general function of this form could be used as a variational trial function. A semiempirical generalization of the nuclear alpha-particle model has been based on this generalization of the independent particle model.[16] The variational trial function to be considered is

$$\Psi = \Phi_0 + \sum_{ij}\sum_{ab} \Phi_{ij}^{ab} c_{ij}^{ab} \tag{29}$$

This wave function treats all orbital pairs in the same manner, and includes the effect of interference between excitations of different orbital pairs. Like the paired electron wave function, Eq. (28), it can be used with localized orbitals. However, in contrast to the paired electron function, Eq. (29) excludes explicit four-particle excitations that arise as products of two-particle excitations. For example, if ψ_{ij} in Eq. (28) contains a term $\phi_a \phi_b c_{ij}^{ab}$, and if ψ_{kl} contains $\phi_c \phi_d c_{kl}^{cd}$, then Ψ contains Φ_{ijkl}^{abcd} with coefficient

$$c_{ijkl}^{abcd} = c_{ij}^{ab} c_{kl}^{cd} \tag{30}$$

Here all indices are distinct as a consequence of the orthogonality constraint imposed on ψ_{ij} and ψ_{kl}, assuming that indices $ijkl$ are all different. Because these higher-order excitations are omitted from Eq. (29), this wave function cannot give an accurate description of the total correlation energy of a system with a large number of particles when there are strong correlation effects within localized particle pairs.

It was originally proposed by Brueckner,[17] in the context of a study of nuclear matter, that the correlation problem could be solved independently for each pair of particles in a many-particle system. The correlation energy is obtained by a formula equivalent to Eq. (27), where the coefficients c_{ij}^{ab} are obtained from the solution of a modified Schrödinger equation for pair (ij). Expressed in terms of differential operators and a two-particle wave function, this form of Brueckner's theory was proposed by Bethe and Goldstone,[18] and the modified Schrödinger equation is referred to as a Bethe–Goldstone equation. It corresponds to the exact solution of a two-particle problem, subject to the constraint that the wave function should be orthogonal to the remaining $N - 2$ orbitals of an N-particle Slater determinant. In Brueckner's theory, the occupied orbitals of this determinant are obtained, in principle, by solution of Eqs. (20b). This introduces an iterative cycle that would, in general, require repeated solution of the Bethe–Goldstone equations. To avoid this iterative procedure, it is convenient to use the Hartree–Fock condition, Eq. (20a), to determine the occupied orbitals or Fermi sea, and then to include specific coefficients c_i^a in the formalism of the Bethe–Goldstone equations.

The proposal by Sinanoğlu,[7,19] for solution of "exact pair" equations, is closely related to the method of Brueckner, Bethe, and Goldstone. The calculations of Geller, Taylor, and Levine on the Be atom,[20] based on Sinanoğlu's formalism, are in fact a variational solution of the Bethe–Goldstone equations, using trial functions that depend explicitly on relative coordinates.

The Bethe–Goldstone equation for pair (ij) can be expressed in the form[21]

$$[\mathscr{H}_{ij}(1) + \mathscr{H}_{ij}(2) + (1 - \Omega_{ij}(1,2))Q(1,2)]\psi_{ij}(1,2) = \varepsilon_{ij}\psi_{ij}(1,2) \tag{31}$$

where

$$\mathscr{H}_{ij} = \mathscr{H}_0 - (i|R|i) - (j|R|j) \tag{32}$$

and $(1 - \Omega_{ij})$ is a projection operator that removes all occupied orbital components ϕ_k, $k \neq i, j$, from the two-particle function upon which it operates. The function ψ_{ij} is constrained to be orthogonal to all ϕ_k, $k \neq i, j$.

It should be emphasized that Eq. (31) is solved independently for each pair (ij), and the equations are coupled only by the constraint of orthogonality to a common set of orbitals ϕ_i. There is no N-particle wave function in this approach to the correlation problem, in contrast to the explicit variational trial functions described by Eqs. (28) and (29). However, Brueckner's theory combines the qualitative virtues of these variational methods. All pairs are treated equivalently, and any given pair is treated to high accuracy, in principle by an exact solution of the constrained two-particle problem.

The Bethe–Goldstone equation for pair (ij) can be expressed in a variational formalism involving N-particle wave functions.[8] The variational trial function, assuming Eq. (20b), is

$$\Psi'_{ij} = \Phi_0 + \sum_{ab} \Phi_{ij}^{ab} c_{ij}^{ab} \tag{33}$$

It can easily be verified that Eqs. (18b) for the coefficients c_{ij}^{ab} in Ψ'_{ij} are equivalent to Eq. (31). In Eq. (33) the Bethe–Goldstone orthogonality constraint is applied explicitly by excluding excitations from orbitals other than ϕ_i and ϕ_j in the trial function. If the Brueckner condition, Eq. (20b), is not satisfied, determinants of the form Φ_i^a and Φ_j^b must be included in Ψ'_{ij}.

4. BETHE–GOLDSTONE EQUATIONS OF *n*th ORDER

If a Fermi sea is defined by the occupied orbitals ϕ_i of a reference state Slater determinant Φ_0, the Bethe–Goldstone equation for pair (ij) is equivalent to variational solution of the two-particle Schrödinger equation, constrained by orthogonality of the wave function to the $N - 2$ orbitals ϕ_k with $k \neq i, j$. Stated in this way, the choice of Φ_0 is a separate consideration, and it is not necessary to restrict the orbitals ϕ_i by either of the conditions usually imposed, Eqs. (20a) or (20b). As indicated by Eq. (33), the Bethe–Goldstone equation is equivalent to a variational calculation with an N-particle trial function.

$$\Psi'_{ij} = \Phi_0 + \sum_{a} \Phi_i^a c_i^a + \sum_{b} \Phi_j^b c_j^b + \sum_{ab} \Phi_{ij}^{ab} c_{ij}^{ab} \tag{34}$$

Here indices i and j are fixed, but the summations over a and b include all members of a complete set of orbitals that are orthogonal to the N occupied orbitals of Φ_0.

The form of Eq. (34) suggests a generalization of the concept of a Bethe–Goldstone equation. An nth-order Bethe–Goldstone equation can be defined as the exact solution of an n-particle Schrödinger equation, with the wave function constrained to be orthogonal to the remaining $N - n$ orbitals of an N-particle Fermi sea. For n specified orbital indices $ijk\ldots$ this is equivalent to a variational calculation with a trial function $\Psi_{ijk\ldots}$ that is a linear combination of Φ_0 and all Slater determinants $\Phi_{ijk\ldots}^{abc\ldots}$ whose indices $ijk\ldots$ are any subset of the specified list. For example, the third-order Bethe–Goldstone equation for the orbital triple ijk is equivalent to diagonalizing the matrix of the N-particle Hamiltonian H over the set of Slater determinants Φ_0, Φ_i^a, Φ_j^b, Φ_k^c, Φ_{ij}^{ab}, Φ_{ik}^{ac}, Φ_{jk}^{bc}, Φ_{ijk}^{abc}, where indices ijk are specified, but abc take on all possible values.

By definition, a sequence of Bethe–Goldstone calculations of increasing order n eventually terminates at order N with an exact solution of the N-particle problem. The increments of energy or of any other physical property of the wave function, taken for successively higher orders of n, form a series whose sum is the exact value. This method becomes a practicable computational technique if these series converge adequately to second or third order.

If diagonal elements of the configuration interaction matrix (matrix of H) are computed relative to the energy mean value D_0 of the reference state, as in Eq. (14), then the energy eigenvalue computed for $\Psi_{ij\ldots}$ will be an energy increment $\Delta E_{ij\ldots}$ with respect to D_0. These quantities will be referred to as *gross* increments. In order to make use of a hierarchy of Bethe–Goldstone equations of increasing order, it is convenient to define a *net* energy increment $e_{ij\ldots}$ as the difference between the corresponding gross increment and the sum of all net increments of lower order whose indices form a subset of the set $ij\ldots$. For example, the net energy increment for pair (23) is defined by

$$e_{23} = \Delta E_{23} - e_2 - e_3 \tag{35}$$

where ΔE_{23} is the directly computed eigenvalue of the configuration interaction matrix for Ψ_{23}, and e_2 and e_3 are net increments obtained from first-order Bethe–Goldstone equations for Ψ_2 and Ψ_3, respectively. If carried to Nth order, this method of bookkeeping gives the exact correlation energy in the form

$$\Delta E = \sum_i e_i + \sum_{ij} e_{ij} + \sum_{ijk} e_{ijk} + \cdots + e_{1\ldots N} \tag{36}$$

Of course, since the Nth-order equation is simply the original Schrödinger equation, this method is of practical value only if high-order terms in Eq. (36) are negligible. Preliminary calculations of third- and fourth-order energy terms, discussed below, indicate that this is true for electronic systems.

The hierarchy of Bethe–Goldstone equations can be used to compute any physical property expressed as the expectation value of a many-electron operator F. Let $F_{\mu\nu}$ be the matrix of this operator in a basis of Slater determinants. If the solution of an nth-order Bethe–Goldstone equation is expressed as

$$\Psi = \sum_\mu \Phi_\mu c_\mu \tag{37}$$

then the gross increment of F is given by the expectation value

$$\Delta F_{ij\ldots} = \sum_\mu \sum_\nu (F_{\mu\nu} - F_{00}\delta_{\mu\nu}) c_\mu^* c_\nu / \sum_\mu c_\mu^* c_\mu \tag{38}$$

The net increment, denoted by $f_{ij\ldots}$, is defined as the difference between $\Delta F_{ij\ldots}$ and the sum of net increments of lower order whose indices form a subset of the set ij. . . .. For example, the net increment of some quantity denoted by f for index triple (234) is given by

$$f_{234} = \Delta F_{234} - f_2 - f_3 - f_4 - f_{23} - f_{24} - f_{34} \tag{39}$$

5. RESULTS OF CALCULATIONS

Variational calculations are carried out on trial wave functions, as indicated by Eq. (34), with Slater determinants Φ_μ selected in the manner described in Section 4. The calculations are organized in essentially the same way as earlier configuration interaction calculations by Boys on Be[22] and by Nesbet and Watson[23] and Watson[24] on He and Be, respectively. The present calculations differ from Boys's earlier work primarily by using an initial matrix

Hartree–Fock calculation[8,25] to determine a reference state and an orthonormal set of orbitals. While no finite set of orbitals can be complete, basis sets are chosen that can be extended systematically to completeness, and enough basis orbitals are chosen to stabilize computed quantities to within a predetermined relative error.

It is stressed in the papers by Boys,[22] and confirmed by later work on atoms and molecules, that the configurations contributing most strongly to the correlation energy must be constructed from orbitals that are as similar as possible to the occupied orbitals which they replace in the reference state. In an orthonormal set these orbitals have spatial extension similar to the occupied orbitals, but orthogonality requires them to have additional nodal surfaces. These perturbing orbitals are quite different from the functions that describe Rydberg series electrons in atoms. The latter spread out rapidly as the number of nodes increases, while the former remain concentrated in the region of the occupied orbitals. The additional nodes in regions of high density cause the perturbing orbitals to have large kinetic energy. For this reason the energy of the perturbing orbitals lies in the continuum of the energy spectrum of the Hartree–Fock one-electron Hamiltonian, and the perturbing orbitals must be thought of as wave packets in the continuum. Such functions, which occur naturally in discrete orbital basis sets orthogonalized to occupied Hartree–Fock orbitals, are a byproduct of the matrix Hartree–Fock method.[8] They occur in an atomic calculation if basis orbitals are included that have exponents at least as large as those characteristic of a given shell, but that have higher powers of r and higher-order spherical harmonic factors. Criteria for the choice of localized perturbing orbitals in molecules, when the occupied orbitals of the reference state are localized, have been discussed by Boys.[26]

Earlier calculations have shown that the correlation energy of a localized electron pair contributed by each perturbing orbital depends primarily on the number of nodal surfaces superimposed on the localized orbital. This has two important implications. First, a consistent orbital basis set should be capable of representing all possible perturbing orbitals with a given number of nodal surfaces. Second, the percentage of the correlation energy obtained in a particular calculation has been found to depend quantitatively on the number of nodes, independent of geometry.[8] For example, basis sets capable of representing perturbing orbitals with all combinations of one, two, or three nodal surfaces should lead to approximately 90, 94, or 97 per cent, respectively, of the correlation energy of a localized orbital pair. This hypothesis is confirmed by the calculations on light atoms reported here.

In the present work, orbital basis sets capable of giving a reasonable approximation to an atomic Hartree–Fock calculation, using the matrix Hartree–Fock method,[8,25] are extended by including additional orbitals in the form of exponential functions exp $(-\zeta r)$, multiplied by arbitrary powers of r and by spherical harmonics. A preliminary series of calculations on several atoms was used to determine optimal values of the exponents ζ. It was found that the best results obtained with sets of exponents different for each angular quantum number l could be duplicated, to within the criterion of 0.001 atomic units of energy used in the present work, by basis sets with a common value of ζ for all values of l. Different values of ζ must still be used to represent perturbations of atomic orbitals with different principal quantum number n. Thus, exponents ζ_K and ζ_L have been determined variationally for perturbations of the K and L shells, respectively, of Li, Be, N, and Ne in their ground states. These exponents are listed in Table 1.

TABLE 1. Exponents used for orbital basis sets.

Z	Atom	ζ_K	ζ_L
3	Li	5.543	1.726
4	Be	7.814	1.419
7	N	14.626	2.734
10	Ne	21.63	4.214

Except for the exponent ζ_L for Li, which corresponds to the special case of the singly-occupied 2s orbital in the Li(^{2}S) reference state, the other exponents in Table 1 depend nearly linearly on the atomic number Z. This fact has been used, in calculations not yet published, to obtain values of ζ_K and ζ_L for the remaining 2p shell atoms by linear interpolation in Table 1.

From earlier configuration interaction calculations, it was estimated that an orbital basis set capable of representing perturbing orbitals with all combinations of up to three nodal surfaces superimposed on occupied Hartree–Fock orbitals should give energy values accurate to within 0.001 atomic units (e^2/a_0) for individual Bethe–Goldstone equations. For K-shell perturbations, this requires basis orbitals with radial and angular factors indicated by quantum numbers (nl) up to $4s$, $4p$, $4d$, $4f$, multiplying exp $(-\zeta_K r)$, in addition to basis orbitals needed to represent the occupied Hartree–Fock orbitals. The computer programs used were designed to allow the inclusion of arbitrary (nl) values, and in some cases higher values of n or l were used when there was any question of the convergence of the energy eigenvalue to 0.001 atomic unit.

One advantage of the systematic definition of gross and net increments of energy, as in Eq. (35), is that these quantities have values defined uniquely in principle, independent of the technique used in a particular approximate method of solving the Bethe–Goldstone equations. This makes it possible to compare the results of independent calculations, and eventually to reach agreement within any stated limit of accuracy on the true values of the net energy increments. These remarks are valid for any expectation value defined through Eq. (38), with net increments given by formulas like Eq. (39).

In the present calculations, the sum of H_{00} and any gross energy increment ΔE is a rigorous upper bound to the ultimate value of this sum. This fact is used to ascertain whether or not a given orbital basis set is superior or inferior to another, and also to examine the convergence of the calculations. Net energy increments of high order are found to be stable in magnitude under considerable variation of the orbital basis. For this reason, truncated basis sets can be used to give reasonable estimates of small net increments, even though the absolute accuracy of larger net increments, or those of lower order, is not adequate when computed with the same orbital basis. In practice small net increments, especially intershell correlation terms, are computed by a hierarchy of variational calculations with a somewhat truncated orbital basis. The intermediate results, net increments of lower order, are used only to define the net increment of immediate interest by formulas such as Eq. (39), which require the gross increment to be computed with accuracy comparable to that of each of the lower order net increments used in the formula. Final values of larger or lower order net increments are obtained by independent variational calculations that use orbital basis sets large enough to give the required accuracy.

Trial calculations, particularly on Be(^1S) and Ne(^1S), showed that results stable to 0.001 atomic unit could be obtained with orbital basis sets containing the "double zeta" matrix Hartree–Fock basis set of Clementi,[27] augmented by two series of basis orbitals with the exponents given in Table 1. Computed net energy increments for Be(^1S) are given in Table 2.[28] Computed

TABLE 2. Correlation energy increments e_{ij} for Be(^1S). Indices 1–4 denote $1s\beta$, $1s\alpha$, $2s\beta$, $2s\alpha$, respectively. The units are Hartree units, e^2/a(Be), in which the reduced mass of an electron in Be is unity.

e_{12}	−0.041827	e_{123}	+0.000030
e_{13}	−0.000813	e_{124}	+0.000030
e_{14}	−0.002119	e_{134}	+0.000428
e_{23}	−0.002119	e_{234}	+0.000428
e_{24}	−0.000813		
e_{34}	−0.045351		
	Σe_{ij}	−0.093042	
	Σe_{ijk}	0.000916	
	Total	−0.092126	

net energy increments for Ne(^1S) are given in Table 3.[29] Details of these calculations are given in the cited publications. Three- and four-particle net correlation energies computed for Ne are found to be small compared with 0.001 atomic unit, as are the three-particle terms for Be shown in Table 2.

The total correlation energy for Be(^1S), including all three-particle terms, is computed to be −0.0921 atomic unit,[28] or 98.1% of the empirical non-relativistic correlation energy.

The sum of the computed pair correlation energies for Ne, shown in Table 3,[29] is -0.3822 atomic unit, or 97.3% of the empirical correlation energy.[30] These results are in good agreement with the level of accuracy expected for the present calculations. The

TABLE 3. Pair correlation energies e_{ij} in $Ne(^1S)$. Units are Hartree atomic units, a.u.(Ne), such that the electron reduced mass in Ne is unity. To convert to cm^{-1}, multiply by 0.2194686×10^6.

i, j	Weight	e_{ij}
$1s\beta, 1s\alpha$	1	-0.039932
$1s\beta, 2s\beta$	2	-0.000469
$1s\beta, 2s\alpha$	2	-0.002099
$1s\beta, 2p_0\beta$	6	-0.001481
$1s\beta, 2p_0\alpha$	6	-0.001834
	K, L subtotal	-0.025026
$2s\beta, 2s\alpha$	1	-0.010834
$2s\beta, 2p_{-1}\beta$	4	-0.003294
$2s\beta, 2p_0\beta$	2	-0.003302
$2s\beta, 2p_{-1}\alpha$	4	-0.010283
$2s\beta, 2p_0\alpha$	2	-0.010317
$2p_{-1}\beta, 2p_0\beta$	4	-0.010906
$2p_{-1}\beta, 2p_1\beta$	2	-0.010927
$2p_{-1}\beta, 2p_{-1}\alpha$	2	-0.016498
$2p_{-1}\beta, 2p_0\alpha$	4	-0.013732
$2p_{-1}\beta, 2p_1\alpha$	2	-0.022837
$2p_0\beta, 2p_0\alpha$	1	-0.025813
	L, L subtotal	0.317269
	Σe_{ij}, all pairs	-0.382227

individual pair correlation energies are in agreement with earlier work expected to be of comparable accuracy: the $1s^2$ and $2s^2$ pair energies of Be computed by Geller and coworkers;[20] the pair energies of Be, including $1s2s$, computed by Kelly;[31] and the $2s^2$ pair correlation energy of Ne computed by McKoy and Sinanoğlu.[32]

The net energy increments e_{ij} for $Ne(^1S)$, listed in Table 3, are all computed independently. The programs used take account only of quantum numbers M_L, M_S, and parity, and the only obvious symmetry in the general Bethe–Goldstone equation is separate sign reversal of all quantum numbers m_l or m_s. This symmetry leads to the weights indicated in Table 3 for individual net energy increments. The near equality of some of these numbers arises from the underlying spherical symmetry of the Hamiltonian. Since the reference state Slater determinant is a ^1S angular momentum eigenfunction, a component of the $N - n$ particle Fermi sea wave function with good quantum numbers L and S can only couple to an n-particle Bethe–Goldstone wave function with the same quantum numbers. If the Slater determinant constructed from $N - n$ orbitals of the Fermi sea is a composite of states with several different (L, S) values, the n-particle Bethe–Goldstone wave function coupled to this Fermi sea must be a composite of the same (L, S) values. For example, since the configuration $1s^22s^22p^4$ has three substates ^3P, ^1D, and ^1S, only three independent parameters should be needed to describe the six $2p^2$ pair correlation energies listed in Table 3. The weights with which these substates occur in any p^4 state are well known, and lead to the coefficients listed in Table 4. The parametrized

TABLE 4. Parametrization of $2p^2$ pair correlation energies e_{ij}. The values of $e(^3P)$, $e(^1D)$, and $e(^1S)$ are listed in Table 5.

	Coefficient of			
i, j	$e(^3P)$	$e(^1D)$	$e(^1S)$	e_{ij}, a.u.(Ne)
$2p_{-1}\beta, 2p_0\beta$	1	0	0	−0.010913
$2p_{-1}\beta, 2p_1\beta$	1	0	0	−0.010913
$2p_{-1}\beta, 2p_{-1}\alpha$	0	1	0	−0.016498
$2p_{-1}\beta, 2p_0\alpha$	$\frac{1}{2}$	$\frac{1}{2}$	0	−0.013706
$2p_{-1}\beta, 2p_1\alpha$	$\frac{1}{2}$	$\frac{1}{6}$	$\frac{1}{3}$	−0.022898
$2p_0\beta, 2p_0\alpha$	0	$\frac{2}{3}$	$\frac{1}{3}$	−0.025691

values of e_{ij}, in the last column of Table 4, agree with the directly computed values shown in Table 3 to within the expected accuracy of the present calculations.

The full list of parameters needed to determine all of the pair correlation energies e_{ij} in Ne is given in Table 5. Except for the

TABLE 5. Pair correlation parameters for Ne(^1S), in a.u.(Ne).

Pair	Value	Pair	Value
$1s^2(^1S)$	−0.039932	$2s2p(^3P)$	−0.003297
$1s2s(^3S)$	−0.000469	$2s2p(^1P)$	−0.017292
$1s2s(^1S)$	−0.003729	$2p^2(^3P)$	−0.010913
$1s2p(^3P)$	−0.001481	$2p^2(^1D)$	−0.016498
$1s2p(^1P)$	−0.002187	$2p^2(^1S)$	−0.044076
$2s^2(^1S)$	−0.010834		

$2p^2$ parameters, only the $1s2p$ and $2s2p$ parameters are not uniquely determined by the data in Table 3. In these cases a weighted average of the computed energies was used to determine the parameters. Because of the close agreement between computed values of $e(2s\beta, 2p_{-1}\beta)$ and $e(2s\beta, 2p_0\beta)$, which when parametrized should be equal and have the value −0.003297 atomic units, and of $e(2s\beta, 2p_{-1}\alpha)$ and $e(2s\beta, 2p_0\alpha)$, both of which should equal −0.010295 atomic units, separate computation of $e(1s\beta, 2p_{-1}\beta)$ and $e(1s\beta, 2p_{-1}\alpha)$ was not carried out. It was considered justified to set these quantities equal to $e(1s\beta, 2p_0\beta)$ and $e(1s\beta, 2p_0\alpha)$, respectively, as indicated in Table 3.

In atomic S states ($L = 0$) the magnetic hyperfine structure depends on a single parameter, a_s, due to the Fermi contact interaction. To compare experimental results it is convenient to define another quantity, the Fermi contact parameter, f, proportional to the electronic spin density at the nucleus. This is

$$f = 4\pi(\rho_+(0) - \rho_-(0)) \tag{40}$$

$$= IJa_s/31.8027\mu_I \tag{41}$$

in units a_0^{-3} if a_s is given in Mc/sec and μ_I is the nuclear magnetic moment in nuclear magnetons. I and J are quantum numbers, respectively, for the nuclear spin and electronic angular momentum. $\rho_\pm(0)$ is the $m_s = \pm\frac{1}{2}$ electronic density at the atomic nucleus.

For an electronic wave function represented by a single Slater determinant, the parameter f is the difference, evaluated at $r = 0$, of the sum of the squared amplitudes of the radial factors of occupied orbitals with $m_s = +\frac{1}{2}$ and of the corresponding sum for $m_s = -\frac{1}{2}$.

Bethe–Goldstone equations have been used to evaluate the Fermi contact parameter for the ^2S ground state of Li and the ^4S ground state of N.[33] The method used, outlined above in Section 4, makes use of Eq. (38) and of the definition of net increments of f exemplified by Eq. (39). Details of the calculations are given in the cited references. Computed net increments of f for Li(^2S) are listed in Table 6. The corresponding results for $N(^4S)$ are listed in Table 7. Results from two sets of basis orbitals are given in these tables. The first set, denoted by HF, is the basis set used for Hartree–Fock calculations,[34] with no additional orbitals. The second set, denoted by DZ + SPDF, is comparable to the basis sets described above for calculations on Be and Ne.

TABLE 6. Fermi contact parameter increments $f(a_0^{-3})$ for Li(^2S). Indices 1, 2, 3 denote $1s\beta$, $1s\alpha$, $2s\alpha$, respectively.

	HF[a]	DZ + SPDF[b]
f_0	2.066865	2.063862
f_1	0.706856	0.632224
f_2	0.0	0.000153
f_3	0.0	−0.000003
f_{12}	0.139291	0.072370
f_{13}	0.077114	0.230334
f_{23}	0.0	−0.023845
f_{123}	−0.079730	−0.082419

[a] Basis orbitals same as in Hartree–Fock calculation; s only.
[b] Basis orbitals described in text; s, p, d, and f.

A comparison of f_{123} for the two different basis sets used for Li(^2S) indicates that it is insensitive to changes in the orbital basis. In the case of N(^4S), because of the large number of third-order equations, they were solved only for the Hartree–Fock orbital basis (s and p orbitals only), and for the single most

TABLE 7. Fermi contact parameter increments $f(a^{-3})$ for $N(^4S)$. Indices 1–7 denote $1s\beta$, $1s\alpha$, $2s\beta$, $2s\alpha$, $2p_{-1}\alpha$, $2p_0\alpha$, $2p_1\alpha$, respectively.

	HF[a]	DZ + SPDF[b]
f_0	0.0	0.0
f_1	−2.728615	−3.125947
f_2	−2.738835	−3.126075
f_3	4.127675	3.673492
f_4	3.383339	2.811017
f_6	0.0	0.0
f_{12}	−0.478375	−0.411986
f_{13}	0.042015	0.064466
f_{14}	0.574009	0.642206
f_{16}	0.055434	0.217955
f_{23}	−0.271872	−0.394888
f_{24}	0.143428	0.142070
f_{26}	−0.053039	−0.153507
f_{34}	1.325324	1.265081
f_{36}	0.052612	0.116482
f_{46}	0.054618	0.142119
f_{56}	0.0	0.0
f_{57}	0.0	0.0
f_{123}	0.288852	
f_{124}	−0.250988	
f_{126}	0.018499	
f_{134}	0.161682	
f_{136}	0.072564	
f_{146}	−0.016403	
f_{234}	0.015826	
f_{236}	−0.008978	
f_{246}	−0.119683	
f_{346}	−0.431617	−0.351496

[a] Basis orbitals same as in Hartree–Fock calculation; *s*, *p* only.
[b] Basis orbitals described in text; *s*, *p*, *d*, and *f*.

important triple (346) in the extended basis. The values of f_{346} are not substantially different. The sum of f_{ijk} used to compute the final value of f, given in Table 8, was taken from the Hartree–Fock basis calculations. Final results for the Fermi contact parameters are listed in Table 8. The third-order terms are expected to be the least accurate, and may account for the remaining disagreement with experiment.

The most striking result of these calculations is a demonstration of the importance of three-particle contributions to the Fermi contact parameter. In contrast, the three-particle correlation energy net increments are all extremely small.[33] Fourth-order effects in N(^4S) were not examined in this work, and there is no assurance that they are negligible. Unpublished calculations show that the three-particle terms arise from the interference between

TABLE 8. Computed Fermi parameters $f(a_0^{-3})$ for Li(^2S) and N(^4S). Data is taken from best available calculation for each net increment.

	Li(^2S)	N(^4S)
f_0	2.066865	0.0
Σf_i	0.632374	0.232487
Σf_{ij}	0.278859	2.276096
Σf_{ijk}	−0.082419	−1.241482
f_{total}	2.895679	1.267101
f_{exp}	2.90960	1.22099

Slater determinants of the form Φ_i^a and those of the form Φ_{jk}^{bc}, rather than from explicit three-particle excitations Φ_{ijk}^{abc}. This is demonstrated by removing the terms Φ_{ijk}^{abc} from the variational trial function. The resulting change in the computed three-particle net increments is found to be negligible. The importance of three-particle terms in the mean value of a one-electron operator, such as the Fermi contact parameter, can be understood qualitatively as an interference between two-particle excitations, the most important terms in the Schrödinger equation because they dominate the correlation energy, and one-particle excitations, which directly influence the value of a one-particle operator. This argument would suggest that the fourth-order net increments of f might be considerably smaller than the third order net increments.

6. BETHE–GOLDSTONE SCATTERING THEORY

The idea of a hierarchy of n-particle Bethe–Goldstone equations can be applied to properties of an electronic wave function other than expectation values. In particular, this formalism has been used to develop a theory of Fermion scattering by a many-Fermion system.[35] Results of calculations are not yet available. Because of the complexity of processes that can occur in many-particle scattering the discussion here will be limited to elastic scattering of a single external electron. Extension of the formalism to describe more complex processes is straightforward.

Let the scattering atom (N electrons) be represented by a Hartree–Fock wave function Φ_0, as in Eq. (1). Then the zeroth-order approximation to a wave function that describes scattering of an external electron is

$$\Phi^k = \det \phi_1(1) \ldots \phi_N(N)\phi_k(N+1) \tag{42}$$

where ϕ_k is a continuum solution of the Hartree–Fock equation

$$\mathscr{H}_0\phi_k = \varepsilon_k\phi_k \tag{43}$$

The operator $\mathscr{H}_0$ is defined by Eq. (11). If the continuum orbital ϕ_k were normalized to make (ϕ_k, ϕ_k) equal unity the integrated electronic density within any finite region would vanish, and there is no self-consistent effect of the external particle on the occupied bound state orbitals ϕ_i, $i \leq N$. The function ϕ_k, for wave number k, determines a set of zeroth-order phase shifts $\delta_l(0)$. For consistency with earlier notation, the symbol det in Eq. (42) means $((N+1)!)^{-\frac{1}{2}}$ times the antisymmetrizing operator.

The first-order Bethe–Goldstone equation with index i, for elastic scattering of a single external particle, is equivalent to a variational calculation with a trial function of the form

$$\Psi_i = \Phi^k + \sum_{ab} \Phi_i^{ab} c_i^{ab} \tag{44}$$

The notation Φ_i^{ab}, consistent with that used in Section 2, above, describes an $(N+1)$-particle Slater determinant in terms of an N-particle reference state. The indices ab refer to members of a complete set of normalized orbitals that are orthogonal to all occupied orbitals ϕ_i of Φ_0. The index k refers to an unnormalized continuum orbital of wave number k, orthogonal to the N occupied orbitals ϕ_i but not to the unoccupied orbitals ϕ_a. Since the function ϕ_k satisfies an integro-differential equation different from Eq. (43), it will in general have different phase shifts. For this reason the continuum function determined by Eq. (43) will be denoted by $\phi_k(0)$, and the continuum function determined variationally from Eq. (44) will be denoted by $\phi_k(i)$. Because this orbital is not normalized, Ψ_i itself is not normalized, but ϕ_k and the coefficients c_i^{ab} contain an arbitrary common numerical factor. The variational equations derived from Eq. (44) can be shown to be equivalent to the Bethe–Goldstone scattering equation derived by Mittleman,[36] to describe electron scattering by alkali atoms.

For simplicity, the analysis in terms of gross and net increments will be described here in terms of phase shifts. It should be clear that the same concepts could equally well be applied to any function of the phase shifts for which additivity of net increments might have a more direct intuitive meaning.

The function $\phi_k(0)$ determines a set of gross phase shifts $\Delta_l(0)$, which by definition are equal to the net phase shifts of order zero, $\delta_l(0)$. The function $\phi_k(i)$, in the variational trial function Ψ_i of Eq. (44), determines a set of gross phase shifts $\Delta_l(i)$. Then the first-order net phase shifts are defined by

$$\delta_l(i) = \Delta_l(i) - \delta_l(0) \tag{45}$$

analogous to Eq. (39) for an expectation value. The intuitive meaning of $\delta_l(i)$ is that it represents the change in phase shift of partial wave l at momentum k due to virtual polarization of the occupied orbital i of the scattering atom, neglecting any interaction or interference with the polarization of other orbitals.

The extension to higher order effects is obvious. The Bethe–Goldstone equation of order n, for a set of n indices $ij\ldots$, is equivalent to a variational calculation with a trial function expressed as a linear combination of Φ^k, Φ_i^{ac}, Φ_j^{bc}, . . ., Φ_{ij}^{abc}, . . ., including all Slater determinants with occupied orbital indices belonging to subsets of $ij\ldots$. The continuum orbital $\phi_k(ij\ldots)$ determines gross phase shifts $\Delta_l(ij\ldots)$. A net phase shift $\delta_l(ij\ldots)$ is defined as the difference between the gross phase shift and the sum of all net phase shifts of lower order with indices belonging to subsets of $ij\ldots$.

By definition, the sum of all net phase shift increments up to order N is the exact phase shift if inelastic processes cannot occur. Obviously an inelastic process would require a trial wave function with more than one continuum function ϕ_k. The assumption that the normalized orbitals ϕ_i, ϕ_a form a complete set means that in principle a function like Φ_i^{ak}, representing a static correlation correction to Φ_0 for the scattering atom, is already included in Eq. (44) and its higher order generalizations. In practice, it would be desirable to include such terms explicitly if they have a significant effect on the phase shifts. Since ϕ_k is not assumed to be orthogonal to the unoccupied orbitals, such terms can be included without changing the structure of the theory.

In Eq. (44), the orbitals ϕ_a and ϕ_b describe polarization of the occupied orbital ϕ_i and distortion of the continuum orbital ϕ_k near the scattering atom. To describe these localized effects, it should be adequate to include orbitals ϕ_a from a discrete complete set of normalized functions that vanish at infinite distance from the atom. The asymptotic behaviour of Ψ'_i is entirely determined by the continuum orbital ϕ_k.

The wave function Ψ'_i is determined by requiring the functional derivative of $(\Psi'_i,(H - E)\Psi'_i)$, taken with respect to Ψ'_i, to vanish. When the explicit expansion, Eq. (44), is used, this variational condition implies two statements. First, the ordinary partial derivative of $(\Psi'_i,(H - E)\Psi'_i)$ with respect to each of the coefficients c_i^{ab} should vanish. This implies

$$((\Phi_i^{cd},H - E)\Psi'_i) = 0, c, \quad d > N \tag{46}$$

a set of inhomogeneous linear equations for the coefficients c_i^{ab}. The second variational statement is that the functional derivative with respect to variations of the continuum function $\phi_k(i)$ should vanish, subject to the constraint that ϕ_k remain orthogonal to the occupied orbitals ϕ_i of Φ_0. This leads to an integro-differential equation for $\phi_k(i)$,

$$\mathscr{H}_0\phi_k(i) + \sum_{ab} c_i^{ab}(i|R|a)\phi_b = (E - H_{00})\phi_k(i) + \sum_j \lambda_j\phi_j \tag{47}$$

which differs from Eq. (43) by inclusion of a polarization term. The coefficients λ_j are Lagrange multipliers needed to make ϕ_k orthogonal to the orbitals ϕ_j, $j \leq N$. If the orbitals ϕ_j are Hartree–

Fock orbitals (eigenfunctions of $\mathscr{H}_0$), then Eq. (47) implies that

$$\lambda_j = \sum_{ab} c_i^{ab}(ij|R|ab) \tag{48}$$

In Eq. (47), H_{00} is the Hartree–Fock energy of the scattering atom. The operator R is defined in Eq. (10). Eqs. (46) and (47) must be solved simultaneously or iteratively.

In analogy with the variational principle of Hulthén,[37] the trial function Ψ_i can be required to satisfy the condition that the functional $(\Psi_i,(H-E)\Psi_i)$ should vanish. Since Eqs. (46) are already required to be satisfied, this is equivalent to requiring that

$$(\Phi^k,(H-E)\Psi_i) = 0 \tag{49}$$

Eq. (49) also follows from Eq. (47). When expressed in terms of matrix elements, Eq. (49) becomes a formula for the parameter E,

$$E - H_{00} = \varepsilon_k + (k|k)^{-1}\sum_{ab}(ik|R|ab)c_i^{ab} \tag{50}$$

Here ε_k is the Hartree–Fock energy of the continuum orbital, $\frac{1}{2}k^2$ in atomic units.

The last term in Eq. (50) represents a correction to the energy of the scattering atom. This can be seen by considering an orbital that vanishes at infinity, so that $(k|k)$ is finite and can be set equal to unity. Then this orbital could be taken to be one of the set ϕ_b, giving an energy correction to $H_{00} + \varepsilon_k$

$$\Delta E_i = \sum_a (ik|R|ak)c_i^{ak} \tag{51}$$

This is identical with the result obtained from Eq. (18a) for the first-order Bethe–Goldstone energy increment of a stationary state described by a Hartree–Fock wave function Φ_0 for the scattering atom. When ϕ_k does not vanish at infinity, it has finite amplitude throughout an infinite volume, and the norm $(k|k)$ becomes infinite. Under these conditions the coefficients c_i^{ab} are finite as are the individual integrals $(ik|R|ab)$. Hence, the last term in Eq. (50) vanishes unless the sum $\sum_{ab}$ diverges. In practice, the complete set of normalized orbitals would be truncated and the sum $\sum_{ab}$ would consist of a finite number of finite terms, precluding any finite energy shift ΔE. If static correlation corrections to the scattering atom are important, the implied energy shift should be taken into account by including terms like Φ_i^{ak} in the wave function.

In addition to the evaluation and manipulation of large numbers of matrix elements of the atomic Hamiltonian, there are special problems arising from the nature of the continuum orbital ϕ_k in applying the present method to scattering theory. Coulombic integrals must be evaluated in which the charge densities contain oscillatory as well as exponentially decreasing functions. In using the method outlined above, a continuum integro-differential equation, Eq. (47), must be solved. However, since this equation has the same formal structure as the Hartree–Fock equations, considerable experience in the necessary numerical methods is available. One advantage of the proposed method is that exchange effects are taken into account in the basic formalism. This has been a source of difficulty in earlier work.

7. ORGANIZATION OF VARIATIONAL BETHE–GOLDSTONE CALCULATIONS

It is not possible here to give a detailed description of the computer programs used in the present work. A general outline of the computational procedure will be given, together with some details of methods used at crucial points in the computations.

There are three relatively independent stages of calculation, outlined in an earlier publication.[25] Stage A consists of the choice of a finite set of basis orbitals and evaluation of all the one- and two-electron integrals that represent matrix elements between these basis orbitals for operators in the many-electron Hamiltonian or in other operators whose expectation values are to be computed. The program used to evaluate these atomic integrals in the present work has been described previously.[25]

Stage B of the overall calculation consists of a matrix Hartree–Fock calculation, in the orbital basis chosen in Stage A. The program used, which carries out calculations by the truncated Hartree–Fock method[8] for open-shell configurations, has also been described previously.[25] The matrix of the Hartree–Fock operator $\mathscr{H}_0$, given by Eq. (10), or some suitable average of this matrix over m_l and m_s values in an open-shell configuration, is diagonalized over the entire set of basis orbitals. This gives in general a large set of orthonormal unoccupied orbitals, orthogonal to the occupied Hartree–Fock orbitals, with orbital energies as widely spread out as possible due to the extremal properties of eigenvalues of an Hermitian matrix. As a consequence of the constraints incorporated into the truncated Hartree–Fock method, these orbitals are all angular momentum and spin eigenfunctions, with well-defined values of quantum numbers l, m_l, and m_s. Within sets of orbitals with given values of these quantum numbers, the radial factors are orthonormal. They are indexed by an integer n in the present work, starting with $l + 1$ and increasing in order of increasing orbital energy parameter (mean value of $\mathscr{H}_0$). Thus the index n, especially for unoccupied orbitals, which must be thought of as wave packets in the continuum, has a somewhat different interpretation than the corresponding principal quantum number of hydrogenic orbitals. For occupied orbitals the index n has its usual significance. In Bethe–Goldstone calculations on light atoms, no further transformation of the orbitals has been carried out. For similar calculations on molecules, and possibly on heavier atoms where orbital subshells have a higher degree of degeneracy, a transformation to localized orbitals might be desirable before beginning the Bethe–Goldstone calculations.[26,38]

Stage C of the overall calculation in the present work involves construction and diagonalization of the hierarchy of configuration interaction matrices ($H_{\mu\nu}$) described above, in Section 3. The one- and two-electron integrals computed in Stage A are transformed to the basis of orthonormal orbitals generated in Stage B. The program used in transforming the two-electron integrals, which requires special organization because of the large number of these quantities, has been described previously.[25] A list of integers, giving the number of values of index n to be used for each angular quantum number l, is supplied from input data, so that not all of the orthonormal orbitals obtained in Stage B need be included in the configuration interaction calculations. Orbitals are included in order of increasing orbital energy values. The programs are arranged so that the integral transformation, followed by a series of Bethe–Goldstone calculations, can be repeated in a loop when a new list of integers, specifying the number of orthonormal radial functions to be used, is supplied from external data read by the program. Each Bethe–Goldstone equation uses configuration interaction matrices whose dimension is governed by the total number of orthonormal orbitals specified for the immediately preceding integral transformation. In this way, a series of calculations can be carried out with systematically increasing orbital sets, without repeating Stages A and B of the calculations. An increase in the original basis set takes place in an outer loop that includes repetition of Stages A and B.

Given lists of transformed integrals, stored in auxiliary memory files in a unique sequence of lists, each of which has its entries in a canonical order so that individual elements can be located by computation,[25] each Bethe–Goldstone calculation is initiated by reading a list of occupied orbital indices $ijk\ldots$ from external data. The configuration interaction matrix corresponding to the trial function $\Psi_{ijk\ldots}$ is then constructed, using Eqs. (7), (8), and (9) to express matrix elements $H_{\mu\nu}$ in terms of integrals taken from the previously stored lists of one- and two-electron integrals.

For each Bethe–Goldstone equation, the lowest eigenvalue and corresponding eigenvector are computed by a rapidly converging iterative method especially designed for use with very large matrices.[13] Matrices up to 1200 by 1200 occurred in the calculations described in Section 5. Mean values of operators other than the Hamiltonian are computed by Eq. (38) after each Bethe–Goldstone calculation. After each Bethe–Goldstone calculation a subsequent calculation with the same orbital set is initiated by reading a new list of occupied orbital indices, without repeating the transformation of one- and two-electron integrals.

Although an efficient general method is available for constructing many-particle eigenfunctions of $\mathbf{L}^2$ and $\mathbf{S}^2$, and of computing matrix elements between these symmetry-adapted functions,[39] in the present work it was concluded, because of the existence of the very rapid method for computation of eigenvectors of large matrices used here, that it would be more efficient to construct configuration interaction matrices directly in a basis of single Slater determinants rather than in a basis of symmetry-adapted functions. In terms of Slater determinants, the formalism presented in Section 3 is used directly in the calculations. Some modification of the formalism is required if the Hartree–Fock state cannot be represented by a single Slater determinant. In particular, Φ_0 must be replaced by a symmetry-adapted function, a linear combination of determinants. This is not necessary for the atomic ground states considered here.

Individual matrix elements $H_{\mu\nu}$ are computed independently of each other. The Slater determinants Φ_μ and Φ_ν are each characterized by a list of pairs of indices $(ai)(bj)$ . . . which denote a determinant $\Phi_{ij\ldots}^{ab\cdots}$ in terms of substitutions of orbitals occupied in the reference state Φ_0. Each orbital index is represented by an encoded computer word containing l, n, m_s, and m_l, in the form of a single positive number. These encoded indices can be ordered and compared by single computer operations. To compute $H_{\mu\nu}$, the index pair lists for μ and ν are compared and reordered, using a simple algorithm, to produce the list of index pairs that represents Φ_μ in the form of a determinant constructed from Φ_ν as reference state, together with a relative sign indicator used to reverse the sign of the matrix element if an odd number of interchanges of orbitals is required in Φ_μ to bring it into this form. This reduced list of index pairs, expressing Φ_μ in terms of excitations from Φ_ν, is used to determine whether the integral vanishes, or to select the appropriate formula, Eqs. (6) or (7) with $\mathscr{H}_0$ replaced by $\mathscr{H}_\nu$, for non-diagonal matrix elements. Diagonal elements are computed from Eq. (15). Matrix elements of $\mathscr{H}_\nu$ are expressed in terms of $\mathscr{H}_0$ by use of Eq. (12). Finally, matrix elements of $\mathscr{H}_0$ are expressed in terms of matrix elements of the operator $\mathscr{H}_{\text{Av}}$ actually used in the Hartree–Fock calculations for open-shell states by adding certain two-electron integrals arising from exchange integrals that involve orbitals occupied in the open shells of Φ_0.[40]

Two-electron integrals of the form indicated in Eq. (6) can be expressed as linear combinations of reduced matrix elements, the generalized Slater parameters $(ai|jb)^k$, which are integrals over the radial factors of the indicated orbital functions only.[9] The coefficients in these linear combinations represent integrals over spherical harmonics and spin summations. They can be expressed in terms of standard vector coupling coefficients by well-known formulas.[9] Only the reduced matrix elements are transformed and stored. The required angular momentum coefficients are computed as needed by a subroutine based on the standard Wigner $3j$ symbol.[41]

References

1. Nesbet, R. K., in *Advances in Quantum Chemistry*, Ed. P.-O. Löwdin, Academic Press, Inc., New York, 1967, pp. 1–24.
2. Nesbet, R. K., *J. Chem. Phys.*, **40,** 3619 (1964).
3. Hartree, D. R., *Proc. Cambridge Phil. Soc.*, **24,** 89, 111 (1928).
4. Fock, V., *Z. Physik*, **61,** 126 (1930); **62,** 795 (1930).
5. Kelly, H. P., this volume, p. 129.
6. Löwdin, P.-O., *Advan. Chem. Phys.*, **2,** 207 (1959).
7. Sinanoğlu, O., *Advan. Chem. Phys.*, **6,** 315 (1963).
8. Nesbet, R. K., *Advan. Chem. Phys.*, **9,** 321 (1965).
9. Condon, E. U., and Shortley, G. H., *The Theory of Atomic Spectra*, Cambridge University Press, New York, 1951, pp. 169–174.
10. Nesbet, R. K., *Phys. Rev.*, **109,** 1632 (1958).
11. Brillouin, L., Actualités sci. et ind. No. 159, Hermann et Cie., Paris, 1934.
12. Møller, C., and Plesset, M. S., *Phys. Rev.*, **46,** 618 (1934).
13. Nesbet, R. K., *J. Chem. Phys.*, **43,** 311 (1965).
14. Goldstone, J., *Proc. Roy. Soc. (London), Ser. A*, **239,** 267 (1957).
15. Hurley, A. C., Lennard-Jones, J. E., and Pople, J. A., *Proc. Roy. Soc. (London), Ser. A*, **220,** 446 (1953).
16. Nesbet, R. K., *Phys. Rev.*, **100,** 228 (1955).
17. Brueckner, K. A., *Phys. Rev.*, **96,** 508 (1954); **97,** 1953 (1955).
18. Bethe, H. A., and Goldstone, J., *Proc. Roy. Soc. (London), Ser. A*, **238,** 551 (1957).
19. Sinanoğlu, O., *J. Chem. Phys.*, **36,** 706, 3198 (1962).
20. Geller, M., Taylor, H. S., and Levine, H. B., *J. Chem. Phys.*, **43,** 1727 (1965).
21. Szasz, L., *Z. Naturforschung*, **14a,** 1014 (1959).
22. Boys, S. F., *Proc. Roy. Soc. (London), Ser. A*, **201,** 125 (1950); **217,** 136 (1953).
23. Nesbet, R. K., and Watson, R. E., *Phys. Rev.*, **110,** 1073 (1958).
24. Watson, R. E., *Phys. Rev.*, **119,** 170 (1960).
25. Nesbet, R. K., *Rev. Mod. Phys.*, **35,** 552 (1963).
26. Boys, S. F., *Rev. Mod. Phys.*, **32,** 296 (1960); *Quantum Theory of Atoms, Molecules and the Solid State*, Ed. P.-O. Löwdin, Academic Press, New York, 1966, pp. 253–262.
27. Clementi, E., *J. Chem. Phys.*, **40,** 1944 (1964).
28. Nesbet, R. K., *Phys. Rev.*, **155,** 51 (1967).
29. Nesbet, R. K., *Phys. Rev.*, **155,** 56 (1967).
30. Clementi, E., *J. Chem. Phys.*, **38,** 2248 (1963).
31. Kelly, H. P., *Phys. Rev.*, **131,** 684 (1963); **136,** B896 (1964).
32. McKoy, V., and Sinanoğlu, O., *J. Chem. Phys.*, **41,** 2689 (1964).
33. Nesbet, R. K., in *Quantum Theory of Atoms, Molecules, and the Solid State*, Ed. P.-O. Löwdin, Academic Press, New York, 1966, pp. 157–165; Colloques Internationaux du CNRS, No. 164 (1967), pp. 87–94.

34. Clementi, E., *IBM J. Res. Develop.*, **9**, Supplement (1965), Tables 4–1, 10–1.
35. Nesbet, R. K., *Phys. Rev.*, **156**, 99 (1967).
36. Mittleman, M. H., *Phys. Rev.*, **147**, 69 (1966).
37. Hulthén, L., *Kgl. Fys. Sallskapets Lund Förhandl.*, **14**, 1 (1944).
38. Edmiston, C., and Ruedenberg, K., *Rev. Mod. Phys.*, **35**, 457 (1963); *Quantum Theory of Atoms, Molecules, and the Solid State*, Ed. P.-O. Löwdin, Academic Press, New York, 1966, pp. 263–280.
39. Nesbet, R. K., *J. Math. Phys.*, **2**, 701 (1961).
40. Nesbet, R. K., *Proc. Roy. Soc. (London), Ser. A*, **230**, 312 (1955).
41. Edmonds, A. R., *Angular Momentum in Quantum Mechanics*, Princeton University Press, Princeton, New Jersey, 1957.

ON THE USE OF THE CLUSTER EXPANSION AND THE TECHNIQUE OF DIAGRAMS IN CALCULATIONS OF CORRELATION EFFECTS IN ATOMS AND MOLECULES

JIŘÍ ČÍŽEK, *Institute of Physical Chemistry, Czechoslovak Academy of Sciences, Prague, Czechoslovakia*

CONTENTS

1. INTRODUCTION

A. General Remarks

During the last decade the mathematical methods of quantum-field theory have been successfully used for the study of correlation effects in both solid-state and nuclear physics.[1] These methods, due to their simplicity, have made it possible to reformulate former theories in a very elegant manner and, simultaneously, made it possible to carry out a number of actual computations. Because of the formal similarity between problems of solid-state and nuclear physics and those of atomic and molecular physics one may expect that in the latter these methods will also be very useful.[2]

The aim of this article is to show how the mathematical methods of quantum-field theory may be used to calculate the wave function describing the non-degenerate ground state of an atomic or molecular system as well as to calculate the mean value of an arbitrary operator. We shall be mainly concerned with the mathematics of this problem, while the possible use of this methodology for the reformulation or even unification of the former theories will only be sketched out. Toward the end of the article we shall briefly present the first results obtained using the suggested methods.

B. Mathematical Methods of the Quantum-Field Theory

The basic concepts used in quantum-field theory are the following:[3]

[1]. The formalism of creation and annihilation operators.
[2]. Particle–hole formalism.
[3]. Wick's theorem.
[4]. Diagram techniques.
[5]. Time-dependent perturbation theory and the adiabatical hypothesis; and eventually, the time-independent perturbation theory using convolutions.

One can say that the first two concepts (e.g. creation and annihilation operators and the particle–hole formalism) represent a "new language". Nevertheless, this "new language" is much simpler and much more useful from certain points of view than the commonly used language. In addition, the particle–hole formalism enables us to establish an analogy between the quantum electrodynamics and the quantum mechanics of many-body systems: the so-called "hole" corresponds to the positron while the "particle" in the excited state corresponds to the electron. Of course, this analogy is only a very formal one, nevertheless, it enables us to apply directly the well-known mathematical method of quantum-field theory to the many-particle problem.

Wick's theorem and diagrammatic techniques represent new mathematical tools which are very powerful, very simple to use, and which enable a simple physical interpretation of certain considerations at the same time.

The new form of the perturbation theory is much simpler than the classical one. Simultaneously, some very difficult problems of classical perturbation theory, as, for example, the problem of unlinked terms, have been completely solved in the new form. Moreover, the new perturbation theory implies that there exists a physical interpretation of the individual terms of the perturbation expansion.

In this article we shall be concerned with the non-perturbational approach to the problem of wave function calculation. Consequently, only the first four concepts will be used.

C. The Formulation of the Problem

Let us consider a system having a non-degenerate ground state consisting of fixed atomic nuclei and of $2n$ electrons. Then, neglecting relativistic and magnetic effects, the Hamiltonian $\hat{H}$ of our problem is given by the following equations:

$$\hat{H} = \hat{Z} + \hat{V} \tag{1}$$

$$\hat{Z} = \sum_{i} \hat{z}_i, \; \hat{V} = \sum_{i<j} \hat{v}_{i,j} \tag{2}$$

where $\hat{z}_i$ is a one-particle operator corresponding to the sum of the kinetic and nuclear field energies and $\hat{v}_{i,j}$ is a two-particle operator of the interelectronic Coulombic repulsion. Our problem is to find the ground-state eigenvector and the corresponding eigenvalue E of the Hamiltonian (1):

$$\hat{H}|\Psi\rangle = E|\Psi\rangle \tag{3}$$

The exact wave function will be written in the form of a cluster expansion. Various forms of cluster expansions have been used in statistical and nuclear physics. The first use of a cluster expansion in atomic and molecular physics is due to Sinanoğlu[4a,b] and the usefulness of this approach has also been discussed by Primas[4c] and da Providencia.[4d] In a cluster-type expansion the exact wave function is expressed in the following form:

$$|\overline{\Psi}\rangle = e^{\hat{T}}|\Phi\rangle \tag{4}$$

where $|\Phi\rangle$ represents the ground-state wave function in the one-particle approximation. Therefore, the operator $\hat{T}$ is a logarithm of the so-called "wave operator".*

We shall now derive a system of equations which will allow us to calculate the individual components of the operator $\hat{T}$. This system may very simply be written in an implicit form. Substituting Eq. (4) into the Schrödinger equation (3) and by multiplying by $\langle\Phi|$ from the left we get

$$E = \langle\Phi|\hat{H}|e^{\hat{T}}\Phi\rangle \tag{5}$$

using the normalization: $\langle\Phi|\overline{\Psi}\rangle = 1$. We observe that the energy expression (5) is not symmetric. It is, however, well known that in the variational procedure a symmetric expression is used. The symmetry of the energy expression represents the main difference between the variational and suggested methods.

Substituting from Eqs. (4) and (5) into the Schrödinger equation (3) and subtracting the expression $\langle\Phi|\hat{H}|\Phi\rangle e^{\hat{T}}|\Phi\rangle$ on both sides we obtain

$$(\hat{H} - \langle\Phi|\hat{H}|\Phi\rangle)e^{\hat{T}}|\Phi\rangle = (\langle\Phi|\hat{H}|e^{\hat{T}}\Phi\rangle - \langle\Phi|\hat{H}|\Phi\rangle)e^{\hat{T}}|\Phi\rangle \tag{6}$$

This is an implicit form of the system of non-linear equations for the components of the operator $\hat{T}$ and the expression $(\langle\Phi|\hat{H}|e^{\hat{T}}\Phi\rangle$

* Nomenclature used by P.-O. Löwdin.

$-\langle\Phi|\hat{H}|\Phi\rangle)$ on the right-hand side represents the correlation energy.

Our next task will be to rewrite this system in an explicit form. This problem has already been partly solved by Coester and Kümmel;[5] they wrote Eq. (6) using the formalism of the creation and annihilation operators in a particle–hole form. They simplified this equation by using the commutation rules for the operator $e^{\hat{T}}$ with one- and two-particle operators. Nevertheless, the final equations obtained by Coester and Kümmel are rather complicated and, at least in the author's opinion, are not very convenient for practical calculations.

Coester and Kümmel did not consider Wick's theorem and diagrammatic techniques. We shall show that by using these concepts the implicit system of equations (6) may be written in an explicit form, which is suitable for practical computations as well as for a general discussion and comparison with the other methods which were previously used for the study of correlation effects.

In addition to the calculation of the wave function in the form of Eq. (4) we shall calculate the following expression

$$\frac{\langle e^{\hat{T}}\Phi|(\hat{L}-\langle\Phi|\hat{L}|\Phi\rangle)|e^{\hat{T}}\Phi\rangle}{\langle e^{\hat{T}}\Phi|e^{\hat{T}}\Phi\rangle} \tag{7}$$

where $\hat{L}$ is an arbitrary spin-independent one- or two-particle operator. The expression (7) represents a difference of mean values of an operator $\hat{L}$ calculated with the exact wave function and with the wave function in one-particle approximation. As will be shown, Wick's theorem and diagrammatic techniques are very useful in this calculation too since they enable us to express the numerator as a product of the denominator and of some additional expression, using only the topological properties of diagrams. We shall be using considerations very similar to those used by Thouless[6] and Kelly[7] in their perturbational calculation of mean values of operators; nevertheless, our approach will be a non-perturbative one similar to that of da Providencia.[8]

Let us further mention that a certain analogy exists between our approach and some approaches used in statistical physics.[9] Nonetheless, this analogy will not be outlined in this article.

The use of diagrammatic methods gives us two possibilities as far as the diagram assignment is concerned: one can assign to each given diagram either a non-antisymmetrized or an antisymmetrized matrix element. Each possibility has its merits: assignment of antisymmetrized matrix elements to diagrams has the advantage of reducing the number of different diagrams which have to be introduced; this simplifies all subsequent considerations. The other possibility is desirable for our spin-independent Hamiltonian since then we can exclude the spin variables from explicit consideration. Indeed, if one assigns to each diagram a corresponding non-antisymmetrized matrix element one can then exclude the spin variables by requiring that the spin orientation be conserved along each oriented particle or hole line of each diagram.

Therefore, we can state that in general the assignment of antisymmetrized matrix elements to diagrams is to be preferred when we are interested in general considerations, while for actual calculations in atomic and molecular physics the assignment of non-antisymmetrized matrix elements is preferable. In the author's paper*[10] only the latter version was used, namely the non-antisymmetrized matrix elements have been assigned to the individual diagrams. We thought it instructive to present here both possibilities so that the advantages and disadvantages of both versions will become apparent.

In ending these remarks let us notice that in nuclear physics only the former version, assigning antisymmetrized matrix elements to diagrams, is useful. This is immediately obvious if we remember that the nuclear forces are spin-dependent, so that one cannot eliminate the spin variables in such a simple way as one can for the Hamiltonian (1) considered in this paper.

D. Approximations Used

In actual computations we shall restrict ourselves to a consideration of the two-particle part of the operator $\hat{T}$, e.g. we shall suppose

$$\hat{T} \cong \hat{T}_2 \tag{8}$$

This means that the form of our wave function will be identical with the wave function used by Sinanoğlu.[4a] Nevertheless, the

* This paper will be hereafter referred to as C.

present method of calculating the wave function differs from Sinanoğlu's variational techniques,[4a,11] some of which are equivalent* to those recently used by Nesbet[12b,12c] and represent a special case of the method described below (compare Reference 13). In order to distinguish these methods we shall refer to the pairs version of Sinanoğlu's theory the "decoupled-pair-many-electron theory", while the method presented in C will be called a "coupled-pair-many-electron theory". We hope that the latter method will be useful in a discussion of the questions associated with a so-called "remainder"[4a] in molecules of "metallic" character.

2. BASIC TERMS AND DEFINITIONS

A. Creation and Annihilation Operators

In this section reference to C will be made frequently and only the most pertinent parts of that paper will be reproduced here.

We introduce a complete set of orthonormal spin-orbitals

$$|A\rangle = |a\rangle|\alpha\rangle| \tag{9}$$

where $|a\rangle$ and $|\alpha\rangle$ designate the space and the spin parts of the spin-orbital $|A\rangle$, respectively. Generally, capital letters will be used to designate spin-orbitals, lower case letters will be associated with orbitals, while Greek letters will be used for spin functions. In addition to this general system of spin-orbitals (9) we shall use spin-orbitals whose space parts are eigenfunctions of the operator

$$(\hat{z} + \hat{u})|a\rangle = \omega_a|a\rangle \tag{10}$$

where $\hat{u}$ is an arbitrary spin-independent one-particle Hermitian operator. Let us note explicitly that Hartree–Fock spin-orbitals fall within this class.

The creation and annihilation operators defined in the system of spin-orbitals (9) will be designated $\hat{X}_A^+$ and $\hat{X}_A$, respectively.

* We have in mind mainly two-particle equations. Three- and more-particle equations are discussed by Nesbet[12b,12c] (cf. also Section 6.A). The relationship between the many-particle equations of Nesbet [12b,12c] and some related equations[14] in nuclear physics will be discussed in a special paper.

These operators satisfy the following anticommutation relations:

$$\hat{X}_A^+\hat{X}_B^+ + \hat{X}_B^+\hat{X}_A^+ = 0$$
$$\hat{X}_A\hat{X}_B + \hat{X}_B\hat{X}_A = 0 \tag{11}$$
$$\hat{X}_A^+\hat{X}_B + \hat{X}_B\hat{X}_A^+ = \langle A|B\rangle$$

The operators $\hat{Z}$ and $\hat{V}$ may then be expressed through the operators $\hat{X}_A^+$ and $\hat{X}_A$ in the following forms:

$$\hat{Z} = \sum_{C_1,B_1} \langle C_1|\hat{z}|B_1\rangle \hat{X}_{C_1}^+\hat{X}_{B_1} \tag{12}$$

$$\hat{V} = \frac{1}{2^c} \sum_{C_1,B_1;C_2,B_2} \langle C_1C_2|\hat{v}|B_1B_2\rangle_c \hat{X}_{C_1}^+\hat{X}_{C_2}^+\hat{X}_{B_2}\hat{X}_{B_1} \tag{13}$$

$$(c = 1, 2)$$

where

$$\langle C_1C_2|\hat{v}|B_1B_2\rangle_2 = \langle C_1C_2|\hat{v}|B_1B_2\rangle_1 - \langle C_1C_2|\hat{v}|B_2B_1\rangle_1 \tag{14}$$

A non-antisymmetrized matrix element is usually denoted $\langle C_1C_2|\hat{v}|B_1B_2\rangle_1 = \langle C_1C_2|\hat{v}|B_1B_2\rangle$. The matrix elements $\langle C_1|\hat{z}|B_1\rangle$ and $\langle C_1C_2|\hat{v}|B_1B_2\rangle_1$ are defined in accordance with Dirac's notation.*

Due to the spin-independency of our Hamiltonian the following relations are valid:

$$\langle C_1C_2|\hat{v}|B_1B_2\rangle_1 = \langle c_1c_2|\hat{v}|b_1b_2\rangle\langle\gamma_1|\beta_1\rangle\langle\gamma_2|\beta_2\rangle \tag{15}$$
$$\langle C_1|\hat{z}|B_1\rangle = \langle c_1|\hat{z}|b_1\rangle\langle\gamma_1|\beta_1\rangle$$

In Eq. (15) the notation introduced in (9) is used.

Later on we shall calculate the average values of the one- or two-particle spin-independent operators $\hat{L}$. Naturally, these operators may be expressed by means of creation and annihilation operators in the same way as the operators $\hat{Z}$ and $\hat{V}$.

B. Hole–Particle Formalism

The symbol $|0\rangle$ will be used to specify the "vacuum state".

Let us suppose that in the complete set of spin-orbitals introduced there exist $2n$ spin-orbitals

$$|A_1'\rangle, |A_2'\rangle, \ldots, |A_{2n}'\rangle \tag{16}$$

* Nesbet[12] uses the following nomenclature:

$$\langle C_1C_2|\hat{v}|B_1B_2\rangle_2 = \langle C_1C_2|\hat{R}|B_1B_2\rangle$$

such that the function

$$|\Phi\rangle = \hat{X}^+_{A_1'} \hat{X}^+_{A_2'} \ldots \hat{X}^+_{A_{2n}'} |0\rangle \tag{17}$$

is a reasonable approximation to the exact ground-state wave function $|\Psi\rangle$. Since we require the ground state to be non-degenerate the spin-orbitals (16) may be formed as products of the following n orbitals:

$$|a_1'\rangle, |a_2'\rangle, \ldots |a_n'\rangle \tag{18}$$

with spin functions corresponding to the positive and negative spin orientations, respectively. The spin-orbitals (16) which span the function $|\Phi\rangle$ will be called non-excited ones while all others, which do not enter the function $|\Phi\rangle$, will be called excited spin-orbitals. The non-excited and excited spin-orbitals will be systematically designated by singly and doubly primed capitals, respectively.

In the following considerations it will be useful to regard the ground state $|\Phi\rangle$ as a "new vacuum state" as well as to interpret the excitation of a particle as the creation of a hole in a non-excited spin-orbital followed by the creation of a particle in an excited spin-orbital. For this purpose a new set of operators is introduced by the following relations:

$$\begin{aligned} \hat{Y}^+_{A''} &= \hat{X}^+_{A''}, \qquad \hat{Y}_{A''} = \hat{X}_{A''} \\ \hat{Y}^+_{A'} &= \hat{X}_{A'}, \qquad \hat{Y}_{A'} = \hat{X}^+_{A'} \end{aligned} \tag{19}$$

The operator $\hat{Y}^+_{A'}$ may now be interpreted as a hole-creating operator while the operator $\hat{Y}_{A'}$ is a hole-annihilating operator. The formal analogy of this approach with quantum electrodynamics is apparent: an electron corresponds to a particle in an excited spin-orbital while a positron corresponds to a hole in a non-excited spin-orbital.

In the following considerations the terms creation and annihilation operators will be associated with the operators $\hat{Y}^+_A$ and $\hat{Y}_A$, respectively. These operators satisfy the same anticommutation relations as the operators $\hat{X}^+_A$, $\hat{X}_A$.

The operator $\hat{T}$ introduced above may be expressed through the operators $\hat{Y}^+_{A''}$, $\hat{Y}^+_{A'}$, in the following form:

$$\hat{T} = \sum_{j=1}^{2n} \hat{T}_j \tag{20}$$

where

$$\hat{T}_j = \left(\frac{1}{j!}\right)^c \sum_{\substack{A_j''\ldots A_1'' \\ A_j'\ldots A_1'}} \langle A_j'' \ldots A_1''|\hat{t}|A_j' \ldots A_1'\rangle_c \prod_{i=1}^{j} \hat{Y}^+_{A_i''} \hat{Y}^+_{A_i'} \qquad (21)$$

$$(c = 1, 2)$$

represents an operator of the j-fold excited state.

To establish a relationship between antisymmetrized and non-antisymmetrized matrix elements of the operator $\hat{T}$ let us define a permutation P;

$$P = \begin{pmatrix} j & \ldots & 1 \\ k_j & \ldots & k_1 \end{pmatrix}$$

Then we get

$$\langle A_j'' \ldots A_1''|\hat{t}|A_j' \ldots A_1'\rangle_2 = \sum_P (-1)^P \langle A_j'' \ldots A_1''|\hat{t}|A_{k_j}' \ldots A_{k_1}'\rangle_1 \qquad (22)$$

C. Wick's Theorem and Contractions

In the following we shall use both Wick's theorem and the generalized Wick's theorem. In order to state these theorems one has to define a normal product of operators, a contraction, and a normal product with contractions. We shall not repeat these definitions or Wick's theorem here—the reader is referred to a book by Schweber[3] or to C. Let us only recall some relations for the contractions of two operators which, together with Eq. (19), will be essential in the following discussion:

$$\cdot\hat{Y}^+_A \cdot\hat{Y}^+_B = \cdot\hat{Y}_A \cdot\hat{Y}_B = \cdot\hat{Y}^+_A \cdot\hat{Y}_B = 0$$

$$\cdot\hat{Y}_A \cdot\hat{Y}^+_B = \langle A|B\rangle \qquad (23)$$

Using Wick's theorem we immediately get the following relations:

$$\hat{X}^+_{C_1}\hat{X}_{B_1} = N[\hat{X}^+_{C_1}\hat{X}_{B_1}] + N[\cdot\hat{X}^+_{C_1}\cdot\hat{X}_{B_1}] \qquad (24)$$

$$\begin{aligned} \hat{X}^+_{C_1}\hat{X}^+_{C_2}\hat{X}_{B_2}\hat{X}_{B_1} = {} & N[\hat{X}^+_{C_1}\hat{X}^+_{C_2}\hat{X}_{B_2}\hat{X}_{B_1}] \\ & + N[\cdot\hat{X}^+_{C_1}\hat{X}^+_{C_2}\cdot\hat{X}_{B_2}\hat{X}_{B_1}] + N[\cdot\hat{X}^+_{C_1}\hat{X}^+_{C_2}\hat{X}_{B_2}\cdot\hat{X}_{B_1}] \\ & + N[\hat{X}^+_{C_1}\cdot\hat{X}^+_{C_2}\cdot\hat{X}_{B_2}\hat{X}_{B_1}] + N[\hat{X}^+_{C_1}\cdot\hat{X}^+_{C_2}\hat{X}_{B_2}\cdot\hat{X}_{B_1}] \\ & + N[\cdot\hat{X}^+_{C_1}\hat{X}^{\cdot\cdot +}_{C_2}\cdot\hat{X}_{B_2}\cdot\cdot\hat{X}_{B_1}] + N[\cdot\hat{X}^+_{C_1}\cdot\cdot\hat{X}^+_{C_2}\cdot\cdot\hat{X}_{B_2}\cdot\hat{X}_{B_1}] \end{aligned} \qquad (25)$$

omitting the vanishing terms. Using relations (23)–(25) and (19) one can write the operators $\hat{Z}$ and $\hat{V}$ in the following form:

$$\hat{Z} = \sum_{C_1,B_1} \langle C_1|\hat{z}|B_1\rangle N[\hat{X}^+_{C_1}\hat{X}_{B_1}] + \langle\Phi|\hat{Z}|\Phi\rangle \tag{26}$$

$$\hat{V} = \frac{1}{2^c} \sum_{C_1,B_1;C_2,B_2} \langle C_1C_2|\hat{v}|B_1B_2\rangle_c N[\hat{X}^+_{C_1}\hat{X}^+_{C_2}\hat{X}_{B_2}\hat{X}_{B_1}] + \sum_{C_1,B_1} \langle C_1|\hat{g}|B_1\rangle N[\hat{X}^+_{C_1}\hat{X}_{B_2}] + \langle\Phi|\hat{V}|\Phi\rangle \tag{27}$$

$$(c = 1, 2)$$

where

$$\langle C_1|\hat{g}|B_1\rangle = \sum_{A'} \langle C_1A'|\hat{v}|B_1A'\rangle_2 \tag{28}$$

For this matrix element we can further write

$$\langle C_1|\hat{g}|B_1\rangle = \langle c_1|\hat{g}|b_1\rangle\langle\gamma_1|\beta_1\rangle \tag{29}$$

where

$$\langle c_1|\hat{g}|b_1\rangle = \sum_{a'} (2\langle c_1a'|\hat{v}|b_1a'\rangle - \langle c_1a'|\hat{v}|a'b_1\rangle) \tag{30}$$

Again, the one- or two-particle operators, the mean values of which we shall calculate later on, may be expressed in the same way as the operators $\hat{Z}$ and $\hat{V}$. Using relations (26)–(29) the total Hamiltonian may be written as follows:

$$\hat{H} - \langle\Phi|\hat{H}|\Phi\rangle = \sum_{C_1,B_1} \langle C_1|\hat{f}|B_1\rangle N[\hat{X}^+_{C_1}\hat{X}_{B_1}] + \frac{1}{2^c} \sum_{C_1,B_1;C_2,B_2} \langle C_1C_2|\hat{v}|B_1B_2\rangle_c N[\hat{X}^+_{C_1}\hat{X}^+_{C_2}\hat{X}_{B_2}\hat{X}_{B_1}] \tag{31}$$

$$(c = 1, 2)$$

where

$$\langle C_1|\hat{f}|B_1\rangle = \langle C_1|\hat{z}|B_1\rangle + \langle C_1|\hat{g}|B_1\rangle \tag{32}$$

This matrix element may be written using (25) and (27) in the following manner:

$$\langle C_1|\hat{f}|B_1\rangle = \langle c_1|\hat{f}|b_1\rangle\langle\gamma_1|\beta_1\rangle \tag{33}$$

where

$$\langle c_1|\hat{f}|b_1\rangle = \langle c_1|\hat{z}|b_1\rangle + \langle c_1|\hat{g}|b_1\rangle$$

Before closing this section let us emphasize that the transcription of the operators (12) and (13) into the new forms (27) and (28) will result in a significant simplification of our considerations and will be very useful in introducing diagrammatical techniques.

3. INTRODUCTION TO "GRAPHOLOGY"

A. General Remarks

The generalized Wick theorem enables us to express a product of normal products of creation and annihilation operators as a sum of normal products of creation and annihilation operators with contractions. The use of the generalized Wick theorem represents a rather complicated combinatorial problem. In order to simplify the solution of this problem it is very useful to introduce some topological concepts, e.g. concepts of the skeleton and the diagram. The concept of the skeleton will be purely topological while that of the diagram will possess direct quantum-mechanical meaning.

Before defining these concepts let us briefly mention the main scheme which we shall follow. We represent the product of normal products by the diagram consisting of vertices and of both oriented and non-oriented lines. By "vertex" we mean a point in which several lines of a diagram are joined together. In this diagram each operator from the product of normal products of operators is represented by an oriented line leaving or entering some vertex of the diagram. We shall represent each normal product with contractions (playing a role in a generalized Wick's theorem) in such a way that every two lines corresponding to contracted operators will be joined together in the diagram mentioned above. The details of the structure of the diagram must be chosen in such a way that we should be able to write easily the explicit form of the normal product with contractions from the topological properties of the diagram.

We shall deal not only with a product of normal products but with sums of expressions of this kind. For that reason it is necessary to introduce the idea of skeletons which is more general than that of diagrams. The concept of a skeleton will enable us to simplify our considerations concerning the sums of products of normal products.

B. Definition and Properties of Skeletons

In order to develop the formal theory we shall give a number of definitions. Let us begin with the definition of a network of vertices: the topological formation consisting of two vertices interconnected by one dashed line will be called a small network while that consisting of $(j + 1)$ vertices $(j \geqslant 1)$ in which one vertex is connected with each of the remaining j vertices by a full non-oriented line will be referred to as a big network. Using the concept of a network of vertices we define a skeleton: it is a topological formation consisting of one or zero small networks and of an arbitrary number (zero inclusive) of big networks, where full oriented lines are entering (and/or leaving, and/or interconnecting) the vertices of these networks.

We obtain a degenerate skeleton from a skeleton if we replace each big and each small network of vertices by one vertex; the latter will be called, respectively, big and small vertices. Every full oriented line leaving (entering) a vertex in a non-degenerate skeleton will leave (enter) that vertex of a corresponding degenerate skeleton by which the network of vertices of a non-degenerate skeleton, containing the vertex mentioned, was replaced. This means that a degenerate skeleton will consist only of oriented lines and of small and big vertices. Small vertices will be denoted by a small full circle and big ones by a big full circle. It is apparent that the set of non-degenerate skeletons is homomorphically mapped on the set of degenerate skeletons by the above rule.

Further, we shall deal with some properties of skeletons and degenerate skeletons. Two skeletons (degenerate skeletons) are defined to be equivalent if they can be mutually transformed into each other through topological deformation. In the case of degenerate skeletons the topological deformations are allowed to transform only big vertices into big vertices and small vertices into small ones. Two skeletons are said to be quasi-equivalent if the corresponding degenerate skeletons are equivalent.

The topological deformation converting a given skeleton (degenerate skeleton) into itself is called an automorphism. The weight of a skeleton (degenerate skeleton) is the reciprocal value of the number of all different automorphisms, including the identical one, of the given skeleton (degenerate skeleton). The degenerate weight of a skeleton is the weight of a corresponding degenerate skeleton. Of course, two equivalent skeletons (quasi-equivalent skeletons) have the same weight (degenerate weight).

Two oriented lines of a skeleton (degenerate skeleton) are considered to be equivalent if there exists an automorphism transferring these lines one into the other. Two oriented lines of a skeleton are considered to be quasi-equivalent if these lines are equivalent in a corresponding degenerate skeleton. Two vertices of a skeleton or degenerate skeleton are equivalent if an automorphism exists which transforms these vertices mutually one into the other.

Generally, it is difficult to give an explicit formula for the weight of a skeleton. On the other hand, one can derive a rather simple formula for the weight of a degenerate skeleton, which does not have equivalent vertices, using considerations due to Hugenholtz.[3c] All lines of this special degenerate skeleton may be so classified that each class contains mutually equivalent lines. Let us suppose that this classification yields k classes, each containing l_i, $(i = 1, \ldots, k)$, oriented lines. Then the weight of this special degenerate skeleton is given by

$$\prod_{i=1}^{k} (l_i!)^{-1} \tag{34}$$

In the following considerations we define four different types of skeletons, namely the H, T, M, and R skeletons. Non-equivalent skeletons of the same type are distinguished by their arguments:

$$S_H(h),\ S_T(t),\ S_M(m),\ S_R(r)$$

Therefore, equivalent skeletons are denoted by identical symbols. The respective weights and degenerate weights are designated by

$$w_H^c(h),\ w_T^c(t),\ w_M^c(m),\ w_R^c(r) \tag{35}$$

where $c = 1$ corresponds to weight* and $c = 2$ corresponds to degenerate weight.

Further, let a positive integer be associated with any full oriented line in such a way that the following rules are respected:

[1]. Quasi-equivalent lines carry different integers.

* The weight of a skeleton was designated in C by the same symbol as here but without the superscript.

[2]. Let us have two equivalent skeletons. Then a topological deformation exists which transfers the skeletons into each other in such a way that the mutually corresponding full oriented lines in both skeletons carry the same integers.

Therefore, the possibility that different lines are labelled by the same integer is not excluded.

C. *H*, *T*, and *M* Skeletons and Diagrams

The H skeletons are those shown in Fig. 1. The left skeleton in Fig. 1 will be designated $S_H(1)$, the right one $S_H(2)$. The assumed

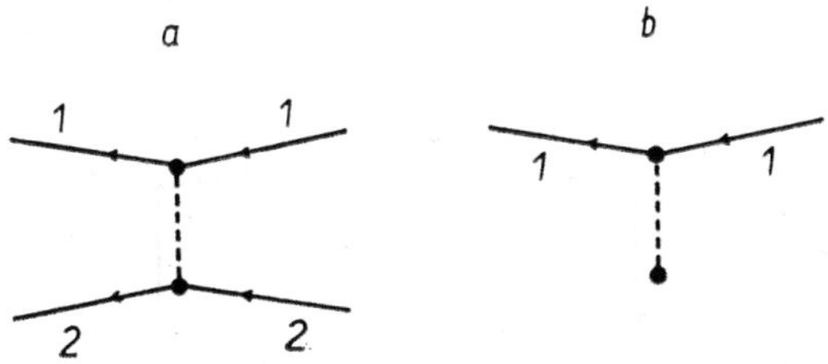

Fig. 1. (a) $S_H(1)$ skeleton, (b) $S_H(2)$ skeleton.

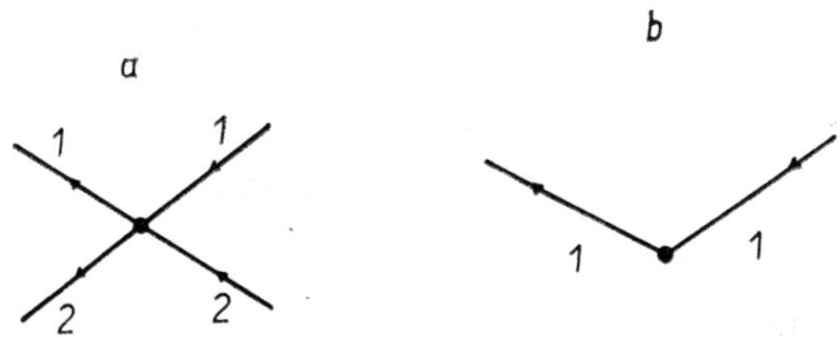

Fig. 2. (a) $S_H(1)$ degenerate skeleton, (b) $S_H(2)$ degenerate skeleton.

numbering of lines is shown as well. The corresponding degenerate skeletons are drawn in Fig. 2.

The weights and degenerate weights associated with H skeletons are:

$$w_H^c(1) = 2^{-c} \qquad w_H^c(2) = 1 \tag{36}$$

$$(c = 1, 2)$$

On labelling the oriented lines of an H skeleton with spin-orbital indices we obtain a so-called H diagram. These diagrams are shown in Fig. 3. To the diagram of Fig. 3a we assign the following two scalar quantities:

$$d_H^c(1, \chi) = \langle C_1 C_2 | \hat{v} | B_1 B_2 \rangle_c \tag{37}$$
$$(c = 1, 2)$$

as well as the following two operators:

$$\hat{D}_H^c(1, \chi) = d_H^c(1, \chi) N[\hat{X}_{C_1}^+ \hat{X}_{C_2}^+ \hat{X}_{B_2} \hat{X}_{B_1}] \tag{38}$$
$$(c = 1, 2)$$

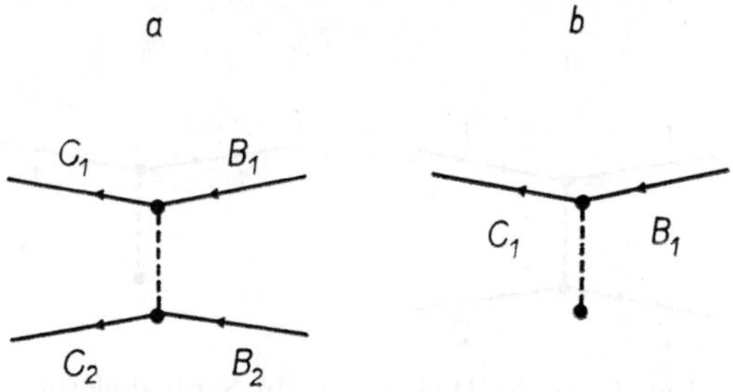

Fig. 3. (a) A diagram formed from the $S_H(1)$ skeleton, (b) a diagram formed from the $S_H(2)$ skeleton.

where χ represents the sequence of numbers C_1, C_2, B_1, B_2.* Similarly the scalar quantities

$$d_H^c(2, \chi) = \langle C_1 | \hat{f} | B_1 \rangle \tag{39}$$
$$(c = 1, 2)$$

and operators

$$\hat{D}_H^c(2, \chi) = d_H^c(2, \chi) N[\hat{X}_{C_1}^+ \hat{X}_{B_1}] \tag{40}$$
$$(c = 1, 2)$$

are assigned to the diagram of Fig. 3b. Here χ represents the numbers C_1, B_1.

Further, to the oriented line labelled $B_i(C_i)$ and entering (leaving) the vertex of the H diagram, the operator $\hat{X}_{B_i}(\hat{X}_{C_i}^+)$ is

* Scalar quantities and operators introduced in C are equal to scalar quantities and operators defined here for $c = 1$.

assigned ($i = 1, 2$). Consequently, (31) may be written in the following form:

$$\hat{H} - \langle\Phi|\hat{H}|\Phi\rangle = \sum_{h=1}^{2} w_H^c(h) \sum_{x} \hat{D}_H^c(h, \chi) \tag{41}$$

$$(c = 1, 2)$$

At this point we should emphasize that the operator assignment to the diagrams as defined in this paper differs from the current usage[3b,3c,3d] in certain respects, the following two being the most striking ones:

[1]. Normally, one assigns to a diagram a simple product of creation and annihilation operators while here a normal product

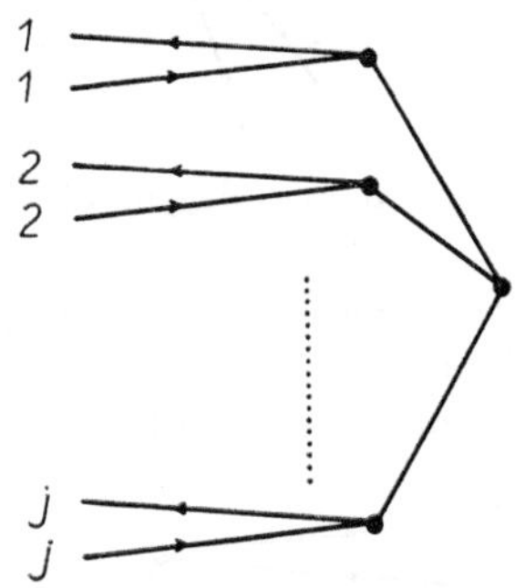

Fig. 4. A general $S_T(j)$ skeleton.

of the same operators is attributed to it. The latter assignment is usually employed in quantum electrodynamics, too (cf. Reference 3a).

[2]. Usually the scalar quantity $-\langle C_1|\hat{u}|B_1\rangle$ is attributed to the H diagram of the second type while here we assign to it the scalar quantity $\langle C_1|\hat{f}|B_1\rangle$.

Let us further observe that, generally speaking, in the theories formulated through non-antisymmetrized matrix elements it is more natural to use the Goldstone[3b] type of diagrams while in those using antisymmetrized quantities Hugenholtz[3c] type diagrams are more appropriate. Furthermore, we find it convenient to draw the interaction lines even in the Hugenholtz-type diagrams since in this way an unambiguous correspondence between the diagrams and the pertinent operators may be achieved.[15] Indeed, if these diagrams are drawn without the interaction lines difficulties in sign assignment arise.

Let us turn our attention to T-type skeletons. A general T-type skeleton as well as the labelling of its oriented lines are

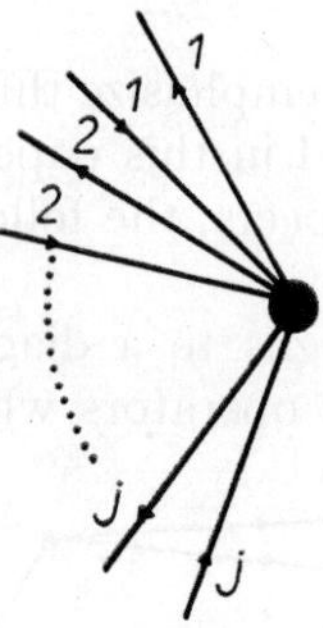

Fig. 5. A general $S_T(j)$ degenerate skeleton.

shown in Fig. 4. These skeletons designated $S_T(j)$ differ in the number of pairs of oriented lines, so that the argument j may be any positive integer. The corresponding degenerate skeleton is

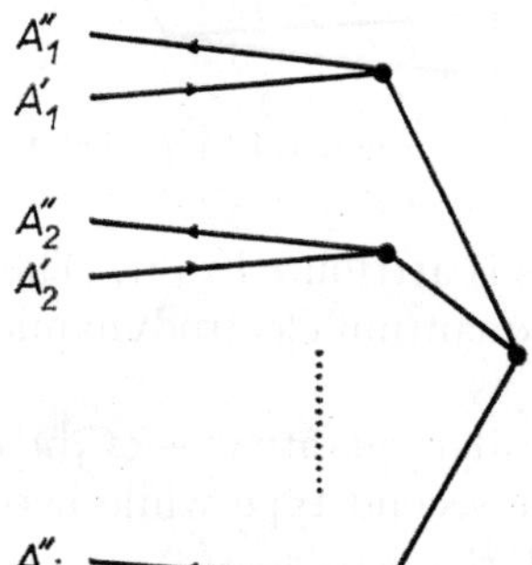

Fig. 6. The diagram formed from the skeleton $S_T(j)$ of Fig. 4.

shown in Fig. 5. The weight and degenerate weight of the $S_T(j)$ skeleton is given by the following relation

$$w_T^c(j) = (j!)^{-c} \tag{42}$$

$$(c = 1, 2)$$

On labelling the oriented lines entering (leaving) the vertices of the T skeleton by the indices of non-excited (excited) spin-orbitals we obtain the T diagram illustrated in Fig. 6. To each T diagram we assign the following two scalar quantities

$$d_T^c(j, \tau) = \langle A_j'' \ldots A_1''|\hat{t}|A_j' \ldots A_1'\rangle_c \tag{43}$$

$$(c = 1, 2)$$

and the operators:

$$\hat{D}_T^c(j, \tau) = d_T^c(j, \tau) \prod_{i=1}^{j} \hat{Y}_{A_i''}^{+} \hat{Y}_{A_i'}^{+} \tag{44}$$

$$(c = 1, 2)$$

where τ represents the sequence of numbers

$$A_1'', A_1', \ldots A_j'', A_j'$$

Furthermore, we shall associate the operator $\hat{Y}_{A'_i}^{+}(\hat{Y}_{A'_i}^{+})$ with every oriented line labelled $A_i'(A_i')$ and entering (leaving) some vertex of the T diagram.

Using the notation specified above, the operator $\hat{T}$ may be expressed in the form

$$\hat{T} = \sum_{t=1}^{2n} w_T^c(t) \sum_{\tau} \hat{D}_T^c(t, \tau) \tag{45}$$

$$(c = 1, 2)$$

A set of T skeletons is called an M skeleton. Generally, a given M skeleton is composed of n_1 skeletons $S_T(t_1)$; n_2 skeletons $S_T(t_2)$, etc. The M skeletons are designated $S_M(m)$, where m represents a sequence of indices t_i of the T skeletons involved, the order of the individual indices being immaterial. Among the M skeletons we also include the so-called empty M skeleton consisting of the empty set of T skeletons.

The weight and degenerate weight of the M skeleton are given by the following formula[3d]

$$w_M^c(m) = \prod_{i=1}^{k} \{[w_T^c(t_i)]^{n_i}/n_i!\} \tag{46}$$

$$(c = 1, 2)$$

In deriving formula (46) it is important to realize that equivalency and quasi-equivalency of T and M skeletons means the same thing. The weight and degenerate weight of the empty M skeleton is equal to unity by definition.

On labelling the oriented lines of an M skeleton with spin-orbital indices in the same way as specified for a T skeleton we obtain the so-called M diagram. Therefore, an M diagram is composed of a certain number of T diagrams. Scalar quantities $d^c_M(m, \mu)$ and operators $\hat{D}^c_M(m, \mu)$ associated with an M diagram are given by the products of the corresponding quantities assigned to the T diagrams constituting a given M diagram (unity and the unit operator are assigned to an empty M diagram by definition). Further, μ specifies the sequence of spin-orbital indices assigned to the oriented lines of an M diagram. Using the concept of M skeleton and M diagram, we can write the wave operator $e^{\hat{T}}$ in the following form:

$$e^{\hat{T}} = \sum_m w^c_M(m) \sum_\mu \hat{D}^c_M(m, \mu) \tag{47}$$

$$(c = 1, 2)$$

4. EQUATIONS FOR MATRIX ELEMENTS OF THE OPERATOR $\hat{T}$

A. Formulation of Auxiliary Theorems

In order to transform the Schrödinger equation (6) into an explicit system of non-linear equations for matrix elements of the operator $\hat{T}$ we have to evaluate the following vector:

$$(\hat{H} - \langle\Phi|\hat{H}|\Phi\rangle)e^{\hat{T}}|\Phi\rangle$$
$$= \sum_h \sum_m w^c_H(h) w^c_M(m) \sum_\chi \sum_\mu \hat{D}^c_H(h, \chi)\hat{D}^c_M(m, \mu)|\Phi\rangle \tag{48}$$

$$(c = 1, 2)$$

We shall carry out this evaluation simultaneously for the non-antisymmetrized version ($c = 1$) as well as for antisymmetrized one ($c = 2$).

For further transformations of Eq. (48) we need two auxiliary theorems that will be presented later on. But before stating these theorems we have to introduce the idea of an R skeleton and R diagram.

An R skeleton is formed from one H and one M skeleton by connecting some lines (none, one, or two) entering the vertices of the H skeleton with lines leaving the vertices of the M skeleton and, further, by connecting some lines (none, one, or two) leaving the vertices of the H skeleton with lines entering the vertices of the M skeleton. The lines connecting the vertices of the H and M skeletons will be called internal lines; all other lines will be called external. The internal hole lines of the R skeleton are those which start from a vertex of the H skeleton and enter a vertex of the M skeleton. The internal lines having the opposite direction—that is, starting at a vertex of the M skeleton and entering a vertex of the H skeleton—will be called internal particle lines. Furthermore, the external hole (particle) lines of an R skeleton are those external lines which enter (leave) a vertex of pertinent M or H skeletons. A sequence of oriented lines will generally be referred to as a path. A closed path will be called a loop. We shall number the lines of an R skeleton independently of the numberings of the corresponding M and H skeletons in the following manner.

Let the R skeleton considered contain j open paths, h internal hole lines, and k internal particle lines. We shall number the external hole lines $1, 2 \ldots j$ and the internal hole lines $j + 1, \ldots, j + h$. Similarly, the external particle lines will be numbered $1, \ldots, j$ and the internal particle lines $j + 1, \ldots, j + k$. The numbering of the external lines is carried out in such a way that the external particle and hole lines associated with the same path have the same number. The weight and the degenerate weight of an R skeleton are given by the definitions stated in Section 3.

By labelling the ith particle (hole) line of the R skeleton with the index of an excited (non-excited) spin-orbital $D_i''(D_i')$ we obtain the R diagram. On dissecting the internal lines of the R diagram we obtain the H and M diagrams. Let $d_H^c(h, \chi)$ and $d_M^c(m, \mu)$ be scalar quantities associated with these H and M diagrams. Then the scalar quantity $d_R^c(r, \sigma)$ assigned to this diagram is given by the expression

$$d_R^c(r, \sigma) = (-1)^{l+h} d_H^c(h, \chi) d_M^c(m, \mu) \qquad (49)$$
$$(c = 1, 2)$$

where l is the number of loops, h is the number of internal hole lines and σ represents the following two sequences σ_1, σ_2

$$\sigma_1 = D_1'', D_1', D_2'', D_2' \ldots D_j'', D_j' \text{ (external lines)}$$

$$\sigma_2 = D_{j+1}'' \ldots D_{j+k}'', D_{j+1}' \ldots D_{j+h}' \text{ (internal lines)}$$

For diagrams lacking external lines we shall write

$$d_R^c(r, \sigma) = d_R^c(r, \sigma_2) \tag{50}$$

$$(c = 1, 2)$$

The operator associated with an R diagram is given by the following equation:

$$\hat{D}_R^c(r, \sigma) = d_R^c(r, \sigma) \prod_{i=1}^{j} \hat{Y}_{D_i''}^{+} \hat{Y}_{D_i'}^{+} \tag{51}$$

$$(c = 1, 2)$$

Again, for diagrams having only internal lines we shall write

$$\hat{D}_R^c(r, \sigma) = d_R^c(r, \sigma_2) \tag{52}$$

$$(c = 1, 2)$$

Let Π^1 designate a set of topologically distinct R skeletons formed from all possible H and M skeletons. The set Π^1 may be divided into several disjunct subsets, each of which contains topologically quasi-equivalent R skeletons. Furthermore, any set of R skeletons containing one representative skeleton from each subset of quasi-equivalent R skeletons will be denoted Π^2.

We are now ready to state the auxiliary theorems mentioned above. A proof of these theorems is given in the Appendix. We can state both theorems simultaneously:

$$w_H^c(h) w_M^c(m) \sum_{\chi} \sum_{\mu} \hat{D}_H^c(h, \chi) \hat{D}_M^c(m, \mu) |\Phi\rangle$$

$$= \sum_{r \in \Pi^c}' w_R^c(r) \sum_{\sigma} \hat{D}_R^c(r, \sigma) |\Phi\rangle \tag{53}$$

$$(c = 1, 2)$$

where the primed summation on the right-hand side of this equation means that the sum extends only over those R skeletons which have the structure $r \in \Pi^c$ and which are formed from one H skeleton and one M skeleton having the structures h and m, respectively.

B. Transformation of the Schrödinger Equation

The Schrödinger may now be given the following form:

$$\sum_{r\in\Pi^c} w^c_R(r) \sum_{\sigma} \hat{D}^c_R(r, \sigma)|\Phi\rangle = (E - \langle\Phi|\hat{H}|\Phi\rangle)e^{\hat{T}}|\Phi\rangle \qquad (54)$$

$$(c = 1, 2)$$

We shall now distinguish connected and disconnected R skeletons. In a connected R skeleton any two vertices are connected by at least one sequence of lines consisting of full oriented, non-oriented, and dashed lines. A disconnected R skeleton is composed of a connected R skeleton and an M skeleton. The weight of the disconnected R skeleton is given by the product of the weights of its constituent connected R skeleton and M skeleton. The scalar quantities and operators assigned to disconnected R diagrams are obtained in an analogous manner. On the basis of the above considerations, the Schrödinger equation may be rewritten in the form

$$e^{\hat{T}}[\sum_{r\in\Delta^c} w^c_R(r) \sum_{\sigma} \hat{D}^c_R(r, \sigma) - (E - \langle\Phi|\hat{H}|\Phi\rangle)]|\Phi\rangle = 0 \qquad (55)$$

$$(c = 1, 2)$$

where Δ^1 is a set of topologically distinct connected R skeletons. The set Δ^1 may be divided into several disjunct subsets each containing topologically quasi-equivalent connected R skeletons. Further we shall designate by Δ^2 any set of connected R skeletons formed in such a way that from *each* subset of the quasi-equivalent connected R skeletons mentioned above we select one representative skeleton.

Equation (55) will be satisfied if the following equations hold

$$\sum_{r\in\Delta_0{}^c} w^c_R(r) \sum_{\sigma_2} d^c_R(r, \sigma_2) = E - \langle\Phi|\hat{H}|\Phi\rangle \qquad (56)$$

$$(c = 1, 2)$$

$$\sum_{r\in\Delta_j{}^c} w^c_R(r) \sum_{\sigma_1\in\Theta^c} x^c(\sigma_1) \sum_{\sigma_2} d^c_R(r, \sigma_1, \sigma_2) = 0 \qquad (57)$$

$$(j \geq 1,\ c = 1, 2)$$

where Δ_j^c is a subset of Δ^c consisting of the skeletons having j open paths. The sets Θ^c are formed in accordance with the following rule.

Let us consider the sequence of indices

$$D''_1, D'_1, \ldots, D''_j, D'_j \tag{58}$$

Further, let us consider two arbitrary permutations of indices $1 \ldots j$:

$$\begin{pmatrix} 1 & \ldots & j \\ k_1 & \ldots & k_j \end{pmatrix} \qquad \begin{pmatrix} 1 & \ldots & j \\ l_1 & \ldots & l_j \end{pmatrix} \tag{59}$$

If the sequence (58) belongs to Θ^1, then the sequence

$$D''_{k_1}, D'_{k_1} \ldots D''_{k_j}, D'_{k_j} \tag{60}$$

also belongs to Θ^1. If the sequence (58) belongs to Θ^2, so does the sequence

$$D''_{k_1}, D'_{l_1} \ldots D''_{k_j}, D'_{l_j} \tag{61}$$

Let P_K and P_L be the parities of the first and second permutations (59), respectively, and further let $P = P_K - P_L$. The quantity $x^c(\sigma_1)$ is equal to unity for $c = 1$ and is equal to $(-1)^P$ for $c = 2$.

Let us note that if we consider a skeleton having j open paths then each set Θ^1 contains $j!$ sequences while each Θ^2 contains $(j!)^2$ sequences.

We shall further specify the form of the Eqs. (57) for the case that the spin-orbitals are eigenfunctions of the operator $(\hat{z} + \hat{u})$. In this case the system of equations (57) will have the following form

$$\sum_{i=1}^{j} (\omega_{Di''} - \omega_{Di'}) \langle D''_1 \ldots D''_j | \hat{t} | D'_1 \ldots D'_j \rangle_c + \sum_{r \in \Delta_j{}^c} w^c_R(r) \sum_{\sigma_1 \in \Theta^c} x^c(\sigma_1) \sum_{\sigma_2} \tilde{d}^c_R(r, \sigma_1, \sigma_2) = 0 \tag{62}$$

$$(c = 1, 2)$$

where the indices $D''_1, D'_1, \ldots, D''_j, D'_j$ form one of the sequences belonging to Θ^c. Further the tilde of $\tilde{d}_R$ indicates that the following scalar quantity is associated with the right diagram of Fig. 3:

$$\begin{aligned} \langle C_1 | \hat{q} | B_1 \rangle &= \langle C_1 | \hat{f} | B_1 \rangle - \langle C_1 | \hat{z} + \hat{u} | B_1 \rangle \\ &= \sum_{A'} \langle C_1 A' | \hat{v} | B_1 A' \rangle_2 - \langle C_1 | \hat{u} | B_1 \rangle \end{aligned} \tag{63}$$

instead of the scalar quantity (39). The matrix elements $\langle B_1|\hat{q}|C_1\rangle$ are identically equal to zero for Hartree–Fock spin-orbitals, as follows immediately from Eq. (63).

If $c = 1$ one can carry out the summation over the spin variables in Eqs. (56), (57), or (62) since our Hamiltonian is spin independent. In doing so one obtains a system of non-linear equations for the matrix elements of the operator $\hat{T}$ in terms of orbitals. We shall not carry out this summation here and the reader is referred to C.

C. Explicit Form of the Equations for Matrix Elements of the Operator $\hat{T}$ in the Approximation $\hat{T} \cong \hat{T}_2$

In order to illustrate the general formalism presented above we shall write down explicitly the system of equations (56) and

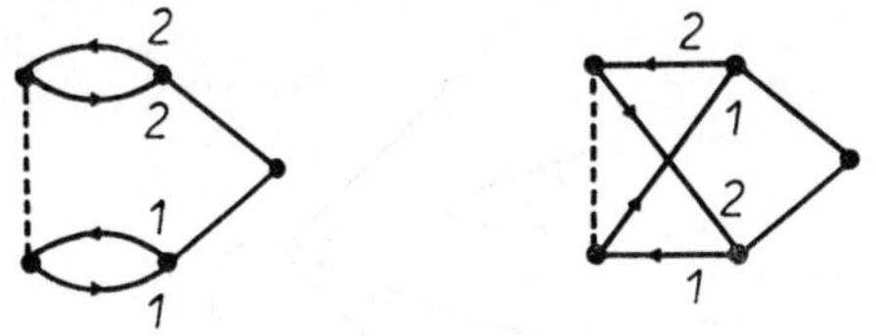

Fig. 7. R skeletons lacking external lines formed from the $S_H(1)$ and $S_T(2)$ skeletons.

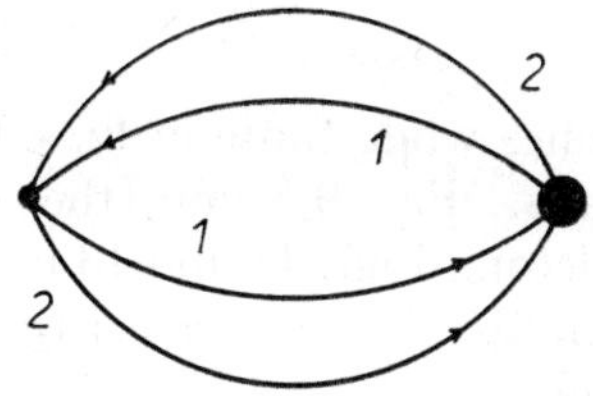

Fig. 8. The degenerate R skeleton corresponding to the skeletons of Fig. 7.

(57) for the case where all components of the operator $\hat{T}$ except $\hat{T}_2$ are disregarded. This approximation, which we call here a "coupled-pair-many-electron theory", was discussed in greater detail in C as was its relation to the work of Sinanoğlu.[4a]

In this special case the M skeletons consist of a set of T skeletons (the empty set being included) each of which has two open paths. The possible R skeletons, having none or two open paths, which can be obtained from these M skeletons are shown in Figs. 7, 9, 11, and 13. The corresponding degenerate skeletons are shown in Figs. 8, 10, 12, and 14. The skeletons shown in Fig. 7 do not

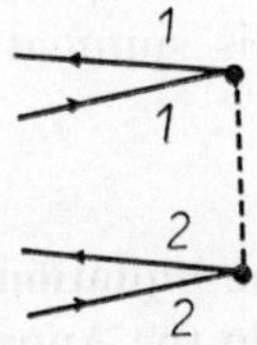

Fig. 9. The R skeleton having two open paths formed from the $S_H(1)$ skeleton and the empty M skeleton.

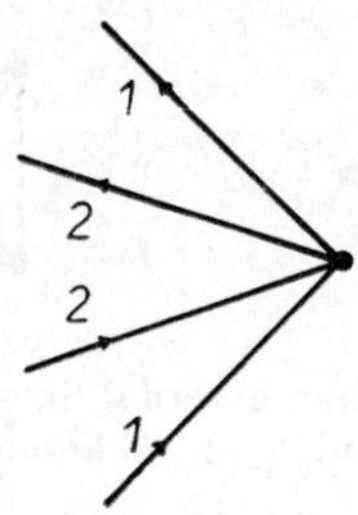

Fig. 10. The degenerate skeleton corresponding to the skeleton of Fig. 9.

have any external lines while those in Figs. 9, 11, and 13 always have two open paths. Fig. 9 shows the R skeleton resulting from the empty skeleton. Figs. 11 and 13 present the R skeletons constructed from the M skeletons consisting of one and two T skeletons, respectively.

The energy difference $(E - \langle\Phi|\hat{H}|\Phi\rangle)$ may now be expressed through the matrix elements over the spin-orbitals as follows:

$$E - \langle\Phi|\hat{H}|\Phi\rangle$$

$$= \frac{1}{2} \sum_{\substack{D_1', D_2' \\ D_1'', D_2''}} \langle D_1'D_2'|\hat{v}|D_1''D_2''\rangle_1 (\langle D_1''D_2''|\hat{t}|D_1'D_2'\rangle_1 - \langle D_1''D_2''|\hat{t}|D_2'D_1'\rangle_1)$$

$$= \frac{1}{4} \sum_{\substack{D_1', D_2' \\ D_1'', D_2''}} \langle D_1'D_2'|\hat{v}|D_1''D_2''\rangle_2 \langle D_1''D_2''|\hat{t}|D_1'D_2'\rangle_2 \quad (64)$$

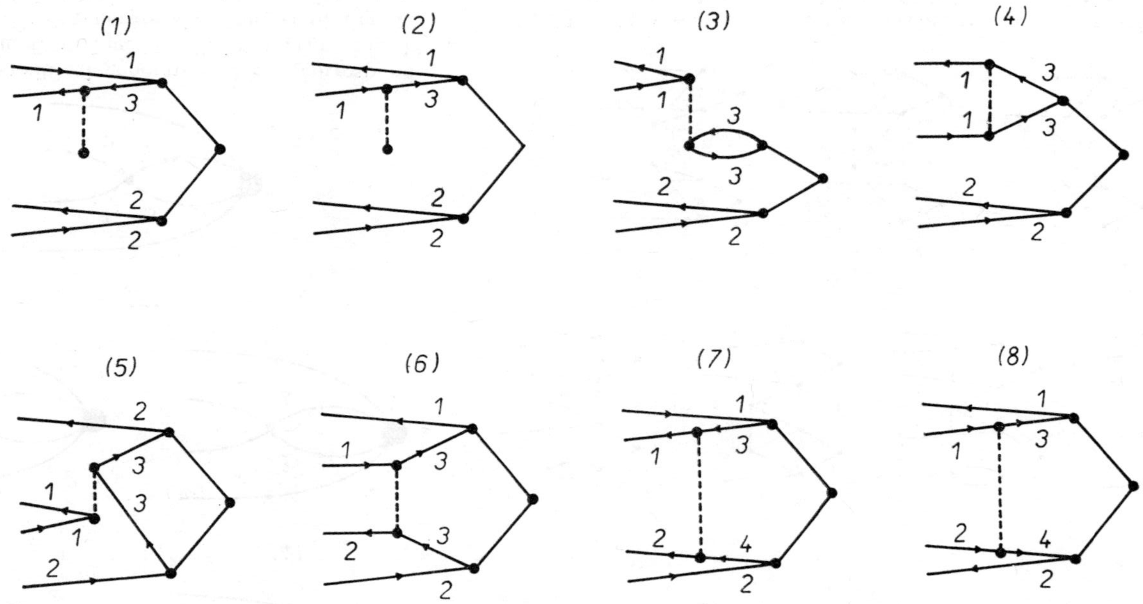

Fig. 11. All possible R skeletons having two open paths formed from the $S_H(1)$ or $S_H(2)$ skeleton and from the M skeleton consisting of one $S_T(2)$ skeleton.

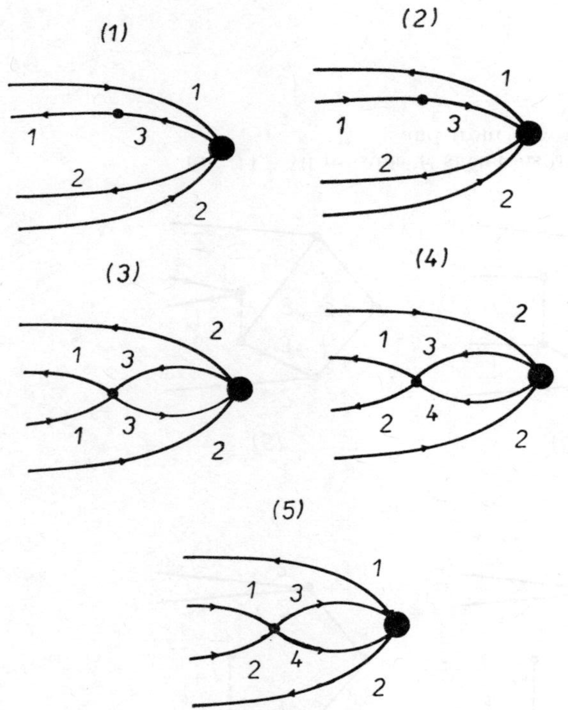

Fig. 12. All possible degenerate skeletons corresponding to skeletons from Fig. 11. Skeleton (1) of Fig. 11 goes over to the degenerate skeleton (1) of Fig. 12, similarly skeleton (2) goes over to the degenerate skeleton (2), skeletons (3), (4), (5), and (6) all go over to the degenerate skeleton (3), skeleton (7) to the degenerate skeleton (4), and finally skeleton (8) to the degenerate skeleton (5).

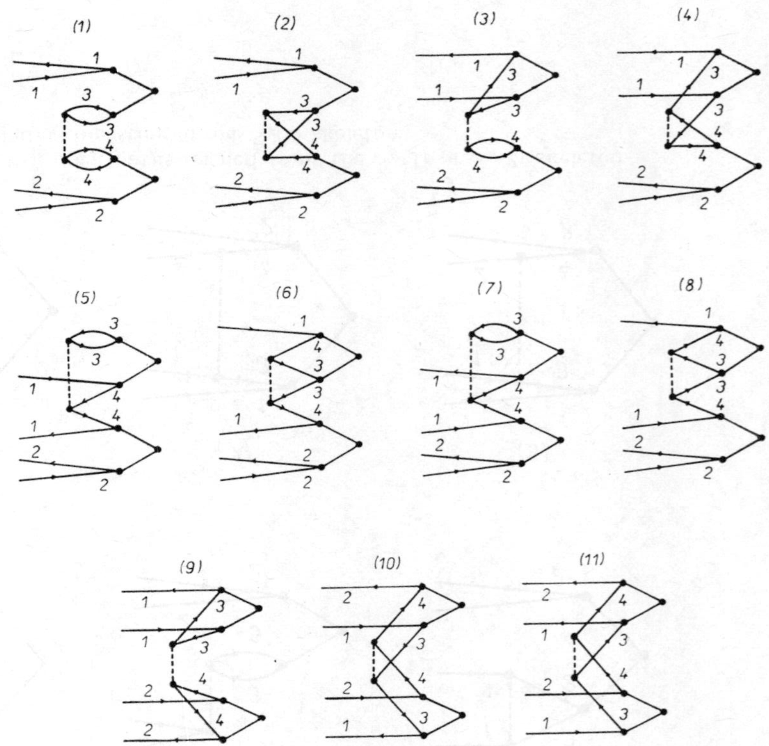

Fig. 13. All possible R skeletons having two open paths formed from the $S_H(1)$ skeleton and the M skeleton consisting of two $S_T(2)$ skeletons.

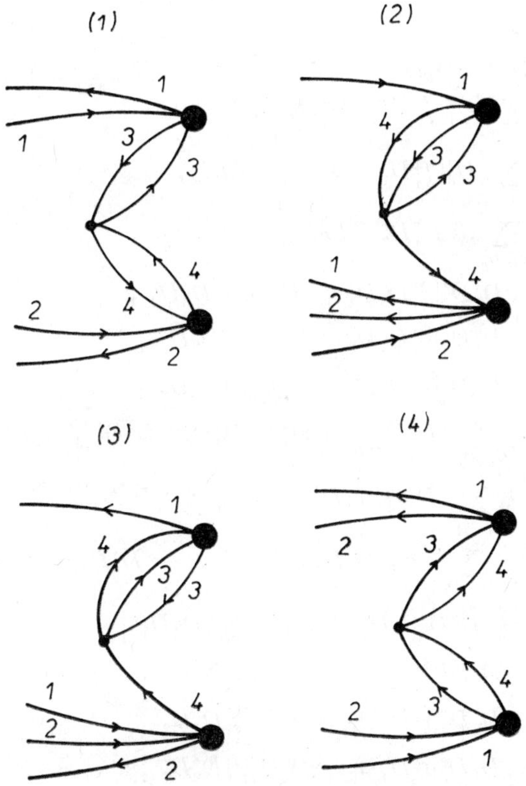

Fig. 14. All possible degenerate skeletons corresponding to skeletons from Fig. 13. Skeletons (1), (2), (3), (4), (9), and (10) of Fig. 13 go over to the degenerate skeleton (1) of Fig. 14, skeletons (5) and (6) go over to degenerate skeleton (2), skeletons (7) and (8) go over to degenerate skeleton (3), and, finally, skeleton (11) goes over to degenerate skeleton (4).

In order to write Eqs. (57) or (62) in the approximation $\hat{T} \simeq \hat{T}_2$ explicitly we introduce

$$\lambda^c(D_1'', D_1';\ D_2'', D_2'),\ c = 1, 2:$$

$$\begin{aligned}
&\lambda^1(D_1'', D_1', D_2'', D_2') \\
&\quad = \tfrac{1}{2}\langle D_1''D_2''|\hat{v}|D_1'D_2'\rangle_1 \\
&\qquad + \sum_{D_3''} \langle D_1''|\hat{f}|D_3''\rangle\langle D_3''D_2''|\hat{t}|D_1'D_2'\rangle_1 \\
&\qquad - \sum_{D_3'} \langle D_3'|\hat{f}|D_1'\rangle\langle D_1''D_2''|\hat{t}|D_3'D_2'\rangle_1 \\
&\qquad + \sum_{D_3', D_3''} [(\langle D_1''D_3'|\hat{v}|D_1'D_3''\rangle_1 \\
&\qquad - \langle D_1''D_3'|\hat{v}|D_3''D_1'\rangle_1)\langle D_3''D_2''|\hat{t}|D_3'D_2'\rangle_1 \\
&\qquad - \langle D_1''D_3'|\hat{v}|D_1'D_3''\rangle_1\langle D_2''D_3''|\hat{t}|D_3'D_2'\rangle_1 \\
&\qquad - \langle D_2''D_3'|\hat{v}|D_3''D_1'\rangle_1\langle D_1''D_3''|\hat{t}|D_3'D_2'\rangle_1] \\
&\qquad + \tfrac{1}{2}\sum_{D_3'', D_4''} \langle D_1''D_2''|\hat{v}|D_3''D_4''\rangle_1\langle D_3''D_4''|\hat{t}|D_1'D_2'\rangle_1 \\
&\qquad + \tfrac{1}{2}\sum_{D_3', D_4'} \langle D_3'D_4'|\hat{v}|D_1'D_2'\rangle_1\langle D_1''D_2''|\hat{t}|D_3'D_4'\rangle_1 \\
&\qquad + \sum_{D_3', D_4';\, D_3'', D_4''} \{[\langle D_3'D_4'|\hat{v}|D_3''D_4''\rangle_1 - \langle D_3'D_4'|\hat{v}|D_4''D_3''\rangle_1] \\
&\qquad \times [\tfrac{1}{2}\langle D_1''D_3''|\hat{t}|D_1'D_3'\rangle_1\langle D_2''D_4''|\hat{t}|D_2'D_4'\rangle_1 \\
&\qquad - \langle D_1''D_3''|\hat{t}|D_3'D_1'\rangle_1\langle D_2''D_4''|\hat{t}|D_2'D_4'\rangle_1 \\
&\qquad - \langle D_3''D_4''|\hat{t}|D_3'D_1'\rangle_1\langle D_1''D_2''|\hat{t}|D_4'D_2'\rangle_1 \\
&\qquad - \langle D_1''D_3''|\hat{t}|D_4'D_3'\rangle_1\langle D_2''D_4''|\hat{t}|D_2'D_1'\rangle_1] \\
&\qquad + \tfrac{1}{2}\langle D_3'D_4'|\hat{v}|D_3''D_4''\rangle_1 \\
&\qquad \times [\langle D_1''D_3''|\hat{t}|D_3'D_1'\rangle_1\langle D_2''D_4''|\hat{t}|D_4'D_2'\rangle_1 \\
&\qquad + \langle D_2''D_3''|\hat{t}|D_4'D_1'\rangle_1\langle D_1''D_4''|\hat{t}|D_3'D_2'\rangle_1 \\
&\qquad + \langle D_1''D_2''|\hat{t}|D_3'D_4'\rangle_1\langle D_3''D_4''|\hat{t}|D_1'D_2'\rangle_1]\}
\end{aligned} \tag{65}$$

$$\begin{aligned}
&\lambda^2(D_1'', D_1';\ D_2'', D_2') \\
&\quad = \tfrac{1}{4}\langle D_1''D_2''|\hat{v}|D_1'D_2'\rangle_2 \\
&\qquad + \tfrac{1}{2}\sum_{D_3''} \langle D_1''|\hat{f}|D_3''\rangle\langle D_3''D_2''|\hat{t}|D_1'D_2'\rangle_2
\end{aligned} \tag{66}$$

$$
\begin{aligned}
&- \tfrac{1}{2} \sum_{D_3'} \langle D_3'|\hat{f}|D_1'\rangle\langle D_1''D_2''|\hat{t}|D_3'D_2'\rangle_2 \\
&+ \sum_{D_3', D_3''} \langle D_1''D_3'|\hat{v}|D_1'D_3''\rangle_2\langle D_3''D_2''|\hat{t}|D_3'D_2'\rangle_2 \\
&+ \tfrac{1}{8} \sum_{D_3'', D_4''} \langle D_1''D_2''|\hat{v}|D_3''D_4''\rangle_2\langle D_3''D_4''|\hat{t}|D_1'D_2'\rangle_2 \\
&+ \tfrac{1}{8} \sum_{D_3', D_4'} \langle D_3'D_4'|\hat{v}|D_1'D_2'\rangle_2\langle D_1''D_2''|\hat{t}|D_3'D_4'\rangle_2 \\
&+ \sum_{D_3', D_4'; D_3'', D_4''} \{\langle D_3'D_4'|\hat{v}|D_3''D_4''\rangle_2 \\
&\times [\tfrac{1}{2}\langle D_1''D_3''|\hat{t}|D_1'D_3'\rangle_2\langle D_2''D_4''|\hat{t}|D_2'D_4'\rangle_2 \\
&- \tfrac{1}{4}\langle D_3''D_4''|\hat{t}|D_3'D_1'\rangle_2\langle D_1''D_2''|\hat{t}|D_4'D_2'\rangle_2 \\
&- \tfrac{1}{4}\langle D_1''D_3''|\hat{t}|D_4'D_3'\rangle_2\langle D_2''D_4''|\hat{t}|D_2'D_1'\rangle_2 \\
&+ \tfrac{1}{16}\langle D_1''D_2''|\hat{t}|D_3'D_4'\rangle_2\langle D_3''D_4''|\hat{t}|D_1'D_2'\rangle_2]\}
\end{aligned} \tag{66}
$$

On the basis of Figs. 9, 11, and 13 we can now write the equations (57) in the following simple form for the case $c = 1$

$$\lambda^1(D_1'', D_1';\ D_2'', D_2') + \lambda^1(D_2'', D_2';\ D_1'', D_1') = 0 \tag{67}$$

For the case $c = 2$ using Figs. 9–14 the Eq. (57) has the form:

$$
\begin{aligned}
&\lambda^2(D_1'', D_1';\ D_2'', D_2') + \lambda^2(D_2'', D_2';\ D_1'', D_1') \\
&\quad - \lambda^2(D_1'', D_2';\ D_2'', D_1') - \lambda^2(D_2'', D_1';\ D_1'', D_2') = 0
\end{aligned} \tag{68}
$$

Permuting indices in (67) we get

$$\lambda^1(D_1'', D_2';\ D_2'', D_1') + \lambda^1(D_2'', D_1';\ D_1'', D_2') = 0 \tag{69}$$

One can see immediately that by subtracting Eq. (69) from (67) we get Eq. (68). This fact follows from general considerations and may also be proved by a direct comparison of Eqs. (59) and (60).

For the orbitals which are eigenfunctions of the operator $(\hat{z} + \hat{u})$ the second and the third terms on the right-hand side of Eqs. (59) and (60) have the form:

$$
\begin{aligned}
&\sum_{D_3''} \langle D_1''|\hat{f}|D_3''\rangle\langle D_3''D_2''|\hat{t}|D_1'D_2'\rangle_c \\
&- \sum_{D_3'} \langle D_3'|\hat{f}|D_1'\rangle\langle D_1''D_2''|\hat{t}|D_3'D_2'\rangle_c \\
&= (\omega_{D_1''} - \omega_{D_1'})\langle D_1''D_2''|\hat{t}|D_1'D_2'\rangle_c \\
&\qquad + \sum_{D_3''} \langle D_1''|\hat{q}|D_3''\rangle\langle D_3''D_2''|\hat{t}|D_1'D_2'\rangle_c \\
&\qquad - \sum_{D_3'} \langle D_3'|\hat{q}|D_1'\rangle\langle D_1''D_2''|\hat{t}|D_3'D_2'\rangle_c
\end{aligned} \tag{70}
$$

$(c = 1, 2)$

Consequently, the following term then appears in expressions (67) or (68)

$$(\omega_{D_1''} + \omega_{D_2''} - \omega_{D_1'} - \omega_{D_2'})\langle D_1''D_2''|\hat{t}|D_1'D_2'\rangle_c \qquad (71)$$
$$(c = 1, 2)$$

Finally, let us note that carrying out the summation over the spin variables in (67) we get the system of equations (60) given in C.

5. THE EVALUATION OF THE MEAN VALUES OF ONE- AND TWO-PARTICLE OPERATORS

A. General Remarks

In contrast to the foregoing sections the operators $\hat{Z}$ and $\hat{V}$ will refer not only to the one-and two-particle parts of the Hamiltonian $\hat{H}$ but will designate arbitrary, spin-independent one- and two-particle operators, respectively. Consequently, $\hat{H}$ will now represent a sum of arbitrary one- and two-particle operators. By properly choosing the matrix elements we can get from $\hat{H}$ the desired one- or two-particle operator we wish to study, e.g. one- or two-particle density matrix operators. We believe that the above convention of operator designation, which will save us from repeating many definitions and equations, will not be misleading in any respect.

In this Section some new types of skeletons and diagrams will have to be defined. Properly, these should have been introduced in Section 3 but we did not wish to overcrowd that Section with definitions not pertinent there. Besides, some notions introduced in Section 4 will be useful to us in defining the skeletons and diagrams needed in the present chapter.

Finally, let us note that the diagrams we are going to introduce in this chapter are directly connected with those introduced by da Providencia.[8] Our approach will differ from that of da Providencia mainly by our consistent use of the topological properties of diagrams and, similarly to the previous Sections, the notion of the automorphism will be employed. Furthermore, for the spinless operators, which represent an important case in the atomic and molecular calculations, the diagrammatic technique which we shall introduce in this Section will allow us to carry out easily the summation over the spin variables.

B. Supplements to "Graphology"

In this Section we shall define four new types of skeletons, namely the $\bar{T}$, $\bar{M}$, N, and D skeletons designated, respectively, by the symbols

$$S_{\bar{T}}(\bar{t}),\ S_{\bar{M}}(\bar{m}),\ S_N(n),\ \text{and}\ S_D(d) \tag{72}$$

Similarly, as in a previous Section, we shall form from these skeletons both degenerate skeletons and diagrams. All definitions from Section 3.B will be valid here as well.

A general $\bar{T}$-type skeleton and the corresponding degenerate skeleton are shown in Figs. 15 and 16, respectively. Labelling the

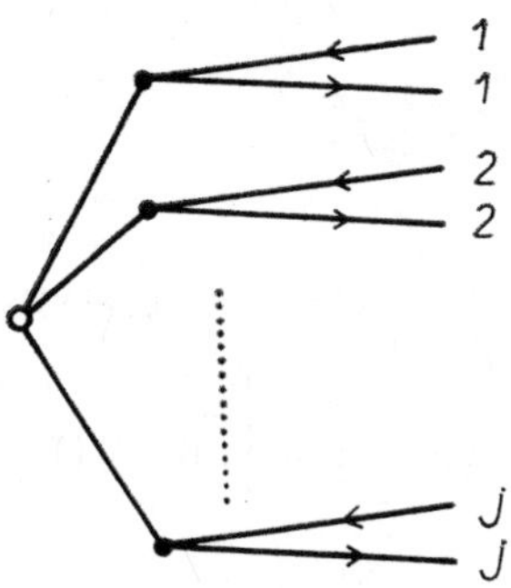

Fig. 15. A general $S_{\bar{T}}(j)$ skeleton.

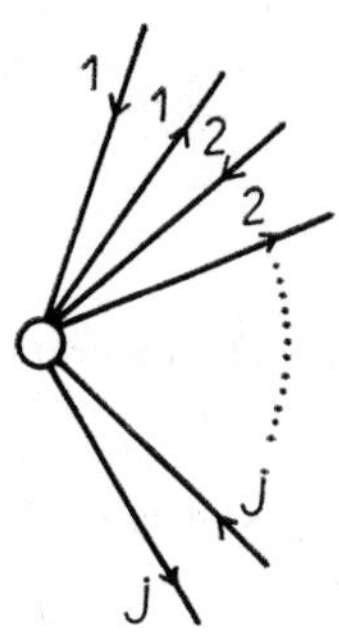

Fig. 16. A general $S_{\bar{T}}(j)$ degenerate skeleton.

oriented lines entering (leaving) the vertices of a $\bar{T}$ skeleton with excited (non-excited) spin-orbital indices we obtain the $\bar{T}$ diagram illustrated in Fig. 17. $\bar{T}$ and T skeletons (degenerate skeletons, diagrams) are distinguished graphically by empty and full central dots, respectively. To a $\bar{T}$ diagram we then assign quantities which are complex conjugates to the quantities (43) and operators which are the Hermitian conjugate of the operators (44). Finally, the $\bar{M}$ skeletons are formed from $\bar{T}$ skeletons in exactly the same way as the M skeletons were obtained from T skeletons, the same also being true for degenerate skeletons and diagrams. Consequently, the following relationship holds:

$$(e^{\hat{T}})^{+} = \sum_{\bar{m}} w^{c}_{\bar{M}}(\bar{m}) \sum_{\bar{\mu}} \hat{D}^{c}_{\bar{M}}(\bar{m}, \bar{\mu}) \tag{73}$$

$$(c = 1, 2)$$

In this formula $\bar{m}$ specifies the structure of the diagram and $\bar{\mu}$ stands for the sequence of spin-orbital indices assigned to the

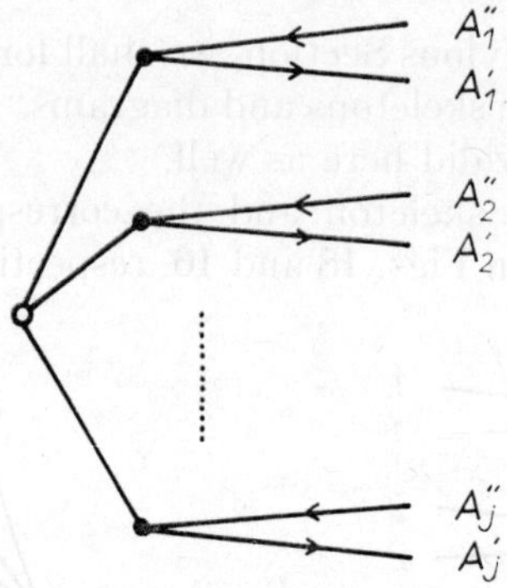

Fig. 17. The diagram formed from the skeleton $S_{\bar{T}}(j)$ of Fig. 15.

oriented lines of the $\bar{M}$ diagram. The dagger designates a Hermitian conjugate.

C. Formulation of Auxiliary Theorems

We shall now turn our attention to the evaluation of both the numerator and denominator of expression (7). For the numerator v_N we get, using (41), (47), (73), and the definitions introduced in this Section, the following expression

$$\begin{aligned} v_N &= \langle e^{\hat{T}}\Phi | \hat{H} - \langle \Phi | \hat{H} | \Phi \rangle | e^{\hat{T}}\Phi \rangle \\ &= \sum_{\bar{m}} \sum_{h} \sum_{m} w^{c}_{\bar{M}}(\bar{m}) w^{c}_{H}(h) w^{c}_{M}(m) \\ &\quad \times \sum_{\bar{\mu}} \sum_{\chi} \sum_{\mu} \langle \Phi | \hat{D}^{c}_{\bar{M}}(\bar{m}, \bar{\mu}) \hat{D}^{c}_{H}(h, \chi) \hat{D}^{c}_{M}(m, \mu) | \Phi \rangle \end{aligned} \tag{74}$$

$$(c = 1, 2)$$

Similarly, for the denominator v_D, we get:

$$v_D = \langle e^{\hat{T}}\Phi | e^{\hat{T}}\Phi \rangle = \sum_{\bar{m}} \sum_{m} w^c_{\bar{M}}(\bar{m}) w^c_M(m) \times \sum_{\bar{\mu}} \sum_{\mu} \langle \Phi | \hat{D}^c_{\bar{M}}(\bar{m}, \bar{\mu}) \hat{D}^c_{\bar{m}}(m, \mu) | \Phi \rangle \tag{75}$$

$$(c = 1, 2)$$

In order to express these formulae explicitly we shall introduce the notion of N and D skeletons, which are analogous to the notion of the R skeleton. The N skeleton is formed from one $\bar{M}$, one H, and one M skeleton is such a way that *each* line leaving (entering) one of these skeletons is joined together with the line entering (leaving) one of the remaining two skeletons. Apparently, the numbers of open paths in the $\bar{M}$ and M skeletons must differ at most by one for $h = 2$ and at most by two for $h = 1$. If all the lines of one $\bar{M}$ and one M skeleton are joined mutually together in such a way that the direction of each line is preserved, the D skeleton results. Evidently both $\bar{M}$ and M skeletons must have the same number of open paths.

All the lines of both N and D skeletons are internal ones. The lines leaving (entering) vertices of $M(\bar{M})$ skeletons are particle lines while all others are hole lines. Labelling the particle (hole) lines of the N and D skeletons with the indices of the excited (non-excited) spin-orbitals we obtain corresponding N and D diagrams, respectively. The set of spin-orbital indices of the N and D diagrams will be designated ν and δ, respectively. Since both N and D diagrams have no external lines we shall assign to them only scalar quantities according to the following rule: On dissecting the internal lines of the $N\{D\}$ diagram we obtain $\bar{M}$, H, and $M\{\bar{M}$ and $M\}$ diagrams. Let $d_M(\bar{m}, \bar{\mu})$, $d^c_H(h, \chi)$, and $d^c_M(m, \mu)\{d^c_M(\bar{m}, \bar{\mu})$ and $d^c_M(m, \mu)\}$ be scalar quantities associated with these $\bar{M}$, H, and M $\{\bar{M}$ and $M\}$ diagrams, respectively. Then the scalar quantities $d^c_N(n, \nu)$ and $d^c_D(d, \delta)$ assigned to these N and D diagrams are given by the following expressions

$$d^c_N(n, \nu) = (-1)^{l+h} d^c_{\bar{M}}(\bar{m}, \bar{\mu}) d^c_H(h, \chi) d^c_M(m, \mu) \tag{76}$$

$$(c = 1, 2)$$

$$d^c_D(d, \delta) = (-1)^{l+h} d^c_{\bar{M}}(\bar{m}, \bar{\mu}) d^c_M(m, \mu) \tag{77}$$

$$(c = 1, 2)$$

where l is the number of closed loops and h is the number of internal hole lines of a given N or D diagram.

In this Section let $\Pi^1(\Gamma^1)$ designate a set of topologically distinct $N(D)$ skeletons. The set $\Pi^1(\Gamma^1)$ may be divided into disjunct subsets each containing topologically quasi-equivalent $N(D)$ skeletons. Furthermore, any set of $N(D)$ skeletons, containing one representative skeleton from *each* subset of quasi-equivalent $N(D)$ skeletons will be denoted $\Pi^2(\Gamma^2)$.

We are now ready to state an auxiliary theorem which will play the same rôle as the auxiliary theorems of Section 3. Namely-the following relationships hold:

$$w^c_{\bar{M}}(\bar{m})w^c_H(h)w^c_M(m)\sum_{\bar{\mu}}\sum_{\chi}\sum_{\mu}\langle\Phi|\hat{D}^c_{\bar{M}}(\bar{m},\bar{\mu})\hat{D}^c_H(h,\chi)\hat{D}^c_M(m,\mu)|\Phi\rangle = \sum_{n\in\Pi^c}{}' w^c_N(n)\sum_{\nu} d^c_N(n,\nu) \quad (78)$$

$$(c = 1, 2)$$

$$w^c_{\bar{M}}(\bar{m})w^c_M(m)\sum_{\bar{\mu}}\sum_{\mu}\langle\Phi|\hat{D}^c_{\bar{M}}(\bar{m},\mu)\hat{D}^c_{\bar{m}}(m,\mu)|\Phi\rangle = \sum_{d\in\Gamma^c}{}' w^c_D(d)\sum_{\delta} d^c_D(d,\delta) \quad (79)$$

where the prime on the summation symbol on the right-hand side of Eqs. (78) and (79) means that the sum extends only over those $N\{D\}$ skeletons having the structure $n\varepsilon\Pi^c\{d\varepsilon\Gamma^c\}$, which are formed from pertinent $S_{\bar{M}}(\bar{m})$, $S_H(h)$, and $S_M(m)\{S(\bar{M}\bar{m})$ and $S_M(m)\}$ skeletons. If no N or D skeleton can be formed from these skeletons, then the prime at the summation symbol means that the corresponding expression (78) or (79) is equal to zero. The proof of these auxiliary theorems is analogous to the proof of the auxiliary theorems of Section 3 given in the Appendix.

D. Final Expression for Mean Values

On the basis of the auxiliary theorems of the preceding section we get

$$v_N = \sum_{n\in\Pi^c} w^c_N(n)\sum_{\nu} d^c_N(n,\nu) \quad (80)$$

$$(c = 1, 2)$$

$$v_D = \sum_{d\in\Gamma^c} w^c_D(d)\sum_{\delta} d^c_D(d,\delta) \quad (81)$$

$$(c = 1, 2)$$

We now define connected and disconnected N skeletons similarly as in Section 4. In this case a disconnected N skeleton consists of a connected N skeleton and of a D skeleton. The weight (degenerate weight) of this disconnected N skeleton is given by the product of the weights (degenerate weights) of its constituent connected N skeleton and D skeleton. In a similar manner a scalar quantity is assigned to each disconnected N skeleton. On the basis of these considerations and using the ideas introduced by Hubbard[3d] we get

$$v_N/v_D = \sum_{n \in \Delta^c} w_N^c(n) \sum_{\nu} d_N^c(n, \nu) \tag{82}$$

$$(c = 1, 2)$$

Here Δ^1 designates a set of topologically distinct, connected N skeletons. The set Δ^2 is defined similarly as in the preceding Section.

For spin-independent operators and the case $c = 1$, we can easily carry out the summation over the spin variables if to each

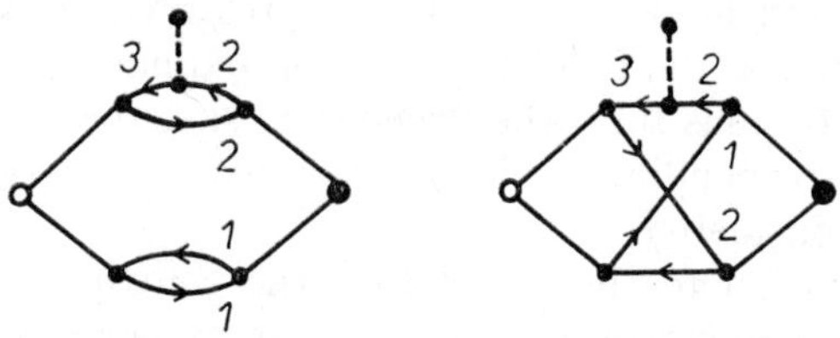

Fig. 18. The N skeletons formed from the $S_T(2)$, $S_H(2)$ and $S_T(2)$ skeletons. Only two of these N skeletons are shown while the other two, which differ only in the direction of all the oriented lines, are omitted.

closed loop of the oriented lines we assign[3d,10] a factor of two, while substituting the orbital indices in place of the spin-orbital ones in formula (82). This formula is also used in all actual computations.

Most simple N skeletons, which should be accounted for in the calculation of a mean value of a general one-particle operator using the approximation $\hat{T} \cong \hat{T}_2$, are those consisting of $S_T(2)$, $S_H(2)$, and $S_T(2)$ skeletons. Two of these are shown in Fig. 18 and the other two may be obtained by reversing the direction of all the oriented lines (we shall call this procedure a "particle–hole inversion" in the following). The degenerate skeletons corresponding to the skeletons of Fig. 18 are shown in Fig. 19.

The further N skeletons which should be considered in the abovementioned case consist of an $S_H(2)$ skeleton and an $\bar{M}$ and an M skeleton formed by two $S_{\bar{T}}(2)$ and two $S_T(2)$ skeletons, respectively. Instead of presenting all these N skeletons we give in Fig. 20 the corresponding degenerate ones except those which differ by particle–hole inversion only.

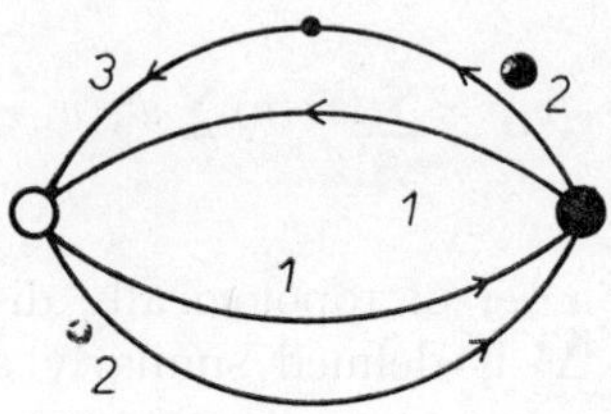

Fig. 19. The degenerate N skeleton corresponding to the skeletons of Fig. 18.

The one-particle density matrices presented in Section 6.B were calculated in the following two approximations:

[1]. Contributions from the degenerate skeleton of Fig. 19 and from the corresponding skeleton obtained by particle–hole inversion are considered.

[2]. In addition to these skeletons the contributions from all degenerate skeletons of Fig. 20 and from corresponding particle–hole inverted skeletons are taken into account. Naturally, in the actual computations we have to draw all the non-degenerate skeletons explicitly so that one can carry out the summation over the spin variables. For example, from the five degenerate skeletons of Fig. 20 we get 34 non-equivalent, non-degenerate skeletons.[16]

Most simple skeletons which will intervene in the calculation of the mean value of a two-particle operator, using the approximation $\hat{T} \simeq \hat{T}_2$, arise in the following four ways:

[1]. from the $S_H(1)$, $S_T(2)$, and empty $\bar{M}$ skeletons,

[2]. from the $S_H(1)$, $S_T(2)$, and empty M skeletons,

[3]. from the $S_H(2)$, $S_{\bar{T}}(2)$, and $S_T(2)$ skeletons, and

[4]. from the $S_H(1)$, $S_{\bar{T}}(2)$, and $S_T(2)$ skeletons.

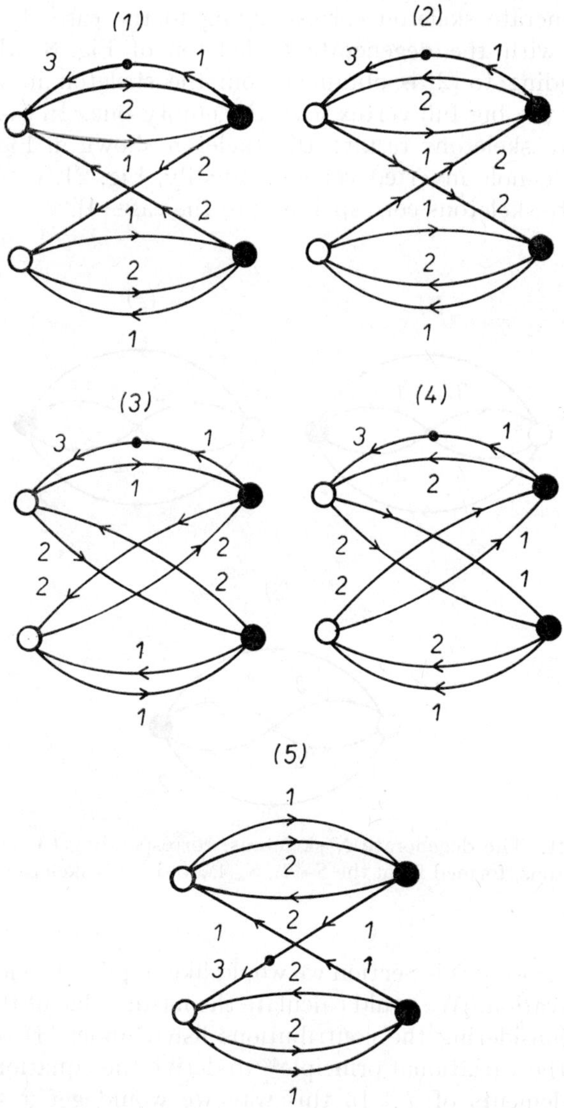

Fig. 20. The degenerate N skeletons corresponding to the N skeletons which are formed from the $S_H(2)$ skeleton and from the $\bar{M}$ and M skeletons, consisting of two $S_{\bar{T}}(2)$ and $S_T(2)$ skeletons, respectively. Only five of these degenerate N skeletons are given; the remaining five may be obtained from those shown here by the reversal of all the oriented lines.

The degenerate skeleton corresponding to the case [1] above is identical with the degenerate R skeleton of Fig. 8, while that corresponding to [2] is obtained from the skeleton in Fig. 8 by replacing the big full vertex with the empty one. In case [3] two degenerate skeletons result: the skeleton shown in Fig. 19 and its particle–hole inverted version. Finally, Fig. 21, presents the degenerate skeletons corresponding to the case [4].

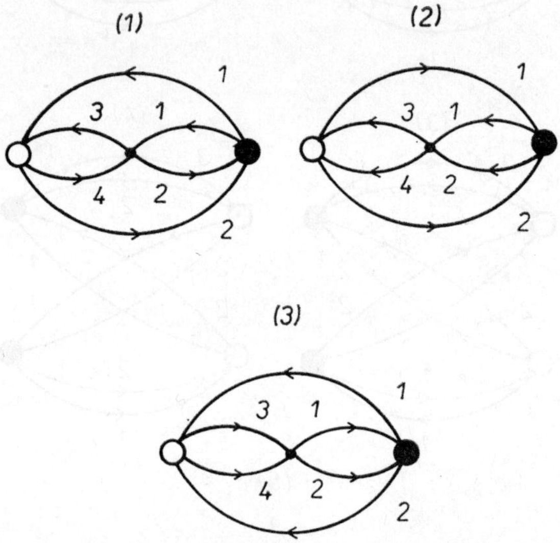

Fig. 21. The degenerate N skeletons, corresponding to N skeletons, formed from the $S_{\hat{T}}(2)$, $S_H(1)$, and $S_T(2)$ skeleton.

Before closing this Section we would like to present one interesting observation. We could calculate the mean value of the Hamiltonian, considering the contributions listed under [1]–[4] above, and use the variational principle[4d] to derive the equations for the matrix elements of $\hat{T}$. In this way we would get a system of equations which is identical with the systems (67) or (66) keeping the linear terms only. Unfortunately, the problem gets rather complicated if we try to keep non-linear terms as well and a simple relationship as in the linear case does not seem to exist.

6. FINAL REMARKS

A. Relationship to Former Theories

The relationship of the present theory with Goldstone's[3b] perturbation theory as well as with the works of Kelly, Sessler,[17,18] and da Providencia[4d] was discussed in the author's former paper.[10] In the latter, the connection of the theory developed here with the Brueckner[19] theory of nuclear matter and the Gell–Mann and Brueckner[20] theory of the electron gas was shown. The physical differences between nuclear and electron gas problems, and the atomic and molecular N-electron problem have been discussed by Sinanoğlu.[4a,11] In this article we shall devote our attention to the relationship of the present theory with the work of Sinanoğlu,[4b,11] Nesbet,[12] and Kelly.[2d]

[1]. First, we shall suppose that the determinant $|\Phi\rangle$ consists of HF orbitals, so that $\langle C_1|\hat{q}|B_1\rangle = 0$ holds for arbitrary C_1 and B_1 (see Eq. 63). Further, neglecting contributions corresponding to degenerate skeletons (1) and (3) of Fig. 14 in Eqs. (68) and restricting the summation over the occupied spin-orbitals in the expression (66) to the spin-orbitals D_1' and D_2' only, one can show that the "coupled-pair-many-electron theory" presented here yields Sinanoğlu's theory, which we call a "decoupled-pair-many-electron theory". These restrictions result in the decoupling of our systems of equations, yielding a set of independent equations for each pair of occupied spin-orbitals. The contributions corresponding to the degenerate skeletons (2) and (4) of Fig. 14 may be written as a product of a pair correlation energy and a corresponding matrix element of the operator $\hat{T}$. Using the terminology of Kelly[2d] we can say that these contributions "violate the Pauli exclusion principle". Kelly[2d] calls these terms "EPV terms of the first type". These terms appear because of the use of the connected cluster formula.

The contributions of degenerate skeletons (1) and (2) of Fig. 12 represent HF one-particle energies while those corresponding to the degenerate skeletons (3) and (5) of Fig. 12 represent "EPV terms of the second type" of Kelly.[2d] The latter appear due to the use of the HF potentials. The contribution corresponding to the degenerate skeleton (4) of Fig. 12 represents a particle–particle interaction.

Let us mention that the original Bethe–Goldstone equation,[21] which was formulated for an infinite system in a form unsuitable for finite systems, contributes only the terms corresponding to the degenerate skeletons (1), (2), and (4) of Fig. 12 and the constant term which corresponds to the degenerate skeleton of Fig. 10.*

[2]. Second, instead of using HF orbitals in our theory we can use an arbitrary unitary transform of them (e.g. localized orbitals).† Then carrying out the same approximations as in the first case above we get, after simple algebraic manipulation, from the system of equations (68) the system of equations (63) from the paper by Nesbet.[12a] Let us mention that Nesbet uses[12b,12c] HF orbitals in his calculations as well. His procedure is quite similar to a configuration interaction procedure described by Sinanoğlu[4a] for calculating pair terms.

B. Actual Calculations

In order to describe briefly the results of actual calculations it is convenient to divide the atomic and molecular systems studied into the following three groups:

[1]. The molecules of benzene, butadiene, and acetylene in the π-electron approximation.

[2]. The π-electron model of cyclic polyenes.

[3]. The beryllium atom.

In all these cases we have used the approximation $\hat{T} \cong \hat{T}_2$. To construct the system of equations (67) and find its solution we have systematically used a program written by Paldus. In this program the Newton–Raphson method, which is a second-order iterational procedure, is used to solve the system of non-linear equations (67). The program is written in a very general way so that in going from one system studied to another it is sufficient to supply new subroutines for the calculation of molecular integrals and for taking advantage of the space symmetry present.

Let us mention that, in all cases studied, full use of the symmetry of the systems was made. The spin symmetry is easy to use

* The question of terminology in this respect is very delicate indeed and controversy may arise even for three-particle equations.[14]

† See also the LO version of Sinanoğlu theory in Reference 4b §27.

if we work with the non-antisymmetrized matrix elements[10] and is permanently built into the program mentioned. The space symmetry for cyclic polyenes and for the beryllium atom required a special subroutine in each case. When the ZDO approximation was used we also exploited alternant symmetry. Systematic use of the symmetry of the systems studied resulted in a significant reduction of the number of the unknown variables in the system of non-linear equations (67). In general, the number of unknown variables in our equations is smaller by one than the order of the corresponding secular problem for CI, including all biexcited states, supposing that the symmetry is respected in both approaches.

Let us now turn our attention to the individual systems [1], [2], and [3] listed above:

[1]. The benzene, butadiene, and acetylene molecules were treated in the π-electron approximation and for each molecule the following calculations were carried out:

(*a*) The wave function and the correlation energy were computed using the CI method, considering all possible configurations which can be obtained from the set of one-electron functions used. This represents an exact solution within a given model and, correspondingly, we use it for comparison with the other calculations.

(*b*) Same as above, except that CI is limited to the biexcited states only.

(*c*) The wave function and correlation energy were calculated using the method of Section 4 (in the approximation $\hat{T} \simeq \hat{T}_2$). In tables in C those calculations are designated by N and in the Table of Reference 22 by II. Further calculations, which we have carried out and will not report here, were also based on the system of equations (67) (or (60) of C), into which different approximations were introduced.

Preliminary results for the benzene molecule have already been published[10] and a detailed paper[23] concerning this molecule, as well as the paper concerning the calculations on butadiene[24] will appear soon. Furthermore, the results concerning the acetylene molecule have been published recently.[22]

In the case of benzene the correlation energy obtained with procedure (*c*) differs from the "exact" correlation energy of calculation (*a*) by less than 1%, nevertheless, a detailed analysis has revealed that this extremely good agreement is due to the cancellation of contributions from neglected terms and a realistic estimate of the accuracy obtained with procedure (*c*) is therefore cca 2%. A similar conclusion can be drawn from the calculations on the butadiene and acetylene molecules. For all the molecules studied we find the results of procedure (*c*) to be significantly better than results of method (*b*).[10,22,24]

In addition, we have calculated the one-particle density matrix for the case of the benzene molecule. Here we have used the following four approaches:

(i) The exact calculation of the one-particle density matrix using the accurate wave function obtained with the procedure (*a*) described above.

(ii) Same as (i) but using the wave function from the procedure (*b*).

(iii) Same as (i) but using the wave function from the procedure (*c*).

(iv) The density matrix calculation using the procedure described at the end of Section **5** using the wave function of procedure (*c*).

The detailed results of these calculations will be published elsewhere.[16] Here we shall restrict ourselves to a general statement that the results of procedures (iii) and (iv) do not differ significantly even though only a few important graphs are retained in method (iv), and that these results are superior to results (ii) in comparison with the exact results (i).

[2]. The π-electron model of cyclic polyenes was extensively studied in a number of papers by Löwdin and coworkers.[25] In a study of these systems, using the method described in this paper, we have run into convergency difficulties when we attempted to solve iteratively the system of equations (67) for larger cyclic polyenes. The size of the polyene for which the method used does not converge depends on the parameterization used.[26] We have found that these difficulties are associated with the stability of the solutions of the HF equations which are used to build up the starting determinant $|\Phi\rangle$. These problems are described in greater detail in a recent paper.[27]

In Reference 27 we classify the instabilities into those of

singlet and triplet types. The HF solution is said to be "triplet unstable" if another solution corresponding to a lower energy exists and if the corresponding determinant $|\Phi'\rangle$ made up from this new solution is no longer an eigenfunction of $\hat{S}^2$. Instabilities of this type are closely connected with the UHF method calculations published recently by Koutecký.[28] If the determinant $|\Phi'\rangle$, in addition to having a lower energy than the starting HF determinant $|\Phi\rangle$, is a pure singlet function we refer to the solution corresponding to $|\Phi\rangle$ as the "singlet unstable" solution. We have found that the HF solution following from the symmetry of the cyclic polyenes is singlet unstable if the number of carbon atoms in the polyene is large enough (the critical size depends again upon the parameterization used).

The fact that the symmetry-adapted HF solutions are singlet unstable implies that the one-particle approximation cannot describe the exact wave function with sufficient accuracy. Consequently, this implies that the use of the approximation $\hat{T} \simeq \hat{T}_2$ in this case is not justified. The problems mentioned in this section are, in fact, much more complex; more detailed investigations in this respect are currently being undertaken.

[3]. Let us make it clear right at the beginning that our calculations on the Be atom cannot compete with the much more complex calculations of Nesbet[12b] insofar as the size of the basis of one-electron functions used is concerned, since our computer facilities are rather limited.* On the other hand, the results of these calculations may be compared directly with those of experiments and, therefore, are more-than-model calculations.

Here we shall present only preliminary results.[29] In these calculations we have used Watson's wave functions[30] and, due to time limitations, we have restricted ourselves to s- and p-type functions only. The necessary integrals for this limited basis set were kindly made available to us by Leclercq,† who has been using the same basis set in his "free geminal" method study[31] of the Be atom. For the sake of comparison he carried[31] out the

* The results[29] reported here were obtained during our (Dr. Paldus and the author) one month stay at CMOA CNRS, Paris. The hospitality and facilities extended to us are greatly appreciated.

† We would very much like to thank Dr. J. M. Leclercq for his kindness.

configuration interaction calculations: (*a*) using the first two Watson *p*-type excited states and, alternatively, (*b*) using the first *s*-type and the first two *p*-type Watson excited one-electron functions, using almost all configurations.* We have carried out the calculations using the method of Section 4 ($\hat{T} \cong \hat{T}_2$) in the same approximations (*a*) and (*b*). As the experimental correlation energy we shall use the value quoted by Nesbet,[12b]

$$\varepsilon_{\text{cor}} = -0.09390 \text{ a.u.}$$

TABLE 1. A comparison of our results for the Be atom (model *a* and *b*) with those of Leclercq (almost complete CI in the given set of orbitals).

	L[a] (a.u.)		PC[b] (a.u.)	
	ε_{cor}	%[e]	ε_{cor}	%[e]
a[c]	−0.06274	66.8	−0.06313	67.2
b[d]	−0.07525	80.1	−0.07563	80.5

[a] Leclercq, see Reference 31.
[b] Paldus and Čížek, see Reference 29.
[c] Using only two excited *p*-type functions (model *a*).
[d] Using only one excited *s*-type function and two excited *p*-type functions (model *b*).
[e] In percent with respect to the exact value of the correlation energy quoted by Nesbet.[12b]

In Table 1 our results are compared with the almost complete CI treatment of Leclercq[31] for both cases (*a*) and (*b*) mentioned above. One can immediately see that in both cases the differences between Leclercq's CI values and our results are always less than 0.4%. This is very promising indeed. Nevertheless, a final conclusion will have to await the completion of the full CI treatment for cases (*a*) and (*b*) above.[29]

* This CI treatment will be extended to include all configurations.[29] Nevertheless, the CI of Leclercq[31] is almost complete and comparison with the results obtained with our methods is interesting.

Let us further mention that for the basis set (*a*) we have also calculated the intrashell correlation energy using the method of Sinanoğlu, using decoupled equations. These results, given in

TABLE 2. A comparison of our results for the Be atom (model *a*) with those computed following Sinanoğlu's method.

	S[a] (a.u.)		PC[b] (a.u.)	
	ε_{cor}	%[f]	ε_{cor}	%[f]
$1s1s$[c]	−0.02061	32.6	−0.02061	32.6
$1s2s$[d]			−0.00157	2.5
$2s2s$[e]	−0.04127	65.4	−0.04095	64.9

[a] Sinanoğlu, see Reference 4b, 11.
[b] Paldus and Čížek, see Reference 29.
[c] Intrashell correlation energy in the $1s$ shell.
[d] Intershell correlation energy between the $1s$ and $2s$ shells.
[e] Intrashell correlation energy in the $2s$ shell.
[f] In percent with respect to the correlation energy obtained in Reference 29, $\varepsilon_{cor} = -0.06313$ a.u.

Table 2, confirm the well-known fact that the decoupled-pair-many-electron theory represents a very good approximation for the Be atom.

Let us now turn our attention to case (*c*) considering all *s*- and *p*-type Watson excited one-electron functions (e.g. four *s* orbitals and five *p* orbitals). In this case it would be very laborious to carry out the complete CI. Therefore, we shall compare our results with those of Weiss[32] and with experiment. Using the basis set (*c*) and the method of Section 4, with $\hat{T} \simeq \hat{T}_2$, we have obtained

$$\varepsilon_{cor} = -0.08572 \text{ a.u.}$$

which corresponds to 91.2% of the experimental value. Weiss, using a larger basis than Watson, obtained 93% of the experimental correlation energy with excited states of the *s*, *p*, *d*, and *f* types, while, when using excited states of *s* and *p* types only, he obtained 89.2%. Since our basis is different from that used by Weiss our results are not directly comparable, nevertheless, this close agreement we find interesting. Some additional results for the basis set (*c*) are given in Table 3; they will be published in

TABLE 3. Intra- and inter-shell correlation energy in the Be atom.

a.u.	TS[d]	K[e]	GTL[b]	N[g]	PC[h]
1*s*1*s*[a]	0.04395	0.04212	0.042083	0.041827	0.03758
1*s*2*s*[b]	0.00648	0.00497	0.00497	0.005864	0.00457
2*s*2*s*[c]	0.04387	0.04357	0.044381	0.045351	0.04357

[a] Intrashell correlation energy in the 1*s* shell.
[b] Intershell correlation energy between the 1*s* and 2*s* shells.
[c] Intrashell correlation energy in the 2*s* shell.
[d] Tuan and Sinanoğlu, see Reference 33.
[e] Kelly, see References 2d and 7.
[f] Geller, Levine, and Taylor, see Reference 34.
[g] Nesbet, see Reference 12b.
[h] Paldus, and Čížek, see Reference 29.

greater detail elsewhere.[29] The results of Table 3 show that for the 2*s* intrashell correlation energy our results are in good agreement with the results of other authors. This is due to the fact that in this case mainly the *p* states intervene as already found by Kelly.[7]

C. On a Generalization of the Theory Presented

We were using a one-determinantal wave function as the initial approximation throughout this article. This implicitly assumes that the starting HF determinant is *singlet stable,*[27,35] otherwise we know that a single Slater determinant wave function does not represent a good starting point and a better zero approximation should be used instead. For this purpose, some of the methods listed, e.g.,

[1]. projected unrestricted HF,
[2]. extended HF, and
[3]. multi-configurational CI,

which are described in detail in articles by Jucys,[36] Lefebvre,[37] Löwdin,[38] and Moser,[39] suggest themselves. How to use these methods as an initial approximation for the method described here or, in other words, how to proceed in the case that the exact wave function is written in the form

$$|\Psi\rangle = e^{\hat{K}}|\Phi'\rangle$$

where $|\Phi'\rangle$ is a multideterminantal wave function, will be the subject of a future communication.[40]

APPENDIX

Proof of the Auxiliary Theorems of Section 3

The proof of the auxiliary theorem used in Section 3 was sketched out in the Appendix of C for the case of $c = 1$. Consequently, we shall be concerned here only with the case $c = 2$; an analogous formula to A.6 of C will be shown to be valid. This lemma has the form:

The number g^2 of all possible quasi-equivalent R skeletons which are obtained by connecting skeletons $S_H(h)$ and $S_M(m)$ in all possible ways is given by the following relationship

$$g^2 = w_R^2(r)/w_H^2(h)w_M^2(m) \tag{A1}$$

A general proof of this lemma is rather complicated and, therefore, we shall prove it only for certain special cases which are, nevertheless, very instructive.

First, let us consider the case for which

(i) $h = 1$,

(ii) a degenerate skeleton belonging to $S_M(m)$ has no equivalent vertices, and

(iii) two particle lines of the H and M skeletons constituting a given R skeleton are joined together.

In (iii) two cases must be distinguished: both particle lines of an $S_M(m)$ skeleton, which connect it together with an $S_H(h)$ skeleton, (*a*) are quasi-equivalent, (*b*) are not quasi-equivalent.

We shall first consider the case (*a*). Let the set of quasi-equivalent lines of the M skeleton, to which both particle lines connecting it with the H skeleton also belong contain j lines.

Then, in accordance with formula (34), the degenerate weight of this M skeleton is

$$w_M^2(m) = (j!)^{-2}C \tag{A2}$$

where C is a constant number given by the structure of the residue of this skeleton. Further, for the degenerate weight $w_R^2(r)$ of the R skeleton we get

$$w_R^2(r) = [4(j-2)!j!]^{-1}C \tag{A3}$$

where the factor 1/2 comes from two external lines leaving the H diagram, another factor 1/2 comes from two internal lines of the

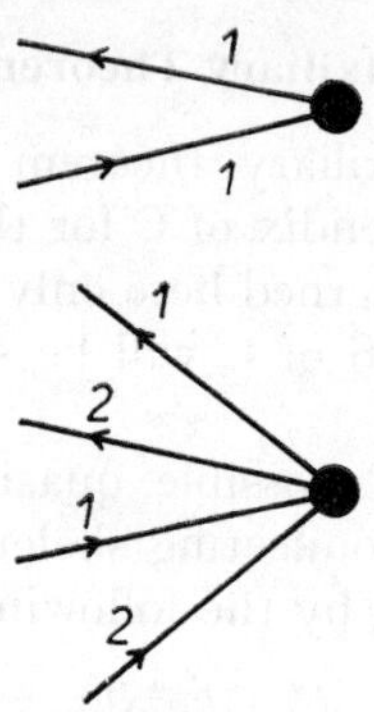

Fig. 22. The degenerate M skeleton which is used as an example in the Appendix on the derivation of Eqs. (A5) and (A7).

R diagram and, finally, the factor $[(j-2)!j!]^{-1}C$ corresponds to the residue of the $S_M(m)$ skeleton.

On the other hand, it is immediately obvious that there are altogether $j(j-1)$ possible ways of joining together the H and M skeletons mentioned in order to obtain a skeleton quasi-equivalent to $S_R(r)$. Indeed, this is in agreement with formula (A1) using the weight from expression (A2), (A3), and (36). The above considerations are graphically illustrated in Figs. 22 and 23.

Let us now consider the case (b). Supposing the number of the quasi-equivalent lines of the M skeleton in two disjunct subsets—each of which contains one of the particle lines joining the H

and M skeletons together—to be j and k, respectively, we get for the degenerate weight of the M skeleton:

$$w_M^2(m) = (j!)^{-2}(k!)^{-2}D \tag{A4}$$

where constant number D depends on the residue of the M skeleton. Further, for the degenerate weight of the R skeleton we get

$$w^2{}_R(r) = [2(k-1)!(j-1)!k!j!]^{-1}D \tag{A5}$$

where the factor 1/2 comes from the two external lines leaving the H skeleton and the factor $[(k-1)!k!(j-1)!j!]^{-1}D$ corresponds to the residue of the $S_M(m)$ skeleton. Again, we can easily

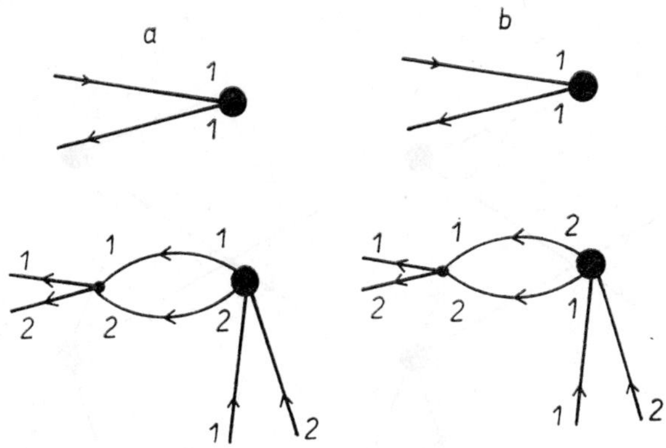

Fig. 23. Examples of degenerate R skeletons referred to in the derivation of formulae (A3) ($C = 1, j = 2$).

see that there are $(2kj)$ possible ways of joining together the H and M skeletons in order to get a quasi-equivalent R skeleton which agrees with (A1) supposing the degenerate weights to be given by formulae (A4), (A5), and (36). The considerations are graphically illustrated in Figs. 22 and 24.

Finally, let us consider the case where

(i) $h = 2$,

(ii) an M skeleton consists of two equivalent T skeletons, each having j particle and j hole lines, and

(iii) an R skeleton is obtained by joining one hole line of the respective H and M skeletons.

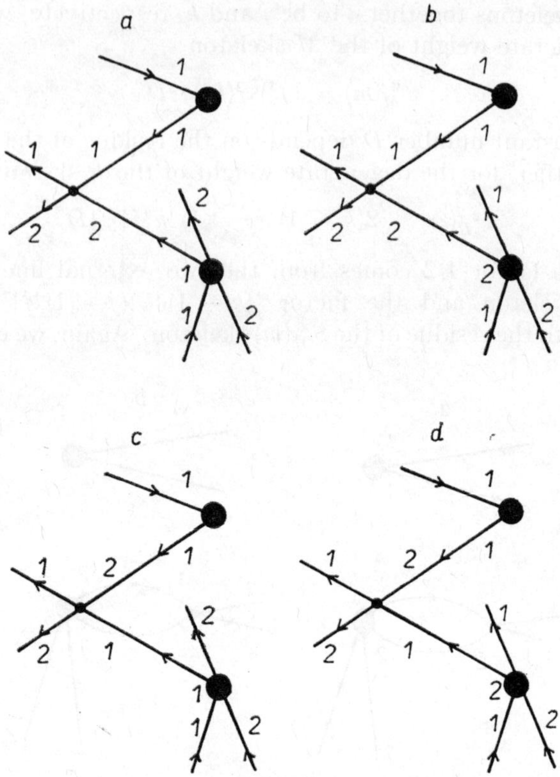

Fig. 24. Examples of degenerate R skeletons referred to in the derivation of formula (A5) ($D = 1$, $k = 1$, $j = 2$).

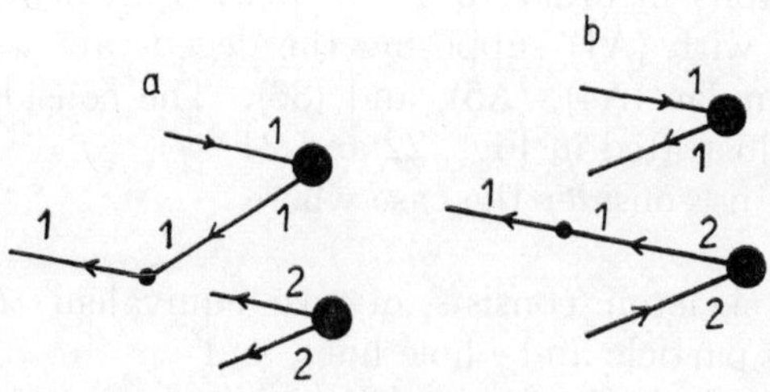

Fig. 25. Examples of degenerate R skeletons referred to in the derivation of formula (A7). The figures are drawn for $j = 1$.

The degenerate weights of the M and R skeletons are

$$w_M^2(m) = (1/2)(j\,!)^{-4} \tag{A6}$$

and

$$w_R^2(r) = [(j\,!)^3(j-1)\,!]^{-1} \tag{A7}$$

respectively. In this case the H and M skeletons may be joined together in $(2j)$ different ways in order to get a quasi-equivalent R skeleton, which is again in agreement with (A1), (A6), (A7), and (36). These considerations are illustrated in Fig. 25.

Acknowledgements

The author would like to express his thanks to Drs. R. Lefebvre, C. Moser, and to all the participants of the Frascati Summer School for many helpful discussions and critical comments concerning the subject of this paper. Especially, it is a pleasure to thank the directors of the School (R. L. and C. M.) for the warm hospitality which created a scientifically fruitful atmosphere during the fortnight at Frascati.

A part of this work originated during the author's visits to the Centre de Mécanique Ondulatoire Appliquée; and his thanks go to Professor R. Daudel and to all the members of the CMOA staff for the interest they showed in the subject of this paper, as well as for their hospitality.

I am very indebted to my mentor, Professor J. Koutecký, for all the support and understanding he has devoted to my work during the past years and for many stimulating discussions and much helpful advice.

Without the help of my friend Dr. J. Paldus this work could hardly have been accomplished. He patiently discussed many problems in detail and his criticisms were very useful in completing this paper. Needless to mention, without his collaboration Section 6.B would not have originated. For all this I would like to cordially thank him.

My thanks are also due to Mrs. L. Šroubková and Mr. J. Vojtík for their help in checking the formulas as well as to Mrs. B. Buchtová, Mrs. E. Ĉutová, Mrs. J. Hochmannová, and Mrs. V. Zedlerová for typing the manuscript and drawing the figures.

References

1*a*. Pines, D., *The Many Body Problem. A Lecture Note and Reprint Series,* W. A. Benjamin, Inc., New York, 1962.
1*b*. Nozières, P., *Le probleme à N corps,* Dunod, Paris, 1963.
2*a*. Tolmachev, V. V., Bulletin of the The Leningrad University, Physics and Chemistry, **1,** 12 (1962) (in Russian).
2*b*. Tolmachev, V. V., *Lietuvos Fizikos Rinkinys,* **3,** 47 (1963) (in Russian).
2*c*. Tolmachev, V. V., *The Field Form of the Perturbation Theory Applied to Many Electron Problems of Atoms and Molecules,* University of Tartu, 1963 (in Russian). For other references see Tolmachev's article in the present volume.
2*d*. Kelly, H. P., *Phys. Rev.,* **131,** 684 (1963).
2*e*. Kelly, H. P., *Perturbation Theory and its Application in Quantum Mechanics,* Ed. C. H. Wilcox, John Wiley and Sons, New York, 1966. For other references see Kelly's article in the present volume.
2*f*. Linderberg, J., and Öhrn, Y., *Proc. Roy. Soc. London, Ser. A,* **285,** 445 (1965).
2*g*. Öhrn, Y., and Linderberg, J., *Phys. Rev.* **139,** A1063 (1965).
2*h*. Linderberg, J., *The Application of Field-Theoretical Methods to Molecular Theory.* (Lecture notes prepared for Summer School on Quantum Chemistry in Hungary, September 1967.)
3*a*. Schweber, S. S., *Introduction to Relativistic Quantum Field Theory,* Row, Peterson and Co., Evanston, Illinois, 1961.
3*b*. Goldstone, J., *Proc. Roy. Soc. London, Ser. A,* **239,** 267 (1957).
3*c*. Hugenholtz, N. M., *Physica,* **23,** 481 (1957).
3*d*. Hubbard, J., *Proc. Roy. Soc. London, Ser. A,* **240,** 531 (1957).
3*e*. Bloch, C., *Nucl. Phys.,* **7,** 451 (1958).
4*a*. Sinanoğlu, O., *J. Chem. Phys.,* **36,** 706 (1962).
4*b*. Sinanoğlu, O., *Advan. Chem. Phys.,* **6,** 315 (1964) and references therein; for other references see Sinanoğlu's article in the present volume.
4*c*. Primas, H., *Modern Quantum Chemistry Istanbul Lectures,* Ed. O. Sinanoğlu, Academic Press, New York, Part II, p. 45.
4*d*. da Providencia, J., *Nucl. Phys.,* **61,** 87 (1965).
5*a*. Coester, F., and Kümmel, H., *Nucl. Phys.,* **17,** 477 (1960).
5*b*. Kümmel, H., *Lectures on the Many Body Problem,* Ed. E. R. Caianiello, Academic Press, New York, 1962, p. 265.
6. Thouless, D. J., *The Quantum Mechanics of Many Body Systems,* Academic Press, New York, 1961, p. 47.
7. Kelly, H. P., *Phys. Rev.,* **136,** B896 (1964).
8. da Providencia, J., *Nucl. Phys.,* **46,** 401 (1963).
9*a*. Mayer, J., and Mayer, M., *Statistical Mechanics,* John Wiley and Sons, New York, 1940.
9*b*. Brout, R., and Carruthers, P., *Lectures on the Many Body Problem,* Interscience Publishers, New York, 1963, Chap. 1.

10. Cížek, J., *J. Chem. Phys.*, **45**, 4256 (1966), hereafter referred to as C.
11. Sinanoğlu, O., *J. Chem. Phys.*, **36**, 3198 (1962).
12*a*. Nesbet, R. K., *Advan. Chem. Phys.*, **9**, 321 (1965) and references therein.
12*b*. Nesbet, R. K., *Phys. Rev.*, **155**, 51 (1967).
12*c*. Nesbet, R. K., *Phys. Rev.*, **155**, 56 (1967).
For other references see Nesbet's article in the present volume.
13. Čížek, J., *Chem. Phys. Letters*, to be published.
14*a*. Bethe, H. A., *Phys. Rev.*, **138**, B804 (1965).
14*b*. Brandow, B. H., *Phys. Rev.*, **152**, 863 (1966).
15. Brandow, B. H., "Linked cluster expansion for the nuclear many body problem," Appendix B, to be published in *Rev. Mod. Phys.*
16. Čížek, J., and Vokatý, E., to be published.
17. Kelly, H. P., and Sessler, A. M., *Phys. Rev.*, **132**, 2091 (1963).
18. Kelly, H. P., *Phys. Rev.*, **134**, A1450 (1964).
19*a*. Brueckner, K. A., *Phys. Rev.*, **97**, 1353 (1955).
19*b*. Brueckner, K. A., *Phys. Rev.*, **100**, 36 (1955).
19*c*. Brueckner, K. A., and Gammel, J. L., *Phys. Rev.*, **109**, 1023 (1958).
For other references see Brueckner's article in this volume.
20. Gell-Mann, M., and Brueckner, K. A., *Phys. Rev.*, **106**, 364 (1957).
21. Bethe, H. A., and Goldstone, J., *Proc. Roy. Soc. (London), Ser. A*, **238**, 551 (1957).
22. Čížek, J., and Pellegati, A., *Intern. J. Quantum Chem.*, **1**, 653 (1967).
23. Čížek, J., Paldus, J., and Šroubková, L., to be published.
24. Paldus, J., and Čížek, J., to be published.
25*a*. Pauncz, R., de Heer, J., and Löwdin, P. O., *J. Chem. Phys.*, **36**, 2247 (1962).
25*b*. Swalen, J. D., and de Heer, J., *J. Chem. Phys.*, **40**, 378 (1964), and references therein.
26. Paldus, J., and Čížek, J., to be published.
27*a*. Čížek, J., and Paldus, J., *J. Chem. Phys.*, **47**, 3976 (1967).
27*b*. Paldus, J., Seminar at the Summer School in Frascati.
28. Koutecký, J., *J. Chem. Phys.*, **46**, 2443 (1967).
29. Paldus, J., and Čížek, J., to be published.
30. Watson, R. E., *Phys. Rev.*, **119**, 170 (1960).
31. Leclercq, J. M., "Etude de l'approximation des geminales libres; application a l'etat fondamental de l'atome de beryllium." Thèse, 3 ème cycle, Université de Paris.
32. Weiss, A. W., *Phys. Rev.*, **122**, 1826 (1961).
33. Tuan, D. T., and Sinanoğlu, O., *J. Chem. Phys.*, **41**, 2677 (1964).
34. Geller, M., Taylor, H. S., and Levine, H. B., *J. Chem. Phys.*, **43**, 1727 (1965).
35. Löwdin, P. O., *Rev. Mod. Phys.*, **35**, 496 (1963).
36. Jucys, A., article in this volume.
37. Lefebvre, R., article in this volume.
38. Löwdin, P. O., article in this volume.
39. Moser, C., article in this volume.
40. Čížek, J., and Paldus, J., to be published.

EFFECTIVE OPERATORS FOR CONFIGURATIONS OF EQUIVALENT ELECTRONS

B. R. JUDD, *The Johns Hopkins University, Baltimore, Maryland*

CONTENTS

1. INTRODUCTION

A characteristic feature of term analyses for configurations of equivalent electrons is the existence, in the central-field approximation, of more experimental data than radial integrals. For example, the configuration f^4 comprises 47 terms (each one labelled by definite values of S and L, the quantum numbers of the total spin and total orbital angular momentum); yet the term energies are expressed through only four Slater integrals $F^{(k)}$. Of course, the spin–orbit interaction complicates matters by splitting each term into levels, but, from an analytical point of view, it multiplies the number of experimental data while adding only one more parameter, the spin–orbit coupling constant ζ, to the analysis.

As a result of this disparity between the number of data and the number of parameters, a study of equivalent electrons is particularly well suited to probe the effects that are neglected in the first stage of the central-field approximation. Interactions between configurations perturb the levels and produce various results. The displacements of the levels of a particular configuration under examination may in part be accommodated by adjustments in the existing parameters, but in general significant discrepancies remain when the observed levels are fitted to the single-configuration theory. These discrepancies are direct evidence of configuration interaction, and they form the subject of the present study.

One way to approach the problem is to consider each perturbing configuration in turn, estimate its effect on the terms of the configuration under examination, and continue until the discrepancies are reduced below some acceptable level. This method, attractive enough at first sight, has a number of disadvantages. The difficulty of obtaining sufficiently accurate inter-configuration radial integrals is an obvious drawback; but, more importantly, the method ignores the possibility of obtaining information about the nature of participating configurations from the discrepancies themselves. It is here that the concept of effective operators is particularly useful. Certain classes of configurations, imposing their influence to a certain order of perturbation theory, can be represented by operators acting solely within the configuration under study. Of course, the strengths of these operators depend on the perturbing configurations; but if the prefacing coefficients are treated as variable parameters a number of advantages are gained. In the first place, a fit with the experimental data immediately reveals whether a certain class of perturbations can or cannot satisfactorily account for the discrepancies; and the values of the parameters obtained from the fitting procedure become available for further study in the same way that the major parameters $F^{(k)}$ and ζ do. Furthermore, the whole process of analysis achieves an agreeable unity: the search for a fit provides effective operators and a number of parameters whose precise values can be regarded as a separate problem, one to be solved with the aid of Hartree–Fock eigenfunctions, perhaps; but for many purposes—for example, for predicting the energies of missing levels of the configuration—this extension is irrelevant.

The effective operators that have to be introduced within a given configuration to reproduce the influence of other configurations often involve the simultaneous coordinates of more than two electrons. This introduces an entirely new feature into atomic spectroscopy: the need to classify many-electron operators, or, strictly speaking, the sums of many-electron operators. We are familiar with operators of the type

$$F = \sum_i f_i$$

and

$$G = \sum_{i \neq j} g_{ij}$$

where f_i is a single-electron operator acting on electron i, and g_{ij} refers to electrons i and j. In addition to quantities of these kinds, it is necessary to study operators such as

$$H = \sum_{i \neq j \neq k} h_{ijk}$$

and their generalizations. Since every operator F, G, H, . . . must be symmetric with respect to the interchange of two electrons, the problem is analogous to classifying the states of a many-boson system. Just as Racah,[1] in his classic paper of 1942, introduced tensor operators to parallel the properties under rotation of single-electron eigenfunctions, so we have to extend a different aspect of Racah's work—that of classifying many-particle eigenfunctions—to the problem of classifying many-electron operators. These methods depend on the theory of continuous groups, and so we must begin with an account of the appropriate features of this theory. It is impossible to do more than give a rough sketch of the ideas involved. Fortunately, the actual mathematical operations are comparatively simple and can be formulated without the need for many preliminaries.

2. GROUPS

A. Permutation Symmetry

A simple and direct introduction to group-theoretical concepts is provided by configurations of two equivalent electrons, l^2. Since the total electronic eigenfunction must be antisymmetric with respect to the interchange of the two electrons, we must combine a symmetric orbital function

$$\psi_1(m)\psi_2(m') + \psi_1(m')\psi_2(m)$$

with an antisymmetric spin function

$$\alpha_1\beta_2 - \beta_1\alpha_2$$

and an antisymmetric orbital function

$$\psi_1(m)\psi_2(m') - \psi_1(m')\psi_2(m)$$

with a symmetric spin function such as $\alpha_1\alpha_2$, $\alpha_1\beta_2 + \beta_1\alpha_2$, or $\beta_1\beta_2$. The symbols m and m' are the eigenvalues of l_z, and $\psi_i(m)$ is proportional to the spherical harmonic $Y_{lm}(\theta_i, \phi_i)$. We represent the symmetry between electrons 1 and 2 by placing two cells side by side:

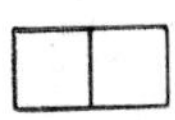

Antisymmetry, by contrast, is represented by placing one cell beneath the other:

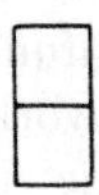

These two patterns are simple examples of *Young Tableaux*. To obtain a totally antisymmetric function, we have to combine our spin and orbital functions (assumed ordered in that way) either corresponding to:

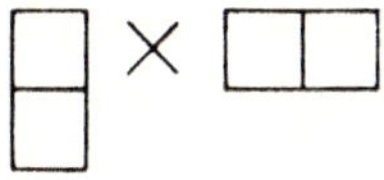

or to

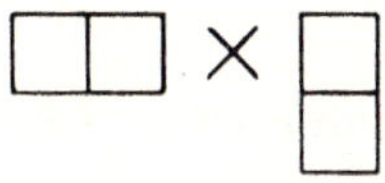

The first gives the singlet states ($S = 0$) of l^2; the second gives the triplet states ($S = 1$). Both tableaux in the above products involve the coordinates of the same two electrons, and the products are called *inner*.

B. Unitary Transformations

Let us consider now an entirely different kind of operation. Suppose the $2l + 1$ orbital functions $\psi(m)$, where $-l \leqslant m \leqslant l$, undergo transformations of the type

$$\psi(m) \rightarrow \sum_{m'} c_{mm'}\psi(m') \tag{1}$$

If we impose no restrictions on the coefficients, the transformations constitute the general linear group in $2l + 1$ dimensions, $GL(2l + 1)$. However, it is convenient to preserve the orthonormality of the eigenfunctions, and this means that the matrices $[c_{mm'}]$ must be unitary. The group is now the unitary group $U(2l + 1)$. The generators of this group are not difficult to find. For the transformation (1), we need to annihilate the state $\psi(m)$ and create a linear combination of the states $\psi(m')$. Now, the annihilation and creation operators for an electron state are labelled not only by m, but also by the spin orientation. Although the states being considered here are spin-free, we must allow for the associated spin orientation by taking

$$\sum_{m'} c_{mm'}(a^\dagger_{\frac{1}{2}m'}a_{\frac{1}{2}m} + a^\dagger_{-\frac{1}{2}m'}a_{-\frac{1}{2}m})$$

as the operator appropriate to (1); for then we get the desired result for either $\alpha\psi(m)$ or $\beta\psi(m)$. The use of annihilation and creation operators is particularly convenient, because we can preserve the Pauli exclusion principle in transformations of many-electron systems simply by imposing the anticommutation relations

$$\begin{aligned} a_\xi^\dagger a_\nu^\dagger + a_\nu^\dagger a_\xi^\dagger &= 0 \\ a_\xi^\dagger a_\nu + a_\nu a_\xi^\dagger &= \delta(\nu, \xi) \\ a_\xi a_\nu + a_\nu a_\xi &= 0 \end{aligned} \tag{2}$$

where ξ and ν are abbreviations for the sets of subscripts.

C. Representations

Any set of functions that transform among themselves under the action of the generators of a group are said to form a representation of the group; and if it is impossible to subdivide the functions into independent linear combinations that separately transform among themselves, the representation is said to be *irreducible*. As examples of irreducible representations of $U(2l+1)$, we may cite the eigenfunctions of l^2 for which: (*a*) $S = M_S = 1$; (*b*) $S = 1$, $M_S = 0$; (*c*) $S = 1$, $M_S = -1$; or (*d*) $S = M_S = 0$. Any one of these four collections of eigenfunctions transforms within itself under the operations of the basic generators

$$E_{m'm} = a_{\frac{1}{2}m'}^\dagger a_{\frac{1}{2}m} + a_{-\frac{1}{2}m'}^\dagger a_{-\frac{1}{2}m}$$

as can be proved by verifying that $E_{m'm}$ commutes with the spin $\mathbf{S}$, which can be written as

$$\mathbf{S} = \sum_{\xi,\nu} a_\xi^\dagger \langle \xi | \mathbf{s} | \nu \rangle a_\nu \tag{3}$$

No further subdivision is possible, and the four representations are thus irreducible.

To obtain labels for the irreducible representations of a group, we have to find the eigenvalues of the commuting generators. In our case, the last can be found by selecting from the $E_{m'm}$ those for which $m = m'$. It is straightforward to confirm that

$$E_{mm}|\phi\rangle = [\delta(m, m'') + \delta(m, m')]|\phi\rangle$$

where

$$|\phi\rangle = [\psi_1(m')\psi_2(m'') + \psi_1(m'')\psi_2(m')][\alpha_1\beta_2 - \beta_1\alpha_2]$$

The sequence of the $2l + 1$ eigenvalues of E_{mm}, beginning with $m = l$ and running down to $m = -l$, are enclosed in brackets and called a *weight*. Thus the weight of $|\phi\rangle$ when $l = 2$, $m' = 1$, and $m'' = -2$ is [01001]. It is clear that when $m \neq m'$, a weight comprises $2l - 1$ zeros and two ones. On the other hand, if $m' = m''$, a weight comprises $2l$ zeros and a two, such as [00200]. Quite generally, the weight $[m_1 m_2 \ldots]$ is said to be *higher* than the weight $[m_1 m_2 \ldots]$ if the first non-vanishing term in the sequence $m_1' - m_1$, $m_2' - m_2$, . . . is positive. It is conventional to label an irreducible representation by its highest weight, a procedure that never leads to ambiguities.[2] Thus, the irreducible representation formed by the singlet functions $|\phi\rangle$ is labelled by [200 . . . 0]. On the other hand, all of the triplet functions involve the orbital part

$$\psi_1(m')\psi_2(m'') - \psi_1(m'')\psi_2(m')$$

which vanishes when $m' = m''$. Consequently, no weights of the type [0 . . . 020 . . . 0] exist, and the highest weight (for any M_S) is [110 . . . 0].

We can now draw a striking correspondence: if we interpret the highest weight $[m_1 m_2 \ldots]$ as giving the number of cells in successive rows of a tableau, the two representations [20 . . . 0] and [110 . . . 0] found above correspond exactly with the tableaux describing the symmetry properties of the orbital eigenfunctions of l^2. This connection between permutation symmetry and unitary transformations rests, of course, on the particular mathematical form used to express the properties in question. However, it is of a completely general nature. For a thorough analysis, see the account by Weyl.[3]

To get the tableaux corresponding to the permutation symmetry in the spin space, we have merely to find the eigenvalues ω_1 and ω_2 of the two commuting operators

$$\begin{aligned} \Omega_1 &= \sum_m a^\dagger_{\frac{1}{2}m} a_{\frac{1}{2}m} \\ \Omega_2 &= \sum_m a^\dagger_{-\frac{1}{2}m} a_{-\frac{1}{2}m} \end{aligned} \qquad (4)$$

of the group $U(2)$, and interpret the highest of the weights $[\omega_1\omega_2]$ as specifying a tableau with ω_1 cells in its first row and ω_2 cells in its second. For example, we find

$$\Omega_1|\phi\rangle = |\phi\rangle, \qquad \Omega_2|\phi\rangle = |\phi\rangle$$

corresponding to the (highest) weight [11].

D. Many-Electron Eigenfunctions

The construction of many-electron eigenfunctions from constituent parts of definite symmetry introduces us to the idea of *outer* products. Consider, for example, the $l(2l + 1)$ orbital states of l^2 for which $S = 1$. We may assign electrons 1 and 2 to these states. Equally well, we may assign electrons 3 and 4 to these same states, thereby obtaining a second set of $l(2l + 1)$ functions. By combining a function of the first set with one of the second, we construct $l^2(2l + 1)^2$ simple products, each of which involves all four electrons. These products form a representation of $U(2l + 1)$. An analysis reveals that this representation can be decomposed into the three irreducible representations $[11110 \ldots 0]$, $[2110 \ldots 0]$, and $[220 \ldots 0]$. In other words,

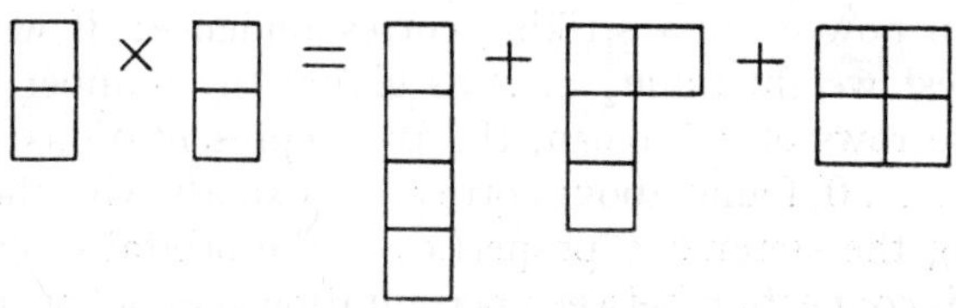

Since we frequently want to build many-electron eigenfunctions from simpler constituents, it is important to have a direct method for obtaining such products. This has been provided by Littlewood.[4] The tableaux appearing in the decomposition of

$$[\xi_1\xi_2 \ldots \xi_p] \times [\eta_1\eta_2 \ldots \eta_q]$$

are those that can be built by adding to the tableau $[\xi_1\xi_2 \ldots \xi_p]$ η_1 cells containing the same symbol α, then η_2 cells containing the symbol β, etc., subject to the two conditions:

[1]. After the addition of each set of cells labelled by a common symbol, we must have a permissible tableau (i.e. one in which no

row of cells overjuts the row above it) with no two identical symbols in the same column;

[2]. If the total set of added symbols is read from right to left in the consecutive rows of the final tableau, then, at any point of the reading, the number of symbols α read up to that point $\geqslant$ the number of symbols β read up to that point $\geqslant$ the number of symbols γ read up to that point, etc. For example,

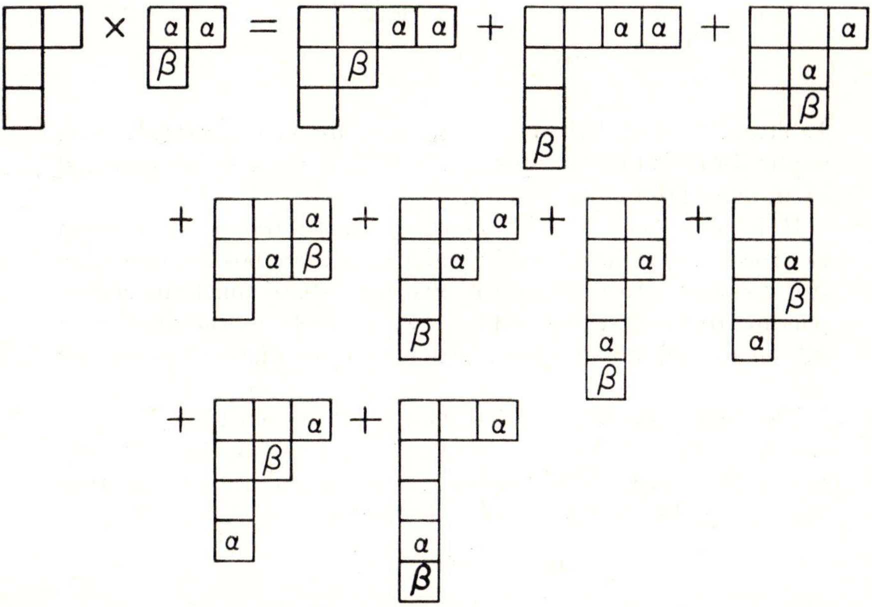

(5)

The condition that the complete many-electron eigenfunction be totally antisymmetric imposes conditions on the tableaux that represent the symmetries in the orbital and spin spaces. We have already seen for l^2 that the representation [2] of $U(2l+1)$ must be combined as an inner product with [11] of $U(2)$. The general rule can be formulated by inspecting special cases. The representation [211] of $U(2l+1)$ possesses as an eigenfunction corresponding to its highest weight any one of the four determinantal product states

$$\{l^+l^-(l-1)^{\pm}(l-2)^{\pm}\}$$

Using Eqs. (4), we see that the weights of $U(2)$ are [31], [22], [22], and [13]. The highest weight is [31], which thus describes the representation of $U(2)$. Now [31] is simply the tableau obtained by reflecting [211] in a diagonal line:

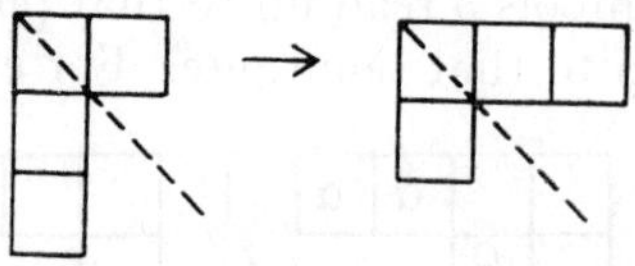

Tableaux related by such reflections are called *adjoint*. The required rule is that the spin and orbital tableaux to be combined as an inner product be adjoint.

It is obvious that spin tableaux cannot possess more than two rows; consequently orbital tableaux cannot possess more than two columns. An attempt to construct orbital functions corresponding to the first five and last two tableaux on the right-hand side of Eq. (5) would merely yield functions that are identically zero.

The total spin S associated with a given spin tableau $[\omega_1\omega_2]$ can be found at once by noting from Eqs. (3) and (4) that $S_z = \frac{1}{2}(\Omega_1 - \Omega_2)$. The maximum eigenvalue of S_z is thus $\frac{1}{2}(\omega_1 - \omega_2)$, and so we arrive at the result

$$S = \tfrac{1}{2}(\omega_1 - \omega_2)$$

E. Subgroups

From all the matrices $[c_{mm'}]$ of Section 2.B, we can select those that correspond to rotations in ordinary three-dimensional space. The transformations are now those of $R(3)$, the rotation group in three dimensions. The generators of $R(3)$ are the three components of **L**, which can, of course, be expressed as linear combinations of the generators of $U(2l+1)$. For example,

$$L_z = \sum_m mE_{mm}$$

Finding how a representation of $U(2l+1)$ decomposes into representations of $R(3)$ can be completely circumvented by combining the knowledge of the spin S associated with the adjoint tableau with the table of Russell–Saunders terms for $l = 0, 1, 2,$ and 3 given by Condon and Shortley.[5] For $l = 4, 5,$ and 6, the extensive tables of Shudeman[6] are available. Thus, for $l = 3$, we obtain the *branching rules*

$$[11] \rightarrow PFH$$

$$[221] \rightarrow P^4D^5F^7G^6H^7I^5K^5L^3M^2NO$$

Evidently $R(3)$ is a subgroup of $U(2l + 1)$; i.e. $U(2l + 1) \supset R(3)$. The search for groups X satisfying $U(2l + 1) \supset X \supset R(3)$ was initiated by Racah.[7] The object is to introduce new representations Ξ that will, hopefully, remove any ambiguity in a state such as $|f^5[221]\Xi LM_LM_S\rangle$. The group obtained by limiting the unitary matrices $[c_{mm'}]$ to those that are orthogonal gives $X = R(2l + 1)$. This corresponds to selecting from the $E_{mm'}$ the $l(2l + 1)$ operators

$$F_{mm'} = E_{mm'} - (-1)^{m+m'}E_{-m'-m} \tag{6}$$

Littlewood's branching rules for $U(2l + 1) \rightarrow R(2l + 1)$ can be expressed as follows (see Jahn[8]). A representation $[\lambda]$ of $U(2l + 1)$ decomposes into a sum of representations (η) of $R(2l + 1)$ according to the scheme

$$[\lambda] \rightarrow \sum_{\eta,\delta} g_{\delta\eta\lambda}(\eta)$$

where: (*a*) the coefficients $g_{\delta\eta\lambda}$ are those appearing in the outer-product decomposition

$$[\delta] \times [\eta] = \sum_{[\lambda]} g_{\delta\eta\lambda}[\lambda]$$

and (*b*) $[\delta]$ runs over all partitions into even parts, i.e. $[0]$; $[2]$; $[4]$, $[22]$; $[42]$, $[222]$; $[44]$, $[422]$, $[2222]$; . . . For example,

$$[421] \rightarrow (100)(210)^2(300)(221)(311)(320)(410)(421)$$

because (deleting all zeros in the representations $[\delta]$ and $[\eta]$)

$$[42] \times [1] \quad = [421] + \cdots$$
$$[22] \times [21] = [421] + \cdots$$
$$[4] \quad \times [21] = [421] + \cdots$$

etc. Representations $W[=(\eta)]$ of $R(2l+1)$ are described by l weights, corresponding to the l eigenvalues of $F_{mm}(l \geqslant m > 0)$. When $[\eta]$ contains more than l non-zero symbols, we may use the modification rules of Murnaghan[9] or simply compare eigenvalues of E_{mm} and F_{mm}. For example, if the eigenvalues of $E_{mm}(m = 3, 2, \ldots -3)$ are [2211110], those of $F_{mm}(m = 3, 2, 1)$ are (210). This follows immediately from Eq. (6). Hence, when $[\eta] = [221111]$ and $l = 3$, then (η) is to be interpreted as (210).

Having obtained the branching rules for $U(2l+1) \rightarrow R(2l+1)$, and knowing those for $U(2l+1) \rightarrow R(3)$, we may readily obtain the branching rules for $R(2l+1) \rightarrow R(3)$ by subtraction.

A knowledge of the branching rules for $U(2l+1) \rightarrow R(2l+1)$ permits us to obtain the decomposition of the products $W \times W'$. We have only to select the appropriate products $[\delta] \times [\eta]$ and set up a chain calculation in which the branching rules are combined with a process of subtraction.

For $l = 3$, an apparently unique event occurs. The exceptional group G_2 of Cartan[10] can be inserted in the sequence of groups as follows:

$$U(7) \supset R(7) \supset G_2 \supset R(3)$$

TABLE 1. Classification of the states of f^N $(0 \leqslant N \leqslant 4)$.

N	$[\lambda]$	W	U	${}^{2S+1}L$	N	$[\lambda]$	W	U	${}^{2S+1}L$
0	[0]	(000)	(00)	${}^{1}S$	4	[1111]	(111)	(00)	${}^{5}S$
1	[1]	(100)	(10)	${}^{2}F$				(10)	${}^{5}F$
2	[11]	(110)	(10)	${}^{3}F$				(20)	${}^{5}DGI$
			(11)	${}^{3}PH$		[211]	(211)	(10)	${}^{3}F$
	[2]	(200)	(20)	${}^{1}DGI$				(11)	${}^{3}PH$
		(000)	(00)	${}^{1}S$				(20)	${}^{3}DGI$
3	[111]	(111)	(00)	${}^{4}S$				(21)	${}^{3}DFGHKL$
			(10)	${}^{4}F$				(30)	${}^{3}PFGHIKM$
			(20)	${}^{4}DGI$			(110)	(10)	${}^{3}F$
	[21]	(210)	(11)	${}^{2}PH$				(11)	${}^{3}PH$
			(20)	${}^{2}DGI$		[22]	(220)	(20)	${}^{1}DGI$
			(21)	${}^{2}DFGHKL$				(21)	${}^{1}DFGHKL$
		(100)	(10)	${}^{2}F$				(22)	${}^{1}SDGHILN$
							(200)	(20)	${}^{1}DGI$
							(000)	(00)	${}^{1}S$

Branching rules for the decomposition of representations W of $R(7)$ into irreducible representations U of G_2 are given elsewhere.[11] The group-theoretical classification of the states of $f^N(0 \leqslant N \leqslant 4)$ are set out in Table 1; a complete tabulation for all N has been made by Wybourne.[12] It is to be stressed that W and U are not *good* quantum numbers; a state ^2F of f^3 that exists in nature is some linear combination of the two states $(210)(21)^2F$ and $(100)(10)^2F$.

3. OPERATORS

A. Scalars

Any operator that is introduced to reproduce the effects of configuration interaction in l^N must be a scalar under the operations of $R(3)$. This follows at once when it is recalled that the perturbation producing the interaction is itself a scalar. In generalizing the operator $E_{m'm}$ to products of the type

$$E_{m_1m_2}E_{m_3m_4}E_{m_5m_6}\cdots$$

in which pairs, triples, etc. of electrons are annihilated and created, a drawback becomes apparent: it is not self-evident what linear combinations of these operators should be taken to ensure that they are scalars in $R(3)$. As a first step to remedy this defect, we introduce the single-electron tensor operators v_q^k for which

$$(lm'|v_q^{(k)}|lm) = (-1)^{l-m'}(2k+1)^{\frac{1}{2}}\begin{pmatrix} l & k & l \\ -m' & q & m \end{pmatrix}$$

We can now construct a scalar from many-electron operators such as

$$a_\xi^\dagger a_\eta^\dagger a_\zeta^\dagger\langle\xi_1\eta_2\zeta_3|(v_q^{(k)})_1(v_{q'}^{(k')})_2(v_{q''}^{(k'')})_3|\sigma_1\tau_2\omega_3\rangle a_\omega a_\tau a_\sigma$$

by the simple expedient of coupling the ranks k, k', and k'' to zero resultant by the usual rules of angular-momentum theory. From the triangular conditions, it is clear that every rank k, k', and k'' must not exceed $2l$; moreover, we must exclude zero ranks for otherwise a many-electron operator reduces to one acting on fewer electrons.

The coupling of the ranks to zero is a straightforward matter for two-electron and three-electron operators, but for more complex cases a number of couplings can give zero resultant. The problem, that of finding and classifying the zero angular-momentum operators that can be formed from N angular momenta $k, k', k'', \ldots$, all satisfying inequalities of the type $0 < k < 2l + 1$, is identical to finding and classifying the S states in the configurations

$$(p + d + f + g + \ldots + 2l)^N$$

of spin-free bosons. That the particles represented by $p, d, f, \ldots$ should be bosons follows from the condition that the complete operator must be symmetric under the interchange of two electrons.

B. Single-Boson States

There exists a wide choice of groups for classifying the states of configurations of the type $(\Sigma k)^N$. However, it is essential to use the same groups as have already been introduced in classifying the many-electron eigenfunctions; for otherwise we cannot take advantage of the Wigner–Eckart theorem to investigate the properties of the matrix elements of the tensor products of the $\mathbf{v}^{(k)}$. But this statement does not go far enough. The transformations induced by the generators A_μ of a group $\mathcal{G}$ on the tensors

$$V_q^{(k)} = \sum_{\zeta,\nu} a_\zeta^\dagger \langle \zeta | v_q^{(k)} | \nu \rangle a_\nu$$

must parallel the transformations induced by the generators B_μ of the group $\mathcal{G}$ on the single-boson states $|k, q\rangle$. The generators A_μ are operators such as $F_{mm'}$ for $R(2l + 1)$, and involve fermion creation and annihilation operators. By contrast, the generators B_μ involve the boson creation and annihilation operators $b_\xi^\dagger$ and b_ν; the analogues of Eqs. (2) are

$$\begin{aligned} b_\xi^\dagger b_\nu^\dagger - b_\nu^\dagger b_\xi^\dagger &= 0 \\ b_\xi^\dagger b_\nu - b_\nu b_\xi^\dagger &= -\delta(\nu, \xi) \\ b_\xi b_\nu - b_\nu b_\xi &= 0 \end{aligned} \tag{7}$$

This parallelism can be established for $R(2l + 1)$ and its subgroups. Corresponding to the generators $F_{mm'}$ of Eq. (6), we take

$$f_{mm'} = e_{mm'} - (-1)^{m+m'} e_{-m'-m}$$

in which

$$e_{mm'} = -\Sigma[1 + (-1)^{k+k'}]\sqrt{\{(2k+1)(2k'+1)\}}b^{\dagger}_{kq}b_{k'q'}$$
$$\times \begin{pmatrix} k & l & l \\ q & m' & m'' \end{pmatrix}\begin{pmatrix} k' & l & l \\ q' & m & m'' \end{pmatrix}$$

the sum running over k, k', q, q', and m''. It is straightforward to show that the coefficients x and x' in the following equations are equal:

$$f_{mm'}|kq\rangle = \Sigma x(mm'kq;\ k'q')|k'q'\rangle$$
$$[F_{mm'}, V^{(k)}_q] = \Sigma x'(mm'kq, k'q')V^{(k')}_{q'} \tag{8}$$

By studying the eigenvalues of $f_{mm'}$ (or $F_{mm'}$), it can be shown that the odd k states ($p, f, h, \ldots$ bosons) belong to the representation $(110\ldots0)$ of $R(2l+1)$; the even k states (d, g, $i, \ldots$ bosons) belong to $(20\ldots0)$. If the s boson is also included, it is found to form by itself the scalar representation $(00\ldots0)$.

C. Many-Boson States

The single-boson classification scheme is set out in Table 2 for $l = 3$. The symbol $[\Lambda]$ stands for a representation of $U(4l^2 + 4l + 1)$, the group of unitary matrices having rows and columns

TABLE 2. Classification of single-boson states.

$[\Lambda]$	$[\lambda]$	W	U	k
[1]	[11]	(110)	(10)	f
			(11)	$p\ h$
	[2]	(200)	(20)	$d\ g\ i$
		(000)	(00)	s

labelled by the $(2l+1)^2$ boson states $|kq\rangle$. The presence of an s boson in [1] means that the many-boson classifications derived from this starting point correspond to the configurations

$$(s + p + d + f + g + h + i)^N \tag{9}$$

Since $\mathbf{v}^{(0)}$ is a pure number, the S states of a given configuration of this type correspond not only to N-particle scalar operators, but to all scalar operators involving fewer than N particles as well. This is no real inconvenience, and it has the advantage that the representation $[\lambda] = [2]$ of Table 2 remains intact.

The two-boson system belongs to $[\Lambda] = [2]$, since the antisymmetric representation [11] must be rejected. There are four outer products $[\lambda] \times [\lambda']$ to consider: $[11] \times [11]$, $[11] \times [2]$, $[2] \times [11]$, and $[2] \times [2]$. Now, from Section 2.D

$$[11] \times [11] = [1111] + [211] + [22]$$

To decide which representations belong to $[\Lambda] = [2]$, we may proceed in several ways:

(1) The dimension of [11] is 21. From n functions it is possible to form $\frac{1}{2}n(n+1)$ symmetric products—in this case, 231. The dimensions of [1111], [211], and [22] are 35, 210, and 196 (see Table 1 or the extensive tables of Flowers[13]). It follows that [1111] and [22] must belong to $[\Lambda] = [2]$. Dimensional arguments of this kind sometimes lead to ambiguities, however.

(2) The separation of $[11] \times [11]$ into symmetric and antisymmetric parts is an example of what Littlewood[14] calls a *plethysm*; in his notation,

$$[11] \otimes [2] = [1111] + [22]$$

$$[11] \otimes [11] = [211]$$

A number of detailed methods for finding $[\lambda] \otimes [\eta]$ have been described by Littlewood and by Smith and Wybourne.[15] They are of great value when simple arguments are not sufficient. Unfortunately, the symbol $\otimes$ has come to be used in the physics literature for outer products, so some new notation seems to be called for.

Similar methods can be used for $[2] \times [2]$. The products $[11] \times [2]$ and $[2] \times [11]$ divide equally between $[\Lambda] = [2]$ and $[\Lambda] = [11]$. Results for the two-boson system are assembled in Table 3. To keep the ranks k of the two contributing bosons in view, the subscripts g (even) and u (odd) are added to the representations $[\lambda]$.

As Wybourne has pointed out, the states of the N-boson system (9) are contained in the reduction of the plethysm

$$([2] + [11]) \otimes [N]$$

where $[N]$ is a tableau comprising N cells in a single row. According to Littlewood,[14] this can be written as

$$\sum_M \{[2] \otimes [M]\} \times \{[11] \otimes [N - M]\}$$

Moreover, $[2] \otimes [M]$ comprises all tableaux corresponding to the partition of $2M$ into even parts. For example,

$$[2] \otimes [4] = [8] + [62] + [44] + [422] + [2222]$$

The composition of $[11] \otimes [M]$ can be rapidly found by taking the adjoint tableaux. Thus

$$[11] \otimes [4] = [11111111] + [221111] + [2222] + [3311] + [44]$$

In this way the classification of the N-boson system can be completed. Tableaux containing more than $2l + 1$ rows can be

TABLE 3. Classification of 2-boson and 3-boson states.

$[\Lambda]$		$[\lambda]$		
[2]	$[1111]_{uu}$	$[211]_{ug}$	$[22]_{gg}$	
	$[22]_{uu}$	$[31]_{ug}$	$[4]_{gg}$	
[3]	$[111111]_{uuu}$	$[21111]_{uug}$	$[2211]_{ugg}$	$[222]_{ggg}$
	$[2211]_{uuu}$	$[222]_{uug}$	$[321]_{ugg}$	$[42]_{ggg}$
	$[33]_{uuu}$	$[3111]_{uug}$	$[33]_{ugg}$	$[6]_{ggg}$
		$[321]_{uug}$	$[411]_{ugg}$	
		$[42]_{uug}$	$[51]_{ugg}$	

rejected for the same reasons that spin tableaux with more than two rows were rejected in Section 2.D. The complete classifications for $N = 2$ and 3 are given in Table 3 for arbitrary l. The branching rules for the reduction $U(7) \to R(7)$ are assembled in Table 4 for those representations $[\lambda]$ of Table 3 that contain at least one S state. Those representations W of Table 4 that contain at least one S state are reduced to representations U of G_2 in Table 5. The number $n(S)$ of S states is included in these tables;

TABLE 4. Branching rules for the reduction $U(7) \to R(7)$.

$n(S)$	[λ]	W
1	[0]	(000)
1	[2]	(000)(200)
1	[1111]	(111)
2	[22]	(000)(200)(220)
2	[4]	(000)(200)(400)
1	[21111]	(111)(210)
4	[222]	(000)(220)(222)
3	[3111]	(111)(211)(311)
3	[321]	(110)(200)(211)(220)(310)(321)
1	[411]	(110)(211)(310)(411)
6	[42]	(000)$(200)^2$(220)(310)(400)(420)
1	[51]	(110)(200)(310)(400)(510)
3	[6]	(000)(200)(400)(600)

TABLE 5. Branching rules for the reduction $R(7) \to G_2$.

$n(S)$	W	U
1	(000)	(00)
1	(111)	(00)(10)(20)
1	(220)	(20)(21)(22)
2	(222)	(00)(10)(20)(30)(40)
2	(311)	(20)(21)(22)(30)(31)(40)
2	(321)	(11)(20)$(21)^2$(22)(30)$(31)^2$(32)(40)(41)
1	(400)	(40)
1	(411)	(30)(31)(32)(40)(41)(50)
3	(420)	(22)(31)(32)(40)(41)(42)
1	(600)	(60)

TABLE 6. Number of S states contained in representations of G_2.

$n(S)$	U
1	(00)
1	(22)
1	(40)
1	(42)
1	(60)

Table 6 gives these numbers for representations U for which $n(S) > 0$.

D. Interpretation of Tables

Tables 3, 4, 5, and 6 enable us to rapidly find the S states of the configurations

$$(s + p + d + f + g + h + i)^N$$

where $N = 2$ or 3. In the case of $N = 2$, seven S states exist. This is what we would expect, since for f electrons there are just seven scalar products

$$\sum_{i \neq j} (\mathbf{v}_i^{(k)} \cdot \mathbf{v}_j^{(k)})$$

corresponding to $k = 0, 1, \ldots, 6$. The assignments $[\lambda]WU$ correspond to particular linear combinations of these scalar operators. They are all two-electron operators except the one for which $k = 0$; this is simply a number.

There are many more possibilities for $N = 3$. In forming the scalars

$$\sum_{h \neq i \neq j} (\mathbf{v}_h^{(k)} \cdot \mathbf{v}_i^{(k')} \cdot \mathbf{v}_j^{(k'')}) \tag{10}$$

where

$$(\mathbf{v}_h^{(k)} \cdot \mathbf{v}_i^{(k')} \cdot \mathbf{v}_j^{(k'')}) = \sum_{q,q',q''} \begin{pmatrix} k & k' & k'' \\ q & q' & q'' \end{pmatrix} (v_q^{(k)})_h (v_{q'}^{(k')})_i (v_{q''}^{(k'')})_j$$

we note first that the three ranks k, k', and k'' cannot all be odd, since neither $[111111]_{uuu}$, $[2211]_{uuu}$, nor $[33]_{uuu}$ contains an S

state. This is to be expected, since if $k = k'$, the associated two-boson operator can be symmetric only if the resultant is even; and in this case a final resultant of zero is impossible to form when k'' is odd. On the other hand, if no equal pair can be selected from k, k', and k'', they must have the values 1, 3, and 5, and the triangular conditions are violated. With two ranks odd and one even, 17 scalars exist. This is consistent with the number of independent symmetric operators characterized by the triads $(kk'k'') = (110)$, (112), (132), (134), (154), (156), (330), (332), (334), (336), (352), (354), (356), (550), (552), (554), (556). Next, for one odd rank and two even, only five scalars exist. Finally, three even ranks give 13 scalars, making 35 in all.

These 35 scalar operators are not all true three-electron operators. If they are taken in the simple form in which each scalar is characterized by a single triad $(kk'k'')$, the ones that are not three-electron operators can be rejected merely by dropping all scalars for which at least one rank (k, k', or k'') is zero. It might seem at first sight that it would be impossible to eliminate the unwanted operators if the linear combinations labelled by $[\lambda]WU$ are not formed. However, the mixing takes place at the $[\lambda]$ level, not at the W; so if we are prepared to abandon the $[\lambda]$ classification, the true three-electron operators can still be described by the labels W and U. The reason is simply that the source of the spurious operators is the presence of s in Table 2; but since it corresponds to an entire representation W, its rejection still leaves the remaining two W representations intact. We could regard these two as forming a representation [1] of $U(48)$; but it would be slightly more awkward to perform the plethysms.

In the present case, seven scalars involve a zero rank, leaving 28 true three-electron operators. Of these, 14 correspond to two odd ranks and one even; five to one odd rank and two even; and nine to three even ranks. This number, 28, represents the total number of scalar three-electron operators that can be formed for the case of f electrons. Not all of them are useful, however. Some possess matrix elements that are proportional to those of other operators; and it may happen too that a given operator has zero matrix elements for all states of the f shell. Again, to a given order of perturbation theory, not all of them may be called on to act as effective operators. The analysis, to this point, tells us what operators and how many are available; but their properties remain to be discussed.

E. Hermiticity

Before going on to discuss the origin of the operators in perturbation expansions, it is convenient to raise the question of their Hermiticity. A symmetric N-particle scalar operator must be some linear combination of operators of the type

$$\Xi = (\ldots(((\mathbf{v}_g^{(k)}\mathbf{v}_h^{(k')})^{(K)}\mathbf{v}_i^{(k'')})^{(K')}\mathbf{v}_j^{(k''')})^{(K'')}\ldots)^{(0)}$$

This form is suitable for calculating matrix elements by means of fractional parentage techniques. Setting Ξ between states ψ and ψ' of l^N, and taking the coefficients of fractional parentage to be real, we obtain the result

$$(\psi|\Xi|\psi') = (-1)^{k+k'+k''+k'''+\ldots}(\psi'|\Xi|\psi)$$

No intermediate ranks K, K', or K'', etc. appear in the phase factor. This means that any scalar operator built from single-electron tensors $\mathbf{v}^{(k)}, \mathbf{v}^{(k')}, \ldots$ is Hermitian if the sum $k + k' + \ldots$ is even, and anti-Hermitian if the sum is odd. This result is independent of the particular type of coupling that may be used to produce a scalar resultant.

4. PERTURBATION THEORY

A. Effective Operators

In the central-field approximation, the configurations are the eigensolutions of the zeroth-order Hamiltonian, H_0. The perturbation V, which must be combined with H_0 to give the total Hamiltonian, splits each configuration up into terms. To first order, the eigenvalues of the terms are obtained by diagonalizing the matrices of V for each configuration separately. If the configurations are labelled by $n = 1, 2, 3, \ldots$, and if the individual eigenfunctions are distinguished by the additional labels t, then the eigenvalues are determined by diagonalizing matrices whose elements are $\langle nt|V|nt'\rangle$. To proceed to second order, we first pick the eigenfunctions

$$|nx\rangle = \sum_t a^n_{xt}|nt\rangle$$

that diagonalize the first-order matrices. Following the conventional method,[16] we find that to second order the term energies are

$$E_n + \langle nx|V|nx\rangle + \sum_{n'y}^{n'\neq n} \langle nx|V|n'y\rangle\langle n'y|V|nx\rangle/(E_n - E_{n'})$$

where E_n is the appropriate eigenvalue of H_0. However, we can obtain an equivalent result—though only to second order—by diagonalizing within each configuration not V, but V', where

$$V' = V + \sum_{n'y}^{n'\neq n} [V|n'y\rangle\langle n'y|V]/(E_n - E_{n'})$$

This diagonalization procedure gives eigenfunctions $|nu\rangle$, which are linear combinations of the $|nx\rangle$ with the same n. The term energies to second order become simply

$$E_n + \langle nu|V'|nu\rangle$$

To second order, the alternative method scarcely appears exceptional. However, several advantages appear as soon as one proceeds to third order. In the conventional method, one would continue to use the eigenfunctions $|nx\rangle$. Although V cannot directly link a bra $\langle nx'|$ to a ket $|nx\rangle$ when $x' \neq x$, yet the introduction of the intermediate states $|n'y\rangle\langle n'y|$ makes this possible. Consequently, energy denominators of the type

$$\langle nx|V|nx\rangle - \langle nx'|V|nx'\rangle$$

appear; and the interpretation by means of effective operators of the terms containing these denominators presents awkward problems. On the other hand, if the eigenfunctions $|nu\rangle$ are used, these terms vanish. However, one should not be content with this rather modest advantage, but instead find a new expansion V'' to replace V', such that to third order the term energies are

$$E_n + \langle nw|V''|nw\rangle$$

At this point a new feature arises: there exists a certain freedom in the choice of V''. Any of the various forms for V''

produce the same eigenvalues to third order, but may differ with regard to higher-order contributions. The expression

$$V'' = V + \sum_{n'y}^{n' \neq n} [V|n'y\rangle\langle n'y|V]/(E_n - E_{n'})$$

$$+ \sum_{n'y}^{n' \neq n} \sum_{n''z}^{n'' \neq n} [V|n'y\rangle\langle n'y|V|n''z\rangle\langle n''z|V]/(E_n - E_{n'})(E_n - E_{n''})$$

$$- \tfrac{1}{2} \sum_{n'yw}^{n' \neq n} [V|nw\rangle\langle nw|V|n'y\rangle\langle n'y|V]/(E_n - E_{n'})^2$$

$$- \tfrac{1}{2} \sum_{n'yw}^{n' \neq n} [V|n'y\rangle\langle n'y|V|nw\rangle\langle nw|V]/(E_n - E_{n'})^2$$

has the advantage that it is Hermitian. However, systematic procedures for generating effective operators to arbitrary orders in perturbation theory do not necessarily exclude non-Hermitian forms. The generalization of the *linked-cluster expansion* to degenerate levels,[17] for example, is ideal for introducing effective operators; but the possibility of having to handle non-Hermitian operators is something of a drawback.

B. Two-Electron Excitations

Suppose that the ground configuration of an atom comprises a number of l electrons outside closed shells. It is convenient to use $a^\dagger$ and a to refer to the open-shell electrons, $c^\dagger$ and c for the electrons in closed shells, and the pair $d^\dagger$ and d for electrons in unoccupied orbitals. There are now various forms that we can use for V in the structure

$$V|n'y\rangle\langle n'y|V$$

For example, suppose we are interested in the effect on the configuration l^N of the excited configuration $l^{N-2}l'l''$. When the second-quantized expressions for V are used, it is essential that the operator on the left contains a sequence of the type $a_\alpha{}^\dagger a_\beta{}^\dagger d_\gamma d_\varepsilon$, and the operator on the right a sequence of the type $d_\xi{}^\dagger d_\eta{}^\dagger a_\zeta a_\lambda$; for otherwise the states $|n'y\rangle$ and $\langle n'y|$ will not be connected to the states of l^N. But, for any choice of subscripts γ, ε, ξ, and η, only one state at most of the configuration n' can enter in such a way as to produce a non-vanishing result; and this means that the closure theorem

$$\sum_{n'y} |n'y\rangle\langle n'y| = 1$$

can be used. The annihilation and creation operators now combine to give the sequence

$$a_\alpha^\dagger a_\beta^\dagger d_\gamma d_\varepsilon d_\xi^\dagger d_\eta^\dagger a_\zeta a_\lambda$$

Since d_γ and d_ε, acting on a state of l^N, cannot annihilate an l' or l'' electron, their transference to the right leads to the result

$$a_\alpha^\dagger a_\beta^\dagger a_\zeta a_\lambda[\delta(\varepsilon, \xi)\delta(\gamma, \eta) - \delta(\varepsilon, \eta)\delta(\gamma, \xi)] \tag{11}$$

Of course, only the skeleton of the mathematics has been given here. Radial integrals of the type $R^k(ll, l'l'')$ enter as well as various n-j symbols. However, the final sequence of annihilation and creation operators shows that the effective operator acting solely within l^N is a two-electron operator; and since V is scalar, this operator must be so too. We have thus obtained a striking result in a very simple way: the changes in the energies of the terms of l^N produced by the configuration $l^{N-2}l'l''$ can be represented, to second order in perturbation theory, by effective two-body scalar operators. According to Section 3.D, there are $2l$ of these, and, in general, they are all required.

This result was recognized by Racah[18] in 1952. For d electrons, there are only two two-electron scalar operators that can differ from those describing the ordinary Coulomb interaction between the electrons. These are the pair

$$\sum_{i\neq j} (\mathbf{v}_i^{(1)} \cdot \mathbf{v}_j^{(1)}), \qquad \sum_{i\neq j} (\mathbf{v}_i^{(3)} \cdot \mathbf{v}_j^{(3)})$$

A particular combination of these operators is proportional to Casimir's operator for $R(5)$, and its eigenvalues differ only slightly for low-lying terms of a configuration d^N. The first operator is proportional to

$$2\sum_{i\neq j} (\mathbf{l}_i \cdot \mathbf{l}_j)$$

which has the eigenvalues $L(L + 1) - 2l(l + 1)$. If, then, two-electron excitations are important, we would expect to be able to account for them fairly well for the low levels by adding the operator $\alpha\mathbf{L}^2$ to the Hamiltonian. The success that such a step had[19] led to the belief that the origin of the perturbations was understood. It was shown in 1960 by Jørgensen and Trees,[20] however, that the excitation of two holes could also be described by the same theory, and that the excitation of two p electrons from $3p^63d^N$ to $3p^43d^{N+2}$ is more important in this case than the excitation of two d electrons. The Feynman diagrams corresponding to these two mechanisms are drawn out in Fig. 1, according

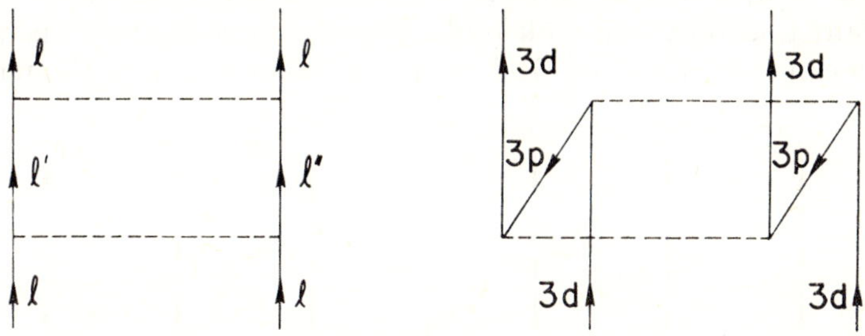

Fig. 1. Feynman diagrams for two second-order mechanisms that can be represented by effective two-electron scalars. The situation at various stages of the perturbation process can be found by examining the labels on the particle-lines that are severed by a horizontal cut. Lines with arrows pointing down represent hole states.

to the prescription of Goldstone.[21] The delta functions in expressions such as (11) above ensure that the diagrams consist of a single connected whole.

C. Single-Electron Excitations

If the states $|n'y\rangle$ belong to the configuration $l^{N-1}l'$, the structure $V|n'y\rangle\langle n'y|V$ gives rise to sequences of operators of the type

$$a_\alpha^\dagger a_\beta^\dagger a_\gamma d_\varepsilon d_\xi^\dagger a_\eta^\dagger a_\zeta a_\lambda$$

which reduces to

$$a_\alpha^\dagger a_\beta^\dagger a_\eta^\dagger a_\zeta a_\gamma a_\lambda \delta(\varepsilon, \xi) + a_\alpha^\dagger a_\beta^\dagger a_\zeta a_\lambda \delta(\varepsilon, \xi)\delta(\gamma, \eta)$$

The second term represents a scalar two-electron operator. The first represents a scalar three-electron operator. The delta function $\delta(\varepsilon, \xi)$ ensures the linking of the two parts of the Feynman diagram; the two possible forms are drawn out in Fig. 2. As explained in Section 3.D, there are 28 three-electron scalars when $l = 3$. However, only a small number of these are required in the present case. When the Coulomb interaction is expanded by means of the addition theorem for spherical harmonics, the states β and γ must be connected by a spherical harmonic of even rank, owing to parity. The same is true of η and ζ. This means that only one rank (say k') of the three-electron operator (10) can take all values. But even this condition can be tightened.

Fig. 2. Feynman diagrams for two second-order mechanisms that can be represented by effective three-electron operators.

For if specific ranks k and k'' are chosen, it clearly cannot make any difference to the final result if k is assigned to the pair $\beta\gamma$ and k'' to the pair $\eta\zeta$ or vice versa. In other words, the three-electron operator must by symmetric under the interchange of two ranks. This is an entirely new type of symmetry. However, the 3-j symbol in the definition of (10) changes sign under an interchange of two columns if $k + k' + k''$ is odd. It follows that all three-electron operators with k' odd cancel out in pairs, so we may take k' to be even. Instead of 28 scalars, we are left with only 9.

This cancellation for $k + k' + k''$ odd can be understood from another point of view. Since V is Hermitian, the operator

$$V|n'y\rangle\langle n'y|V$$

must be so too. According to Section 3.E, the sum $k + k' + k''$ of the ranks of the component tensors of a Hermitian three-electron scalar operator must be even. It follows that the operators for which $k + k' + k''$ is odd must vanish to second order in perturbation theory. All two-electron scalar operators must necessarily be Hermitian, of course.

Since the matrix elements of $\mathbf{V}^{(k)}$ with even (non-zero) rank k change sign under conjugation (the substitution of hole states for electron states), all diagonal matrix elements must vanish for the half-filled shell. The operator

$$(\mathbf{V}^{(k)} \cdot \mathbf{V}^{(k')} \cdot \mathbf{V}^{(k'')})$$

which contains an odd number of such tensors $\mathbf{V}^{(k)}$, must similarly have null diagonal matrix elements for the half-filled shell when k, k', and k'' are all even. This means that the contribution coming from the three-electron part can be expressed, under these conditions, as the negative of the contribution from the two-electron and zero-electron parts. As Racah realized,[22] this lies at the root of the great success Trees[19] had with the spectra of several transition-metal ions containing the half-filled d shell. The addition of the term $\alpha L(L+1)$ reduces the mean deviation by factors of roughly five for Mn II d^5s and Fe III d^5s.

D. The Wigner–Eckart Theorem

It is at this point that group theory plays its role, and allows us to reduce the number of three-electron scalars even further. Suppose we consider the linear combinations of these operators that correspond to the irreducible representations of groups. To find these representations for f electrons, we first note that with three even ranks, only $[222]_{ggg}$, $[42]_{ggg}$, and $[6]_{ggg}$ of Table 3 contribute. The representations W contained in these tableaux and themselves containing at least one S state can be found from Tables 4–6: they are

$$(000)^3(220)^2(222)(400)^2(420)(600)$$

As already explained, these representations include those used in labelling the operators for which at least one rank (k, k', or k'') is zero. To find these superfluous representations, we refer to the zero-electron and two-electron operators classified in Tables 3 and 4. In this way, we find that we have to subtract off $(000)^2(220)(400)$. The remaining representations are listed with their subgroup structure in Table 7. The rank K $(=0)$

emphasizes that we are dealing solely with scalars. The nine linear combinations of the three-electron operators corresponding to a specific WUK are assigned representations in this table.

To use the Wigner–Eckart theorem, a matrix element is visualized in the form

$$\langle \Gamma a | \mathrm{O}(\Gamma' x) | \Gamma'' b \rangle$$

where Γ, Γ', and Γ'' are irreducible representations of a group $\mathscr{G}$. The symbols a and b specify a particular state and x a particular

TABLE 7. Three-electron scalar operators classified for f electrons.

t_i	WUK
t_1	(000)(00)0
t_2	(220)(22)0
t_3	(222)(00)0
t_4	(222)(40)0
t_5	(400)(40)0
t_6	(420)(22)0
t_7	(420)(40)0
t_8	(420)(42)0
t_9	(600)(60)0

operator of those spanning the representations in question. A vital quantity is $c(\Gamma\Gamma'\Gamma'')$, the number of times Γ occurs in the reduction of the product $\Gamma' \times \Gamma''$. If $c(\Gamma\Gamma'\Gamma'') = 0$, the matrix element vanishes; if $c(\Gamma\Gamma'\Gamma'') = 1$, the set of matrix elements for all a, x, and b is proportional to any other set labelled by the same representations Γ, Γ', and Γ''.

Applying this theorem to the operators t_i of Table 7, we see immediately that $c(W(600)W') = 0$ for all representations W and W' used in classifying states of f^N, since their leading weight (w_1 or w_1') can at most be two (see Section 2.E), and the method of forming products effectively imposes a triangular condition on the triad ($w_1\, 6\, w_1'$). Thus the matrix elements of t_9 are always zero, and this operator disappears from the analysis. Next, it can be shown that $c(W(400)W') \leqslant 1$ for all representations W

and W' used in classifying states of f^N. Thus the matrix elements of t_5 must always be proportional to those of the two-electron scalar transforming like (400). If this operator has already been included in the parametrizing procedure, it is unnecessary to also include t_5, which can therefore be dropped. Finally, it can be shown that the matrix elements of the total scalar t_1 can be reproduced in f^3 by combining the two-electron scalar (000) with the zero-particle scalar (000). Thus t_1 can also be disregarded, and the nine three-electron operators are reduced to six. The details of this simplification—in particular, the explicit construction of the operators t_i from the basic three-electron operators—have been described elsewhere.[23]

The six three-electron operators, when combined with the six two-electron operators and the zero-electron operator, make thirteen in all. With this number, the energies of the terms of the configurations f^N can be rigorously fitted to second order in perturbation theory. In practice, the spin–orbit splitting necessitates the addition of a fourteenth parameter. The experimental situation for the rare-earth and actinide elements is not sufficiently advanced to make a good test of the theory possible; but Shadmi[24] has found that the three-electron operators greatly improve the fit for $3d$ electrons. In these cases, only two (rather than six) parameters are required for the three-electron operators.

5. THIRD-ORDER PERTURBATION THEORY

A. Distinct Intermediate Configurations

As far as the experimental situation goes, there is no immediate call to extend the theory to third order. However, there is considerable interest in the purely theoretical aspects of the problem, since the treatment to second order is really too simple to indicate what happens in higher orders. No attempt is made here to treat the third-order processes in a comprehensive manner; yet it is felt that the examples give a reasonably accurate impression of the kind of analysis that has to be made.

As a first example, we consider a class of interactions in which two distinct intermediate configurations enter. In the notation of Section 4.A, the symbols n, n', and n'' stand for l^N, $l^{N-1}l'$, and $l^{N-1}l''$, with the electron l' assumed distinct from the electron l''. The most obvious new feature is the appearance of effective four-electron operators. The corresponding Feynman diagrams are drawn out in Fig. 3. Since there are many ways of coupling four angular momenta to a zero resultant, a form such as

$$\sum_{g \neq h \neq i \neq j} (\mathbf{v}_g^{(k)} \cdot \mathbf{v}_h^{(k')} \cdot \mathbf{v}_i^{(k'')} \cdot \mathbf{v}_j^{(k''')})$$

does not uniquely define a four-electron scalar. At first sight, it might appear attractive to introduce some intermediate rank K, and so specify a coupling scheme such as $((kk')K(k''k''')K)0$. This device is to be avoided when dealing with ranks k, k', k'', and k''' that may not all be different, since only certain linear combinations of the four-electron operators may be permitted if they are to be totally symmetric with respect to electron interchange. (For the same reason, intermediate angular momenta are not given when specifying states of equivalent electrons.) In fact, the natural way to define the four-electron scalars is to use the group-theoretical approach.

Just as with the three-electron scalars of Section 4.C, two ranks (say k and k''') must be even. Each of the six diagrams of Fig. 3 can be represented as a particular sum over operators of the type

$$T = a_\alpha^\dagger a_\beta^\dagger a_\varepsilon^\dagger a_\theta^\dagger \langle \alpha_1 \beta_2 \varepsilon_3 \theta_4 | (v_q^{(k)})_1 (v_{q'}^{(k')})_2 (v_{q''}^{(k'')})_3 (v_{q'''}^{(k''')})_4 | \zeta_1 \xi_2 \eta_3 \lambda_4 \rangle a_\lambda a_\eta a_\xi a_\zeta$$

However, when the detailed synthesis of the diagrams is made, it is found that the operators T occur with associated factors that permit

$$T\left[\frac{1}{E'E''} + \frac{(-1)^{k''}}{(E'+E'')E''} + \frac{(-1)^{k'}}{E'(E'+E'')} + \frac{(-1)^{k''}}{(E'+E'')E'} + \frac{(-1)^{k'+k''}}{E''E'} + \frac{(-1)^{k'}}{E''(E'+E'')}\right]$$

to be separated out. The energies E' and E'' are the energies of the excitations $l \to l'$ and $l \to l''$; each of the terms in the above expression corresponds to one Feynman diagram in Fig. 3. For the expression not to vanish, both k' and k'' must be even.

The diagrams of Fig. 3 do not exhaust all possible four-electron operators. Cases in which three of the four particle-lines are involved in only one interaction remain. It turns out, however,

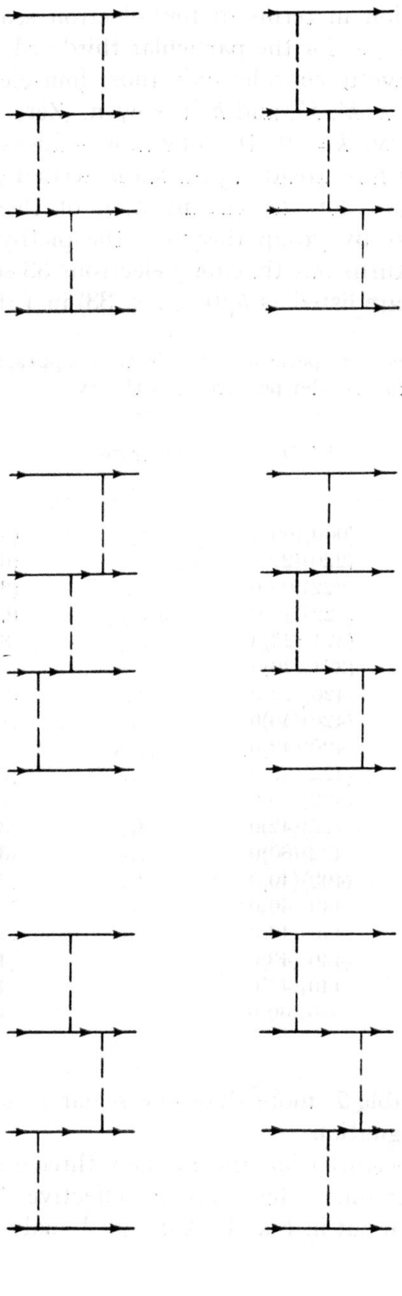

Fig. 3. Feynman diagrams for six third-order mechanisms that can be represented by effective four-electron scalar operators

that their description in terms of four-electron scalars requires even ranks only. Thus, for the particular third-order mechanism under study, we have to consider only those four-electron scalars for which all ranks k, k', k'', and k''' are even. Zero ranks can be excluded, for if, say, $k = 0$, the operator reduces to a three-electron scalar that has already appeared in second order.

The starting point for the classification of these particular four-electron scalars by group theory is the plethysm $[2] \otimes [4]$ of Section 3.C. It turns out that for f electrons 33 scalars can be constructed: they are listed as $p_i (0 < i \leqslant 33)$ in Table 8. Unlike

TABLE 8. Scalar operators for f electrons appearing in third-order perturbation theory.

Operator	*WUK*	Operator	*WUK*
p_1, p_2	(000)(00)0	p_{28}	(620)(42)0
p_3, p_4, p_5	(220)(22)0	p_{29}	(620)(60)0
p_6	(222)(00)0	p_{30}	(620)(61)0
p_7	(222)(40)0	p_{31}, p_{32}	(620)(62)0^2
p_8	(321)(22)0	p_{33}	(800)(80)0
p_9	(321)(40)0		
p_{10}, p_{11}	(420)(22)0	t_{10}	(000)(00)0
p_{12}, p_{13}	(420)(40)0	t_{11}	(111)(00)0
p_{14}, p_{15}	(420)(42)0	t_{12}, t_{13}	(220)(22)0
p_{16}	(422)(22)0	t_{14}	(222)(00)0
p_{17}	(422)(40)0	t_{15}	(222)(40)0
p_{18}	(422)(42)0	t_{16}	(311)(22)0
p_{19}	(422)(60)0	t_{17}	(311)(40)0
p_{20}, p_{21}, p_{22}	(400)(40)0	t_{18}	(321)(22)0
p_{23}	(440)(40)0	t_{19}	(321)(40)0
p_{24}	(440)(42)0	t_{20}	(400)(40)0
p_{25}	(440)(43)0	t_{21}	(420)(22)0
p_{26}	(440)(44)0	t_{22}	(420)(40)0
p_{27}	(600)(60)0	t_{23}	(420)(42)0

the scalars t_i of Table 7, more than one scalar p_i sometimes has the same *WU* designation.

In third-order perturbation theory, new three-electron scalars appear. Four Feynman diagrams of effective three-electron operators are drawn out in Fig. 4. A detailed analysis shows that

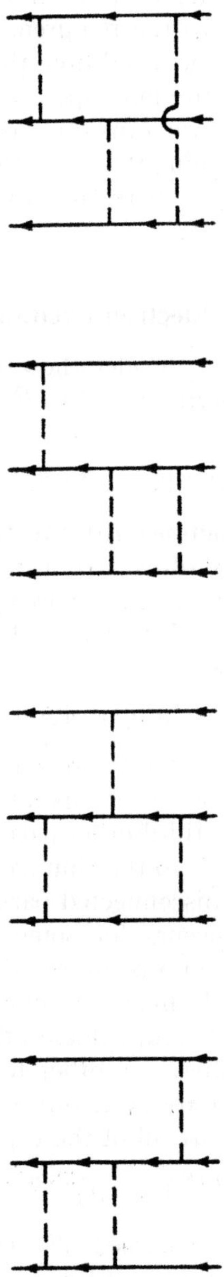

Fig. 4. Feynman diagrams for four third-order mechanisms that can be represented by effective three-electron scalar operators.

the requirement of Section 4.C that the three ranks k, k', and k'' be all even can be relaxed to the requirement that $k + k' + k''$ be even. This is merely the condition that the operators be Hermitian. In addition to the nine operators t_i (for f electrons), there are those for which two ranks are odd and one is even. They can be found by examining the representations $[\lambda]$ of Table 3 that are subscripted by *uug*. There are 14 in all: they are listed as t_i $(10 \leqslant i \leqslant 23)$ in Table 8.

B. Single-Electron Excitations

As a second example, we consider those third-order contributions that come from the interaction $l^N \leftrightarrow l^{N-1}l'$ and which involve the factors of the type

$$R^{(k)}(ll, ll')R^{(k')}(ll, ll)R^{(k'')}(ll, ll')/E'^2$$

In this expression, the functions $R^{(k)}$ are the usual Slater integrals,[5] and E' is the energy of excitation for $l \rightarrow l'$. In contrast with the previous example, terms involving such factors come from all three of the third-order parts of V'' (defined in Section 4.A). Sequences of the type

$$a_\alpha^\dagger a_\beta^\dagger a_\varepsilon d_\gamma \, . \, a_\xi^\dagger a_\eta^\dagger a_\nu a_\lambda \, . \, d_\gamma^\dagger a_\rho^\dagger a_\sigma a_\tau$$

arise when the intermediate states are removed, indicating the possible existence of five-electron operators. However, such terms vanish when all three third-order parts of V'' are combined. This cancellation corresponds to the elimination of certain graphs comprising more than one disconnected part.

A similar cancellation occurs for some disconnected graphs corresponding to four-electron operators. Some of the operators that remain are of the kind already encountered in Section 5.A, and correspond to even (non-zero) values of the ranks k, k', k'', and k'''. Rather surprisingly, there are other four-electron operators for which two of the ranks are even (say k and k'') and two are odd (say k' and k'''). They are all of the type

$$((\mathbf{v}_g^{(k)}\mathbf{v}_h^{(k')})^{(K)} \, . \, (\mathbf{v}_i^{(k'')}\mathbf{v}_j^{(k''')})^{(K)}) + ((\mathbf{v}_g^{(K)}\mathbf{v}_h^{(k')})^{(k)} \, . \, (\mathbf{v}_i^{(k'')}\mathbf{v}_j^{(k''')})^{(k)})$$

in which K is even and never exceeds $2l$. These conditions on K limit the number of operators rather severely. This is not altogether an advantage, since it presents an awkward problem when the group-theoretical classification of the operators is studied. For f electrons, the total number of four-electron scalars possessing two even and two odd ranks is equal to the number of S states in the decomposition of

$$\{(200) \otimes [2]\} \times \{(110) \otimes [2]\}$$

which works out at 106. We can thus construct 106 scalar operators having precise WU designations; but only certain linear combinations of these are required for the present problem. In other words, we cannot freely parametrize the 106 operators, as we might have been led to do from our previous experience. Of course, higher-order terms in perturbation theory will almost certainly produce four-electron operators without restrictions on K; but this is beside the point. As far as third-order terms go, we cannot profitably use the group-theoretical description for this particular class of operators.

6. CONCLUSION

In spite of the fact that only two of the many possible third-order mechanisms have been considered, several general features are clear. Perhaps the most striking is the increase in complexity in going from second-order to third-order perturbation theory. In second order, the use of groups is extremely valuable in reducing the number of parameters that are needed to represent the effects of configuration interaction. It is difficult to see how this simplification could have been brought about without the explicit introduction of group-theoretical concepts. In third order, however, the number of operators that are necessary to reproduce the influence of excited configurations has increased enormously; and, as has been seen in Section 5.B, we cannot expect that they can all be replaced by an equivalent set of operators having well-defined values of WU. Of course, contributions from different orders of perturbation theory do not necessarily fall off in a simple way as the order is increased, so that it may in fact be more convenient to introduce classes of operators in a different sequence from that implied by successive orders of perturbation theory. It may, for example, be more appropriate to complete the parametrization of all Hermitian three-electron operators before proceeding to four-electron operators. Alternatively, it may prove better to rely more on the physics of the situation. For example, the radial eigenfunction of an electron l must certainly expand as higher terms of the configuration l^N are studied, and this simple process is reproduced in terms of numerous many-electron operators when the effective operator approach is employed. By assuming a gradual expansion of the eigenfunction, the number of parameters would be enormously reduced.

Throughout the analysis, no mention has been made of the role played by the central field: the actual perturbation V of Section 4 should, of course, be the difference between the inter-electronic Coulomb interactions and the energy arising from the central-field potential. Since this last is a one-electron scalar, its effects in a given order of perturbation theory can be represented by those effective operators that appear in a lower order. Although modifying the size of the effective operators, it does not introduce ones of a different tensorial character. Since we are not concerned here with detailed numerical evaluation, no account has been presented of the effect of the central potential.

Certain theoretical problems are of considerable interest for their own sake. It is not obvious how the parities of the ranks k of the single-particle tensors $\mathbf{v}^{(k)}$ that form the many-electron operators are determined from the type of perturbation mechanism under study. The use of graphs may be of considerable value here, in view of the topological equivalence between the Feynman graphs and the angular-momentum diagrams of Jucys and his collaborators.[25] In this connection, it would be of great interest to search for new rules for a linked cluster expansion that would ensure that the effective operators are Hermitian.

Acknowledgements

The final form of this article has benefited from discussions with those present at the Summer Institute at Frascati. The preparatory work was partially supported by the United States Atomic Energy Commission.

References

1. Racah, G., *Phys. Rev.*, **62,** 438 (1942).
2. Weyl, H., *Zeits. Math.*, **23,** 271 (1925).
3. Weyl, H., *Gruppentheorie und Quantenmechanik,* S. Hirzel Verlag, Leipzig, 1931; translated by H. P. Robertson, *The Theory of Groups and Quantum Mechanics,* reprinted by Dover Publications, New York.
4. Littlewood, D. E., *The Theory of Group Characters,* Oxford University Press, Oxford, 1950, p. 94.
5. Condon, E. U., and Shortley, G. H., *The Theory of Atomic Spectra,* Cambridge University Press, Cambridge, 1935.
6. Shudeman, C. L. B., *J. Franklin Inst.*, **224,** 501 (1937).
7. Racah, G., *Phys. Rev.*, **76,** 1352 (1949).
8. Jahn, H. A., *Proc. Roy. Soc.* (*London*), *Ser. A*, **201,** 516 (1950).
9. Murnaghan, F. D., *The Theory of Group Representations,* Johns Hopkins Press, Baltimore, 1938.
10. Cartan, E., *Sur la Structure des Groupes de Transformation finis et continus,* thesis, Nony, Paris, 1894.
11. Judd, B. R., *Operator Techniques in Atomic Spectroscopy,* McGraw-Hill Book Co., New York, 1963.
12. Wybourne, B. G., *Spectroscopic Properties of Rare Earths,* John Wiley and Sons, New York, 1965.
13. Flowers, B. H., *Proc. Roy. Soc.* (*London*), *Ser. A*, **210,** 497 (1952).
14. Littlewood, D. E., *The Theory of Group Characters,* Oxford University Press, Oxford, 1950, p. 290.
15. Smith, P. R., and Wybourne, B. G., *J. Math. Phys.*, **8,** 2434 (1967), I am indebted to Professor Wybourne for communicating his results on plethysms prior to publication.
16. Condon, E. U., and Shortley, G. H., *The Theory of Atomic Spectra,* Cambridge University Press, Cambridge, 1935, p. 34.
17. Sandars, P. G. H., this volume, p. 365. See also Bloch, C., and Horowitz, J., *Nucl. Phys.*, **8,** 91 (1958), and des Cloiseaux, J., *Nucl. Phys.*, **20,** 321 (1960).
18. Racah, G., *Phys. Rev.*, **85,** 381 (1952).
19. Trees, R. E., *Phys. Rev.*, **83,** 756 (1951); **84,** 1089 (1951).
20. Jørgensen, C. K., and Trees, R. E., *Phys. Rev.*, **123,** 1278 (1961).
21. Goldstone, J., *Proc. Roy. Soc.* (*London*), *Ser. A*, **239,** 267 (1957).
22. Racah, G., Lectures at the Collège de France, unpublished, Paris, 1964.
23. Judd, B. R., *Phys. Rev.*, **141,** 4 (1966). For the case of *d* electrons, see Feneuille, S., *Compt. Rend.*, **262,** 23 (1966). The same writer has also examined several mixed configurations of the type $(l + l')^N$: J. de Phys., **28,** 61, 315, 497 (1967).
24. Shadmi, Y., private communication.
25. Sandars, P. G. H., *La Structure Hyperfine Magnétique des Atomes et des Molécules,* Editions du Centre National de la Recherche Scientifique, Paris, 1967, p. 111.

APPLICATIONS OF MANY-BODY DIAGRAM TECHNIQUES IN ATOMIC PHYSICS*

HUGH P. KELLY, *Physics Department, University of Virginia, Charlottesville, Virginia, U.S.A.*

CONTENTS

1. INTRODUCTION

Although the many-body perturbation methods were first developed to attack the problems of infinite nuclear matter[1–6] and of the infinite electron gas,[7] these methods are also applicable to finite systems such as the nucleus or an atom.

For finite systems, there is no longer translational invariance, and the problem is complicated by the fact that the appropriate basis states for the perturbation expansion are no longer plane wave states. One of the major considerations in applying the many-body methods to a finite system such as an atom is the

* Work supported in part by the U.S. Atomic Energy Commission. Document ORO-2915-83.

appropriate choice of the single-particle basis states; and much of the numerical effort in calculations may be spent in evaluating the set of single-particle solutions. However, once these states have been calculated, one may readily apply the perturbation expansion to a wide range of physical problems. A discussion of basis states will be given in more detail in succeeding sections.

The linked cluster perturbation expansion, which was first demonstrated by Brueckner[1–3] and later proved to all orders by Goldstone,[4] will be reviewed in Section 2. Section 3 contains a discussion of the choice of single-particle basis states and also a discussion of the explicit evaluation of perturbation terms for atoms. Applications of these methods to atomic correlation energy calculations and to transition probabilities are given in Section 4. Multiple perturbations are discussed in Section 5. Examples of multiple perturbations are given by polarizabilities, shielding factors, and hyperfine interactions when the unperturbed initial state is not the exact non-relativistic solution. In Section 6 it is demonstrated that the same techniques which have been used to evaluate the terms of the many-body perturbation expansion may also be used to calculate terms in a Green's function perturbation expansion. These terms may then be used to obtain the phase shifts for elastic scattering. A brief discussion of inelastic scattering is also given.

2. REVIEW OF MANY-BODY PERTURBATION THEORY

A. The Brueckner–Goldstone Linked Cluster Expansion

We consider the problem of N identical fermions interacting through two-body potentials v_{ij}. We may also have a number of one-body potentials acting upon these particles. The basic problem is to solve the Schrödinger equation

$$H\psi = E\psi \tag{1}$$

where ψ and E are the exact non-relativistic wave function and energy. The Hamiltonian

$$H = \sum_{i=1}^{N} T_i + \sum_{i<j}^{N} v_{ij} \tag{2}$$

where T_i is the sum of the kinetic energy operator for the ith particle and all one-body potentials acting on the ith particle. For example, for atoms of atomic number Z,

$$T_i = -\frac{\nabla_i^2}{2} - \frac{Z}{r_i} \tag{3}$$

Atomic units are used throughout ($e = m = \hbar = 1$). In order to simplify the problem initially, we may approximate the effective potential for the ith particle due to the $N - 1$ other particles by an effective potential V_i. For the moment we do not specify any particular choice for V_i except that it must be Hermitian. A frequent choice of V_i is the Hartree–Fock potential. However, the choice of V_i will be discussed in detail later. The term Σv_{ij} in Eq. (2) is now approximated by $\sum_{i=1}^{N} V_i$. We now have an approximate Hamiltonian

$$H_0 = \sum_{i=1}^{N} (T_i + V_i) \tag{4}$$

Eq. (1) now becomes

$$H_0\Phi_0 = E_0\Phi_0 \tag{5}$$

where Φ_0 is an approximation for ψ. Although the subscript on Φ_0 and E_0 implies that we are interested in the ground state, we may also use Eq. (5) for excited states. For closed-shell states, the state Φ_0 is a determinant containing N single-particle solutions φ_n of the equation

$$(T + V)\varphi_n = \varepsilon_n\varphi_n \tag{6}$$

When Φ_0 approximates the ground state, the states φ_n in Φ_0 are the N solutions of Eq. (6) which are lowest in energy. For open-shell states, Φ_0 is in general a linear combination of determinants. However, for the ground states of many open-shell atoms Φ_0 is a single determinant when the choice $M_L = \pm L$, $M_S = \pm S$ is made.[8] The single-particle states φ_n which are occupied in Φ_0 are called unexcited states and all others are called excited states. The unoccupied, unexcited states are called holes and occupied, excited states are called particles. The present treatment corresponds to that of Goldstone.[4]

In the matrix element representation, we have

$$H_0 = \sum_n \varepsilon_n \eta_n^+ \eta_n \tag{7}$$

where η_n^+ and η_n are the creation and annihilation operators for the state φ_n. They satisfy the usual Fermi–Dirac anticommutation relations. In Eq. (7), the sum n runs over all states φ_n which are solutions of Eq. (6), and the combination $\eta_n^+ \eta_n$ is the number operator for the state n.

We also have

$$H' \equiv H - H_0 = \sum_{pqmn} \langle pq|v|mn\rangle \eta_p^+ \eta_q^+ \eta_n \eta_m - \sum_{pm} \langle p|V|m\rangle \eta_p^+ \eta_m \tag{8}$$

In the summation $\sum_{pqmn}$ only distinct matrix elements are included. For example $\langle pq|v|mn\rangle$ is not distinct from $\langle qp|v|nm\rangle$; however, $\langle pq|v|mn\rangle$ is distinct from $\langle pq|v|nm\rangle$. Alternatively, we need not restrict ourselves to distinct matrix elements. In that case $\sum_{pqmn}$ must be multiplied by 1/2. In this paper, we shall consider only distinct matrix elements.

We shall now use time-dependent perturbation theory in the interaction representation. We assume that the perturbation H' is slowly turned on beginning at $t = -\infty$ and is brought to full strength at $t = 0$. This may be expressed by multiplying H' by $e^{\alpha t}$, where α will be allowed to approach zero. We assume that ψ will be adiabatically derived from Φ_0 in this manner.

Consider

$$\psi(t) = e^{-iH_0 t} U(t) \Phi_0 \tag{9}$$

where $U(t)$ is defined by Eq. (9). Now $\psi(t)$ must satisfy the time-dependent Schrödinger equation

$$i \frac{\partial \psi}{\partial t} = H\psi = (H_0 + H' e^{\alpha t})\psi \tag{10}$$

this leads to the following equation for $U(t)$

$$i \frac{\partial U}{\partial t} = e^{iH_0 t} H' e^{\alpha t} e^{-iH_0 t} U \tag{11}$$

subject to the boundary condition $U(t = -\infty) = 1$. The solution to Eq. (11) is given by

$$U_\alpha(t) = \sum_{n=0}^{\infty} (-i)^n \int_{0>t_1>t_2 \cdots >t_n} H'(t_1)H'(t_2 \cdots H'(t_n)dt_1 \cdots dt_n \tag{12}$$

where

$$H'(t) \equiv e^{iH_0t}H'e^{\alpha t}e^{-iH_0t}$$

the subscript α has been attached to U to indicate the dependence on α. The first few terms of Eq. (12) are explicitly written

$$U_\alpha(t) = 1 + (-i)\int_{-\infty}^{t} H'(t_1)dt_1 + (-i)^2\int_{-\infty}^{t} dt_1 H'(t_1)\int_{-\infty}^{t_1} dt_2 H'(t_2) + \cdots \tag{13}$$

Carrying out these time integrations for $U_\alpha(0)\Phi_0$, we get

$$U_\alpha(0)\Phi_0 = \left[1 + \frac{1}{E_0 - H_0 + i\alpha} H' + \frac{1}{E_0 - H_0 + 2i\alpha} H' \frac{1}{E_0 - H_0 + i\alpha} H' + \cdots \right] \Phi_0 \tag{13a}$$

From Eq. (9) $\psi(t = 0) = U_\alpha(0)\Phi_0$. We now choose the normalization $\langle \Phi_0 | \psi \rangle = 1$, which we obtain by dividing $U_\alpha(0)\Phi_0$ by $\langle \Phi_0 | U_\alpha(0) | \Phi_0 \rangle$. We now have

$$\psi = \lim_{\alpha \to 0} U_\alpha(0)\Phi_0 / \langle \Phi_0 | U_\alpha(0) | \Phi_0 \rangle \tag{14}$$

The energy correction

$$\Delta E \equiv E - E_0 \tag{15}$$

is obtained from

$$(H_0 + H')\psi = (E_0 + \Delta E)\psi \tag{16}$$

by taking the scalar product from the left with Φ_0.

The result is

$$\Delta E = \langle \Phi_0 | H' | \psi \rangle \tag{17}$$

By use of Wick's theorem,[9] $U_\alpha(0)\Phi_0$ becomes a sum of terms which may be represented by Feynman diagrams or graphs. A particle in an excited state is represented by a line directed upwards and a hole in an unexcited state is represented by a line directed downwards. The direction of increasing time is upwards. the graphical representation of matrix elements is illustrated in Fig. 1.

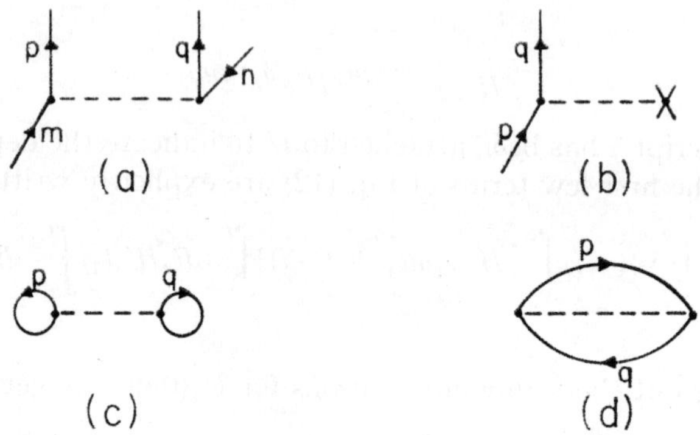

Fig. 1. Diagrams associated with matrix elements: (a) $\langle pq|v|mn\rangle$. Particles in the excited state φ_m and the unexcited state φ_n interact through v and scatter into excited states φ_p and φ_q leaving a hole in φ_n. (b) $-\langle q|V|p\rangle$. A particle in the excited state φ_p is scattered by V into φ_q. (c) $\langle pq|v|pq\rangle$. (d) $-\langle pq|v|qp\rangle$.

The rules for obtaining $U_\alpha(0)\Phi_0$ are the following:

[1]. Draw all distinct diagrams with no free lines at the bottom.

[2]. For each diagram multiply the v and V matrix elements and the $e^{\pm iH_0t}$ and $e^{\alpha t}$ factors. Include a factor $(-1)^{h+l}$ where h is the number of internal hole lines and l is the number of closed loops. In Fig. 1(c), $h = 2$ and $l = 2$. In Fig. 1(d), $h = 2$ and $l = 1$.

[3]. Attach the pairs of η^+ and η operators corresponding to the free (external) lines of each open loop with η^+ to the left of η.

[4]. Carry out the time integrations, noting that the diagrams are time-ordered.

An unlinked part of a diagram is defined as any part of a diagram which is completely disconnected from the rest and has no external lines attached. A linked diagram is defined as one containing no unlinked parts. For a diagram containing unlinked parts and also other pieces, there are other diagrams which differ only by having the interactions in the unlinked parts in different positions relative to the other pieces. The time-orderings of the interactions are kept fixed in the separate parts. Sums over the different relative positions of the interactions are equivalent to carrying out the time integrations with only the time-ordering restrictions in each part separately. The result is a factorization into a product of the separate disconnected parts of the diagrams. Alternatively, we may first carry out the time integrations and then demonstrate the factorizations algebraically. As a result of the factorization, $U_\alpha(0)\Phi_0$ becomes a product of linked terms times the sum of all possible unlinked terms; and the sum of all unlinked terms is $\langle\Phi_0|U_\alpha(0)|\Phi_0\rangle$. In Eq. (14) the denominator is then cancelled, and we have

$$\psi = \lim_{\alpha\to 0} \sum_L \frac{1}{E_0 - H_0 + ni\alpha} H' \cdots \frac{1}{E_0 - H_0 + 2i\alpha} H' \frac{1}{E_0 - H_0 + i\alpha} H'\Phi_0 \qquad (18)$$

where $\sum_L$ means that only linked diagrams are to be included. Then Φ_0 cannot occur as an intermediate state in Eq. (18). If $E_0 - H_0$ does not vanish for intermediate states, then we may set $\alpha = 0$. When Φ_0 is a non-degenerate ground state, $E_0 - H_0$ will not vanish for intermediate states. However, $E_0 - H_0$ will also not vanish in many cases when Φ_0 is a degenerate state provided the perturbation H' does not lead to the degenerate intermediate states. An example of this is found in L–S coupling. Consider an open-shell state Φ_0 specified by L, S, M_L, and M_S. If H' commutes with L and S, then the perturbation H' will not lead to those intermediate states of different M_L and M_S which are degenerate with Φ_0 and we may still set $\alpha \to 0$. We then obtain the well-known "linked cluster" result

$$\psi = \sum_L \left(\frac{1}{E_0 - H_0} H'\right)^n \Phi_0 \qquad (19)$$

$$\Delta E = \sum_{L'} \left\langle \Phi_0 | H' \left(\frac{1}{E_0 - H_0} H'\right)^n |\Phi_0 \right\rangle \qquad (20)$$

where $\sum_{L'}$ indicates a sum over connected diagrams with no external lines.

First-order corrections to Φ_0 are given in Fig. 2. The diagram of Fig. 2(a) corresponds to

$$\frac{1}{\varepsilon_p + \varepsilon_q - \varepsilon_k - \varepsilon_{k'}} \langle kk'|v|pq\rangle \eta_k^+ \eta_{k'}^+ \eta_q \eta_p |\Phi_0\rangle \tag{21}$$

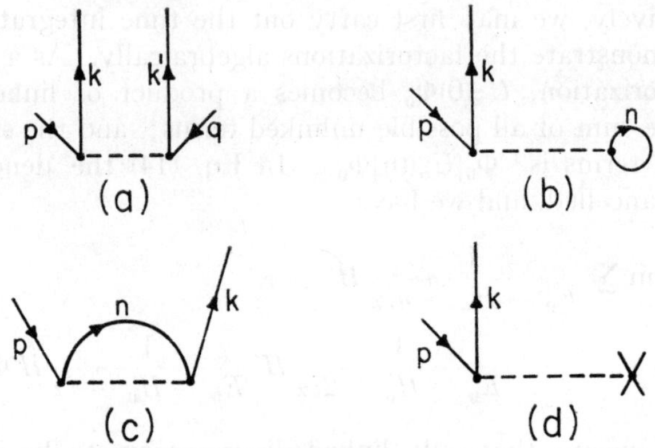

Fig. 2. First-order corrections to Φ_0. (a) Two-body correlation in which electrons in states p and q are excited into k and k' through the interaction $\langle kk'|v|pq\rangle$. (b) Electron in φ_p excited into φ_k through the interaction $\langle kn|v|pn\rangle$ with the unexcited state φ_n. (c) Exchange interaction $-\langle nk|v|pn\rangle$. (d) Interaction $-\langle k|V|p\rangle$. Diagrams (b), (c), and (d) sum to zero when V is the Hartree–Fock potential V_{HF}.

The diagram of Fig. 2(b) is given by

$$\sum_{\substack{n=1\\ n\neq P}}^{N} \frac{1}{\varepsilon_p - \varepsilon_k} \langle kn|v|pn\rangle \eta_k^+ \eta_p |\Phi_0\rangle \tag{22}$$

The diagram of Fig. 2(c) is

$$-\sum \frac{1}{\varepsilon_p - \varepsilon_k} \langle nk|v|pn\rangle \eta_k^+ \eta_p |\Phi_0\rangle \tag{23}$$

and Fig. 2(d) is

$$-\frac{1}{\varepsilon_p - \varepsilon_k} \langle k|V|p\rangle \eta_k^+ \eta_p |\Phi_0\rangle \tag{24}$$

When the potential V is chosen to be the Hartree–Fock potential V_{HF}, the diagrams of Figs. 2(b), (c), and (d) cancel. V_{HF} is defined by

$$\langle a|V_{\mathrm{HF}}|b\rangle = \sum_{n=1}^{N} (\langle an|v|bn\rangle - \langle an|v|nb\rangle) \tag{25}$$

where a and b are any states of the orthonormal set. In general, for open-shell systems one must use a "restricted" Hartree–Fock

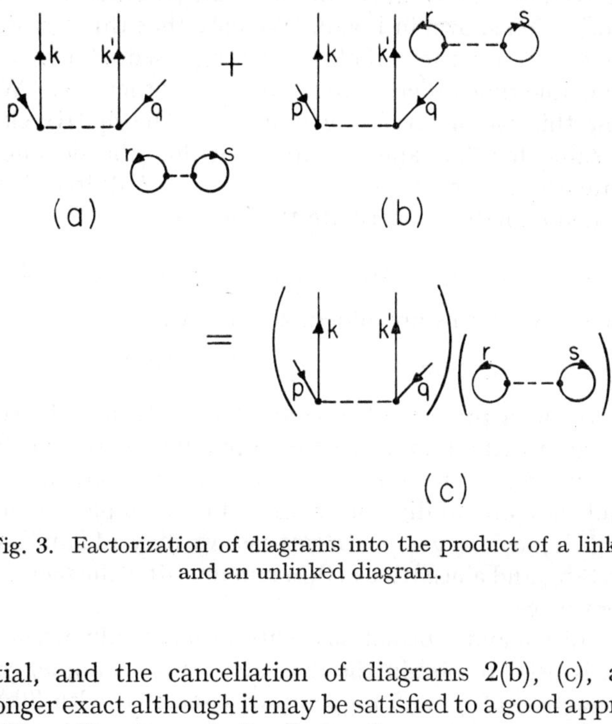

Fig. 3. Factorization of diagrams into the product of a linked and an unlinked diagram.

potential, and the cancellation of diagrams 2(b), (c), and (d) is no longer exact although it may be satisfied to a good approximation. When V_{HF} is used, the first-order energy correction is

$$\langle\Phi_0|H^1|\Phi_0\rangle = \sum_{p<q}^{N} (\langle pq|v|pq\rangle - \langle pq|v|qp\rangle) - \sum_{p=1}^{N} \langle p|V_{\mathrm{HF}}|p\rangle = -\tfrac{1}{2}\sum_{p=1}^{N} \langle p|V_{\mathrm{HF}}|p\rangle \tag{26}$$

B. Exclusion Principle Violating Terms

When we examine the Brueckner–Goldstone expansion, we find that there are terms which are said to violate the exclusion principle in intermediate states. These terms are found to be quite significant in atomic calculations and they represent important physical effects. These same terms are also obtained in the familiar Rayleigh–Schrödinger expansion when certain terms are grouped together. Exclusion principle violating (EPV) terms arise from the factorization of diagrams and from the interactions with the "passive unexcited states" and with the potential V. Consider the factorization shown in Fig. 3. We note that this factorization is possible because the unlinked part represented by $\langle rs|v|rs\rangle$ has all possible time orderings relative to the other part. We may also show this factorization algebraically. The matrix elements are the same for 3(a) and for 3(b). In 3(a) the denominators contribute the factors $(\varepsilon_p + \varepsilon_q - \varepsilon_k - \varepsilon_{k'} + 2i\alpha)^{-1}(i\alpha)^{-1}$ and in 3(b) the denominators contribute the factors

$$(\varepsilon_p + \varepsilon_q - \varepsilon_k - \varepsilon_{k'} + i\alpha)^{-1}, \quad (\varepsilon_p + \varepsilon_q - \varepsilon_k - \varepsilon_{k'} + 2i\alpha)^{-1}$$

When these two terms are added, the result is

$$(\varepsilon_p + \varepsilon_q - \varepsilon_k - \varepsilon_{k'} + i\alpha)^{-1}(i\alpha)^{-1}$$

which gives us a product of disconnected diagrams. If we write out the annihilation and creation operators corresponding to Fig. 3(b), we get a zero result unless p, q, r, and s are all included in Φ_0 and they are all different states. For example, we obtain a zero result if $p = r$ or $q = s$. In this case we may add and subtract diagram 3(b), and a negative of 3(b) remains after the factorization has taken place.

This addition and subtraction occurs automatically when Wick's theorem is used to obtain the diagrams. For example, consider the contraction of η^+ and η operators associated with 3(b) when $p = r$ and $q = s$. We have the following matrix elements and operators. (The denominators have been omitted.)

$$\langle pq|v|pq\rangle\langle kk'|v|pq\rangle\eta_p^+\eta_q^+\eta_q\eta_p\eta_k^+\eta_{k'}^+\eta_q\eta_p \tag{27}$$

The operators may be contracted in four different ways and the corresponding diagrams are shown in Fig. 4. In this theory, contractions are defined as follows:

$$C(\eta_n \eta_n^+) = 1; \qquad \varphi^n \text{ an excited state}$$

$$C(\eta_n^+ \eta_n) = 1; \qquad \varphi^n \text{ an unexcited state}$$

All other contractions are zero. We could also have obtained the usual result[9] for contractions by defining $b_n = \eta_n^+$, $b_n^+ = \eta_n$ for unexcited states φ_n. The operator b_n creates a hole and b_n^+ destroys a hole. In Fig. 4, all four diagrams are equal in magnitude.

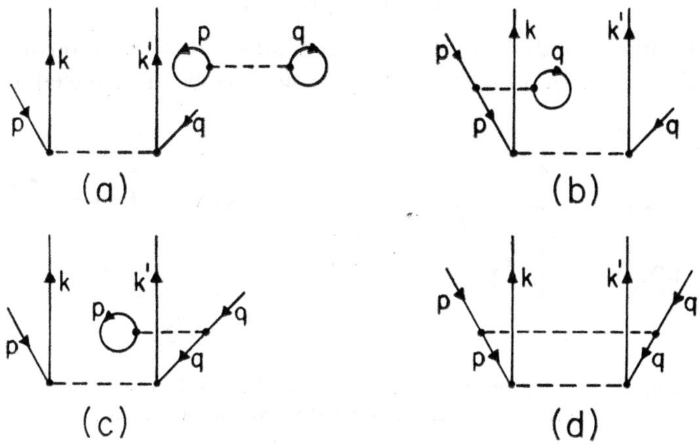

Fig. 4. The four diagrams resulting from the contractions of Eq. (27).

However, a plus sign is associated with (a) and (d) and a minus sign is associated with (b) and (c). Diagrams (b) and (c) are ignored when the Hartree–Fock potential is used. That is, there are other diagrams which cancel 4(b) and 4(c). Diagram 4(a) is used in the factorization of the unlinked part and only diagram 4(d) remains. It is described as a linked diagram with one hole–hole interaction. It is seen to be an exclusion principle violating (EPV) diagram because holes are created in states φ_p and φ_q and then one more hole is created in each of these states before the first two holes are filled. Similarly, in the next order of perturbation theory there is a diagram with two hole–hole interactions.

We may also obtain the terms of Fig. 4 by considering the usual formulation of Rayleigh–Schrödinger perturbation theory as given, for example, by Messiah.[10] The cancellation of unlinked clusters is carried out in the method of Brueckner.[3] The second-order contribution to $|\psi\rangle$ is

$$|\psi^{(2)}\rangle = \frac{Q_0}{E_0 - H_0}(H' - E^{(1)})\frac{Q_0}{E_0 - H_0}H'|\Phi_0\rangle \tag{28}$$

where

$$E^{(1)} = \langle\Phi_0|H'|\Phi_0\rangle \tag{29}$$

and

$$Q_0 = 1 - |\Phi_0\rangle\langle\Phi_0| \tag{30}$$

In the many-body case, the intermediate states are determinants in which one or more of the φ_n in Φ_0 has been replaced by an

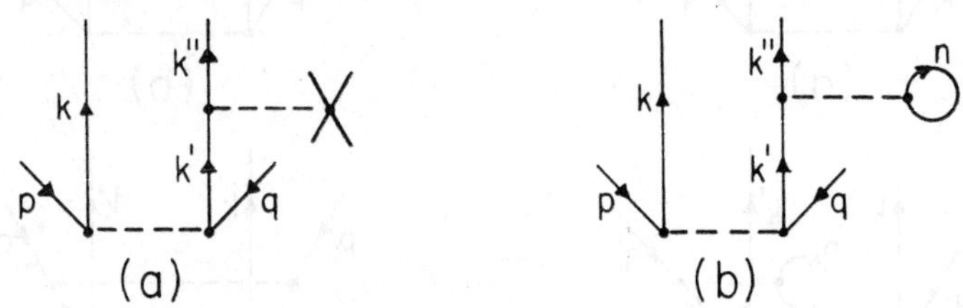

Fig. 5. (a) Scattering by the potential V. (b) Diagram causing partial cancellation of the diagram shown in (a).

excited single-particle state φ_k. Consider the determinant $\Phi_{pq}^{kk'}$ in which φ_p and φ_q have been replaced by φ_k and $\varphi_{k'}$. There is a contribution to $|\psi^{(2)}\rangle$ given by

$$|\Phi_{pq}^{kk'}\rangle \frac{1}{\varepsilon_p + \varepsilon_q - \varepsilon_k - \varepsilon_{k'}}(\langle\Phi_{pq}^{kk'}|H'|\Phi_{pq}^{kk'}\rangle - E^{(1)})\frac{1}{\varepsilon_p + \varepsilon_q - \varepsilon_k - \varepsilon_{k'}}\langle\Phi_{pq}^{kk'}|H'|\Phi_0\rangle \tag{31}$$

When the matrix elements of $\langle\Phi_{pq}^{kk'}|H'|\Phi_{pq}^{kk'}\rangle$ and $E^{(1)}$ are written out, there is a cancellation of the "unlinked" terms and there is, among other terms, $-\langle pq|v|pq\rangle$ which corresponds to the sum of diagrams 4(b), (c), and (d).

We may also obtain second-order EPV diagrams from the net interaction with the passive unexcited states as shown in Fig. 5(b)

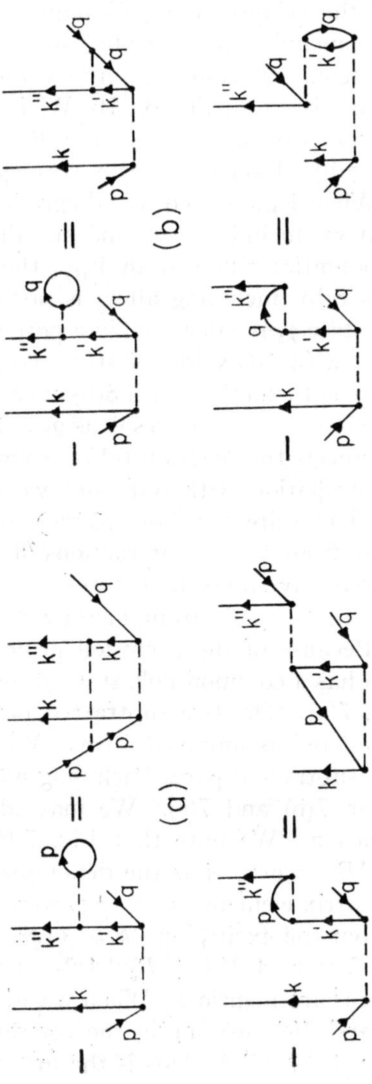

Fig. 6. Diagrams resulting from net interaction of a particle with the Hartree–Fock potential V_{HF} and with all passive unexcited states. (a) and (b) are hole–particle EPV diagrams. (c) and (d) are the corresponding exchange terms.

and with the potential V as shown in Fig. 5(a). If V is the Hartree–Fock potential, the diagram of Fig. 5(a) is cancelled by the diagrams of Fig. 5(b) and the corresponding exchange terms only if the summation over n includes all unexcited states. Since φ_p and φ_q are no longer occupied, the cancellation is incomplete and the net result is the terms shown in Fig. 6. By Wick's theorem, one includes all the diagrams on the right in Fig. 6, and all unexcited states n in Fig. 5(b). Exchange terms corresponding to 5(b) are also included. When V_{HF} is used, the diagrams of Fig. 5 cancel exactly (provided we include all n) and the diagrams of Fig. 6 remain. If the potential differs from V_{HF}, then the net cancellation of 5(a) with 5(b) (including all n) is not exact. If, for example, the excited state φ_k is calculated in a potential which includes all interactions with unexcited states except for φ_q, then the term of 5(a) does not exactly cancel 5(b) even though all unexcited states n are included. The net result is now a term like 5(b) with $n = q$. This cancels the diagram 6(b). There is a corresponding exchange cancellation with 6(d) and we need only consider 6(a) and 6(c). These are EPV hole–particle diagrams.[11] Corresponding terms arise from the net interactions of k with the potential V and with the passive unexcited states n.

Higher order EPV diagrams result from factorizations of the type shown in Fig. 7. Because of the exclusion principle (since each part of the diagram has a common hole state q), one obtains the result shown in Fig. 7(a). The two subtracted parts on the right should not be considered as unlinked terms. When Wick's theorem is used, the two subtracted parts (including minus signs) are obtained as shown in 7(b) and 7(c). We may also note a further type of factorization. We note that Fig. 7(b) includes the factor $A^{-1}(A + B)^{-1}B^{-1}$ where A is the denominator which we would obtain if the matrix element $\langle kk'|v|pq\rangle$ were alone and B is the denominator from the excitation $\langle k''k'''|v|qr\rangle$. Fig. 7(c) has the corresponding factor $A^{-1}(A + B)^{-1}A^{-1}$. The matrix elements for both diagrams are identical. When we add the contributions of both 7(b) and 7(c), we obtain the common matrix elements times the factor $A^{-1}A^{-1}B^{-1}$. This is the first term in an expansion of the denominator:

$$(\varepsilon_p + \varepsilon_q + |\langle qr|v|k''k'''\rangle|^2(\varepsilon_q + \varepsilon_r - \varepsilon_{k''} - \varepsilon_{k'''})^{-1} - \varepsilon_k - \varepsilon_{k'})$$

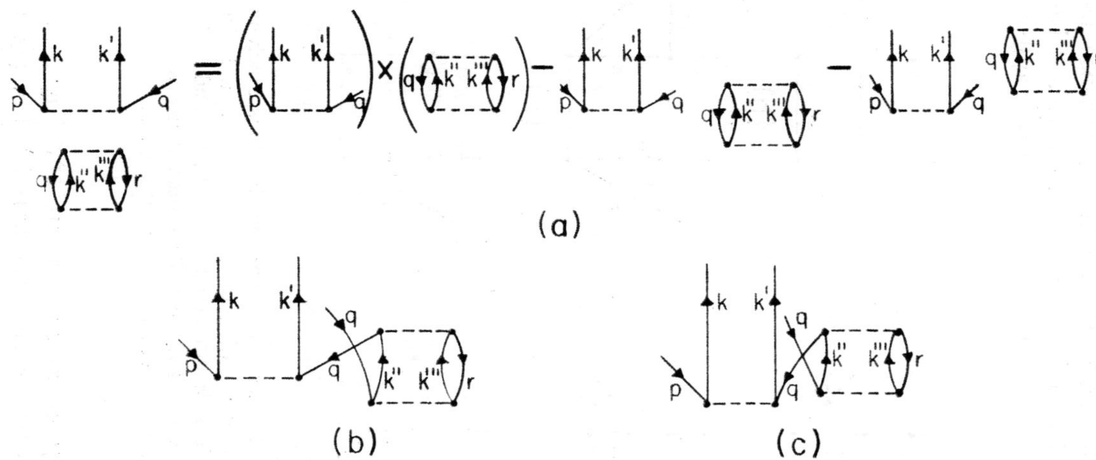

Fig. 7. (a) Factorization of diagrams when disconnected parts have a common hole line. (b) and (c) are the EPV diagrams on the right of (a) drawn in the usual many-body form.

C. Sums of Diagrams

Second-order diagrams for ψ involving only the hole states p and q are shown in Fig. 8. Corresponding exchange diagrams are not shown but should in general be included. Diagrams 8(d) and (f) may not be present when a potential other than V_{HF}

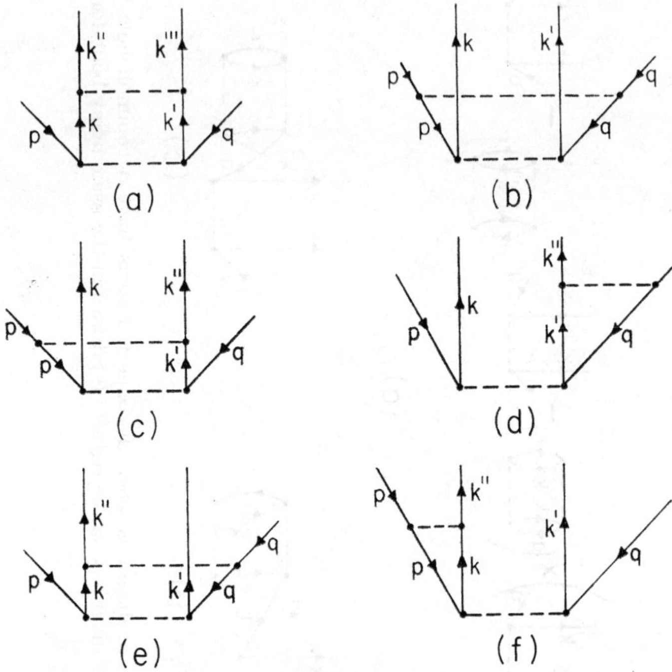

Fig. 8. Second-order wave function diagrams involving only the pair pq. Exchange diagrams are not shown. (a) Ladder diagram. (b) Hole–hole interaction. (c), (d), (e), and (f) occur as in Fig. 6 because V_{HF} includes interactions with all unexcited states including p and q.

is used. All diagrams except 8(a) are EPV diagrams. Diagrams (a) and (b) are positive and diagrams (c), (d), (e), and (f) are negative. However, since the energy denominators are negative, (a) and (b) are negative relative to the first-order term associated with $\langle kk'|v|pq\rangle$. In diagram calculations for beryllium and oxygen,[8] it was found that the diagrams of Fig. 8 were all approximately equal in magnitude. With V_{HF}, these diagrams do not cancel because there are two positive diagrams and four negative diagrams. In this case, taking account of the fact that denominators are negative, the diagrams of Fig. 8 will increase the effect of the first-order term. If the excited states k' had been calculated with a V which includes interactions with all unexcited states except for q, then the diagram of 8(d) would not be included. Likewise, 8(f) would not be included if k had been calculated with a V which did not include interactions with state p. There is an advantage, then, in calculating states in a potential V_{N-1} which contains interactions with $N - 1$ particles rather than with N as in the Hartree–Fock case. There may also be additional terms resulting from the net effect of cancellations between terms involving interactions with V and with the passive unexcited states. This effect is small, however, since the principal effects have been included by the diagrams of Fig. 8.

When the excited states of Fig. 8 are bound states, the most important contributions to the diagrams of Fig. 8 are the diagonal interactions. In 8(a) these are $k'' = k$ and $k''' = k'$. For the diagonal interactions, the sum of diagrams 8(a) through (f) is given by the product of the first-order term

$$(\varepsilon_p + \varepsilon_q - \varepsilon_k - \varepsilon_{k'})^{-1}\langle kk'|v|pq\rangle \tag{32}$$

and the factor

$$\begin{aligned}(\varepsilon_p + \varepsilon_q - \varepsilon_k - \varepsilon_{k'})^{-1}[&\langle kk'|v|kk'\rangle + \langle pq|v|pq\rangle \\ &- \langle pk'|v|pk'\rangle - \langle k'q|v|k'q\rangle - \langle kq|v|kq\rangle - \langle pk|v|pk\rangle]\end{aligned} \tag{33}$$

Exchange terms should also be included. We note that the terms $-\langle k'q|v|k'q\rangle$ and $-\langle pk|v|pk\rangle$ are only included when V_{HF} is used. These interactions are repeated in each order and in each order are accounted by the factor of Eq. (33). These diagonal terms are summed geometrically to give the first-order term of Eq. (32) with the denominator $(\varepsilon_p + \varepsilon_q - \varepsilon_k - \varepsilon_k')$ replaced by

$$\begin{aligned}D = [\varepsilon_p + \varepsilon_q - \langle pq|v|pq\rangle] - [&\varepsilon_k + \varepsilon_{k'} + \langle kk'|v|kk'\rangle \\ &- \langle pk'|v|pk'\rangle - \langle k'q|v|k'q\rangle - \langle kq|v|kq\rangle \\ &- \langle pk|v|pk\rangle]\end{aligned} \tag{34}$$

The terms in the first bracket give the pair energy for the p–q pair, and the terms in the second bracket give the approximate energy for the pair kk'. In the first bracket $\langle pq|v|pq\rangle$ is subtracted from $\varepsilon_p + \varepsilon_q$ since the p–q potential interaction is included both in ε_p and ε_q. In the second bracket $\langle kk'|v|kk'\rangle$ is the potential interaction between k and k'. The subtracted terms represent the fact that the excited states were calculated in a potential which included interactions with these unexcited states but the excited states are now propagating in the absence of these unexcited states. Diagrams as shown in Figs. 7(b) and (c) and similar higher-order terms may be summed to give an exact p–q pair energy and an approximate k–k' pair energy.[12,13] The result is that the first bracket of Eq. (34) is increased by

$$E_{\text{corr}}(p, q) + E_{\text{corr}}(p, r \neq q) + E_{\text{corr}}(r \neq p, q) \tag{35}$$

where $E_{\text{corr}}(p, q)$ is the correlation energy of the p–q pair and $E_{\text{corr}}(p, r \neq q)$ is the correlation energy of p with all unexcited electrons except for q. $E_{\text{corr}}(r \neq p, q)$ is the correlation energy of q with all unexcited states except for p. There is also a term $-E_{\text{corr}}(k, k')$ added inside the second bracket. This term accounts for the fact that correlations among other pairs of unexcited states which involve excitations into states k, k' are eliminated by the Pauli principle when p and q are excited into k and k'. Effects involving E_{corr} are usually small and are treated as a perturbation and not as a shift in the denominators. Similar methods may also be used for excited states in the continuum.[8] It is also possible to derive Eqs. (34) and (35) in a non-perturbative way.[12]

3. SINGLE-PARTICLE STATES AND EVALUATION OF DIAGRAMS

A. The Potential

In any calculation of many-body diagrams for a finite system it is first necessary to obtain the (supposedly complete) set of single-particle states φ_n which are determined by

$$(T + V)\varphi_n = \varepsilon_n\varphi_n \tag{36}$$

In general, these solutions will include both bound states and continuum states. As was discussed in Section 2, if the Hartree–Fock potential V_{HF} is chosen, then there will be no single excitations in the first-order corrections to Φ_0. For open-shell systems, in general, we cannot choose an exact V_{HF} but must use a "restricted" Hartree–Fock solution and in this case there will be a contribution in first order from single excitations. However, if V is still a good approximation to V_{HF}, the contributions from single excitations are expected to be small.

One criterion which leads to V_{HF} is to carry out the minimization

$$\frac{\delta}{\delta\varphi_n}\langle\Phi_0|H|\Phi_0\rangle = 0 \tag{37}$$

subject to orthonormality conditions on the φ_n. The single-particle states φ_n then satisfy the Hartree–Fock equation[14]

$$-\frac{1}{2}\nabla^2\varphi_n(\mathbf{r}) - \frac{Z}{r}\varphi_n(\mathbf{r}) + \left(\sum_{j=1}^{N}\int d\mathbf{r}'\,\frac{\varphi_j^*(\mathbf{r}')\varphi_j(\mathbf{r}')}{|\mathbf{r}-\mathbf{r}'|}\,\varphi_n(\mathbf{r}) - \sum_{j=1}^{N}\delta(m_{sn}, m_{sj})\int d\mathbf{r}'\,\frac{\varphi_j^*(\mathbf{r}')\varphi_n}{|\mathbf{r}-\mathbf{r}'|}(\mathbf{r}')\varphi_j(\mathbf{r})\right) = \varepsilon_n\varphi_n(\mathbf{r}) \tag{38}$$

The terms under the summations represent the Hartree–Fock potential V_{HF} acting on φ_n. The negative term, which is the exchange part of the potential, only contributes when φ_j and φ_n have parallel spins. The potential V_{HF} may also be represented by matrix elements as in Eq. (25). We note that Eq. (38) actually defines a set of coupled equations for the N solutions which are the unexcited states for Φ_0. These N solutions must be solved self-consistently to determine V_{HF}. However, since V_{HF} is then determined, there is no self-consistency requirement for the remaining solutions which are the excited states.

In general, for open-shell atoms V_{HF} is not spherically symmetric and the solutions of Eq. (38) do not have the desired form

$$\varphi_n(\mathbf{r}) = R_{n'l}(r)Y_{lm}(\theta, \varphi)\chi_s(m_s) \tag{39}$$

In such cases we assume the functional form of Eq. (39) and minimize $\langle\Phi_0|H|\Phi_0\rangle$ with respect to all $R_{n'l}$ subject to the orthonormality constraints. The resulting equations are called "restricted" Hartree–Fock equations. Since the $R_{n'l}(r)$ are independent of m_l and m_s, Φ_0 may be constructed as an eigenstate of L^2, L_Z, S^2, and S_Z.

One of the results of using V_{HF} is that for neutral atoms all the excited states may lie in the continuum.[11] This is understood by observing that when Eq. (38) is used to solve for an unexcited state φ_n, there is a cancellation of the direct and exchange term for $j = n$; and so the unexcited states actually have interactions with the nucleus and with $N - 1$ other electrons. However, this cancellation does not take place for the excited states which are then calculated in the field of the nucleus and N other electrons. For neutral atoms, the excited states are calculated in a potential for which the nucleus is exactly screened at large distances, and so it is not necessary that there should be any bound excited states. However, this must be tested for each atomic calculation which uses V_{HF}. We note that the excited states of V_{HF} for neutral atoms in a first approximation bear a relationship to negative ion states. However, the unexcited states which define V_{HF} are not affected by the excited-state particle, unlike the true negative ion case.

In testing for bound excited states in a given local potential $V(r)$, we make use of the Schrödinger equation at zero energy for $P(r) = rR(r)$ for a given l.

$$\frac{d^2P}{dr^2} = \frac{l(l+1)}{r^2} P + 2V(r)P - 2\varepsilon P \tag{40}$$

If ε is negative, it partially cancels a negative $V(r)$, and $P(r)$ is more likely to "turn over" for $\varepsilon = 0$ than for $\varepsilon < 0$. In order to find the total number of bound states for a given $V(r)$, we set $\varepsilon = 0$ in Eq. (40) and observe the number of nodes of $P(r)$ excluding the one at $r = 0$. This test was applied to V_{HF} for beryllium[11] and for oxygen,[8] and in both cases it was found that all excited states are in the continuum.

In testing for bound states of V_{HF} by setting $\varepsilon = 0$, we note that the exchange part of V_{HF} is energy-dependent and failure of the wave function to turn over for $\varepsilon = 0$ does not rigorously ensure that it will not turn over at a lower energy. However, it seems unlikely that the difference in the exchange part of V_{HF} at zero energy and at an energy close to zero should be sufficient to cause a bound state.

This approach is related to the use of Levinson's Theorem[15] which states that the number of bound states n_l of fixed l for

$V(r)$ is given by

$$\pi n_l = \delta_l(0) - \delta_l(\infty) \tag{41}$$

where $\delta_l(0)$ and $\delta_l(\infty)$ are the phase shifts at zero and infinite energy. Levinson's theorem is restricted to local potentials satisfying the conditions

$$\int_0^\infty dr |V(r)| < \infty$$

and

$$\int dr\, r^2 |V(r)| < \infty$$

In all cases, the continuum states should be included; and the existence of bound excited states increases the amount of effort in calculating second-order terms. In addition, use of $V_{\mathbf{HF}}$ yields the best Φ_0. However, in Section 2 it was pointed out that there is a serious disadvantage in using $V_{\mathbf{HF}}$ because of the unbalance in the cancellation of the higher-order diagrams for a given pair pq as discussed in connection with Fig. 8. The perturbation expansion may then converge much more slowly than if the potential V had been chosen with $N - 1$ interactions with unexcited states. The slow convergence with $V_{\mathbf{HF}}$ was found in the correlation energy calculations for beryllium[11] and will be discussed more fully in the following section.

We may construct V in many ways, all of which involve interactions with $N - 1$ unexcited states. One approach is to calculate all single-particle states in the Hartree–Fock field of the atom minus one of the outer electrons. The potential may be chosen such that the outer unexcited states are Hartree–Fock states. However, the excited states in this case approximate the physical single-particle excitations of the atom. The first many-body calculations for Be with the $(1s)^2(2s)^2$ configuration used $V_{\mathbf{HF}}$. A second calculation used a potential in which one $2s$ electron was removed. In this case the Schrödinger equation was

$$-\frac{\nabla^2}{2}\varphi_n(\mathbf{r}) - \frac{Z}{r}\varphi_n(\mathbf{r}) + 2\int d\mathbf{r}'\,\frac{\varphi_{1s}^*(\mathbf{r}')\varphi_{1s}(\mathbf{r}')}{|\mathbf{r}-\mathbf{r}'|}\,\varphi_n(\mathbf{r})$$
$$-\int d\mathbf{r}'\,\frac{\varphi_{1s}^*(\mathbf{r}')\varphi_n(\mathbf{r}')}{|\mathbf{r}-\mathbf{r}'|}\,\varphi_{1s}(\mathbf{r}) + \int d\mathbf{r}'\,\frac{\varphi_{2s}^*(\mathbf{r}')\varphi_{2s}(\mathbf{r}')}{|\mathbf{r}-\mathbf{r}'|}\,\varphi_n(\mathbf{r}) = \varepsilon_n\varphi_n(\mathbf{r}) \tag{42}$$

If φ_{1s} and φ_{2s} in Eq. (42) are Hartree–Fock orbitals for beryllium, then the equation for the $n = 2s$ solution is the 2s Hartree–Fock equation. The solution $n = 1s$ differs from the Hartree–Fock φ_{1s}, the difference being approximately one digit in the third decimal place.

In the oxygen many-body calculations,[8] the $l = 1$ states were calculated in the field of the $(1s)^2(2s)^2(2p)^3$ configuration analogous to Eq. (42) for Be. The $l = 0$ states were calculated in the configuration $(1s)^2(2s)(2p)^4$. That is, the potential for $l = 0$ states had one 2s electron removed. These potentials were chosen so that they gave the Hartree–Fock 2s and 2p solutions for the $l = 0$ and $l = 1$ cases, respectively. The resulting Φ_0 contains Hartree–Fock 2s and 2p orbitals, and only the 1s orbitals differ from oxygen Hartree–Fock solutions. It was found that $\langle\Phi_0|H|\Phi_0\rangle$ in this case differs by only 0.0000215 a.u. from E_{HF} which equals -74.80936 a.u. It was also found that the differences between φ_{1s} and the Hartree–Fock φ_{1s} are approximately one digit in the third decimal place. However, ε_{1s} (-21.726 a.u.) does differ significantly from the Hartree–Fock ε_{1s} (-20.669 a.u.). This may be understood by considering the differences in the two potentials. This difference turns out to be insignificant when we consider the higher-order terms of the perturbation expansion. When we include the net effect of potential and passive particle interactions on the 1s hole lines, we find that these may be summed to shift the ε_{1s} value to -20.6705 which is extremely close to the Hartree–Fock value.[8]

For larger atoms one could choose the potential for each l value such that it reproduces the Hartree–Fock solution for the outermost unexcited state of that l. However, in some cases one might prefer a different choice depending on the physics of the problem. For example, consider the iron atom with a ground-state configuration $(1s)^2(2s)^2(2p)^6(3s)^2(3p)^6(3d)^6(4s)^2$. We might choose our potential for s states in analogy with Eq. (44) so that it reproduces the Hartree–Fock 4s solution. For $l = 1$, we could choose our potential by the removal of one 3p electron from the ground-state configuration in such a way that the Hartree–Fock 3p state is reproduced and similarly for the $l = 2$ potential. However, it may be preferable for atoms like iron to calculate the $l = 1$ states with one 4s electron removed from the ground configuration so that these states will give accurate $4s \rightarrow 4p$ transition matrix elements. This is because we may expect important perturbation contributions from those excited states which have the same principal quantum numbers as some of the unexcited states. Although we may in principle use any orthogonal set of states, it is highly desirable that convergence of the perturbation expansion be as rapid as possible.

B. Sums over Excited States for Evaluation of Diagrams

The single-particle bound states are labeled by the principal quantum number n, orbital angular momentum l, azimuthal quantum number m, and spin projection m_s. Continuum states are labeled k, l, m, m_s, where $\varepsilon = k^2/2$. The states are calculated in a spherically symmetric potential so that

$$\langle \mathbf{r} | k, l, m, m_s \rangle = R(k, l;\ r) Y_{lm}(\theta, \varphi) \chi_s(m_s) \tag{43}$$

We assume that the atom is enclosed in a large spherical volume of radius R_0 which tends to infinity. At distances far from the atom,

$$R(k, l;\ r) - \cos\,[kr + \delta_l + (q/k) \ln 2kr - \tfrac{1}{2}(l + 1)\pi]/kr \tag{44}$$

where $V \rightarrow q/r$ as $r \rightarrow \infty$. The normalization factor for R of Eq. (44) is $(2/R_0)^{\frac{1}{2}}k$. Also, since R must vanish at $r = R_0$,

$$kR_0 + \delta_l + (q/k) \ln 2kR_0 - \tfrac{1}{2}(l + 1)\pi = (n + \tfrac{1}{2})\pi \tag{45}$$

where n is an integer (not to be confused with the principal quantum number). In the limit as $R_0 \rightarrow \infty$,

$$\Delta n = (R_0/\pi) \Delta k \tag{46}$$

and $\sum_k \rightarrow (R_0/\pi) \int_0^\infty dk$. In calculating any observable each single-particle state occurs twice and we may replace $\sum_k$ by $(2/\pi) \int_0^\infty dk$ if we drop the normalization factor $(2/R_0)^{\frac{1}{2}}$ from $R(kl;\ r)$. Sums over l, m, and m_s still must be carried out. In practice, only a limited number of l values are included

The sums over bound excited states are carried out explicitly for those bound states which have been calculated (typically ten or so for each l value). For $V(r) \to q/r$ and if $q \geq 1$, then there will be an infinite number of bound excited states. However, the infinite sums are readily carried out by use of the n^{-3} rule, according to which

$$\lim_{\substack{n\to\infty \\ m \text{ fixed}}} n^3 |\langle mn|v|ab\rangle|^2 = \text{constant} \tag{47}$$

This result has been carefully checked numerically,[8,12] and it is expected since for hydrogen-like atoms of large principal quantum number n

$$R_{nl} \approx 2\left(\frac{Z}{n}\right)^{3/2} \frac{(2Zr)^l}{(2l+1)!}\left[1 = \frac{2rZ}{2l+2} + \cdots\right] \tag{48}$$

For example, in perturbation calculations for the energy we evaluate terms of the form

$$\sum_{m=2}^{\infty} \sum_{n=2}^{\infty} |\langle mpnp|v|2s2s\rangle|D^{-1} \tag{49}$$

which contribute to the correlation energy among two $2s$ electrons. We might evaluate up to $n = 9$ or 10 explicitly, then use Eq. (47) to extrapolate $|\langle mpnp|v|2s2s\rangle|^2$ up to $n = N_f \approx 15$, and then approximate the remainder by

$$C \int_{N_f+1}^{\infty} n^{-3} dn = C/[2(N_f + 1)^2] \tag{50}$$

For greater accuracy, the Riemann zeta function ζ (3) is used, where

$$\zeta(s) = \sum_{n=1}^{\infty} n^{-s} \tag{51}$$

Any desired accuracy may be achieved by calculation of a sufficient number of excited states. The sums in Eq. (49) must be made for each value of m, and an extrapolation is made for $m \to \infty$ just as for n.

4. CORRELATION ENERGIES AND TRANSITION PROBABILITIES

A. Correlation Energies

In many-particle calculations there is much interest in the correlation energy E_{corr}, which is defined as the difference between the exact non-relativistic energy E and an exact Hartree–Fock solution[16]

$$E_{\text{corr}} = E - E_{\text{HF}} \tag{52}$$

The exact non-relativisitic energy E may be obtained from the total binding energy E_{exp} minus the relativistic contributions to the energy.[17]

TABLE 1. Second-order energies for Be calculated with normal and modified denominators.[a]

pq	$l = l'$	$E_2(ll';\ pq)$	$E_2^{(m)}(ll';\ pq)$
$1s^+1s^-$	0		−0.01152
$1s^+1s^-$	1	−0.02255	−0.02077
$1s^+1s^-$	2		−0.00345
$2s^+2s^-$	0	−0.00241	−0.00192
$2s^+2s^-$	1	−0.02228	−0.01605
$2s^+2s^-$	2	−0.00383	−0.00326
$1s2s^b$	0		−0.002033
$1s2s^b$	1	−0.00499	−0.004719
$1s2s^b$	2		−0.000612
$1s2s^c$	0		0.000990
$1s2s^c$	1		0.001108
$1s2s^c$	2		0.00030
Total			−0.06194

[a] From Reference 11. All energies in a.u.

[b] Direct terms. Results include factor 4 since there are four $1s$–$2s$ pairs.

[c] Exchange terms. A factor 2 is included since there are two $1s$–$2s$ pairs of parallel spin.

When V_{HF} is used, E_{corr} is given by the sum of all second-order and higher-order energy diagrams. When a restricted Hartree–Fock potential is used, there will be small contributions due to the single-particle excitations shown in Figs. 2(b), (c), and (d). When we do not use a Hartree–Fock potential, there are additional first-order one-particle excitations from diagrams 2(b), (c), and (d). These latter terms are not included in E_{corr} but are added to $\langle\Phi_0|H|\Phi_0\rangle$ in order to bring it to the Hartree–Fock value E_{HF}.

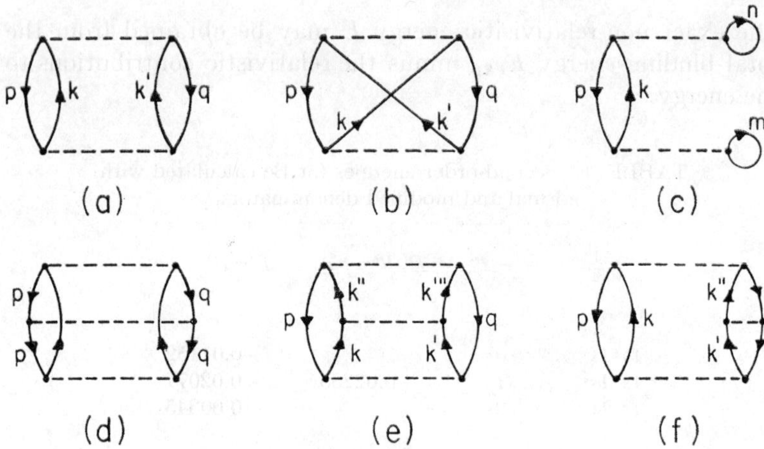

Fig. 9. Energy diagrams. (a) Second-order direct diagram. (b) Second-order exchange diagram. (c) Second-order diagram with single excitation. When V_{HF} is used, diagram (c) cancels with other similar diagrams involving exchange interactions and interactions with V_{HF}. (d) Third-order diagonal hole–hole interaction. (e) Particle–particle interaction or ladder diagram. (f) Hole–particle interaction.

Many-body perturbation theory has been used to calculate correlation energies for beryllium[11,12] and for oxygen.[8] The first calculations for beryllium used V_{HF} and the results are given in Table 1. A number of correlation energy diagrams are given in Fig. 9. Since V_{HF} is used for beryllium, diagrams like Fig. 9(c) vanish when the exchange terms and the interactions with V_{HF} are included. The second-order correlation diagrams are 9(a) and its exchange, Fig. 9(b). In this case, all the excited states are in the continuum. For excited states k, k' of orbital angular momentum l and l', the diagram of Fig. 9(a) is given by

$$E_2(ll'; pq) = \left(\frac{2}{\pi}\right)^2 \int_0^\infty dk \int_0^\infty \frac{dk' \langle pq|v|kl, k'l'\rangle\langle kl, k'l'|v|pq\rangle}{\varepsilon_p + \varepsilon_q - \dfrac{k^2}{2} - \dfrac{k'^2}{2}} \tag{53}$$

Sums over m and m_s must also be carried out, and the total value for E_2 is obtained by summing over l, l', p, and q. The exchange diagram of Fig. 9(b) is given by Eq. (53) with $\langle kl, k'l'|v|pq\rangle$ replaced by $-\langle k'l', kl|v|pq\rangle$. Since the excited states are in the continuum, only the hole–hole interactions of Fig. 8 are readily summed geometrically. A "modified" second-order calculation $E_2{}^m$ may be made which includes these hole–hole interactions.

$$E_2^{(m)}(ll';\ pq) = \left(\frac{2}{\pi}\right)^2 \int_0^\infty dk \int_0^\infty \frac{dk' \langle pq|v|kl, k'l'\rangle\langle kl, k'l'|v|pq\rangle}{D(k, k')} \tag{54}$$

where

$$D(k, k') = \varepsilon_p + \varepsilon_q - \langle pq|v|\widetilde{pq}\rangle - \frac{k^2}{2} - \frac{k'^2}{2} \tag{55}$$

and $\langle pq|v|\widetilde{pq}\rangle \equiv \langle pq|v|pq\rangle - \langle pq|v|qp\rangle$. For Be with $|pq\rangle = |1s1s\rangle$, k and k' for the most significant contributions ranged from 0.0 to 10.0 a.u. For $|pq\rangle = |2s2s\rangle$, the significant values for k and k' ranged between 0.0 and 1.0. The results of Table 1 are quite far from the experimental value $E_{\text{corr}} = -0.0944$ a.u. This discrepancy indicates the necessity for including the higher-order interactions of Fig. 8.

Consider the ratio of the third-order hole–particle diagrams of Fig. 9(f) to the second-order diagram 9(a). It has been found that in many cases this ratio may be accurately approximated by

$$a(l;\ q) = -\left(\frac{2}{\pi}\right) \int_0^\infty dk' \langle k''q|v|k'q\rangle D^{-1}(kk') \frac{\langle kk'|v|pq\rangle}{\langle kk''|v|pq\rangle} \tag{56}$$

The ratio $a(l;\ q)$ depends on the l values of k, k', k'' and also is a function of k''. However, it was found in the beryllium calculations that $a(l;\ q)$ depended weakly on k'' provided k'' was chosen in the middle of the range of important k'' values. This furthermore indicates that the hole–particle diagrams may be summed geometrically to a good approximation. The ratio of the exchange diagram corresponding to Fig. 9(f) to the second-order term is given by $b(l;\ q)$ and involves replacing $\langle k''q|v|k'q\rangle$ by $-\langle qk''|v|k'q\rangle$ in Eq. (56). If we use D given by Eq. (55), we have included the diagonal hole–hole interactions and our second-order diagram is then the "modified" diagram of Eq. (54). In an analogous way, the ratio of the third-order ladder diagram of Fig. 9(e) to the second-order diagram is given approximately by

$$t(l;\ pq) = \left[\left(\frac{2}{\pi}\right)^2 \int_0^\infty dk \int_0^\infty dk' \langle k''k'''|v|kk'\rangle D^{-1}(k, k') \times \langle kk'|v|pq\rangle\right] / \langle k''k'''|v|pq\rangle \tag{57}$$

where k'' and k''' are again continuum values in the middle of the range of important values. Strictly speaking,

$$D = \varepsilon_p + \varepsilon_q - \frac{k^2}{2} - \frac{k'^2}{2}$$

if we do not use "modified" diagrams.

In Be for $pq = 2s2s$, the sum of the third-order energy diagrams corresponding to the diagrams of Fig. 8 is given by

$$E_2^{(m)} \times (4a + 2b + t)$$

The hole–hole interaction is already included if we use D given by Eq. (55). The interactions of Fig. 8 are repeated in fourth and higher orders and are summed geometrically to give

$$E_2^{(m)} \times (1 - 4a - 2b - t)^{-1} \tag{58}$$

In these calculations, the EPV diagrams of Figs. 7(a) and (b) were approximately included by modifying Eq. (58) to

$$E_2^{(m)}(1 - \gamma S)[1 - (4a + 2b + t)(1 - \gamma S)]^{-1} \tag{59}$$

where γ is an average reciprocal denominator D^{-1} used in $E_2^{(m)}$ and S is approximately the total $2s$–$2s$ correlation energy.

The only diagrams involving a single pair pq which must still be considered are ladder diagrams like Fig. 9(e) in which l values of the excited states are changed by the middle interaction.

Higher-order interactions in which l values do not change are included by multiplying by the same factor which multiplies $E_2^{(m)}$ in Eq. (59). A summary of the $2s$–$2s$ Be correlation energy contributions including the higher-order terms is given in Table 2. $E_{corr}(l)$ is the contribution from intermediate states with given l values and is given by Eq. (59). The ladder diagrams with only one change of l values are given by E_{3lch} and the first diagram of this type occurs in third order. Ladder diagrams with two changes in l values are denoted by E_{4lch}, etc.

TABLE 2. Contributions to the $2s$–$2s$ correlation energy in Be.[a]

Term	Energy (a.u.)
$E_{corr}^{(0)}$ [b]	−0.00370
$E_{corr}^{(1)}$	−0.04256
$E_{corr}^{(2)}$	−0.00532
E_{3l} [c]	0.00970
E_{4l} [c]	−0.00250
ΣE_{nl} [c]	0.00051
Total	−0.04387

[a] From Reference 11.
[b] $E_{corr}(l)$ is the correlation energy from excited states of fixed l.
[c] Ladder-type diagrams with changes in l values of excited states.

The modified second-order calculations among $1s$–$2s$ pairs are given in Table 1 and give −0.00497 a.u. The higher-order terms for these pairs should also be included, in addition to the three-body and four-body diagrams. The sum of the third-order three-body diagrams was calculated to be +0.0007 a.u. The higher-order $1s$–$2s$ diagrams were estimated to contribute approximately −0.0007 a.u., cancelling the three-body contribution. A summary of the Be results is given in Table 3. The theoretical total is

—0.091 a.u. as compared with the experimental result —0.094 a.u. The greatest uncertainty in this total comes from the $2s$–$2s$ result because of the slow convergence.

The $l = 1$ contribution to the $2s$–$2s$ correlation energy was recalculated using the single-particle states calculated in the Be potential of the $(1s)^2(2s)$ configuration as described in the previous section.[13] There was a complete change in the nature of the convergence of the perturbation expansion. In fact, the second-order result was approximately 10 per cent larger than the final result. Most of the second-order result (approximately 90 per cent)

TABLE 3. Be correlation energy in a.u.[a]

Electrons	Correlation energy
$1s$–$1s$	—0.04212
$1s$–$2s$	—0.00497
$2s$–$2s$	—0.04387[b]
Total	—0.09096

[a] From Reference 11.

[b] An improved value —0.04488 is given in Reference 13.

came from $2s2s \rightarrow 2p2p$ excitations. The final value for $E_{\text{corr}}(2s, 2s)$ due to $l = 1$ excitations only was found to be —0.04357 a.u. as compared with the value —0.04256 a.u. calculated with the states of V_{HF}. The value —0.04357 is expected to be more accurate due to the rapid convergence of the perturbation expansion. When this new value is used with the other values in Tables 1, 2, and 3, the total calculated beryllium correlation energy becomes —0.092 a.u.

The correlation energy calculations for the ^3P ground state of oxygen are now discussed.[8] As explained in Section 2, it is often possible to use the many-body theory of Brueckner and Goldstone without modification in calculations for open-shell systems with degenerate ground states. However, a general extension of the linked-cluster expansion to include degenerate ground states of systems of fermions has been given by Bloch and Horowitz[18] and also by Brandow.[19]

We assume L–S coupling and start from an unperturbed state Φ_0 specified by the configuration and the quantum numbers L^2, M_L, S^2, and M_S. The operators $L_\pm$, L_Z, $S_\pm$, and S_Z commute with the perturbation $\sum_{l<j} v(|\mathbf{r}_i - \mathbf{r}_j|)$. These same operators commute with $\sum_{i=1}^{N} V_i$ provided V is spherically symmetric and spin independent, and so they also commute with H'. The perturbation can only lead to combinations of excited single-particle states which give a total N-particle state with the same eigenvalues of L^2, L_Z, S^2, and S_Z as Φ_0. Since the degenerate states in M_L and M_S of open-shell atoms cannot be reached through H' in this case, we may use non-degenerate perturbation theory.

In the Brueckner–Goldstone (BG) expansion, Φ_0 is a single determinant. However, for open-shell systems Φ_0 in general must be expressed as a linear combination of determinants. In this case the BG expansion may be readily extended. The operator $U_\alpha(0)$ acts on Φ_0 and then $\langle\Phi_0|U_\alpha(0)|\Phi_0\rangle$ is factored out and cancels this normalization factor in the denominator. The combinations of single-particle states occupied in Φ_0 are excluded as intermediate N-particle states in the perturbation expansion. This extension is not necessary for many open-shell ground states since the choice $M_L = \pm L$, $M_S = \pm S$ often causes Φ_0 to be a single determinant.

The calculations for the ^3P ground state of oxygen used the single determinant

$$\Phi_0(^3P;\ M_L = L, M_S = S) = (+1^+0^+ \ -\ 1^+ + 1^-)$$

where $+1^+$ indicates a $2p$ electron with $m = +1$, $m_s = +\frac{1}{2}$; 0^+ indicates a $2p$ electron with $m = 0$, $m_s = +\frac{1}{2}$, etc. As discussed in Section 3, the oxygen calculations were carried out with single-particle states calculated in the field of the neutral oxygen atom with one electron removed. The $l = 0$ states were calculated with one $2s$ electron removed and $l = 1$ states were calculated with one $2p$ electron removed. The potentials were chosen so that the Hartree–Fock $2s$ and $2p$ orbitals were reproduced. The $l = 2$ and $l = 3$ states were calculated with one $2p$ electron removed. Excited states with $l > 3$ were not included in this calculation. The energy $\langle\Phi_0|H|\Phi_0\rangle$ was calculated by perturbation theory to be 0.0000215 a.u. above $E_{\text{HF}}(-74.80936\text{ a.u.})$.

A summary of the oxygen correlation energy results is given in Table 4. These results only include pair correlations and one-body corrections which correct a restricted Hartree–Fock solution. The total −0.274 a.u. is in reasonable agreement with the "experimental" value −0.258 a.u., which is estimated to be accurate to five percent.[20] The difference between these values gives a very rough estimate of +0.016 for the three-body and higher terms. However, the uncertainties in both the calculated and experimental values make a direct calculation of three-body terms highly desirable.

TABLE 4. Electron correlation energies for oxygen in a.u.[a]

Electrons	Correlation energy
$2p$–$2p$	−0.09058
$2s$–$2p$	−0.10044
$2s$–$2s$	−0.01500
$1s$–$2p$	−0.014996
$1s$–$2s$	−0.00629
$1s$–$1s$	−0.04383
$2p$[b]	−0.00113
$2s$[b]	−0.00161
$1s$[b]	−0.00009
Total	−0.2740

[a] From Reference 8. Only one-body and two-body terms included.
[b] One-body corrections due to use of a restricted Hartree–Fock potential.

The $2p$–$2p$ results were contributed mostly by continuum excitations $2p, 2p \rightarrow kp, k'p$ and $2p, 2p \rightarrow kd, k'd$. In these calculations it was found that the factors a and t were nearly equal in magnitude and of opposite sign. The second-order $2p$, $2p$ results were a few percent larger than the final value. Again we see the advantage in using a potential V_{N-1}. The $2s$–$2p$ results are given in detail in Table 5; the excitations of $2s$ electrons into the two unoccupied $2p$ states and continuum excitations account for most of $E_{\text{corr}}(2s, 2p)$. The total $2s$, $2s$ pair correlation energy is -0.01500 a.u., with the largest contribution -0.00529 a.u. coming from $2s, 2s \rightarrow kd, k'd$ excitations. There would be a contribution to $E_{\text{corr}}(2s, 2s)$ of -0.0607 a.u. in second order from $2s2s \rightarrow 2p2p$ excitations if all the $2p$ states were unoccupied, but these excitations are prevented by the exclusion principle, as pointed out by McKoy and Sinanoğlu.[21] The $1s$–$2p$, $1s$–$2s$,

TABLE 5. Oxygen $2s$–$2p$ correlation energies.[a]

Excitations	Energy in a.u.
$2s^- \rightarrow 2p(0^-, -1^-)$ $2p^\pm \rightarrow kd^\pm$	-0.03675
$2s \rightarrow ks$ $2p \rightarrow k'p$	-0.02623
$2s \rightarrow kp$ $2p \rightarrow k'd$	-0.01097
$2s \rightarrow kd$ $2p \rightarrow k'f$	-0.01589
$2s \rightarrow ns(n \geq 3)$ $2p \rightarrow kp$	-0.00581
$2s \rightarrow ks$ $2p \rightarrow np(n \geq 3)$	-0.00274
$2s \rightarrow np(n \geq 3)$ $2p \rightarrow kd$	-0.00168
Other	-0.00037
Total	-0.10044

[a] From Reference 8. k, k' refer to continuum states. n refers to bound excited states.

and $1s$–$1s$ correlation energies are dominated by continuum excitations. The $1s$–$1s$ correlation energy in oxygen is slightly larger than that calculated for Be, which agrees with the result of Linderberg and Shull[22] that the $1s$–$1s$ correlation energy in two-electron atoms increases slowly with atomic number. The small contributions from single excitations of $2s$ and $1s$ electrons are corrections which account for the fact that both $2s^\pm$ states and $1s^\pm$ states are calculated with exchange interactions with $2p$ electrons. However, the $2s^+$ and $1s^+$ states have three $2p$ exchange interactions and the $2s^-$ and $1s^-$ electrons have exchange interactions with only one $2p$ electron.

B. Transition Probabilities

The BG expansion may be used to obtain accurate wave functions as well as energies. One application has been the calculation of transition probabilities or oscillator strengths between the ground state and excited states of Be; and it was found that these oscillator strengths are strongly affected by electron correlations in the ground state. The oscillator strength f_{ni}^{Z} for an electron transition between an initial state $|i\rangle$ and a final state $|n\rangle$ is given in atomic units by[23]

$$f_{ni}^{Z} = 2\omega_{ni}|\langle n|Z_{op}|i\rangle|^2 \tag{60}$$

where

$$Z_{op} = \sum_{j=1}^{N} Z_j \tag{61}$$

and ω_{ni} is the energy difference between states $|n\rangle$ and $|i\rangle$. We consider the Be transitions $(1s)^2(2s)^2\ {}^1S$ to $(1s)^2(2s)(np)\,{}^1P$. When correlations in states $|n\rangle$ are neglected, they may be written

$$|(2s)(np)\,{}^1P\rangle = \frac{1}{\sqrt{2}}(\eta_{2s}^{+} + \eta_{np}^{+} - \eta_{2s}^{+}\eta_{np^{+}}^{+})|0\rangle \tag{62}$$

where $|0\rangle$ is the "core state" $(1s)^2\ {}^1S$. The excitation energy is

$$\omega_{ni} = \langle 2snp\,{}^1P|H|2snp\,{}^1P\rangle - \langle (2s)^2\,{}^1S|H|(2s)^2\,{}^1S\rangle \tag{63}$$

In the first approximation, neglecting correlations,

$$\omega_{ni} = \varepsilon_{2p} - \varepsilon_{2s} + \langle 2snp|v|np2s\rangle \tag{64}$$

In the first approximation for the oscillator strengths

$$f_{ni}^{Z} = 4\omega_{ni}|\langle np|Z|2s\rangle|^2 \tag{65}$$

Neglecting the lowest-order electron correlations in the excited states, contributions to $\langle np|Z_{op}|i\rangle$ are illustrated diagrammatically in Fig. 10. The neglect of correlations in excited states should be a reasonable approximation except perhaps for the $n = 2$ excited state. Also, we note that our np states are only approximately the same orbitals we would obtain from a new Hartree–Fock calculation for the $(2s)(np)\,{}^1P$ excited states since our states are solutions of Eq. (44). The ground state is connected to each of the excited-state determinants in Eq. (60) through $\langle np|z|2s\rangle$ as shown in Fig. 10(a). The effects of the $2s$–$2s$ correlations in

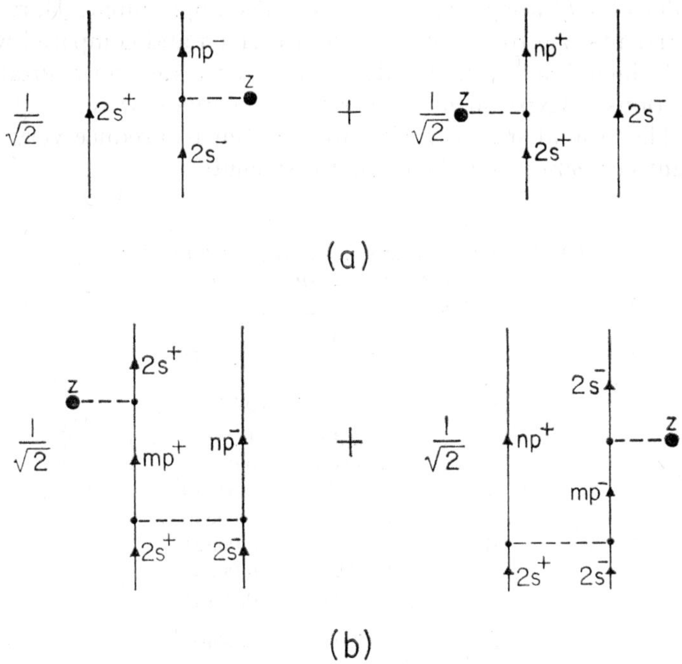

Fig. 10. Diagrams contributing to $\langle n|Z_{op}|i\rangle$. (a) Hartree–Fock approximation. The interaction line is $\langle np|z|2s\rangle$. (b) Including one correlation interaction. These diagrams differ from Goldstone's prescription in that both particle and hole lines are directed upwards. The lines at the bottom represent single-particle states occupied in $|i\rangle$. Lines at the top are single-particle states in $|n\rangle$.

Φ_0 are shown in Fig. 10(b). The sum of diagrams 10(a) and (b) is

$$\langle n|Z_{op}|i\rangle = \sqrt{2}(\langle np|z|2s\rangle + \sum_m \langle 2s|z|mp\rangle D^{-1}\langle mpnp|v|2s2s\rangle) \quad (66)$$

where D is given by Eq. (34). Higher-order non-diagonal terms must also be included. Normalization corrections must also be calculated because it has been assumed that states $|n\rangle$ and $|i\rangle$ are normalized to unity. For Be, $\langle\psi_0|\psi_0\rangle = 1.14$, the difference from 1.00 coming mostly from the $2s$–$2s$ correlations.

Values for f_{ni}^Z are given in Table 6. The approximate Hartree–Fock results due to use of Eq. (65) are given and compared with results including ground-state correlations and normalization corrections. Experimental excitation energies were used for ω_{ni}. The ground-state correlations are seen to produce very important corrections for the oscillator strengths.

TABLE 6. Oscillator strengths for transitions $(2s)^2\,{}^1S \rightarrow 2snp\,{}^1P$.[a]

n	$f_{ni}^z(\text{HF})$[b]	$f_{ni}^z(\text{corr})$[c]
2	2.0293	1.2540
3	0.00541	0.01676
4	0.00530	0.01013
5	0.00318	0.00549
6	0.00194	0.00321
7	0.00125	0.00202
8	0.00084	0.00135
$\sum_{n=9}^{\infty}$	0.00295	0.00467

[a] From Reference 13.
[b] Calculated from Eq. (65) using approximate Hartree–Fock states of Eq. (42).
[c] Correlation and normalization corrections included.

Oscillator strengths of all $1s$ and $2s$ excitations both to bound and continuum states are required in order to check the Thomas–Reiche–Kuhn sum rule

$$\sum_n f_{ni}^Z = N \tag{67}$$

where N is the total number of electrons. Contributions to this sum rule are listed in Table 7. These also include normalization corrections and contributions to $\langle n|Z_{op}|i\rangle$ in which correlations excite an electron pair in $|i\rangle$ and the pair is annihilated by correlations in $|n\rangle$.

The beryllium photoionization cross section σ_k for transitions of a 2s electron into a continuum p state was calculated with the

TABLE 7. Sum-rule evaluation for oscillator Strengths in Be.[a]

Transitions	Hartree–Fock[b]	Correlated
$2s \to np$	2.0499	1.2977
$2s \to kp$	0.6420	0.5969
$1s \to np$	0.2091	0.2029
$1s \to kp$	2.0663	1.9744
$(2s)^2 \to (ns)(2p)$	0.0000	0.0690
Total	4.967	4.141

[a] From Reference 13.

[b] Approximate Hartree–Fock states calculated by Eq. (42).

set of single-particle states of Eq. (42); and σ_k was also calculated including ground-state correlations and normalization corrections.[13]

$$\sigma_k = 4\pi^2\omega_E \frac{\alpha}{k} \sum_{M_i=-L}^{+L} |\langle 2snp\,^1P(M_L)|Z_{op}|(2s)^2\,^1S\rangle|^2 \tag{68}$$

where ω_E is the ionization potential plus $k^2/2$ and α is the fine structure constant. The continuum function in this case has the asymptotic normalization $(2/\pi)^{\frac{1}{2}}$. The calculated correlated cross section was 1.3×10^{-18} cm^2 at $k = 0$ as compared with the approximate Hartree–Fock value 0.85×10^{-18} cm^2.

5. Multiple Perturbations

Consider the atom perturbed by an external potential $\sum_{i=1}^{N} V_{\text{ex}}(r_i)$. Since the exact atomic states in the absence of V_{ex} are not known, it is impossible to carry out a perturbation calculation in terms of V_{ex} alone. However, perturbation theory may still be used to calculate the exact non-relativistic wave function and energy in the presence of V_{ex}. We start from the usual unperturbed state Φ_0 and use the BG expansion with the perturbation

$$H' = \sum_{i<j}^{N} v_{ij} - \sum_{i=1}^{N} V(r_i) + \sum_{i=1}^{N} V_{\text{ex}}(r_i) \tag{69}$$

The perturbation expansion in H' is a double perturbation expansion in the Coulomb correlations and the external potential. There are first-order energy corrections

$$\sum_{n=1}^{N} \langle n | V_{\text{ex}} | n \rangle \tag{70}$$

in addition to the first-order correlation terms; and there are first-order corrections to Φ_0 with single excitations due to V_{ex}. The shift in energy due to V_{ex} is given by the sum of all energy diagrams in which V_{ex} appears one or more times, and there is a corresponding correction for the wave function. These terms may be represented by diagrams in the usual way.

A useful application of multiple perturbation theory is the calculation of atomic polarizabilities and shielding factors. An extensive review of these quantities has been given by Dalgarno.[24] We consider an atom perturbed by a small charge Z' located at a distance r' from the nucleus of charge Z, and r' is chosen along the $\hat{z}$ axis. The interaction energy between Z' and the atom in a.u. is given by

$$V_{\text{int}} = \frac{ZZ'}{r'} - Z' \sum_{i=1}^{N} |\mathbf{r}_i - \mathbf{r}'|^{-1} \tag{71}$$

A Legendre expansion gives

$$V_{\text{int}} = \frac{ZZ'}{r} - Z' \sum_{i=1}^{N} \sum_{k=0}^{N} (r_<^k / r_>^{k+1})_i P_k(\cos \theta_i) \tag{72}$$

where $(r_<^k/r_>^{k+1})_i = r_i/r'^{k+1}$ for $r_i < r'$; and $(r_<^k/r_>^{k+1})_i = r'^k k/r_i^{k+1}$ for $r_i > r'$.

In calculating polarizabilities and shielding factors, it is assumed that $r' > r_i$ for all i. However, in calculating polarization potentials we calculate a shift in energy due to Z' as a function of r', and the case $r_i > r'$ must also be considered.

For $r_i < r'$, we have

$$V_{\text{int}}(r_i < r') = \frac{Z'(Z-N)}{r'} - Z' \sum_{i=1}^{N} \sum_{k=1}^{\infty} (r_i^k/r'^{k+1}) P_k(\cos \theta_i) \quad (73)$$

Using $V_{\text{int}}(r_i < r')$ for V_{ex} in the perturbation expansion, we find

$$\psi_n = \psi_n^{(0)} + Z' \sum_{k=1}^{\infty} \psi_{n,1}^{(k)}/r'^{k+1} + \theta(Z'^2) \quad (74)$$

where $\psi_n^{(0)}$ is the exact non-relativistic wave function for the nth state in the absence of Z', and $\psi_{n,1}^{(k)}$ is $\psi_n^{(0)}$ perturbed once by

$$V_{\text{ex}}^{(k)} = -\sum_{i=1}^{N} r_i^k P_k(\cos \theta_i) \quad (75)$$

In the BG expansion $\psi_n^{(0)}$ is given by all diagrams starting from an unperturbed state Φ_n with the perturbation $(\Sigma v_{ij} - \Sigma V_i)$ but with no interactions with V_{ex}. The term $\psi_{n,1}^{(k)}$ is given by all diagrams in which $V_{\text{ex}}^{(k)}$ acts once and only once and for which Φ_n is the unperturbed state.

The 2^L-pole polarizability $\alpha_2 L$ for $\psi_{(n)}^0$ is defined by

$$\alpha_2 L = 2\langle \psi_n^{(0)} | \sum_{i=1}^{N} r_i^L P_L(\cos \theta_i) | \psi_{n,1}^{(L)} \rangle / \langle \psi_n^{(0)} | \psi_n^{(0)} \rangle \quad (76)$$

The 2^L-pole shielding factor is given by

$$\gamma_2 L = 2\langle \psi_n^{(0)} | \sum_{i=1}^{N} r_i^{-L-1} P_L(\cos \theta_i) | \psi_{n,1}^{(L)} \rangle / \langle \psi_n^{(0)} | \psi_n^{(0)} \rangle \quad (77)$$

Terms contributing to $\alpha_2 L$ and to $\gamma_2 L$ may be represented by diagrams. There are disconnected diagrams which occur in the numerators of Eqs. (76) and (77). When these disconnected diagrams are factored, the result is the product of $\langle \psi_n^{(0)} | \psi_n^{(0)} \rangle$ times a sum of connected diagrams and the factors $\langle \psi_n^{(0)} | \psi_n^{(0)} \rangle$ in the numerator and denominator cancel. In the usual way, there are exclusion principle violating (EPV) diagrams resulting from the factorization. The definition of $\alpha_2 L$ given by Eq. (76) is equivalent to

$$\alpha_2 L = -2E_{n,2}^{(L)}(t) \quad (78)$$

where $E_{n,2}^{(L)}(t)$ is the total contribution to the energy from all linked cluster diagrams with two interactions with $V_{ex}^{(L)}$ and an unlimited number of Coulomb correlation interactions. For the dipole case $L = 1$,

$$\langle n|V_{ex}^{(L)}|n\rangle = 0 \tag{79}$$

Second-order and higher-order terms (due to correlation interactions) are shown in Fig. 11. We have the same diagrams for

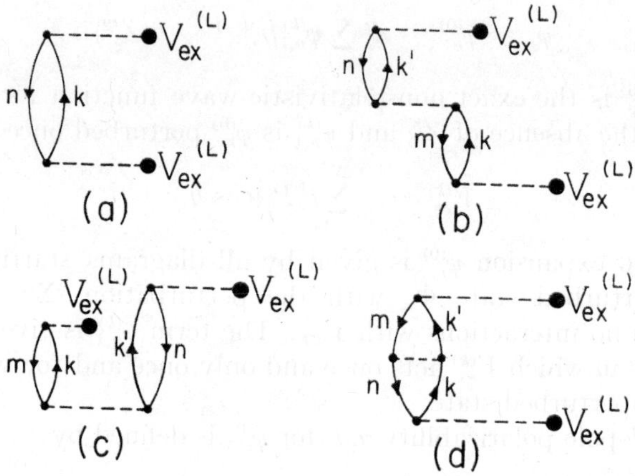

Fig. 11. Diagrams contributing to $E_{n,2}^{L}(t)$ or to $\alpha_2 L$. The heavy dot interactions correspond to $V_{ex}^{(L)}$. (a) Second-order. (b), (c), and (d) are third-order and contain one correlation interaction. The diagram obtained by inverting (c) should also be included in addition to the exchange diagram of (c). When V_{HF} is used, diagram (d) with $m = n$ occurs.

γ_{2L} except that the upper interaction with $V_{ex}^{(L)}$ is replaced by an interaction with

$$\sum_{i=1}^{N} P_L(\cos\theta_i) r_i^{-L-1} \tag{80}$$

Diagrams (b), (c), and (d) are second order in $V_{ex}^{(L)}$ and first order in the correlation perturbation. In diagrams like Fig. 11(c), the effects of higher-order correlation interactions may be included by the methods used for correlation energy diagrams.

Diagrams like 11(d) with $m = n$ occur when V_{HF} is used, due to the net interactions with V_{HF} and with the passive unexcited states. When the excited states are calculated with V_{N-1}, the terms with $m = n$ in diagram (d) do not occur. Because of the diagonal character, diagram (d) with $m = n$ is in general large even though the true correlation effects are small. If good results are obtained in second order with the states of V_{HF}, there has been a fortunate degree of cancellation between diagrams (b), (c), and (d) in Fig. 11. We note that diagram (d) is negative with respect to diagrams (b) and (c).

An amusing illustration of this effect is found when we consider the "Hartree–Fock equation for the hydrogen atom".[25]

$$\left(-\frac{\nabla^2}{2} - \frac{1}{r}\right)\varphi_n(\mathbf{r}) + \int d\mathbf{r}'\varphi_{1s}^*\frac{(\mathbf{r}')\varphi_{1s}(\mathbf{r}')}{|\mathbf{r} - \mathbf{r}'|}\varphi_n(\mathbf{r})$$
$$-\int d\mathbf{r}'\,\frac{\varphi_{1s}^*(\mathbf{r}')\varphi_n(\mathbf{r}')\varphi_{1s}(\mathbf{r})}{|\mathbf{r} - \mathbf{r}'|} = \varepsilon_n\varphi_n(\mathbf{r}) \qquad (81)$$

The lowest solution is the usual hydrogenic 1s state. All other solutions are in the continuum and correspond to triplet states for $e + H$ scattering in the Hartree–Fock approximation.

For hydrogen, the second-order expression for α_d is

$$\alpha_d^{(2)} = -2\sum_{k\neq 1s} |\langle k|r\cos\theta|1s\rangle|^2/(\varepsilon_{1s} - \varepsilon_k) \qquad (82)$$

When the states $|k\rangle$ are the usual hydrogen states, Eq. (82) is the exact expression for α_d which has the value 4.50 a.u.[24] Equation (82) was also evaluated using the excited states of Eq. (81) and the result was found to be 2.5601 a.u. When these states are used, there are higher-order terms like Fig. 11(d) with $m = n = 1s$ plus the corresponding exchange terms and higher-order iterations of these interactions; and these are the only types of higher-order diagrams in the special case of hydrogen. The discrepancy between the second-order result 2.5601 a.u. and the exact value 4.50 a.u. illustrates the importance of these higher-order terms.

These hole–particle interactions may be approximately summed by the same method used to sum the hole–particle interactions of the correlation energy diagrams given in Eq. (56).

The ratio of diagram 11(d) to 11(a) is (with $m = n = 1s$) approximated by

$$a(k') = \left[-\frac{2}{\pi} \int_0^\infty dk \langle k'1s|v|k1s \rangle \left(\varepsilon_{1s} - \frac{k^2}{2} \right)^{-1} \times \langle k|r \cos\theta|1s \rangle \right] / \langle k'|r \cos\theta|1s \rangle \tag{83}$$

where k' is a typical excitation of importance in the calculation of $\alpha_d^{(2)}$. The ratio of the exchange diagram corresponding to 11(d) to diagram 11(a) is approximated by

$$b(k') = \left[\frac{2}{\pi} \int_0^\infty dk \langle k'1s|v|1sk \rangle \left(\varepsilon_{1s} - \frac{k^2}{2} \right)^{-1} \times \langle k|r \cos\theta|1s \rangle \right] \Big/ \langle k'|r \cos\theta|1s \rangle \tag{84}$$

All these higher-order terms may be approximately summed geometrically to give

$$\alpha_d = \alpha_d^{(2)}[1 - a(k') - b(k')]^{-1} \tag{85}$$

In calculating $\alpha_d^{(2)}$, $k' = 0.70$ is approximately in the middle of the range of important k values. The calculation gave

$$a(0.70) = 0.49266 \quad \text{and} \quad b(0.70) = -0.05815$$

Using these values in Eq. (85), the result is $\alpha_d = 4.527$ a.u., in good agreement with the exact value 4.50 a.u.

Dipole and quadrupole polarizabilities and shielding factors have been calculated for beryllium in second order with single-particle states determined by V_{HF} and also with states of V_{N-1}. In the latter case, the higher-order terms were also calculated. These results are summarized in Table 8. It has been shown that the second-order calculations with states determined by V_{HF} are equivalent to the "uncoupled Hartree–Fock" calculations.[27] The dipole shielding factor is known to have the exact value N/Z, 1.00 for beryllium.[24]

In the case of the calculations with the states of V_{N-1} the $2s \rightarrow 2p$ excitations give 11.94 Å^3. These large differences in the second-order calculations indicate that the contributions from diagrams like Fig. 11(d) with $m = n = 2s$ are large. For the second-order dipole shielding with states V_{N-1} the $2s$ excitations contribute 3.814 and $1s$ excitations contribute 0.482 of which $1s \rightarrow kp$ is 0.437. The $2s \rightarrow 2p$ result is 3.890 and the other $2s$ excitations contribute small negative values. The increased importance of $1s$ excitations for the dipole shielding as compared with polarizability calculations is due to matrix elements of $\cos \theta / r^2$ which emphasize the inner regions of the atom.

For the quadrupole polarizability α_q, the second order $2s \rightarrow 3d$ excitations give 6.776 Å^5 and $2s \rightarrow kd$ contribute 4.262 Å.^{5} The $1s$ result is 0.0006 Å.^{5} For the second-order quadrupole shielding γ_∞, the $2s \rightarrow kd$ terms are 0.516 and $2s \rightarrow nd$ are 0.372. The $1s \rightarrow kd$ contribution is 0.170.

TABLE 8. Polarizabilities and shielding factors for Be.

Calculation	α_d(Å^3)[a]	β_∞^b	α_q(Å^3)	γ_∞^d
2nd order using V_{HF}^e	4.54	1.77	9.26	0.67
2nd order using V_{N-1} of Eq. (42)	12.15	4.30	15.1	1.06
Including higher orders	6.93	0.972	14.1	0.75

[a] Dipole polarizability 0.1 Å^3 = 10^{-24} cm^3.
[b] Dipole shielding.
[c] Quadrupole polarizability.
[d] Quadrupole shielding.
[e] All excited states in continuum. From Reference 27.
[f] Includes only second- and third-order terms.

The discrepancies between the two classes of second-order results in Table 8 indicate the necessity of considering higher-order terms for Be. Basic third-order diagrams are shown in Figs. 11(b), (c), and (d). The inverted form of 11(c) and exchange diagrams should be included. Since we are using states of V_{N-1}, Fig. 11(d) with $m = n$ is not included. Each time a correlation interaction $\langle 2p2s|v|2s2p\rangle$ as shown in Fig. 11(b) occurs, the diagram is multiplied by a factor $\langle 2p2s|v|2s2p\rangle/(\varepsilon_{2s} - \varepsilon_{2p})$. These interactions may be summed to all orders geometrically by multiplying by the factor $[1 - \langle 2p2s|v|2s2p\rangle/(\varepsilon_{2s} - \varepsilon_{2p})]^{-1}$

where appropriate. Diagrams such as 11(b) with excited states different from $2p$ must also be calculated. Diagrams like 11(c) are calculated with shifted denominators to account for higher-order correlations. The $(\varepsilon_{2s} - \varepsilon_{2p})$ denominators were shifted by small energies to account for the insertions on hole and particle lines from diagrams as shown in Figs. 12(a)–(d). Diagrams like

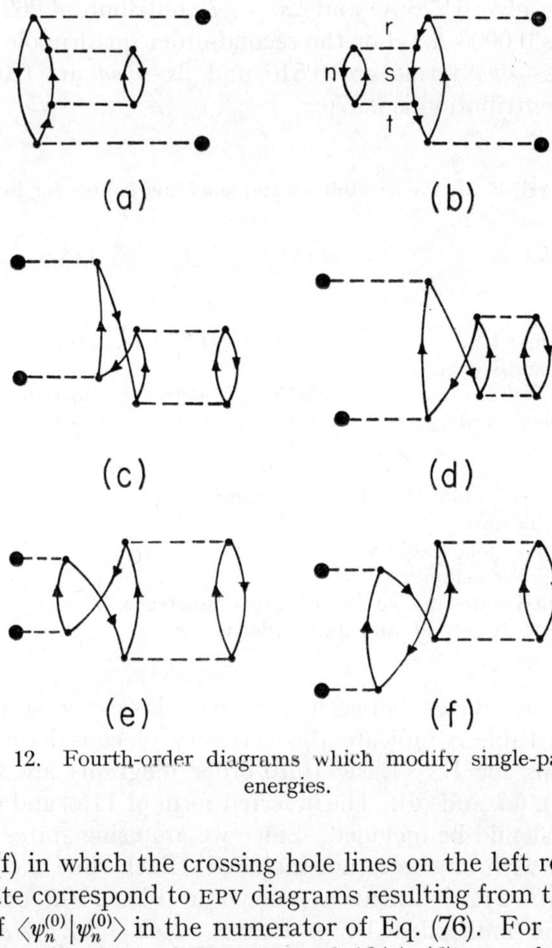

Fig. 12. Fourth-order diagrams which modify single-particle energies.

(e) and (f) in which the crossing hole lines on the left refer to the same state correspond to EPV diagrams resulting from the factorization of $\langle\psi_n^{(0)}|\psi_n^{(0)}\rangle$ in the numerator of Eq. (76). For beryllium, all the contributions of Figs. 11 and 12(a)–(d) were found to be 5.569 Å.[3] The remaining diagrams for α_d are the types shown in

Fig. 13. When the diagrams of Fig. 13 and of Figs. 12(e) and (f) were included, the final result for α_d was 6.93 Å.[3] The coupled Hartree–Fock calculations[28,29] give the result 6.8 Å.[3]

In calculating the Be dipole shielding factor β_∞, the sum of all diagrams of Fig. 11 including the higher iterations was found to be 0.498, which includes a contribution −1.189 from third-order 1*s*–2*s* diagrams. The final result including diagrams of Fig. 13 was found to be 0.972.

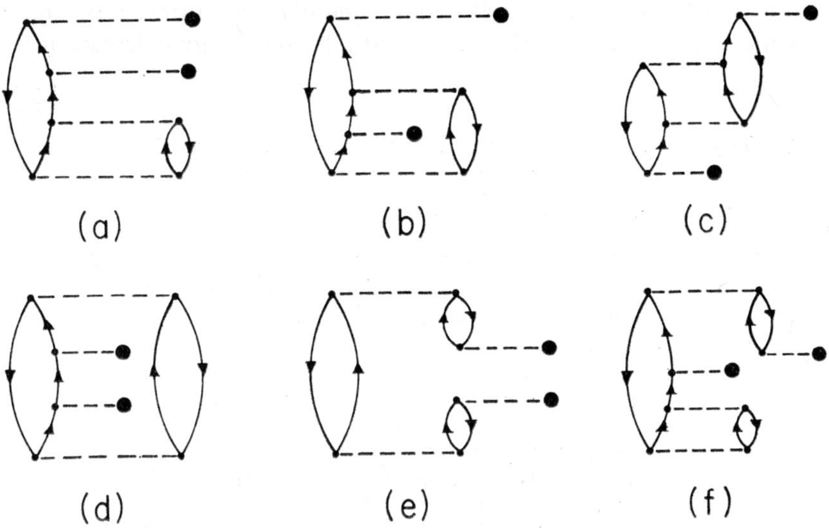

Fig. 13. Basic fourth-order diagrams with two external interactions are given by (a), (b), (c), (d), and (e). These are not obtained by repeating the third-order interactions. (f) A fifth-order diagram modifying (a). Similar diagrams modify (b) and (c). In (b) and (d) one or both interactions with $V_{\mathrm{ex}}^{(L)}$ may be on right-hand particle line.

In the quadrupole polarizability and shielding factor results listed in the third row of Table 8, only the second- and third-order diagrams of Fig. 11 were included. The third-order terms are relatively less important in the quadrupole than in the dipole case because for Be the correlations between two 2*s* electrons are given almost entirely by $l = 1$ excitations.

Although correlation energies are independent of M_L and M_S, polarizabilities and shielding factors depend on M_L and should be averaged over M_L. They are independent of M_S, however, since

$$[\mathbf{S}, V_{\text{ex}}] = 0$$

where V_{ex} is the potential due to an external charge. The dipole polarizability has been calculated for the ground state and a number of low-lying excited states of the neutral oxygen atom by use of the BG expansion.[30] The single-particle states were the same as those used for the oxygen correlation energy calculation.[8]

TABLE 9. Second-order contributions to α_d(Å^3) for oxygen ^3P.[a]

Excitations[b]	$M_L = \pm 1$	$M_L = 0$
$2p \rightarrow nd$	0.47479	0.51131
$2p \rightarrow ns$	0.10015	0.20030
$2s \rightarrow np$[c]	0.23240	0.02097
$1s \rightarrow np$	0.00034	0.00028
Total	0.80768 Å^3	0.73286 Å^3

[a] From Reference 30. Results are independent of M_s.

[b] Both bound state and continuum excitations.

[c] Includes higher-order terms involving net interactions with passive unexcited states and potential.

The effects of shifts in single-particle energies due to net interactions of particle and hole lines with the potential and with passive unexcited states were included. The second-order contributions for the ^3P ground state are summarized in Table 9. Most of the $2p \rightarrow nd$ contribution is from continuum l states. The third-order diagrams shown in Figs .11(b), (c), and (d) contributed -0.1456 Å.^{3} Estimates of the fourth- and higher-order terms indicated that they cancel the reduction from the third-order terms and that the second-order result is approximately correct. In this case the second-order result 0.783 Å^3 compares very favourably with the experimental value (0.77 ± 0.06) Å^3 obtained in a shock tube experiment.[31] Second-order calculated polarizabilities for the ^3P ground state and for low-lying excited states are listed in Table 10.

TABLE 10. Calculated polarizabilities α_d for oxygen.[a]

State	α_d in Å^3
$(2p)^4\ {}^3P$	0.7827
$(2p)^4\ {}^1D$	0.7898
$(2p)^4\ {}^1S$	0.8357
$(2p)^3\ {}^4S\ 3s\ {}^5S$	47.17[b]
$(2p)^3\ {}^4S\ 3s\ {}^3S$	50.63[b]
$(2p)^3\ {}^4S\ 4s\ {}^5S$	800.97[b]
$(2p)^3\ {}^4S\ 4s\ {}^3S$	831.73[b]
$(2p)^3\ {}^4S\ 3p\ {}^5P$	69.50[b]

[a] From Reference 30.
[b] Contributions from outermost electron only.

Consider the case where we wish to determine the shift in energy due to the external charge Z' as a function of r', the distance of Z' from the nucleus. There are the usual first-order terms as given by Eq. (70). There are second-order and higher perturbations due to terms

$$Z' \sum_{k=0}^{\infty} V_p^{(k)} \tag{86}$$

where

$$V_p^{(k)} = -\sum_{i=1}^{N} (r_<^k / r_>^{k+1})_i P_k(\cos\theta_i) \tag{87}$$

Second-order terms should be calculated for each $V_p^{(k)}$ including $k = 0$ since we may have $r' < r_i$. The second-order shift in energy due to $V_p^{(k)}$ gives a polarization potential $V_{\mathrm{polz}}^{(k)}(r')$. The total polarization potential is

$$V_{\mathrm{polz}}(r') = \sum_{k=0}^{\infty} V_{\mathrm{polz}}^{(k)}(r') \tag{88}$$

and it represents that part of the interaction energy due to the distortion of the atom by the perturbing charge. The potential $V_{\mathrm{polz}}(r')$ is an approximation to the second-order term $V_{c2}(r')$ of the optical potential.[31–33]

As $r' \to \infty$,

$$V_{c2}(r') \approx -\alpha_d Z'^2/(2r'^4) \tag{89}$$

Goldberger and Watson[31] note that it has been customary to approximate $V_{c2}(r')$ by $V_p(r')$ which has the asymptotic form of Eq. (89) but contains a "cut-off" to avoid the singularity as $r' \to 0$.

$$V_{c2}(r') \to V_p(r') = -\frac{\alpha_d Z'^2}{2(r'^2 + S^2)^2} \tag{90}$$

where S is a length related to the atomic size. Mittleman and Watson[33] estimate

$$S^4 \approx \frac{\alpha_d}{2} Z^{-\frac{1}{3}} \tag{91}$$

When $V_{\mathrm{polz}}(r')$ is used, there is no need to insert a cut-off since $V_{\mathrm{polz}}(r')$ remains finite as $r' \to 0$ because $r_>$ and $r_<$ are used rather than $r' > r_i$.

The polarization potential has been calculated for the $(2p)^3\,{}^4S\,3s\,{}^5S$ excited state of oxygen by including the $k = 0$ and $k = 1$ terms of Eqs. (87) and (88), and the results are given in Fig. 14. Contributions from the $(1s)^2(2s)^2(2p)^3$ core were omitted. Since the core electrons contribute relatively little to α_d for the $3s\,{}^5S$ state, it is expected that these contributions are small except for small r' where $V_{\mathrm{polz}}(r')$ is relatively small compared to V_{HF}. However, effects of core electrons and higher k values could be easily included. For large r', most of the $k = 0$ and $k = 1$ terms come from the $3s \to 3p$ excitations. However, as $r' \to 0$, continuum excitations give approximately one-half of the $k = 0$ term and 15 per cent of the $k = 1$ term.

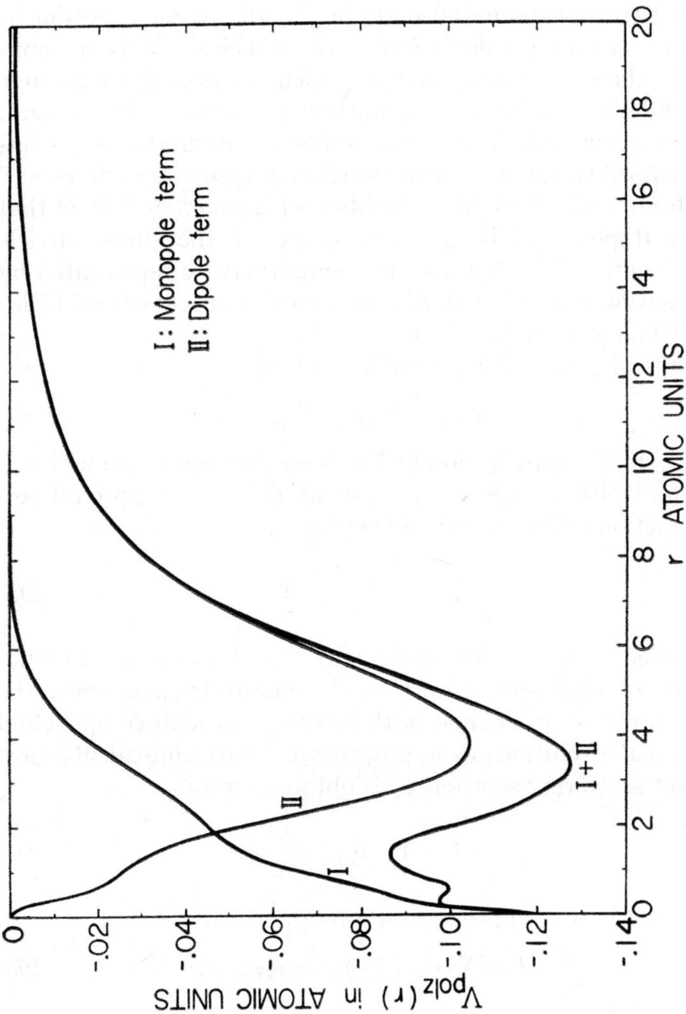

Fig. 14. The polarization potential $V_{\text{polz}}(r)$ from excitations of the $3s$ electron in the $(2p)^3\,{}^4S\,3s\,{}^5S$ state of oxygen. The monopole ($k = 0$) and dipole ($k = 1$) contributions are also given separately.

6. DIAGRAM TECHNIQUES FOR SCATTERING

A. Diagrams

The polarization potential described in the previous section is useful in scattering calculations. Nevertheless, it is a semi-classical, adiabatic approximation which neglects the exchange effects and the motion of the incoming electron. The general optical potential which does not suffer these shortcomings has been discussed for electron–atom scattering by many authors.[31–37] It has been recognized by a number of authors [32, 36, 38, 39] that the optical potential is the self-energy of the single-particle Green's function and that the self-energy may be represented by a perturbation expansion similar to that of Brueckner and Goldstone for bound-state problems.

The optical potential V_{op} may be written

$$V_{\text{op}} = V_{\text{op}}^{(0)} + V'_{\text{op}} \tag{92}$$

where $V_{\text{op}}^{(0)}$ is the contribution to V_{op} from the interaction with the nucleus and with a single-particle potential V which approximates the interactions with the atomic electrons.

$$V_{\text{op}}^{(0)} = -\frac{Ze^2}{r} + V \tag{93}$$

As discussed in the earlier sections, if $V = V_{\text{HF}}$, then an orthogonal set of single-particle states is obtained; and since the excited states are calculated with interactions with N unexcited electrons, these solutions are appropriate for scattering calculations. The exact scattering solution ψ_k is obtained from

$$(-\tfrac{1}{2}\nabla^2 + V_{\text{op}})\psi_k = \frac{k^2}{2}\psi_k \tag{94}$$

and is given in first approximation by φ_k, where

$$(-\tfrac{1}{2}\nabla^2 + V_{\text{op}}^{(0)})\varphi_k = \varepsilon_k\varphi_k \tag{95}$$

with $\varepsilon_k = \frac{1}{2}k^2$.

The first-order contributions to V'_{op}, are given by the net interactions with the potential V and with the passive unexcited states. When we choose $V = V_{\text{HF}}$, the first-order contributions vanish.

The lowest-order contributions to $\langle k|V'_{\rm op}|k\rangle$ are the second-order terms represented by the diagrams of Fig. 15. Diagrams (b) and

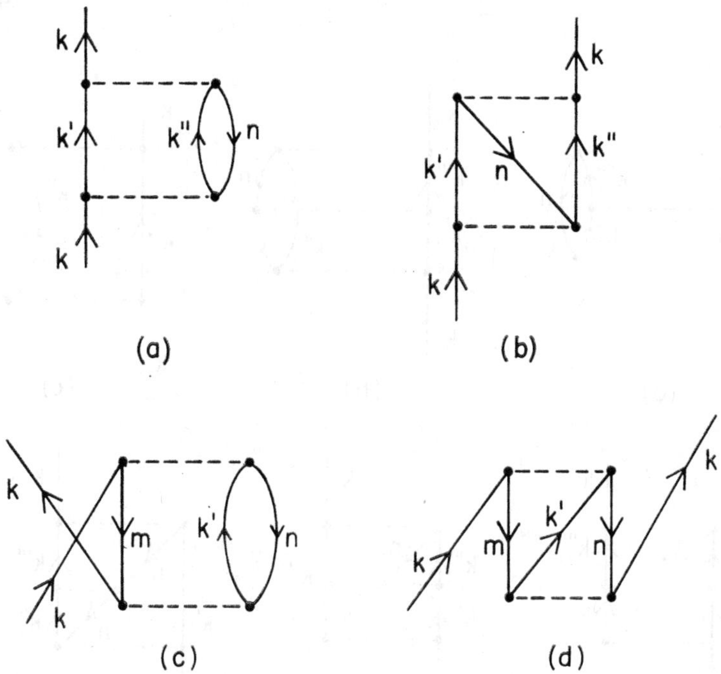

Fig. 15. Second-order diagrams for $\langle k|V'_{\rm op}|k\rangle$. Diagrams (a) and (b) represent correlations between an electron in state k and an atomic electron in state n.

(d) are the exchange diagrams corresponding to (a) and (c), respectively. Diagrams (c) and (d) are EPV diagrams arising from the linked-cluster factorization. These diagrams are present in any problem with two or more ground-state electrons. The contribution of Fig. 15(a) to $\langle k|V'_{\rm op}|k\rangle$ is given by

$$\sum_{k',k'',n} \frac{\langle kn|v|k'k''\rangle\langle k'k''|v|kn\rangle}{\varepsilon_k + \varepsilon_n - \varepsilon_{k'} - \varepsilon_{k''}} \tag{96}$$

where the summations over k' and k'' include only excited states and only unexcited states are included in the sum over n. The contribution of Fig. 15(c) to $\langle k|V'_{\mathrm{op}}|k\rangle$ is

$$-\sum_{k',m,n} \frac{\langle mn|v|kk'\rangle\langle kk'|v|mn\rangle}{\varepsilon_m + \varepsilon_n - \varepsilon_k - \varepsilon_{k'}} \tag{97}$$

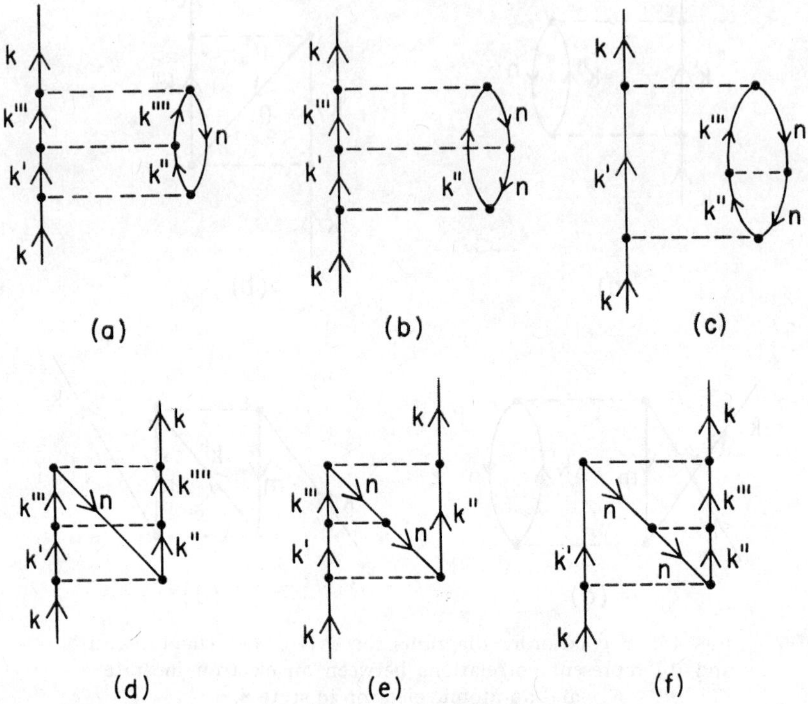

Fig. 16. Third-order diagrams for a single electron pair. Exchange diagrams corresponding to (b), (c), (e), and (f) should also be included. Diagrams (a), (b), and (c) modify the basic diagram of Fig. 15(a) and diagrams (d), (e), and (f) modify 15(b). (a) and (d) are ladder diagrams. (b), (c), (e), and (f) result from use of V_{HF} in calculating single-particle states. Similar interactions modify diagrams of Fig. 15(c) and (d).

The higher-order diagrams involving a single pair of electrons are the ladder diagrams and hole–particle diagrams corresponding to Figs. 15(a) and (b), and these diagrams are shown in third order in Fig. 16. There are also third-order ladder diagrams,

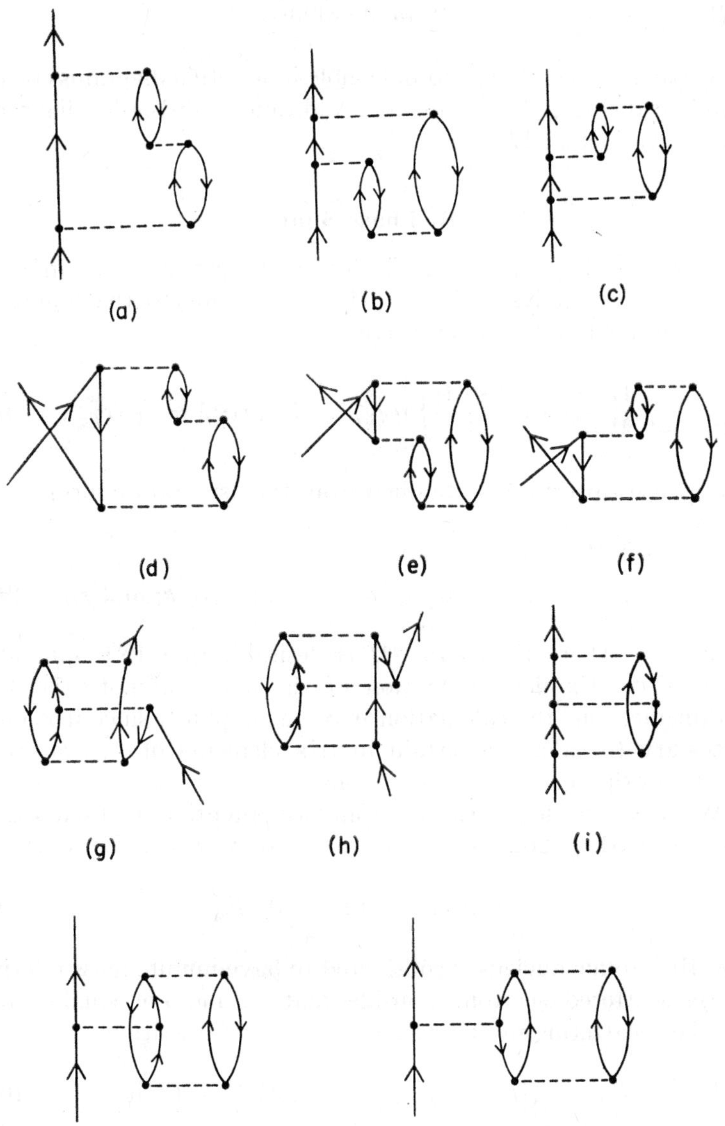

Fig. 17. Additional third-order diagrams for $\langle k | V'_{\text{op}} | k \rangle$. There are also diagrams corresponding to (g), (h), and (i) but with external lines as in (d), (e), and (f). There are also exchange diagrams corresponding to those shown. For atoms with two or more electrons, there are also hole–hole interaction diagrams associated with diagrams (c) and (d) of Fig. 15.

hole–particle diagrams, and hole–hole interaction diagrams corresponding to Figs. (15(c) and (d). Additional third-order diagrams are shown in Fig. 17.

B. Phase Shifts

Assuming V_{op} to be spherically symmetric we may make a partial wave analysis of Eq. (94) which leads to the following equation for the lth partial wave

$$-\frac{1}{2}\left(\frac{d^2}{dr^2} - l\,\frac{(l+1)}{r^2}\right)(r\psi_{k,l}^{\pm}) + V_{op}(r\psi_{k,l}^{\pm}) = \frac{k^2}{2}(r\psi_{k,l}^{\pm}) \qquad (98)$$

The phase shift $\delta_l(k)$ is obtained from the asymptotic form

$$r\psi_{k,l}^{\pm} \to i^l(2/\pi)^{\frac{1}{2}} \exp[\pm i\delta_l(k)] \times \sin(kr - l\pi/2 + \delta_l(k) + (q/k)\ln 2kr)/k \qquad (99)$$

as $r \to \infty$, where $V_{op} \to q/r$. It is desirable that $V_{op}^{(0)} \to q/r$ also as $r \to \infty$. Explicit evaluation of the non-local potential V_{op} is difficult; but the calculation may be simplified since it is only necessary to calculate certain matrix elements of V_{op} in order to obtain the phase shift corrections.

We now consider scattering from two potentials as discussed in Reference 31, p. 202. The scattering matrix $\mathscr{T}$ may be written

$$\langle b|\mathscr{T}|a\rangle = \delta(\mathbf{P}_b - \mathbf{P}_a)T_{ba} \qquad (100)$$

and the atomic nucleus is considered to have infinite mass relative to the scattered electron. Consider states α and β of equal energy E. The scattering matrix element $T_{\beta\alpha}$ may be written

$$T_{\beta\alpha} = \langle (H - E)\chi_\beta|\psi_\alpha^+\rangle = \langle \chi_\beta|V_{op}^{(0)} + V'_{op}|\psi_\alpha^+\rangle \qquad (101)$$

where

$$\chi_\beta = (2\pi)^{-3/2}\exp(i\mathbf{k}_\beta\cdot\mathbf{r}) \qquad (102)$$

and

$$H = -\tfrac{1}{2}\nabla^2 + V_{op}^{(0)} + V'_{op} \qquad (103)$$

Also

$$\begin{aligned}\psi_\alpha^+ &= \chi_\alpha + (E - H + i\eta)^{-1}(V_{op}^{(0)} + V'_{op})\chi_\alpha \\ &= \chi_\alpha + (E - K + i\eta)^{-1}(V_{op}^{(0)} + V'_{op})\psi_\alpha^+\end{aligned} \tag{104}$$

where $K = -\frac{1}{2}\nabla^2$ and $\eta \to 0^+$.

We also consider incoming wave states ϕ_β^- which satisfy the equation

$$\phi_\beta^- = \chi_\beta + (E - K - i\eta)^{-1} V_{op}^{(0)} \phi_\beta^- \tag{105}$$

We use Eq. (105) to write χ_β in terms of ϕ_β^- and then substitute into Eq. (101). The result is

$$T_{\beta\alpha} = \langle \phi_\beta^- | V'_{op} | \psi_\alpha^+ \rangle + \langle \phi_\beta^- | V_{op}^{(0)} | \chi_\alpha \rangle \tag{106}$$

The term $\langle \phi_\beta^- | V_{op}^{(0)} | \chi_\alpha \rangle$ is the exact scattering matrix element when $V'_{op} = 0$. Several derivations of this result have been given by Goldberger and Watson (GW).[31]

For central forces,

$$\langle \mathbf{k}_f, S, \nu' | T | \mathbf{k}, S, \nu \rangle = \delta_{SS'} \delta_{\nu',\nu} T(\hat{\mathbf{k}}_f \cdot \hat{\mathbf{k}}, k_f, k, S) \tag{107}$$

where S', S are the values for the total final and initial spins and ν', ν are total spin projections on the z axis. For fixed S, we may expand

$$T(\hat{\mathbf{k}}_f \cdot \hat{\mathbf{k}}, k_f, k) = \sum_{l=0}^{\infty} \sum_{m=-l}^{+l} Y_{lm}(\hat{\mathbf{k}}_f) Y_{lm}^*(\hat{\mathbf{k}}) T_l(k_f, k) \tag{108}$$

For elastic scattering, $k_f = k$ and we write $T_l(k;\ k)$ as $T_l(k)$. As shown by GW, Reference 31, p. 232,

$$T_l(k) = -[\pi k^2/(d\varepsilon/dk)]^{-1} \exp(i\delta_l) \sin \delta_l \tag{109}$$

where $\delta_l = \delta_l(k)$. Also,

$$T_l(k) = \langle \chi_{k,l} | V_{\text{int}} | \psi_{k,l}^+ \rangle_r \equiv \int_0^\infty dr\, r^2 \chi_{k,l}^* V_{\text{int}} \psi_{k,l}^+ \tag{110}$$

where V_{int} is the potential for the scattered electron, and $\chi_{k,l} = i^l (2/\pi)^{\frac{1}{2}} \times j_l(kr)$. We consider Eq. (106) with α and β referring to k and k_f.

When $V'_{\text{op}} = 0$, ϕ_k^- is an exact scattering solution and $T_{\beta\alpha}$ is given by $\langle\phi_\beta^-|V_{\text{op}}^{(0)}|\psi_\alpha\rangle$. The partial-wave expansion of Eq. (106) is

$$-[\pi k^2/(d\varepsilon/dk)]^{-1} \exp(i\delta_l) \sin \delta_l = \langle\phi_{k,l}^-|V'_{\text{op}}|\psi_{k,l}^+\rangle r - [\pi k^2/(d\varepsilon/dk)]^{-1} \exp(i\delta_l^{(0)}) \sin \delta_l^{(0)} \quad (111)$$

where $\delta_l^{(0)}$ are the phase shifts for $V'_{\text{op}} = 0$. Introducing states $\varphi_{k,l}$ and $\psi_{k,l}$ with the form

$$\varphi_k = RY_{lm}X_s \quad \text{and} \quad R \to \sin[kr + (q/k) \ln 2kr - \tfrac{1}{2}l\pi + \delta_l]/r$$

as $r \to \infty$, Eq. (111) becomes

$$\exp(i\delta_l) \sin \delta_l = \exp(i\delta_l^{(0)}) \sin \delta_l^{(0)} - \frac{2}{k} \exp(i\delta_l^{(0)} + i\delta_l) \times \langle\varphi_{k,l}|V'_{\text{op}}|\psi_{k,l}\rangle \quad (112)$$

We define $\Delta\delta_l$ by

$$\delta_l = \delta_l^{(0)} + \Delta\delta_l \quad (113)$$

When V'_{op} is small relative to $V_{\text{op}}^{(0)}$, we may approximate $\psi_{k,l}$ in Eq. (112) by $\varphi_{k,l}$. Also assuming $\Delta\delta_l$ is small and expanding, we find

$$\Delta\delta_l \approx -\frac{2}{k} \langle\varphi_{k,l}|V'_{\text{op}}|\varphi_{k,l}\rangle \quad (114)$$

Equation (114) was first obtained by Pu and Chang[40] by means of a variational expression for δ_l.

These methods with the approximation of Eq. (114) have been applied to the elastic scattering of electrons by helium[40] and by hydrogen for the *S*-wave triplet states.[26] In the case of helium, only diagrams (a) and (b) of Fig. 15 were included. In the case of hydrogen (a) and (b) are the only second-order diagrams since there is only one unexcited electron.

For hydrogen, the Hartree–Fock values are compared with the accurate variational values of Schwartz[41] in Table 11. The second-order diagram contributions for intermediate $l = 0, 1,$ and 2 states are given in Table 12, and it is noted that there is a significant discrepancy between these values and the variational results. It is apparent, then, that the higher-order hole–particle diagrams

TABLE 11. *S*-wave phase shifts for electron–hydrogen scattering in the triplet state.[2]

k (a.u.)	δ^{a}_{HF}	$\delta^{b}_{\mathrm{var}}$	$\delta_{\mathrm{var}} - \delta_{\mathrm{HF}}$
0.100	2.90760	2.9388(4)	0.0312(4)
0.200	2.67915	2.7171(5)	0.0380(0)
0.300	2.46114	2.4996(8)	0.0385(4)
0.400	2.25731	2.2938(4)	0.0365(3)
0.500	2.07008	2.1046(4)	0.0345(6)
0.600	1.90057	1.9329(8)	0.0324(1)
0.700	1.74883	1.7797(6)	0.0309(3)
0.800	1.61412	1.643(3)	0.0292(0)

[a] Phase shifts in radians calculated with the Hartree–Fock potential.

[b] Phase shifts calculated by Schwartz using the variational principle. See Reference 41. Figures in parentheses indicate the uncertainty in the last digit.

TABLE 12. Phase shift corrections due to second-order terms of the optical potential.[a]

	$\Delta\delta_0(l = 0)^b$	$\Delta\delta_0(l = 1)^b$	$\Delta\delta_0(l = 2)^b$	$\Delta\delta_0^c$	$(\delta_{\mathrm{var}} - \delta_0^{(0)})^d$
$k = 0.20$					
Direct	0.008435	0.033036	0.003770	0.024425	0.0380(0)
Exchange	−0.008237	−0.010437	−0.002142		
$k = 0.40$					
Direct	0.016384	0.041642	0.005870	0.025169	0.0365(3)
Exchange	−0.015907	−0.018926	−0.003894		
$k = 0.60$					
Direct	0.023848	0.045135	0.006945	0.023289	0.0324(1)
Exchange	−0.022920	−0.024654	−0.005065		

[a] Phase shifts for *S*-wave electron–hydrogen triplet scattering.

[b] The l values in parentheses refer to l values of intermediate state k', k'' used for $\langle k | V'_{\mathrm{op}} | k \rangle$ as given by Eq. (96).

[c] Sum of the second-order direct and exchange terms with $l = 0$, 1, and 2 intermediate states.

[d] The "exact" result for $\Delta\delta_0$ based on the accurate variational calculations of Schwartz listed in Table 1. Figures in parentheses indicate uncertainty in the last digit. Values for $\delta_0^{(0)}$ are Hartree–Fock results listed as δ_{HF} in Table 11.

and ladder diagrams of Fig. 16 should be included in these calculations. The ratio of the ladder diagram of Fig. 16(a) to the second-order diagram of Fig. 15(a) is denoted t and may be approximated by the method of Eq. (57) used for the correlation energies. Similarly, the ratio of diagram 16(b) to 15(a) is written a_b, and the ratio of the corresponding hole–particle exchange interaction term to 15(a) is written b_b. The ratio of 16(c) to 15(a) is written a_c and its exchange is written b_c. These quantities are now approximated by the ratios of matrix element techniques of Eqs. (56), (57), (58), (83), and (84). The ratios of diagrams 16(d), (e), and (f) to 15(b) are then given by the same ratios. These types of diagrams (which are the only diagrams for hydrogen) may be summed geometrically to all orders to give the result

$$\langle k|V'_{\rm op}|k\rangle = \langle k|V'_{\rm op}|k\rangle^{(2)}(1 - t - a_b - b_b - a_c - b_c)^{-1} + l\text{-changing ladder diagrams} \quad (115)$$

The factor $(1 - t - a_b - b_b - a_c - b_c)^{-1}$ was calculated to be 1.5915 for $k = 0.40$ and for $l = 1$ intermediate states.[26] The contribution from the l-changing ladder diagrams was estimated to be -0.0042. For the $l = 0$ and $l = 2$ intermediate states, the $l = 1$ factor 1.5915 was also used. Using these values in Eqs. (114) and (115) along with the second-order result $\Delta\delta_0^{(2)}(k = 0.40) = 0.025169$, we obtain the improved value $\Delta\delta_0(k = 0.40) = 0.0359$, which compares favourably with the "exact" value 0.0365 given in Table 12.

C. Inelastic Scattering

Consider diagrams like those of Figs. 15 and 16 and let us suppose that we have bound excited states $k' = p$ and $k'' = q$; that is, we calculate excited intermediate states with V_{N-1}. For a given pair of electrons, then, the ladder diagram becomes the most important third-order diagram. The diagonal ladder with bound excited states are then summed geometrically as usual, and we obtain the second-order term with a new denominator given by

$$(\varepsilon_k + \varepsilon_n - (\varepsilon_p + \varepsilon_q + \langle pq|v|pq\rangle - \langle qp|v|pq\rangle)) \quad (116)$$

The quantity in parentheses approximates the energy of an excited atomic state with $N + 1$ electrons. As discussed by Fano[42] and by Burke,[43] states similar to these give rise to resonances in the scattering phase shifts. Singularities in the optical potential are now given by Eq. (116). There is, in general, an infinite number of bound states and therefore an infinity of resonances. Additional, smaller shifts of Eq. (116) may be produced by summing additional classes of non-diagonal ladder terms and by the net interactions of particle and hole lines with the passive unexcited states and with the potential V_{N-1}.

As the energy of the incoming electron increases, the resonances given by zeros of Eq. (116) become closer and closer as the principal quantum number n of either the state p or q becomes large. For energies ε_k above the inelastic threshold, the denominators of diagrams (a) and (b) of Fig. 15 vanish for excitations for which either k' or k'' is a continuum state k' and the other excitation k'' is a bound state q (or a continuum state if ε_k is sufficiently large). The condition for the vanishing denominator is

$$\frac{k^2}{2} + \varepsilon_n - \frac{k'^2}{2} - \varepsilon_q = 0 \tag{117}$$

This singularity is treated by introducing a small imaginary part η into Eq. (117). We now write

$$\lim_{\eta\to 0^+} \left(\frac{k^2}{2} + \varepsilon_n - \frac{k'^2}{2} - \varepsilon_q + i\eta\right)^{-1} = P\left(\frac{k^2}{2} + \varepsilon_n - \frac{k'^2}{2} - \varepsilon_q\right)^{-1} - i\pi\delta\left(\frac{k^2}{2} + \varepsilon_n - k'^2 2 - \varepsilon_q\right) \tag{118}$$

where P indicates that we take the principal value in integration. The contribution to $\Delta\delta_l$ given by Eq. (114) from the diagram of Fig. 15(a) for a particular bound excited state q is

$$\Delta\delta_l^{(a)}(q) = -\frac{2}{k}\frac{2}{\pi} P\int_0^\infty dk' \langle kn|v|k'q\rangle \left(\frac{k^2}{2} + \varepsilon_n - \varepsilon_q - \frac{k'^2}{2}\right)^{-1} \langle k'q|v|kn\rangle + i\pi\frac{2}{k}\frac{2}{\pi}\frac{1}{k'}\,|\langle kn|v|k'q\rangle|^2 \tag{119}$$

where $k' = (k^2 + 2\varepsilon_n - 2\varepsilon_q)^{\frac{1}{2}}$ in the imaginary term. There is a similar contribution from the exchange term.

Following Mott and Massey,[44] if we have a complex phase shift

$$\delta_l = \lambda_l + i\mu_l \tag{120}$$

where λ_l and μ_l are real, we have the following expressions for the elastic and inelastic scattering cross sections

$$\sigma_{\text{el}} = \frac{2\pi}{k^2} \sum_{l=0}^{\infty} (2l + 1) \exp(-2\mu_l)[\cosh 2\mu_l - \cosh 2\lambda_l] \tag{121}$$

$$\sigma_{\text{in}} = \frac{\pi}{k^2} \sum_{l=0}^{\infty} (2l + 1)(1 - \exp(-4\mu_l)) \tag{122}$$

For $4\mu_l$ small, we may expand $\exp(-4\mu_l)$ in Eq. (122) to obtain

$$\sigma_{\text{in}} \approx \frac{\pi}{k^2} \sum_{l=0}^{\infty} (2l + 1)4\mu_l \tag{123}$$

If we take the lowest-order contributions to μ_l obtained from Eq. (119) and its exchange, we find

$$\sigma_{\text{in}} \approx \frac{16\pi}{k^3k'} \sum_{l=0}^{\infty} (2l + 1)(|\langle kn|v|k'q\rangle|^2 - \langle nk|v|k'q\rangle \times \langle k'q|v|kn\rangle) \tag{124}$$

Equation (124) may also be derived in lowest-order perturbation theory using the usual "Golden-Rule" formula for transition probabilities. Although Eq. (124) is physically valid, in higher orders of perturbation theory, for $k^2/2$ above the inelastic threshold, we may encounter infinite matrix elements due to the long-range character of the Coulomb potential; and these higher-order terms require further investigation.

In the preceding sections we have seen that the many-body diagram techniques may be applied to a wide range of problems in atomic physics. One of the initial computational difficulties is the generation of a suitable basis set of single-particle states. However, once the states are obtained, it is often relatively easy to apply the many-body techniques to the calculation of many different properties for a single atom. Additional properties to be investigated include time-dependent problems, long-range interactions between atoms, hyperfine structure, relativistic effects, phase shifts for open-shell atoms, and molecular properties.

References

1. Brueckner, K. A., *Phys. Rev.*, **97,** 1353 (1955).
2. Brueckner, K. A., *Phys. Rev.*, **100,** 36 (1955).
3. Brueckner, K. A., *The Many-Body Problem,* John Wiley and Sons, Inc., New York, 1959.
4. Goldstone, J., *Proc. Roy. Soc.* (*London*), *Ser. A,* **239,** 267 (1957).
5. Brueckner, K. A., and Gammel, J. L., *Phys. Rev.*, **109,** 1023 (1958).
6. Brueckner, K. A., and Masterson, Jr., K. S., *Phys. Rev.*, **128,** 2267 (1962).
7. Gell-Mann, M., and Brueckner, K., *Phys. Rev.*, **106,** 364 (1957).
8. Kelly, H. P., *Phys. Rev.*, **144,** 39 (1966).
9. Wick, G. C., *Phys. Rev.*, **80,** 268 (1950).
10. Messiah, A., *Quantum Mechanics Vol. II,* North Holland Book Publishing Co., Amsterdam, 1965.
11. Kelly, H. P., *Phys. Rev.*, **131,** 684 (1963).
12. Kelly, H. P., *Phys. Rev.*, **134,** A1450 (1964).
13. Kelly, H. P., *Phys. Rev.*, **136,** B896 (1964).
14. Slater, J. C., *Quantum Theory of Atomic Structure, Vol. II,* McGraw-Hill Book Co., Inc., New York, 1960.
15. Levinson, N., *Danske Videnskab, Selskab, Mat.-fys. Medd.*, **25,** No. 9 (1949).
16. Löwdin, P. O., *Advan. Chem. Phys.*, **2,** 207 (1959).
17. Froman, A., *Phys. Rev.*, **112,** 870 (1958).
19. Brandow, B., *Rev. Mod. Phys.*, to be published (1967).
20. Clementi, E., *J. Chem. Phys.*, **38,** 2248 (1963).
21. McKoy, V., and Sinanoğlu, O., *J. Chem. Phys.*, **41,** 2689 (1964).
22. Linderberg, J., and Shull, H., *J. Mol. Spectry.*, **5,** 1 (1960).
23. Bethe, H. A., *Intermediate Quantum Mechanics,* W. A. Benjamin, Inc., New York, 1964.
24. Dalgarno, A., *Advan. Phys.*, **11,** 281 (1962).
25. Kelly, H. P., *Phys. Letters,* **25A,** 6 (1967).
26. Kelly, H. P., *Phys. Rev.*, **160,** 44 (1967).
27. Kelly, H. P., and Taylor, H. S., *J. Chem. Phys.*, **40,** 1478 (1964).
28. Dalgarno, A., and McIntyre, H. A., *Proc. Phys. Soc.* (*London*), **85,** 47 (1965).
29. Cohen, H. D., and Roothaan, C. C. J., *J. Chem. Phys.*, **43,** 534 (1965).
30. Kelly, H. P., *Phys. Rev.*, **152,** 62 (1966).
31. Goldberger, M. L., and Watson, K. M., *Collision Theory,* John Wiley and Sons, Inc., New York, 1964, p. 853.
32. Fetter, A. L., and Watson, K. M., *Advances in Theoretical Physics, Vol. I,* Ed. K. A. Brueckner, Academic Press, New York, 1965, p. 115.
33. Mittleman, M. H., and Watson, K. M., *Phys. Rev.*, **113,** 198 (1959).
34. Watson, K. M., *Phys. Rev.*, **89,** 575 (1953).
35. Lippmann, B. A., Mittleman, M. H., and Watson, K. M., *Phys. Rev.*, **116,** 920 (1959).
36. Bell, J. S., and Squires, E. J., *Phys. Rev. Letters,* **3,** 96 (1959).
37. Mittleman, M. H., in Ref. 32, p. 283.

38. Namiki, M., *Progr. Theoret. Phys.* (*Kyoto*), **23**, 629 (1960).
39. Thouless, D. J., *The Quantum Mechanics of Many-Body Systems*, Academic Press Inc., New York, 1961.
40. Pu, R. T., and Chang, E. S., *Phys. Rev.*, **151**, 31 (1966).
41. Schwartz, C., *Phys. Rev.*, **124**, 1468 (1961).
42. Fano, U., *Phys. Rev.*, **124**, 1866 (1961).
43. Burke, P. G., *Advan. Phys.*, **14**, 521 (1965).
44. Mott, N. F., and Massey, H. S., *The Theory of Atomic Collisions*, Oxford University Press, London, England, 1965, p. 185.

ON THE HARTREE–FOCK METHOD IN MULTI-CONFIGURATION APPROXIMATION

A. P. JUCYS, *Institute of Physics and Mathematics of the Academy of Sciences of Lithuanian SSR, V. Kapsukas State University Vilnius, Vilnius, Lithuanian SSR*

CONTENTS

1. INTRODUCTION

The methods of quantum-mechanical calculations of atomic structures in which operator matrix elements connecting different configurations are neglected constitute single-configuration approximation. To a certain extent such treatment is justified, and the quantum theory of atoms has scored large successes with the help of such an approximation. However, the further development of the theory requires taking into account the non-diagonal matrix elements connecting different configurations. This has led to the multi-configuration approximation, at first called the inclusion of configuration interaction or superposition of configurations. The term mixing of configurations appeared later.

The first to pay attention to this problem were Bacher[1] and Ufford[2] in the early days of the formulation of the methods of quantum-mechanical calculations of atomic structures. Bacher considered the energy of the $3s3d\,^1D$ configuration of MgI taking into account the perturbation by the configuration $3p^2$. Ufford investigated the term intervals of TiII and ZrII superposing four configurations. In these calculations the non-diagonal matrix elements connecting different configurations were estimated, the approximation was rather poor.

The first to calculate the non-diagonal matrix elements of the energy operator with the help of radial orbitals of the self-consistent field were Hartree and Swirles (now Jeffreys).[3] They considered the energy levels of the ground configurations of the oxygen atom and their ions taking into account the matrix element between configurations

$$1s^22s^22p^q - 1s^22p^{q+2}(1-2) \tag{1}$$

In the case of $q = 2$ for the ratio between inter-multiplet separations

$$\frac{{}^1D - {}^1S}{{}^3P - {}^1D} \tag{2}$$

they obtained 0.76 instead of 1.5 in single-configuration approximation and the experimental value of 1.13. This result, which is similar to the results of Bacher, led to the conclusion that the effect of superposition of configurations is sufficient to account for the deviation of the ratio (2) from the experimental value.

A little later Jucys[4] performed similar calculations for CI with the same result. In these calculations the Hartree–Fock radial orbitals were used instead of the Hartree orbitals which were used by Hartree and Swirles. In both cases calculations were performed under the assumption that

$$P_2(nl|r) = P_1(nl|r) \tag{3}$$

$P_i(nl\ r)$ being the radial orbital due to Hartree of the configuration i. Jeffreys[5] and Jucys[6] repeated the calculations using the exact radial orbitals for both configurations. The results did not change essentially.

It was considered that other higher configurations might improve the results. Jucys[7] incorporated the configurations $1s^22s^22p3p$ and $1s^22s2p^23s$ with (1). The results for the ground configuration were almost the same. However, it was encouraging that the results obtained for the inter-multiplet separations for $1s^22s^22p3p$ were very good as compared with single-configuration approximation.

The development of electronic computing techniques makes it possible to carry out the calculations far more easily and to apply the multi-configuration approximation to a much greater number of atomic systems, taking into account many more configurations.

The recalculation of the two-configuration approximation (1) for oxygen with the help of exact Hartree–Fock radial orbitals by Froese[8] has given almost the same results. In the calculations of Watson,[9] Weiss,[10] and Brown and Fontana[11] a great number of configurations have been taken into account.

In the applications of the multi-configuration approximation a great variety of radial orbitals are used. In most cases they represent the single-configuration approximation for each configuration separately.

In the pioneer work of Hartree and coworkers[12] the problem of taking into account the effect of superposition of configurations on the Hartree–Fock radial orbitals themselves has been solved. This was done using the example of oxygen making the assumption (3) before applying the variation principle. Jucys solved the same problem without using the assumption mentioned.[13] General consideration of the problem has been given by Slater.[14]

The same problem using analytical radial orbitals was solved by Boys,[15] and by Taylor and Parr.[16] The latter used a method similar to that of the classical work of Hylleraas,[17] as has been observed by Shull and Löwdin.[18]

The general conclusion is that it is quite reasonable to use the radial orbitals evaluated in multi-configuration approximation. In what follows we are going to outline the Hartree–Fock method generalized to the multi-configuration approximation. It comprises the calculation of physical quantities as well as the evaluation of radial orbitals in multi-configuration approximation.

2. THE EXPRESSION FOR ENERGY IN MULTI-CONFIGURATION APPROXIMATION

The wave function of the atom in multi-configuration approximation is written in the form

$$\Psi(A^j/X) = \frac{1}{\sqrt{\sum_i \lambda_i^2}} \sum_i \Psi(A_i/X) \tag{4}$$

Here $\Psi(A_i/X)$ is the antisymmetric wave function of the atom corresponding to the configuration i. A_i indicates the set of quantum numbers. As long as we are working in the LS-coupling scheme, it contains LSM_LM_S. The latter being the same in all configurations owing to the diagonality of the electrostatic energy operator with respect to these quantum numbers. A_i may differ in the additional characteristics of the term and in one or two sets of nl. A^j indicates the same sets of quantum numbers as A_j, the superscript, instead of subscript, showing that the corresponding state is described in the multi-configuration approximation. The wave function (4) represents this improved description of the configuration j. For this case λ_j is maximal, the other configurations being responsible for corrections only. Each j having its own set of the values of λ_i.

The wave function (4) leads to the following expression for the energy (neglecting the magnetic interactions)

$$E = \frac{1}{\sum_i \lambda_i^2}\left[\sum_i \lambda_i^2 E_{ii} + 2\sum_{j>i} \lambda_i \lambda_j E_{ij}\right] \tag{5}$$

Here E_{ij} is the energy matrix element connecting configurations i and j, that is

$$E_{ij} = E_{ji} = \langle A_i|H|A_j\rangle \tag{6}$$

the operator in atomic units of Hartree has the form

$$H = \sum_i H_i + \sum_{j>i} \frac{1}{r_{ij}} \tag{7}$$

$$H_i = -\tfrac{1}{2}\Delta_i + \frac{Z}{r_i} \tag{8}$$

Δ_i being the Laplace operator and Z the number of the corresponding element in the Mendeleeff system of elements.

The matrix elements E_{ij} are expressed in terms of generalized Slater integrals

$$I(nl, n'l)_{ij} = -\frac{1}{2}\int_0^\infty P_i(nl|r)\left[\frac{d^2}{dr^2} + \frac{2Z}{r} - \frac{l(l+1)}{r^2}\right] P_j(n'l|r)dr \tag{9}$$

$$R_k(nln'l', n''l''n'''l''')_{ij} = \int_0^\infty \frac{1}{r} P_i(nl|r)P_j(n'l'|r)Y_k(n''l'', n'''l'''|r)_{ij}dr$$
$$= \int_0^\infty \frac{1}{r} P_i(n''l''|r)P_j(n'''l'''|r)Y_k(nl, n'l'|r)_{ij}dr \tag{10}$$

$$Y_k(nl, n'l'|r)_{ij} = Z_k(nl, n'l'|r_{ij}) + U_{k+1}(nl, n'l'|r)_{ij}, \tag{11}$$

$$Z_k(nl, n'l'|r)_{ij} = \frac{1}{r^k} \int_0^r r'^k P_i(nl|r')P_j(n'l'|r)dr' \tag{12}$$

$$U_k(nl, n'l'|r)_{ij} = r^k \int_r^\infty \frac{1}{r'^k} P_i(nl|r')P_j(n'l'|r')dr' \tag{13}$$

R_k reduces to the traditional Slater integrals $F_k(nl, n'l')$ and $G_k(nl, n'l')$ when $i = j$. In this case we have

$$I(nl, nl) = I(nl) \tag{14}$$

$$R_k(nlnl, n'l'n'l') = F_k(nl, n'l') \tag{15}$$

$$R_k(nln'l', nln'l') = G_k(nl, n'l') \tag{16}$$

The expressions for E_{ij} in terms of radial integrals may be found by the methods of Racah.[20] For constructing the expressions for diagonal matrix elements of the electrostatic energy operator the tables of reduced matrix elements of U^k and V^{1k} of Karazija and coworkers[21] may be very useful. Besides that, the part of these last-mentioned expressions giving the electrostatic interaction inside the shells of equivalent electrons may be taken from the tables of Nielson and Koster,[22] which contain the reduced matrix elements of U^k for calculating the direct electrostatic interaction between shells as well. General considerations of the methods of expressing the energy matrix elements may be found in the papers of Shore[23] and Fano.[24]

In the expressions for non-diagonal matrix elements connecting different configurations the overlap integral appears. Its definition is

$$N(nl, nl)_{ij} = \int_0^\infty P_i(nl|r)P_j(nl|r)dr \tag{17}$$

It equals unity for $i = j$, and a little less than unity otherwise.

The method we are outlining does not need to include continuum states in (4). These may be handled with the methods of Fraga.[25]

As an example we take the four-configuration approximation of helium-like atomic systems

$$1s^2 - 2p^2 - 2s^2 - 3d^2 \tag{18}$$

The expressions for the energy operator matrix elements are

$$E_{11} = 2I(1s) - F_0(1s, 1s) \tag{19}$$

$$E_{22} = 2I(2p) + F_0(2p, 2p) + \tfrac{2}{5}F_2(2p, 2p) \tag{20}$$

$$E_{33} = 2I(2s) + F_0(2s, 2s) \tag{21}$$

$$E_{44} = 2I(3d) + F_0(3d, 3d) + \tfrac{2}{7}F_2(3d, 3d) + \tfrac{2}{7}F_4(3d, 3d) \tag{22}$$

$$E_{12} = \frac{1}{\sqrt{3}} G_1(1s, 2p)_{12} \qquad E_{13} = G_0(1s, 2s)_{13}$$

$$E_{14} = \frac{1}{\sqrt{5}} G_2(1s, 3d)_{14} \qquad E_{23} = \frac{1}{\sqrt{3}} G_1(2p, 2s)_{23} \tag{23}$$

$$E_{24} = \frac{1}{\sqrt{15}} G_1(2p, 3d)_{24} + \frac{9}{7\sqrt{15}} G_3(2p, 3d)_{24}$$

$$E_{34} = \frac{1}{\sqrt{5}} G_2(2s, 3d)_{34}$$

These formulae are written for the state ^1S which occurs in all four configurations.

3. HARTREE–FOCK EQUATIONS IN MULTI-CONFIGURATION APPROXIMATION

Hartree–Fock equations are derived from the variation principle which consists of making some functional stationary with respect to variations of one-electron wave functions. In Fock's traditional variation method radial orbitals are varied only, owing to the fact that the functional is expressed in terms of radial integrals. In the original Fock theory the functional is the expression for the energy, without magnetic interactions, with additional terms added accounting for the orthonormalization conditions.

In the multi-configuration approximation the Fock functional according to Kybartas and coworkers[26] is brought to the form

$$E' = \frac{1}{\sum_j a_{ij}^2} [\sum_j a_{ij}^2 E'_{jj} + 2 \sum_{k>j} a_{ij} a_{ik} E'_{jk}] \tag{24}$$

Here

$$E'_{jk} = E_{jk} + \sum_{nl,n'l} (1 + \delta_{jk}\delta_{nn'} - \delta_{nn'}) \varepsilon_{(nl,n'l)_{jk}} N(nl, n'l)_{jk} \tag{25}$$

$$a_{ij} = \lambda_j/\lambda_i \qquad (a_{ii} = 1,\ a_{ij} \ll 1, \text{ if } j \neq i) \tag{26}$$

$\varepsilon(nl, n'l)_{jk}$ are Lagrange multipliers. As is easily seen, $N(nl, nl)_{jk}$ $(j \neq k)$ drop out of the functional, because these integrals are uniquely defined by the normalization conditions inside the configurations i and j.

In most cases the radial orbitals are evaluated by varying (25) with $j = k$ for all configurations to be taken into account in the problem under consideration. Then (24) is varied with respect to a_{ij} only. Such an approach may be compared with the ordinary Ritz method, as was observed by Jucys and Vizbaraite.[27] According to the idea of Hartree and coworkers[12] the functional (24) is to be made stationary with respect to both radial orbitals and coefficients a_{ij} simultaneously. Such an approach in Reference 27 is compared with the generalized Ritz method. This approach constitutes the foundation of the Hartree–Fock method in the multi-configuration approximation, the main feature of which we are setting forth.

Generalizing the method of Hartree and Hartree[28] we write the variation of (24) with respect to radial orbitals in the following form

$$\delta E' = \int_0^\infty \sum_{inl} \frac{\partial E'}{\partial P_i(nl|r)} \delta P_i(nl|r) \mathrm{d}r \tag{27}$$

To equate this expression to zero requires all multipliers at the variations of radial orbitals to vanish. That is

$$\frac{\partial E'}{\partial P_i(nl|r)} = 0 \quad (\text{for all } P_i(nl|r)) \tag{28}$$

This is the most general form of Hartree–Fock equations. A little more explicit form of these equations follows from (24). Then we have

$$\frac{\partial E'_{ii}}{\partial P_i(nl|r)} + 2_j \sum_{j(\neq i)} a_{ij} \frac{E'_{ij}}{\partial P_i(nl|r)} = 0 \tag{29a}$$

$$\frac{\partial E'_{kk}}{\partial P_k(nl|r)} + 2 \sum_{j(\neq k)} \frac{a_{ij}}{a_{ik}} \frac{\partial E'_{kj}}{\partial P_k(nl|r)} = 0 \qquad (k \neq i) \tag{29b}$$

Omitting the sums in (29) one obtains the ordinary Hartree–Fock equations (in single-configuration approximation). It must be accentuated that in obtaining (29b) we divided by a_{ik}^2; this is justified, owing to the fact that a_{ik} does not vanish. On the contrary, the configuration k drops out of the multi-configuration approximation under consideration.

Each radial integral of the expression for the energy matrix elements containing a radial orbital subjected to variation yields one term in the explicit form of Hartree–Fock equations. For these contributions we have

$$\frac{\partial I(nl, n'l)_{ij}}{\partial P_i(nl|r)} = -\tfrac{1}{2}(1 + \delta_{ij}\delta_{nn'}) \times \left[\frac{d^2}{dr^2} + \frac{2Z}{r} - \frac{l(l+1)}{r^2}\right] P_j(n'l|r) \; \partial N(nl, n'l)_{ij} \tag{30}$$

$$\partial P_i(nl|r) = (1 + \delta_{ij}\delta_{nn'}) P_j(n'l|r) \tag{31}$$

$$\frac{\partial R_k(nln'l', n''l''n'''l''')_{ij}}{\partial P_i(nl|r)} = (1 + \delta_{nn''} + \delta_{ij}\delta_{nn'} + \delta_{ij}\delta_{nn'}\delta_{nn''}\delta_{nn'''}) \times \frac{Y_k(n''l'', n'''l'''|r)_{ij}}{r} P_j(n'l'|r) \tag{32}$$

In the particular case one obtains the corresponding formulae to those of Hartree and Hartree.[28]

The variation of (24) with respect to the parameters a_{ij} yields the system of equations

$$\sum_k (E_{jk} - E\delta_{jk}) a_{jk} = 0 \qquad (j = 1, 2, \ldots) \tag{33a}$$

The corresponding secular equation is

$$\begin{vmatrix} E_{11} - E & E_{12} & E_{13} \ldots \\ E_{21} & E_{22} - E & E_{23} \ldots \\ \ldots & & \end{vmatrix} = 0 \tag{33b}$$

Equations (29) will now be solved together with Eqs. (33). Equations (29) are solved by the usual methods of solution of Hartree–Fock equations. After each approximation the energy matrix elements are calculated and (33) solved. This gives a new set of parameters a_{ij} with $a_{ii} = 1$. Each value of i gives a separate set of solutions of (29) and (33).

A particular i indicates the configuration for which the calculation is improved by the corresponding multi-configuration approximation. The other configurations bring corrections into the results for configuration i. The configuration i may be called the configuration under investigation and the other ones the correcting configurations.

For the four-configuration approximation (18) with $i = 1$ Eqs. (29) have this explicit form

$$\left[\frac{d^2}{dr^2} + \frac{2Z - 2Y_0(1s, 1s|r)}{r} - \varepsilon_{1s,1s}\right] P(1s|r)$$
$$- \frac{2}{\sqrt{3}} a_{12} \frac{Y_1(1s, 2p|r)}{r} P(2p|r) - 2a_{13} \left[\frac{Y_0(1s, 2s|r)}{r}\right.$$
$$\left. - \varepsilon_{1s,2s}\right] P(2s|r) - \frac{2}{\sqrt{5}} a_{14} \frac{Y_2(1s, 3d|r)}{r} P(3d|r) = 0 \tag{34}$$

$$\left[\frac{d^2}{dr^2} + \frac{2Z - 2Y_0(2p, 2p|r) - \frac{4}{5}Y_2(2p, 2p|r)}{r}\right.$$
$$\left. - \frac{2}{r^2} - \varepsilon_{2p,2p}\right] P(2p|r) - \frac{2}{\sqrt{3}} \frac{1}{a_{12}} \frac{Y_1(1s, 2p|r)}{r} P(1s|r)$$
$$- \frac{2}{\sqrt{3}} \frac{a_{13}}{a_{12}} \frac{Y_1(2p, 2s|r)}{r} P(2s|r) - \frac{2}{\sqrt{15}} \frac{a_{14}}{a_{12}} \left[2 \frac{Y_1(2p, 3d|r)}{r}\right.$$
$$\left. + \frac{9}{7} \frac{Y_3(2p, 3d|r)}{r}\right] P(3d|r) = 0 \tag{35}$$

$$\left[\frac{d^2}{dr^2}+\frac{2Z-2Y_0(2s,2s|r)}{r}-\varepsilon_{2s,2s}\right]P(2s|r)$$

$$-2\frac{1}{a_{13}}\left[\frac{Y_0(1s,2s|r)}{r}-\varepsilon_{1s,2s}\right]P(1sr)$$

$$-\frac{2}{\sqrt{3}}\frac{a_{12}}{a_{13}}\frac{Y_1(2p,2s|r)}{r}P(2p|r)$$

$$-\frac{2}{\sqrt{5}}\frac{a_{14}}{a_{13}}\frac{Y_2(2s,3d|r)}{r}P(3d|r)=0 \tag{36}$$

$$\left[\frac{d^2}{dr^2}+\frac{2Z-2Y_0(3d,3d|r)-\frac{4}{7}Y_2(3d,3d|r)-\frac{4}{7}Y_4(3d,3d|r)}{r}\right.$$

$$\left.-\frac{6}{r^2}-\varepsilon_{3d,3d}\right]P(3d|r)-\frac{2}{\sqrt{5}}\frac{1}{a_{14}}\frac{Y_2(1s,3d|r)}{r}P(1s|r)$$

$$-\frac{2}{\sqrt{15}}\frac{a_{12}}{a_{14}}\left[2\frac{Y_1(2p,3d|r)}{r}+\frac{9}{7}\frac{Y_3(2p,3d|r)}{r}\right]P(2p|r)$$

$$-\frac{2}{\sqrt{5}}\frac{a_{13}}{a_{14}}\frac{Y_2(2s,3d|r)}{r}P(2s|r)=0 \tag{37}$$

Here for conciseness we have omitted the subscripts indicating configurations to which the corresponding radial orbitals belong, because it is evident by itself, owing to the simplicity of the configurations.

As may be seen from the equations given above the calculations even for the simplest case of the theory of many-electron atoms ought to be very complicated. They become more difficult as the number of configurations taken into account and the number of electrons in the atom under investigation increase. For this reason in the early stages of the practical application of the multi-configuration approximation efforts have been made to seek simplification of the corresponding Hartree–Fock equations.

4. SIMPLIFICATION OF HARTREE–FOCK EQUATIONS AND SOME RESULTS

The Hartree–Fock equation in multi-configuration approximation symbolically may be written in the form

$$H+F+K=0 \tag{38}$$

Omission of K, which we call the configuration term, gives the usual Hartree–Fock equation (single-configuration approximation). Further omission of F, which consists of exchange terms and of terms containing $Y_k(nl, nl|r)$ with $k \neq 0$, yields the Hartree equation.

The essential part of Eq. (38) is H. F and K present the further refinement of the theory. It is important to establish which of F and K, and in which cases, is the more substantial. Then we can omit the one which is less important and conserve the other.

At first it must be observed that $\partial E_{ij}/\partial P_i(nl)$ is of the same order of magnitude as the corresponding contributions to the exchange term. The configuration term in (29a) contains $a_{ij} \ll 1$ which are absent in the exchange term. For this reason in (29a) K is less significant than F, and may be omitted as compared with F. Then we are left with the ordinary Hartree–Fock equations for the configuration under investigation.

In Eqs. (29b) the multipliers a_{ij}/a_{ik} are of the order of unity except the one with $j = i$, which is much more than unity. For this reason all terms of the sum with $j \neq i$ can be omitted as compared with that in which $j = i$. This last case is more significant than the exchange term F. Consequently, the exchange term in (29b) can be neglected as well. Thus, for the correcting configurations we are left with the Hartree equations with the additional (configuration) term, this last-mentioned term being very important. This term appears only for those radial orbitals which are contained in the non-diagonal energy operator matrix element connecting corresponding correcting configurations with the configuration under investigation. Other radial orbitals of the correcting configuration may be taken from the configuration under investigation.

For each two-configuration approximation one solves Eq. (29b) together with the two-row equation (33b) and the system of two equations (33a). With the help of the radial orbitals so obtained Eqs. (33) are solved in full. The energy, neglecting magnetic interactions, is obtained as a solution of (33b). The other quantities may be evaluated with the help of the wave function (4), the coefficients of which are determined from Eq. (33a).

Hartree–Fock equations for the configuration under investigation are the same for all two-configuration approximations. They are solved only once for all two-configuration approximations. For each correcting configuration Hartree equations with the configuration term for those electronic shells which differ in the number of electrons from the configuration under investigation must be solved, the other radial orbitals being taken from the configuration under investigation.

In the problem of the ground state of the helium atom, besides the Hartree–Fock equation for the ground state, it is necessary to solve these additional Hartree equations with the configuration term

$$\left[\frac{d^2}{dr^2} + \frac{2Z - 2Y_0(nl, nl|r)}{r} - \frac{l(l+1)}{r^2} - \varepsilon_{nl,nl}\right] P(nl|r)$$
$$- \frac{2}{\sqrt{2l+1}} \frac{1}{a_{1s,nl}} \frac{Y_1(1s, nl|r)}{r} P(1s|r) - \sum_{n'>n} \varepsilon_{nl,n'l} P(n'l|r) = 0 \tag{39}$$

for $nl = 2p, 2s, 3d, \ldots$ The non-diagonal Lagrange multipliers are involved for securing the orthogonality of radial orbitals with the same quantum number l and different n. The correcting configurations are arranged in the order of diminishing effect. For this reason $2p$ is placed before $2s$, $3d$ before $3p$, and so on. In (39) the configurations of the type $nln'l$ are not included. Those radial orbitals must be taken from the configurations nl^2 and $n'l^2$.

It is interesting that the configuration term displaces the radial orbitals of correcting configurations from their position of single-configuration approximation into the direction of the radial orbital of the configuration under investigation. Therefore the radial orbitals of correcting configurations overlap with that of the configuration under investigation. Consequently, the concept of the electronic shell is conserved. For this reason Linderberg and Shull[29] taking $2p^2$, for example, as correcting the K shell ($1s^2$) call it $K2p^2$ as compared with $L2p^2$ of single-configuration approximation for the L shell.

The convergence of the method is rapid, especially with respect to the correcting configurations lying near the configuration under investigation. The results of Vizbaraite and coworkers[30] show that six-configuration approximation

$$1s^2 - 2p^2 - 2s^2 - 3d^2 - 3p^2 - 2p3p \tag{40}$$

yields for the energy of the helium atom —2.9000 Hartree. Kaminskas and Kaveckis[31] applied a 15-configuration approximation

$$1s^2 - nl^2 - \ldots (n = 2, 3, 4, 5, \qquad l = 0, 1, 2, 3, 4) \tag{41}$$

with the result —2.90242 Hartree against the result —2.86167 of the ordinary Hartree–Fock method and the experimental —2.90376.

In the case of the many-electron atom the multi-configuration approximation is to be applied on the base of the two-electron state approximation as is observed by Jucys and coworkers.[32] The simplest atom of this kind is lithium. In a five-configuration approximation

$$(1s^2 - 2p'^2 - 2s'^2 - 3d'^2 - 3p'^2)nl \tag{42}$$

Jucys and coworkers[33] obtained for the energy of two lower states of lithium ($nl = 2s,\ 2p$) —7.472 and —7.404 Hartree, as compared with —7.433 and —7.365 Hartree from the ordinary Hartree–Fock method, and the experimental values of —7.478 and —7.410. In the case of the ground state of beryllium applying the (5 + 2)-configuration approximation

$$(1s^2 - 2p'^2 - 2s'^2 - 3d'^2 - 3p'^2)(2s^2 - 2p''^2) \tag{43}$$

they obtained —14.659 Hartree against —14.577 by the ordinary Hartree–Fock method and the experimental value of —14.667. These last results are to be a little changed due to the improved estimation of the theoretical value of the ordinary Hartree–Fock method, which must be —14.573 as given by Watson.[34] This fact reduces the result —14.659 practically to —14.655.

The utility of multi-configuration approximation based on the two-electron state approximation is seen from the results given above as compared with those of Brown and Fontana (Reference 11), where 208 configurations have been used to obtain the same result for the configuration $1s^22p$ of lithium we have obtained with five configurations. This approach is a variety of the method called pair correlation, and the two-electron wave function in multi-configuration approximation is a variety of geminals.

The method outlined here yields a further advance of the former theory in which the superposition of configurations was not compatible with the concept of the shell structure of atomic electrons, as observed by Daudel and coworkers.[35] The results for lithium and beryllium (Reference 33) show that the concept of shell structure for these atoms is conserved as is in helium, where two-electron correlation coincides with the ordinary multi-configuration approximation.

Calculations connected with the application of multi-configuration approximation are rather difficult, when a great number of electronic shells are present. But there is no need to apply this approximation to its full extent. In most cases it is quite enough to perform the calculations for the shells in improved calculations in which one is interested. The calculations of Strockytė and Jucys[36] show that such an approach gives correct values for the electron affinity of alkali metals. Here the improved calculations were performed for the outer shell ns^2 of the negative ion only. The energy of the inner shells being the same in both cases, with a good approximation, and for the ns state of the neutral atom the conventional method of calculation is valid.

The procedure of evaluation of radial orbitals may be further simplified using the method of Hartree in multi-configuration approximation.[37] In this method the Hartree functionals representing the above-stated two-configuration approximations must be made stationary. The part of kinetic energy and the potential energy with respect to the nucleus in the Hartree functional is the same as in the Hartree–Fock treatment. For obtaining the part of inter-electronic electrostatic interaction according to Jucys and Vizbaraite[38] the product of radial orbitals $R(nl|r) = P(nl|r)/r$ must be used instead of antisymmetric wave functions. At this point we emphasize that utilizing the product of spin-orbitals or orbitals for this purpose is not justified, because the latter is not uniquely defined in the case of an incomplete shell containing more than one electron or more than one hole. However, such a functional may by no means be treated as the actual energy of the atomic system under consideration.

In the Hartree method in multiconfiguration approximation

the Hartree equations for the configuration under investigation coincide with the conventional Hartree equations for this configuration. Those for correcting configurations are the corresponding Hartree equations with the configuration terms, these contain non-diagonal $Y_0(nl, n'l'|r)_{ij}$ instead of $Y_k(nl, n'l'|r)_{ij}$ involved in the corresponding Hartree–Fock equations. In the case of the ground state of helium in Eqs. (39) the expression $Y_1(ls, nl|r)/\sqrt{2l+1}$ is to be substituted by $Y_0(1s, nl|r)$ for obtaining the corresponding Hartree equations in two-configuration approximations.

After solving the Hartree equations in two-configuration approximations all the physical quantities needed ought to be calculated in the same way as using the solutions of corresponding Hartree–Fock equations. That is, Eqs. (33) must be solved in full and parameters obtained are to be used in (4).

References

1. Bacher, R. F., *Phys. Rev.*, **43,** 264 (1933).
2. Ufford, C. W., *Phys. Rev.*, **44,** 732 (1933).
3. Hartree, D. R., and Swirles, Bertha, *Proc. Cambridge Phil. Soc.*, **33,** 239 (1937).
4. Jucys, A., *Proc. Roy. Soc. (London), Ser. A*, **173,** 59 (1939), Teorinis Ionu C^{4+} ir C^{++} ir Neutralaus C Tyrimas. Kaunas, 1941.*
5. Jeffreys, Bertha, *Proc. Cambridge Phil. Soc.*, **38,** 290 (1942).
6. Jucys, A., *Journal of Physics USSR*, **11,** 49 (1947), Vilniaus Universiteto Mokslo Darbai, Gamtos ir Matematikos mokslu serija, **1,** 13 (1947).
7. Jucys, A., *Zh. Eksperim. i Teor. Fiz.*, **19,** 565 (1949).*
8. Froese, Charlotte, *Proc. Phys. Soc. (London)*, **90,** 39 (1967).
9. Watson, R. E., *Quarterly Progress Report of Solid-State and Molecular Theory Group*, Massachusetts Institute of Technology, October 1956, p. 38.
10. Weiss, A. W., *Phys. Rev.*, **122,** 1826 (1961).
11. Brown, R. T., and Fontana, P. R., *J. Chem. Phys.*, **45,** 4248 (1966).
12. Hartree, D. R., Hartree, W., and Swirles, Bertha, *Phil. Trans. Roy. Soc. London*, **238,** 229 (1939).
13. Jucys, A., *Zh. Eksperim. i Teor. Fiz.*, **23,** 129 (1952).*

* Translated into English for Air Force Cambridge Research Laboratories, Office of Aerospace Research, L. G. Hanscom Field, Bedford, Massachusetts, 01731, by the American Meteorological Society, Translation Service, 872 Massachusetts Avenue, Cambridge, Mass. 02139.

14. Slater, J. C., *Phys. Rev.*, **91,** 528 (1953).
15. Boys, S. F., *Proc. Roy. Soc. (London), Ser. A*, **206,** 489 (1951); **A207,** 181, 197 (1951); *Phil. Trans. Roy. Soc. London,* **245,** 95 (1952); Bernal, M. J. M., and Boys, S. F., *Phil. Trans. Roy. Soc. London,* **245,** 116 (1952); Boys, S. F., and Sahni, R. C., *Phil. Trans. Roy. Soc. London,* **246,** 463 (1953).
16. Taylor, G. R., and Parr, R. G., *Proc. Natl. Acad. Sci.*, **38,** 154 (1952).
17. Hylleraas, E. A., *Z. Physik.*, **48,** 469 (1928).
18. Shull, H., and Löwdin, Per-Olov, *J. Chem. Phys.*, **25,** 1035 (1956).
19. Hartree, D. R., *The Calculation of Atomic Structures*, John Wiley and Sons, New York, 1957.
20. Racah, G., *Phys. Rev.*, **63,** 367 (1943).
21. Karazija, R., Vizbaraite, J., Rudzikas, Z., and Jucys, A., *Tables for the Calculation of Matrix Elements of Atomic Quantity Operators*, Computing Center of the Academy of Sciences USSR, Moscow, 1967.
22. Nielson, C. W., and Koster, G. F., *Spectroscopic Coefficients for the p^n, d^n, and f^n Configurations,* The Massachusetts Institute of Technology, Cambridge, Massachusetts, 1963.
23. Shore, B. W., *Phys. Rev.*, **139,** 1042A (1965).
24. Fano, U., *Phys. Rev.*, **140,** 67A (1965).
25. Fraga, S., *J. Math. Phys.*, **6,** 18 (1965).
26. Kybartas, V., Kaveckis, V., and Jucys, A., *Zh. Eksperim. i Teor. Fiz.*, **29,** 623 (1955); *Soviet. Phys. JETF Eng. Transl.*, **2,** 481 (1956).
27. Jucys, A., and Vizbaraite, J., *Lietuvos TSR Mokslu Akad. Darbai, Ser. B,* **3(26),** 3 (1961).
28. Hartree, D. R., and Hartree, W., *Proc. Roy. Soc. (London), Ser. A,* **154,** 588 (1936).
29. Linderberg, J., and Shull, H., *J. Mol. Spectry.*, **5,** 1 (1960).
30. Vizbaraite, J., Šironas, V., Kaveckis, V., and Jucys, A., *Opt. i Spektroscopiya,* **1,** 277 (1956).
31. Kaminskas, V., and Kaveckis, V., *Lietuvos Fiz. Rinkinys,* 7 (1967). In press.
32. Batarūnas, J., Kaveckis, V., and Jucys, A., *Lietuvos TSR Mokslu Akad. Darbai, Ser. B,* **3,** 9 (1955); **2,** 3 (1956).
33. Jucys, A., Vizaraite, J., Batarūnas, J., and Kaveckis, V., *Lietuvos TSR Mokslu Akad. Darbai, Ser. B,* **2,** 3 (1958).
34. Watson, R. E., *Phys. Rev.*, **119,** 170 (1960).
35. Daudel, R., Odiot, S., and Brion, H., *J. Phys. Radium,* **15,** 804 (1954).
36. Strockyte, T., and Jucys, A., *Lietuvos TSR Mokslu Akad. Darbai, Ser. B,* **1,** 11 (1958); **2,** 17 (1958); **3(19),** 3 (1959).
37. Vizbaraite, J., Strockyte, T., and Jucys, A., *Lietuvos TSR Mokslu Akad. Darbai, Ser. B,* **3(26),** 19 (1961).
38. Jucys, A., and Vizbaraite, J., *Lietuvos TSR Akad. Darbai, Ser. B,* **3(26),** 11 (1961).

ON THE APPLICATION OF THE EXTENDED METHOD OF CALCULATION TO THE ATOMIC ELECTRONS

A. P. JUCYS AND V. A. KAMINSKAS, *Institute of Physics and Mathematics of the Academy of Sciences of Lithuanian SSR, V. Kapsukas State University Vilnius, Vilnius, Lithuanian SSR*

CONTENTS

1. INTRODUCTION

The term the extended method of calculation is understood as the use of as many radial orbitals as there are electrons in the shell of equivalent atomic electrons. The corresponding calculations are very complicated, because the wave function of the whole atom is a sum with a great number of summands. The difficulties with respect to the angle and spin variables are overcome by finding the correspondence between the radial integrals of the ordinary method of calculation and the linear combinations of those of the extended method of calculation as given by Jucys.[1]

The expressions for the matrix elements of operators in terms of radial integrals of the extended method of calculation are found by expanding the permanents of overlap integrals. In these one or two rows being substituted by one- or two-electron radial integrals for one- or two-electron operators correspondingly. As the first aim of this note is to indicate how it is to be done in practice.

For the time being the calculations have been carried out to the full extent mainly for the *s* electrons in helium-,[2,3] lithium-,[3,4] beryllium-,[5] and boron-like[6] atoms and ions. In the case of the shell of p electrons, calculations are performed for the configurations $1s^2 2p^N (N = 2, \ldots, 6)$.[7] To give some results for the configurations $1s^2 2s^2 2p^N$ constitutes the second aim of this note.

As far as we are interested in the properties of $2p$ electrons we are going to apply the extended method of calculation to these electrons only, and to conserve the ordinary method of calculation for s electrons. This will influence the absolute energy values, the differences between the energies of different states of the $2p$ shell not being affected noticeably.

2. THE PROCESS OF EXPRESSING THE MATRIX ELEMENTS OF OPERATORS

As is seen from the paper of Jucys[1] in the extended method of calculation the permanent of overlap integrals plays an important role

$$S_N(l) = \begin{vmatrix} (l^{(1)}l^{(1)}) & (l^{(1)}l^{(2)}) & \ldots & (l^{(1)}l^{(N)}) \\ (l^{(2)}l^{(1)}) & (l^{(2)}l^{(2)}) & \ldots & (l^{(2)}l^{(N)}) \\ & \ldots & & \\ (l^{(N)}l^{(1)}) & (l^{(N)}l^{(2)}) & \ldots & (l^{(N)}l^{(N)}) \end{vmatrix} \tag{1}$$

$$(l^{(i)}l^{(j)}) = (l^{(j)}l^{(i)}) = \int_0^\infty P(l^{(i)}|r)P(l^{(j)}|r)dr \tag{2}$$

Here

$$P(l|r) = rR(l|r) \tag{3}$$

is the radial orbital after Hartree,[8] $R(l|r)$ being the proper radial orbital. N is the number of electrons in the shell, the fundamental quantum number is omitted.

Equation (1) may be expanded in the form

$$S_N(l) = \sum_P P(l^{(1)}l^{(1)})(l^{(2)}l^{(2)}) \ldots (l^{(N)}l^{(N)}) \tag{4}$$

where the summation runs over all permutations P of the first (or second) radial orbitals in overlap integrals.

For the one-electron operator radial integral according to the Eq. (14) of Reference 1 we have the correspondence

$$NR(ll) \to \frac{1}{S_N(1)} \sum_{i,P} P(l^{(1)}l^{(1)}) \ldots R(l^{(i)}l^{(i)}) \ldots l(^{(N)}l^{(N)}) \tag{5}$$

Here i runs from 1 to N. As an example we take the radial integral of the kinetic energy operator and the electrostatic energy operator with respect to the nucleus for the configuration p^3. Then we have

$$3I(pp) \rightarrow \frac{I_p}{S_3} \tag{6}$$

$$\begin{aligned} I_p = {} & I(p^{(1)}p^{(1)})[1 + (p^{(2)}p^{(3)})^2] \\ & + I(p^{(2)}p^{(2)})[1 + (p^{(1)}p^{(3)})^2] + I(p^{(3)}p^{(3)})[1 + (p^{(1)}p^{(2)})^2] \\ & + 2I(p^{(1)}p^{(2)})[(p^{(1)}p^{(2)}) + (p^{(1)}p^{(3)})(p^{(2)}p^{(3)})] \\ & + 2I(p^{(1)}p^{(3)})[(p^{(1)}p^{(3)}) + (p^{(1)}p^{(2)})(p^{(2)}p^{(3)})] \\ & + 2I(p^{(2)}p^{(3)})[(p^{(2)}p^{(3)}) + (p^{(1)}p^{(2)})(p^{(1)}p^{(3)})] \end{aligned} \tag{7}$$

$$\begin{aligned} S_3 = {} & 1 + (p^{(1)}p^{(2)})^2 + (p^{(1)}p^{(3)})^2 + (p^{(2)}p^{(3)})^2 \\ & + 2(p^{(1)}p^{(2)})(p^{(1)}p^{(3)})(p^{(2)}p^{(3)}) \end{aligned} \tag{8}$$

For the two-electron operator radial integral according to (14)[1] we have the correspondence

$$\begin{aligned} & \frac{N(N-1)}{2} R(ll, ll) \rightarrow \\ & \quad \frac{1}{S_N(1)} \sum_{j>i,P} P(l^{(1)}l^{(1)}) \ldots R(l^{(i)}l^{(i)}, l^{(j)}l^{(j)}) \ldots \\ & \qquad \ldots (l^{(N)}l^{(N)}) \end{aligned} \tag{9}$$

For the electrostatic energy operator radial integral of p^3 one obtains

$$3R(pp, pp) \rightarrow \frac{R_{pp}}{S_3} \tag{10}$$

$$\begin{aligned} R_{pp} = {} & R(p^{(1)}p^{(1)}, p^{(2)}p^{(2)}) + R(p^{(1)}p^{(2)}, p^{(1)}p^{(2)}) \\ & + R(p^{(1)}p^{(1)}, p^{(3)}p^{(3)}) + R(p^{(1)}p^{(3)}, p^{(1)}p^{(3)}) \\ & + R(p^{(2)}p^{(2)}, p^{(3)}p^{(3)}) + R(p^{(2)}p^{(3)}, p^{(2)}p^{(3)}) \\ & - 2[R(p^{(1)}p^{(1)}, p^{(2)}p^{(3)}) + R(p^{(1)}p^{(2)}, p^{(1)}p^{(3)})](p^{(2)}p^{(3)}) \\ & + 2[R(p^{(2)}p^{(2)}, p^{(1)}p^{(3)}) + R(p^{(1)}p^{(2)}, p^{(2)}p^{(3)})](p^{(1)}p^{(3)}) \\ & + 2[R(p^{(1)}p^{(2)}, p^{(3)}p^{(3)}) + R(p^{(1)}p^{(3)}, p^{(2)}p^{(3)})](p^{(1)}p^{(2)}) \end{aligned} \tag{11}$$

Here

$$R(pp, pp) = F_0(pp, pp) + \alpha F_2(pp, pp) \tag{12}$$

$\alpha = -1/5, -2/25, 0$ for the terms ^4S, ^2D, and ^2P correspondingly.

For the interaction operator radial integral between two shells according to Eq. (23)[1] we have the correspondence

$$N_1N_2R(l_1l_1, l_2l_2) \rightarrow \frac{1}{S_{N_1}(l_1)S_{N_2}(l_2)} \times \sum_{i,P_1,j,P_2} P_1P_2(l_1^{(1)}l_1^{(1)}) \ldots R(l_1^{(i)}l_1^{(i)}, l_2^{(j)}l_2^{(j)}) \ldots (l_2^{(N_1)}l_2^{(N_2)}) \tag{13}$$

When the shell of l_1 electrons is treated in the ordinary method of calculation and that of l_2 electrons in the extended method we have

$$N_2R(l_1l_1, l_2l_2) \rightarrow \frac{1}{S_{N_2}(l_2)} \sum_{j,P_2} P_2(l_2^{(1)}l_2^{(1)}) \ldots R(l_1l_1, l_2^{(j)}l_2^{(j)}) \ldots (l_2^{(N_2)}l_2^{(N_2)}) \tag{14}$$

instead of (13). As an example we find the expression for the electrostatic interaction between the shells s^2 and p^3, the first one being considered in the ordinary method of calculation and the second in the extended method. Then (14) yields

$$3R(ss, pp) \rightarrow \frac{R_{sp}}{S_3} \tag{15}$$

R_{sp} is obtained from (7) by the substitution

$$I(p^{(i)}p^{(j)}) \rightarrow R(ss, p^{(i)}p^{(j)}) \tag{16}$$

Here

$$R(ss, pp) = F_0(ss, pp) - \tfrac{1}{3}G_1(sp, sp) \tag{17}$$

The summation of permanents of overlap integrals is performed by an electronic computer. In order to evaluate the corresponding radial integrals, each summand of overlap integral permanent is multiplied by

$$\frac{R(l^{(i)}l^{(j)})}{(l^{(i)}l^{(j)})} \quad \text{or} \quad \frac{R(l^{(i)}l^{(i)}, l^{(j)}l^{(j)})}{(l^{(i)}l^{(i)})(l^{(j)}l^{(j)})}$$

for (5) or (9) correspondingly and the summation is performed over all permutations of the first radial orbitals, as indicated above.

The radial integrals $R(l^{(i)}l^{(j)})$ and $R(l^{(i)}l^{(i)}, l^{(j)}l^{(j)})$ are to be calculated by the conventional routine.

3. SOME RESULTS OF CALCULATIONS

Now we are going to report some results for the configurations $1s^2 2s^2 2p^N (N = 2, \ldots, 6)$ obtained with the help of hydrogen-like radial orbitals

$$P(1s|r) = \sqrt{4\alpha^3}\, re^{-\alpha r} \tag{18}$$

$$P(2s|r) = \sqrt{\frac{12\beta^5}{\alpha^2 - \alpha\beta + \beta^2}}\, r\left(1 - \frac{\alpha + \beta}{3} r\right) e^{-\beta r} \tag{19}$$

$$P(2p|r) = \sqrt{\frac{4\gamma^5}{3}}\, r^2 e^{-\gamma r} \tag{20}$$

As mentioned above, the ordinary method of calculation is applied for s electrons and the extended method for p electrons. Then one obtains N radial orbitals $P(2p|r)$ with different values of γ.

The initial values of parameters were taken from the paper of Kirilowa and coworkers[9] and the Fock functional (the expression for non-relativistic energy) was made stationary. The parameters α and β underwent only an insignificant change. The γ's, as in the configurations $1s^2 2p^N$, differ from one another considerably.

In Table 1 the values of γ are given for neon in different degrees

TABLE 1. The values of parameters γ for neon and its ions.

N	LS	γ_1	γ_2	γ_3	γ_4	γ_5	γ_6	γ
6	^1S	4.95	3.77	2.86	2.24	1.65	1.58	2.86
5	^2P	4.99	3.49	2.70	2.24	1.76		3.06
4	^1S	4.87	3.34	2.51	2.19			3.23
	^1D	4.87	3.34	2.55	2.24			3.25
	^3P	4.86	3.34	2.59	2.26			3.26
3	^2P	4.67	3.09	2.55				3.43
	^2D	4.65	3.11	2.59				3.45
	^4S	4.64	3.12	2.67				3.48
2	^1S	4.36	2.83					3.59
	^1D	4.33	2.94					3.63
	^3P	4.31	3.01					3.66
1	^2P	3.83						3.83

of ionization. It is interesting to notice that the relation

$$\gamma = \frac{\Sigma i \gamma_i}{\Sigma i} + 0.8\, \frac{N - 1}{N + 1} \tag{21}$$

approximately holds. This means that in constructing γ from the ordinary method of calculation with the aid of the extended one the weights increase with decreasing values. This follows from the fact that differences between adjacent values diminish together with these values.

In Table 2 the inter-multiplet ratio's for several atoms and ions

TABLE 2. Inter-multiplet separation ratios.

	Ratio	Ordinary method	Extended method	Observed
CI	$\frac{{}^1D - {}^1S}{{}^3P - {}^1D}$	1.5	1.35	1.13
NII	$\frac{{}^1D - {}^1S}{{}^3P - {}^1D}$	1.5	1.42	1.14
OIII	$\frac{{}^1D - {}^1S}{{}^3P - {}^1D}$	1.5	1.45	1.14
NI	$\frac{{}^2D - {}^2P}{{}^4S - {}^2D}$	0.67	0.64	0.50
OII	$\frac{{}^2D - {}^2P}{{}^4S - {}^2D}$	0.67	0.65	0.51
FIII	$\frac{{}^2D - {}^2P}{{}^4S - {}^2D}$	0.67	0.65	0.51
OI	$\frac{{}^1D - {}^1S}{{}^3P - {}^1D}$	1.5	1.46	1.11
FII	$\frac{{}^1D - {}^1S}{{}^3P - {}^1D}$	1.5	1.47	1.12
NeIII	$\frac{{}^1D - {}^1S}{{}^3P - {}^1D}$	1.5	1.48	1.18

are given. It is seen from here that the extended method of calculation brings a correct change in these ratios. However, this alteration is too small to reach the experimental values.

It is very likely that for obtaining the right theoretical values of inter-multiplet separations the extended method of calculation must be combined with the multi-configuration approximation based on the two-electron state model. Then we should have the approximation

$$l^{(1)}l^{(2)}\ldots l^{(N)} - l^{(1)}l^{(2)}\ldots l^{(N-2)}l'^{2} \tag{22}$$

A necessary condition here is that $l' \neq l$, because the extended method covers the correlation with the same l. In the approximation (22) the radial orbitals of $N - 2\,l$ electrons of the correcting configurations may be taken from the configuration under investigation. The radial orbital of l' electrons must be found by the methods described by Jucys.[10]

References

1. Jucys, A. P., *Intern. J. Quantum Chem.*, **1**, 311 (1967).
2. Eckart, C., *Phys. Rev.*, **36**, 878 (1930); Trefftz, E., Schlüter, A., Dettmar, K.-H., and Jorgens, K., *Z. Astrophys.*, **44**, 1 (1957); Skvorcowa, G. S., Shugurov, V. K., and Eringis, K. K., *Lietuvos TSR Mokslu Akad. Darbai, Ser. B*, **4**(**20**), 27 (1961).
3. Hurst, R. P., Gray, J. D., Brigman, G. H., and Matsen, F. A., *Mol. Phys.*, **1**, 189 (1958).
4. Brigman, G. H., and Matsen, F. A., *J. Chem. Phys.*, **27**, 829 (1957).
5. Brigman, G. H., Hurst, K. P., Gray, J. D., and Matsen, F. A., *J. Chem. Phys.*, **29**, 251 (1958); Eringis, K. K., Vizbaraitė, J. J., and Jucys, A. P., *Lietuvos Fiz. Rinkinys*, **6**, 343 (1966).
6. Brigman, G. H., and Silverman, J. N., *J. Chem. Phys.*, **44**, 3136 (1966).
7. Eringis, K. K., Vizbaraitė, J. J., and Jucys, A. P., *Lietuvos Fiz. Rinkinys*, **1**, 291 (1961); Eringis, K. K., and Jucys, A. P., *Lietuvos, Fiz Rinkinys*, **5**, 335 (1965); Ali, M. A., and Cook, D. B., *J. Chem. Phys.*, **45**, 3873 (1966).
8. Hartree, D. R., *The Calculation of Atomic Structures*, John Wiley and Sons, New York, 1957.
9. Kirillowa, I. V., Veselov, M. G., and Bratzew, V. F., *Opt. i Spectroscopiya*, **15**, 3 (1963).
10. Jucys, A. P., this issue, p. 191.

THE CORRELATION ENERGY OF A NON-UNIFORM ELECTRON GAS

K. A. BRUECKNER, *Department of Physics, University of California at San Diego, La Jolla, U.S.A.*

CONTENTS

1. INTRODUCTION

The correlation energy of many-electron systems is of greatest interest and importance in atoms and molecules, for which modern computing techniques allow determination of the Hartree–Fock energy with great precision. In such systems, the electron density is high, since the characteristic atomic dimension (in the Fermi–Thomas approximation) scales as

$$\begin{aligned} R_{\text{atomic}} &\sim a_0/Z^{1/3} \\ a_0 &= \hbar^2/me^2 \\ &= \text{Bohr radius} \end{aligned} \tag{1}$$

Then the atomic volume scales as

$$\Omega_{\text{atomic}} \sim \frac{4\pi}{3}\frac{a_0^3}{Z} \tag{2}$$

and the volume per electron

$$\Omega_{\text{electron}} \equiv \frac{4\pi}{3} r_s^3 a_0^3 \approx \frac{4\pi}{3}\frac{a_0^3}{Z^2} \tag{3}$$

Thus the scale parameter r_s varies as

$$r_s = Z^{-2/3} \tag{4}$$

Thus for Z considerably greater than unity, the electron density (or the average) is high. This result does not hold for the outer electrons; nevertheless, even for those the scale parameter r_s is of the order of or less than unity. This feature of atoms (and usually molecules) allows one to study the correlation problem as one of a high density non-uniform electron gas.

2. THE UNIFORM ELECTRON GAS

As a first step in analysing the problem, we consider the uniform electron gas and then introduce the effects of non-uniformity. The zeroth-order energy, the kinetic energy of the degenerate Fermi gas, is given by (we set $\hbar = 1$)

$$\frac{E_0}{N} = \frac{3}{10m}\, p_0^2 \tag{5}$$

with the Fermi momentum

$$p_0 = \left(\frac{9\pi}{4}\right)^{1/3} \frac{1}{a_0 r_s} \tag{6}$$

Thus

$$\begin{aligned} E_0/N &= \frac{3}{5}\left(\frac{9\pi}{4}\right)^{2/3} \frac{1}{r_s^2} \text{ Rydbergs} \\ &= \frac{2.22}{r_s^2} \text{ Rydbergs} \end{aligned} \tag{7}$$

The first-order energy results from the direct and exchange interaction of the electrons. We assume that the electrostatic energy is cancelled by a positive uniform charge background. The exchange energy then is (summing over electrons with parallel spin)

$$E_1 = -2\,\frac{4\pi e^2}{\Omega} \sum_{p_i} \sum_{p_j} n_i n_j \frac{1}{(\mathbf{p}_i - \mathbf{p}_j)^2} \tag{8}$$

which is easily evaluated to give

$$\begin{aligned} E_1/N &= -\frac{3}{2\pi}\left(\frac{9\pi}{4}\right)^{1/3} \frac{1}{r_s} \text{ Rydbergs} \\ &= -\frac{0.916}{r_s} \text{ Rydbergs} \end{aligned} \tag{9}$$

The above results show that the energy is given by a series in r_s (and, as we shall see later, in $\ln(r_s)$), which appears to converge rapidly for r_s much less than unity. This result of course follows from the perturbation expansion in the Coulomb interaction which is in powers of e^2 and hence in terms of the scaling parameter r_s.

The first two terms of the energy give the Hartree–Fock energy of the uniform gas. The first correlation correction arises from the second-order energy, which is

$$E_2 = 4 \times \frac{1}{2} \times \left(\frac{4\pi e^2}{\Omega}\right)^2 \sum_{p_i}\sum_{p_j}\sum_{q} \frac{1}{(\varepsilon_i + \varepsilon_j - \varepsilon_{i+q} - \varepsilon_{j+q})} \left[\frac{1}{q^2} - \frac{1}{2}\frac{1}{(\mathbf{q} + \mathbf{p}_i + \mathbf{p}_j)^2}\right] \tag{10}$$

with the restrictions on the sums

$$\begin{aligned} p_i &\leqq p_0 \\ p_j &\leqq p_0 \\ [|\mathbf{p}_i + \mathbf{q}|] &\geqq p_0 \\ [|\mathbf{p}_j + \mathbf{q}|] &\geqq p_0 \end{aligned} \tag{11}$$

In Eq. (10) the first term is the non-exchange contribution and the second (with the factor of $\frac{1}{2}$) the exchange term. The exchange term can (with difficulty) be evaluated analytically and gives a constant term in the energy. The most interesting term is the first; its structure may be easily determined for small values of q where the integrand becomes large. We measure all momenta in units of p_0 and retain only leading terms in q. In this limit the restrictions of Eq. (11) allow us to make the replacement

$$\int d\mathbf{p}_i = \int d\Omega_i p_0^2 \int_{p_0 - q\mu_i}^{p_0} dk_i \tag{12}$$

with

$$\mu_i = \mathbf{p}_i \cdot \mathbf{q}/p_i q \tag{13}$$

and also to replace p_i and p_j elsewhere by p_0. The result is

$$E_2 = -2(4\pi e^2)^2 \Omega \frac{1}{(2\pi)^9} p_0^3 \int \frac{dq}{q} \int d\Omega_i \int d\Omega_j \frac{\mu_i \mu_j}{(\mu_i + \mu_j)} \tag{14}$$

which gives

$$E_2/N = -\frac{4}{\pi^2}(1 - \ln 2)\int \frac{dq}{q} \text{ Rydbergs} \qquad (15)$$

The limits on the integral are, of course, zero and infinity. The integral is, however, well behaved for large q if the full dependence on q is included. The divergence at small q is real and results from the failure of perturbation theory to describe the long-range screening of the Coulomb interaction. To determine the screening properly, higher order corrections must be taken into account.

The higher order terms in perturbation theory are in the form of a power series if r_s multiplied by dimensionless integrals. The dominant contribution in each order therefore arises from the most singular integral which also arises from the Coulomb interaction for the same momentum transfer acting the maximum number of times. The corresponding perturbation diagrams are the well-known ring diagrams, the first few terms of which are

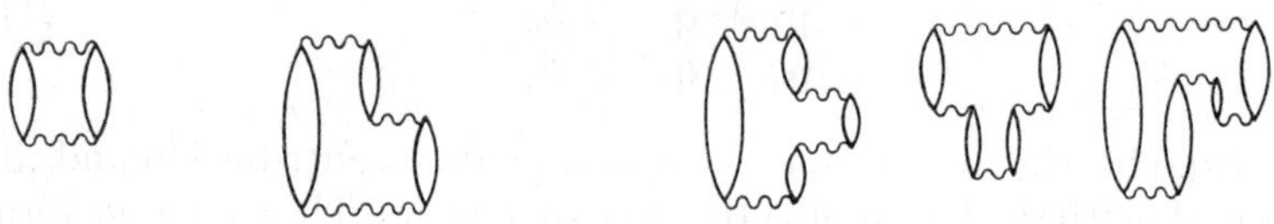

Fig. 1. Time ordered ring diagrams.

shown in Fig. 1. The corresponding matrix element in order n may be shown to have the form

$$\int \frac{dq}{q}\left(-\frac{r_s}{q^2}\right)^{n-2} \qquad (16)$$

so that the perturbation sum is of the form

$$\int \frac{dq}{q}\sum_{n=2}^{\infty} C_n\left(-\frac{r_s}{q^2}\right)^{n} \qquad (17)$$

with C_n a numerical constant determined by the sum of all ring diagrams of a given order.

The sum of Eq. (17) has been evaluated by three different methods. The first (Gell–Mann and Brueckner[1]) evaluated the coefficients C_n directly. The second method (Sawada[1]) made use of the result that the ring diagrams were precisely equal to those generated by perturbation theory applied to the truncated Hamiltonian

$$H_{\mathrm{T}} = \sum_q \left\{ \sum_i w_i b_i^* b_i + \frac{1}{2} \frac{4\pi e^2}{q^2} [\sum_i (b_i^* + b_i)]^2 \right\} \tag{18}$$

with the b_i's boson operators creating and annihilating particle–hole pairs, i.e.

$$b_i = a_{i+q} a_i, \; |k_i + q| > k_F, \; |k_i| < k_F \tag{19}$$

and ω_i the excitation energy of the pair.

The commutation rules of the b_i's are those for bosons within the ring diagram approximation. The simple form of H_{T} allows the Hamiltonian to be diagonalized and the ground-state energy determined. The plasmon dispersion relation and the excitation spectrum may also be immediately obtained.

The most direct procedure for obtaining the energy makes use of modern perturbation theory for which all ring diagrams for a given order result from the various possible time orderings for a single Feynman diagram. The time-ordered diagrams of Fig. 1 may then be replaced by those of Fig. 2. Each wavy line in Fig. 2

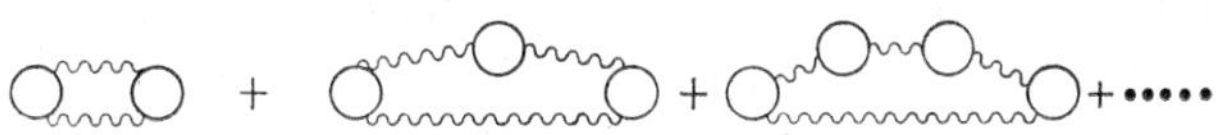

Fig. 2. Feynman ring diagrams.

corresponds to a matrix element of the interaction momentum transfer q,

$$V(q) = \frac{4\pi e^2}{q^2} \tag{20}$$

For each closed loop is introduced the propagator product

$$F_0(q, w) = 2 \int \frac{d^3 p\, dw'}{(2\pi)^4 i} G_0(p, w') G_0(\mathbf{p} + \mathbf{q}, w + w') \tag{21}$$

with

$$G_0(p, w) = \frac{1}{w + \mu - \varepsilon_p + i\eta \operatorname{sgn}(\varepsilon_p - \mu)} \tag{22}$$

and

$$\mu = \text{Zeroth-order chemical potential}$$
$$= p_0^2/2m \tag{23}$$

The sum then is

$$E_{\text{corr}} = -\Omega \sum_{n=2}^{\infty} \frac{1}{n} \int \frac{d\mathbf{q}dw}{(2\pi)^4 i} [F_0(q, w)V(q)]^n$$
$$= \Omega \int \frac{d\mathbf{q}dw}{(2\pi)^4 i} \{\ln[1 - F_0(q, w)V(q)] + F_0(q, w)V(q)\} \tag{24}$$

The function $F_0(q, w)$ is, after carrying out the w' integration,

$$F_0(q, w) = -\frac{2}{(2\pi)^3} \int_{\substack{|p+q| > p_0 \\ p < p_0}} \mathbf{dp} \left[\frac{1}{w - (\varepsilon_{p+q} - \varepsilon_p)} - \frac{1}{w + (\varepsilon_{p+q} - \varepsilon_p)}\right] \tag{25}$$

which for small q is easily evaluated to give

$$F_0(q, w) = -\frac{p_0 m}{\pi^2}\left(1 - \frac{wm}{ip_0 q} \tan^{-1} \frac{ip_0 q}{wm}\right) \tag{26}$$

The w integral in Eq. (24) runs below the real w axis for w negative and above the real w axis for ω positive; thus the contour of integration may be rotated to lie along the imaginary axis from $-i\infty$ to $+i\infty$ (note that the poles of the integrand lie on the real w axis). Then making the variable change

$$w = \frac{ip_0 qu}{m} \tag{27}$$

we bring Eq. (24) to the form

$$E_{\text{corr}} = -\Omega \frac{po}{m} \int_{-\infty}^{\infty} \frac{du}{2\pi} \int \frac{qd\mathbf{q}}{(2\pi)^3} \left\{\ln\left[1 + \frac{4mp_0e^2}{\pi q^2} R(u)\right] - \left[\frac{4mp_0e^2}{\pi q^2} R(u)\right]\right\} \tag{28}$$

with

$$R(u) = \left(1 - u \tan^{-1}\left(\frac{1}{u}\right)\right) \tag{29}$$

The integral of Eq. (28), which gives the small q contribution, together with the remainder of the large q contribution to the second-order energy and the non-divergent second-order exchange energy were evaluated by Gell–Mann and Brueckner (1957) with the result

$$\frac{E_{\text{corr}}}{N} = (-0.096 + 0.0622 \ln (r_s)) \text{ Rydbergs} \tag{30}$$

This result gives the correlation energy exact to terms of order r_s and $r_s \ln r_s$. These arise from higher order terms in the perturbation series not included in the ring diagrams.

The result of Eq. (30) can be used to estimate the correlation energy of atoms, if corrections of higher order on r_s or depending on the non-uniformity of the electron density are assumed to be small. A rough estimate may be obtained by using the approximation relation

$$\overline{\ln r_s} \approx -\frac{2}{3} \ln Z \tag{31}$$

so that

$$\frac{E_{\text{corr}}}{N} \cong -(0.096 + 0.043 \ln Z) \text{ Rydbergs} \tag{32}$$

This result is confirmed by actual integration of Eq. (30) over atomic densities. Experimental correlation energies from oxygen to potassium are between 0.070 and 0.085 Rydbergs per electron, which is about 60% of the result of Eq. (32). Thus the uniform electron gas result gives a very inaccurate estimate for the atom. Thus we need to determine the correlation energy correction arising from the non-uniformity of the electron density. The correction to Eq. (30) from higher order terms in r_s has been estimated by using various interpolation formulae for the correlation energy believed to be reasonably accurate for intermediate values of r_s, and does not appear to account for the discrepancy between theory and experiment.

3. THE NON-UNIFORM ELECTRON GAS

The following theory is taken from Ma and Brueckner (1967). The energy E and the average density $\rho(\mathbf{r})$ may be obtained by minimizing the energy functional $E[\rho]$ keeping the total number of electrons fixed. We break up $E[\rho]$ into two parts:

$$E[\rho] = E_{\mathrm{HF}}[\rho] + E_{\mathrm{c}}[\rho] \tag{33}$$

$E_{\mathrm{HF}}[\rho]$ gives the Hartree–Fock energy and density and $E_{\mathrm{c}}[\rho]$ is the correlation energy functional. When the density variation is slow, we expect that $E_{\mathrm{c}}[\rho]$ may be expanded in powers of the density gradient. If only the first two terms are kept, one has

$$E_{\mathrm{c}}[\rho] = \int d^3x\{A(\rho(\mathbf{x})) + B(\rho(\mathbf{x}))|\nabla\rho(\mathbf{x})|^2\} \tag{34}$$

Our problem is to evaluate the coefficient $B(\rho)$, $A(\rho)$ already being known from the electron gas results.

Consider an electron gas perturbed by a weak static field

$$\phi(\mathbf{r}) = \sum_{k\neq 0} \phi_k e^{i\mathbf{k}.\mathbf{r}} \tag{35}$$

The perturbing Hamiltonian is then

$$\mathscr{H}' = \sum_{k\neq 0} \phi_k \hat{\rho}_{-k} \tag{36}$$

with $\hat{\rho}_k$ the density operator. The energy to second order in ϕ is

$$E = E^{(0)}(n) + \sum_{k\neq 0} \phi_k\phi_{-k} \sum_m \frac{\langle 0|\hat{\rho}_k|m\rangle\langle m|\hat{\rho}_{-k}|0\rangle}{E^{(0)} - E^{(m)}} \tag{37}$$

with n, $|0\rangle$, $E^{(0)}$, $|m\rangle$, $E^{(m)}$ respectively the density, exact ground and excited states and their energies of a uniform electron gas.

The perturbed density to first order in ϕ is

$$\rho_k = 2\sum_m \langle 0|\hat{\rho}_k|m\rangle \frac{1}{E^{(0)} - E^{(m)}} \langle m|\hat{\rho}_{-k}\phi_k|0\rangle \tag{38}$$

$$\rho_0 = n$$

which we write as

$$\rho_k = \phi_k \mathscr{F}(k, 0) \tag{39}$$

with $\mathscr{F}(k, 0)$ the density response function of a uniform electron gas. Using Eqs. (38) and (39), we may then rewrite Eq. (37) as

$$E = E^{(0)}(n) + \tfrac{1}{2} \sum_{k\neq 0} \phi_k\phi_{-k}\mathscr{F}(k, 0) \tag{40}$$

This result and the result of Eq. (39) may be obtained by minimizing the functional

$$E[\rho] = E^{(0)}(n) + \sum_{k\neq 0} \phi_{-k}\rho_k - \tfrac{1}{2}\sum_{k\neq 0} \rho_k\rho_{-k}\mathscr{F}^{-1}(k, 0)$$

Similarly, the Hartree–Fock energy and density may be obtained by minimizing the functional

$$E_{\mathrm{HF}}[\rho] = E^{(0)}_{\mathrm{HF}}(n) + \sum_{k\neq 0} \phi_{-k}\rho_k - \tfrac{1}{2}\sum_{k\neq 0} \rho_k\rho_{-k}\mathscr{F}^{-1}_{\mathrm{HF}}(k, 0) \qquad (41)$$

with $\mathscr{F}_{\mathrm{HF}}$ the density response function in the Hartree–Fock approximation. Thus the correlation energy functional for the weakly perturbed gas is

$$\begin{aligned} E_{\mathrm{c}}[\rho] &= E[\rho] - E_{\mathrm{HF}}[\rho] \\ &= E^{(0)}_{\mathrm{c}}(n) - \tfrac{1}{2}\sum_{k\neq 0} \rho_k\rho_{-k}[\mathscr{F}^{-1}(k, 0) - \mathscr{F}^{-1}_{\mathrm{HF}}(k, 0)] \qquad (42) \end{aligned}$$

The density response function is in general given by the series of diagrams shown in Fig. 3 in which the shaded closed loop F does

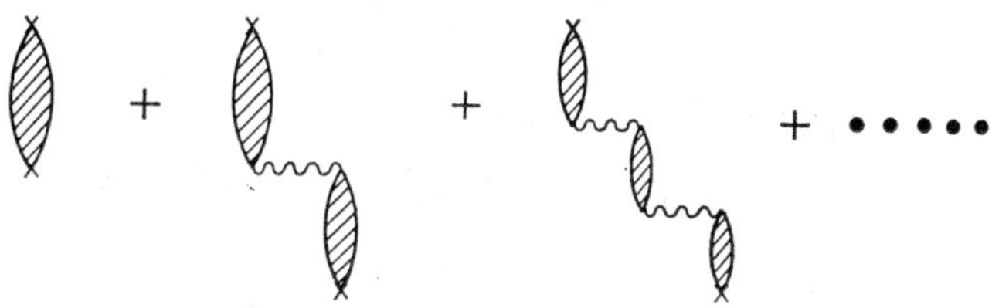

$$\mathscr{F} = F + FVF + FVFVF + \cdots\cdots$$

Fig. 3. Density response function.

not contain any isolated interaction line. The series is readily summed analogous to the closely related sum for the ground-state energy, and gives

$$\mathscr{F} = F/(1 - VF) \qquad (43)$$

Thus

$$\mathscr{F}^{-1} = F^{-1} - V \qquad (44)$$

and, returning to Eq. (42)

$$E_{\mathrm{c}}[\rho] = E^{(0)}_{\mathrm{c}}(n) - \tfrac{1}{2}\sum_{k\neq 0} \rho_k\rho_{-k}(F^{-1}(k, 0) - F^{-1}_{\mathrm{HF}}(k, 0)) \qquad (45)$$

Now we suppose that the response functions are known as power series in k^2 i.e.

$$\begin{aligned} F(k, 0) &= a^{-1} + bk^2 + \ldots \\ F^{-1}(k, 0) &= a - a^2bk^2 + \ldots \end{aligned} \tag{46}$$

The constant a^{-1} is given by

$$\begin{aligned} a^{-1} &= \lim_{k \to 0} F(k, 0) \\ &= -\frac{dn}{d\mu} \end{aligned} \tag{47}$$

and the chemical potential by

$$\mu = \frac{dE^{(0)}(n)}{dn} \tag{48}$$

thus

$$a = \frac{d^2E^{(0)}(n)}{dn^2} \tag{49}$$

The correlation energy then is

$$\begin{aligned} E_{\mathrm{c}}[\rho] = E_{\mathrm{c}}^{(0)}(n) &+ \tfrac{1}{2}\sum_{k \neq 0} k^2 \rho_k \rho_{-k}(a^2b - a_{\mathrm{HF}}^2 b_{\mathrm{HF}}) \\ &+ \frac{d^2}{dn^2} E_{\mathrm{c}}^{0}(n) \tfrac{1}{2} \sum_{h \neq 0} \rho_k \rho_{-k} \end{aligned} \tag{50}$$

Returning to the density expansion, we conclude that the coefficient $B(n)$ is given by

$$B(n) = \tfrac{1}{2}(a^2b - a_{\mathrm{HF}}^2 b_{\mathrm{HF}}) \tag{51}$$

The leading term $A(n)$ can, to first order in the gradient correction, be evaluated at the Hartree–Fock density.

We separate b into the Hartree–Fock and the correlation part, i.e.

$$b = b_{\mathrm{HF}} + b_{\mathrm{c}} \tag{52}$$

Eq. (51) then becomes

$$B(n) = \tfrac{1}{2}[(a^2 - a_{\mathrm{HF}}^2)b_{\mathrm{HF}} + a^2 b_{\mathrm{c}}] \tag{53}$$

The ground-state energy differs from the Hartree–Fock energy by terms of order e^4 (or r_s^2) while b_c will be shown to be of order e^2. Thus to leading order in e^2 (or in the limit of high density) we may retain only the second term in Eq. (53) and replace a by a_{HF} which is

$$a_{HF} = -\frac{d^2E_{HF}^{(0)}}{dn^2} = -\frac{\pi^2}{p_0 m}(1 + \theta(r_s)) \tag{54}$$

The result is

$$B(n) = \tfrac{1}{2}\left(\frac{\pi^2}{p_0 m}\right)^2 b_c(1 + \theta(r_s)) \tag{55}$$

The problem now is to evaluate b_c to the leading order in e^2 (or r_s).

The density response function $F(k, 0)$ is determined by the sum of all diagrams perturbed twice by a static external field, containing no isolated interaction line. The Hartree–Fock contribution results from the sum of all diagrams with a single closed loop but with the interaction acting an indefinite number of times. To the order of Eq. (55), the gradient correction results from the interaction of free electrons perturbed twice by the external field and with at least two electron–hole pairs excited. These may be obtained from the ground-state energy diagrams of Fig. 2 perturbed twice by an external field, as shown in Fig. 4. In Fig. 4 the double wavy lines represent the sum of repeated

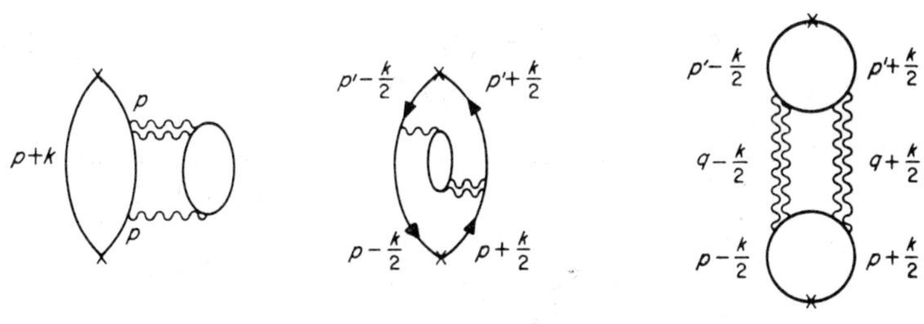

Fig. 4. Gradient corrections.

Coulomb interactions, i.e. the Coulomb interaction modified by the dielectric function

$$\tilde{V}(q, w) = \frac{V(q)}{1 - F(q, w)V(q)} \tag{56}$$

the appearance of $\tilde{V}$ rather than V in the precise form indicated in Fig. 4 is readily understood from examination of the possible ways of introducing the perturbing external field into the ground-state diagrams. In Fig. 4 we have not included the second-order exchange diagram which, because of its lower degree of singularity for small momentum transfers (i.e. its insensitivity to changes in the screening) does not contribute to b_c to the order considered.

We now evaluate the contributions to order k^2 from diagrams (a), (b), and (c) in Fig. 4.

A. Diagram (a). The matrix element is

$$F^{(1)}(k, 0) = 2 \times 2 \times \int dp G^2(p) G(p + k) \Sigma(p) \tag{57}$$

with

$$\Sigma(p) = -\int dq V(q) \tilde{V}(q) G(p + q) F_0(q) \tag{58}$$

For convenience of notation we represent the four-dimensional integrals by

$$\int dp \rightarrow \int \frac{d^3p\, dw}{(2\pi)^4 i} \tag{59}$$

and use the four vectors

$$\begin{aligned} p &\rightarrow (\mathbf{p}, w) \\ q &\rightarrow (\mathbf{q}, w') \end{aligned} \tag{60}$$

The factor of 2×2 in Eq. (57) arises from the spin and from the two types of equivalent diagrams represented by (a) of Fig. 4.

In Eq. (57) the term $G^2(p)$ represents the propagation of a single particle (or hole), and can be obtained by differentiation of $G(p)$, i.e.

$$\begin{aligned} G^2(p) &= -\frac{\partial}{\partial \varepsilon} \frac{1}{\varepsilon + \mu - \varepsilon_p + i\eta \operatorname{Sgn}(\varepsilon_p - \mu)} \\ &= -\frac{\partial G(p)}{\partial \varepsilon} \end{aligned} \tag{61}$$

This is to be distinguished from the derivative of G with respect to μ, since this represents a limit approached in which the particles (or holes) need not propagate at the same energy and momentum values. This difference can be made explicit, since

$$G(p) = \frac{1}{\varepsilon + \mu - \varepsilon_p + i\eta \operatorname{Sgn}(\varepsilon_p - \mu)} \tag{62}$$

may be written

$$\begin{aligned} G(p) &= P\frac{1}{\varepsilon + \mu - \varepsilon_p} - \pi i\delta(\varepsilon + \mu - \varepsilon_p)\operatorname{Sgn}(\varepsilon_p - \mu) \\ &= P\frac{1}{\varepsilon + \mu - \varepsilon_p} - \pi i\delta(\varepsilon + \mu - \varepsilon_p)\operatorname{Sgn} \varepsilon \end{aligned} \tag{63}$$

thus

$$G(p) = \frac{1}{\varepsilon + \mu - \varepsilon p + i\eta \operatorname{Sgn} \varepsilon} \tag{64}$$

From these forms it follows that

$$\begin{aligned} -\frac{\partial G}{\partial \varepsilon} &= [\varepsilon + \mu - \varepsilon p + i\eta \operatorname{Sgn}(\varepsilon_p - \mu)]^{-2} \\ -\frac{\partial G}{\partial \mu} &= [\varepsilon + \mu - \varepsilon p + i\eta \operatorname{Sgn} \varepsilon]^{-2} \end{aligned} \tag{65}$$

and

$$\frac{\partial G}{\partial \mu} = \frac{\partial G}{\partial \varepsilon} + 2\pi i\delta(\varepsilon)\delta(\varepsilon_p - \mu) \tag{66}$$

Thus in Eq. (57) we use Eq. (66) and carry out the integral over the four components of p. The result is

$$F^{(1)}(k, 0) = -4\int dp\, \frac{\partial G}{\partial \mu} G(p + k)\Sigma(p) + \Sigma(p_0, 0)\,\frac{\partial F_0(k, 0)}{\partial \mu} \tag{67}$$

with

$$F_0(k, w) = 2 \int_{\substack{|p+k|>p_0 \\ p<p_0}} \frac{\mathbf{dp}}{(2\pi)^3}\left(\frac{1}{w - \varepsilon_{p+k} + \varepsilon_p} - \frac{1}{w + \varepsilon_{p+k} - \varepsilon_p}\right) \tag{68}$$

To obtain the expansion of $F^{(1)}$, we use the expansion of $G(p+k)$

$$G(p+k) = G(p) + k\cdot\nabla pG(p) + \tfrac{1}{2}(k\cdot\nabla p)^2G(p) + \cdots$$
$$= G(p) - \left(\frac{k\cdot p}{m} + \frac{k^2}{2m}\right)\frac{\partial G(p)}{\partial\mu} + \frac{(k\cdot p)^2}{2m^2}\frac{\partial^2 G(p)}{\partial\mu^2} \tag{69}$$

For an isotropic medium, we drop the term linear in **k** and replace $(k\cdot p)^2$ by $p^2k^2/3$. The expansion of $F_0(k, 0)$ gives

$$\frac{\partial}{\partial\mu}F_0(k,0) = -\frac{m^2}{\pi^2 p_0} - \frac{k^2m^2}{12\pi^2 p^3} \tag{70}$$

Thus, collecting results, we find

$$b_c^{(1)} = -\frac{2}{m}\int dp\,\frac{\partial G}{\partial\mu}\left(-\frac{\partial G}{\partial\mu} + \frac{2}{3}\,\varepsilon_p\,\frac{\partial^2 G}{\partial\mu^2}\right)\Sigma(p) - \frac{m^2}{12\pi^2 p_0^3}\Sigma(p_0, 0) \tag{71}$$

Using the relations

$$\frac{\partial G}{\partial\mu}\frac{\partial G}{\partial\mu} = -\frac{1}{6}\frac{\partial^3 G}{\partial\mu^3}$$
$$\frac{\partial G}{\partial\mu}\frac{\partial^2 G}{\partial\mu^2} = -\frac{1}{12}\frac{\partial^4 G}{\partial\mu^4} \tag{72}$$

we may also rewrite Eq. (71) as

$$b_c^{(a)} = \frac{1}{m}\int dp\left\{-\frac{\partial^3 G}{\partial\mu^3} + \frac{1}{9}\,\varepsilon_p\,\frac{\partial^4 G}{\partial\mu^4}\right\}\Sigma(p) - \frac{m}{24\pi^2\mu p_0}\Sigma(p_0, 0) \tag{73}$$

B. Diagram (b). The contributions to the k^2 term from diagram (b) of Fig. 4 are given in Fig. 5.

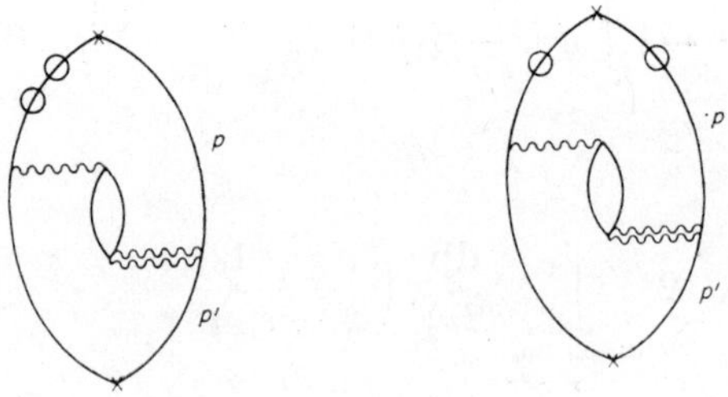

Fig. 5. Contributions from diagram (b) of Fig. 4.

Each circle represents a factor $k \cdot \nabla_p$. The numerical coefficient for the first term arises from 2 for spin states, 1/2 from the second derivative in the k expansion, $(-\frac{1}{2})^2$ for the coefficient of k, and 4 for the equal contributions from the different terms in the expansion. The coefficient for the second term arises from 2 for spin states, $\frac{1}{2} \times (-\frac{1}{2})$ from the coefficient of k, and 2 from the two equal contributions from the expansion. The result is

$$-\int dp[k \cdot \nabla(k \cdot \nabla G(p))G(p) - (k \cdot \nabla G(p))(k \cdot \nabla G(p))]\Lambda^{(b)}(p) \quad (74)$$

with

$$\Lambda^{(b)}(p) = \int dp' \tilde{V}(p-p')F_0(p-p')V(p-p')G(p')G(p') \quad (75)$$

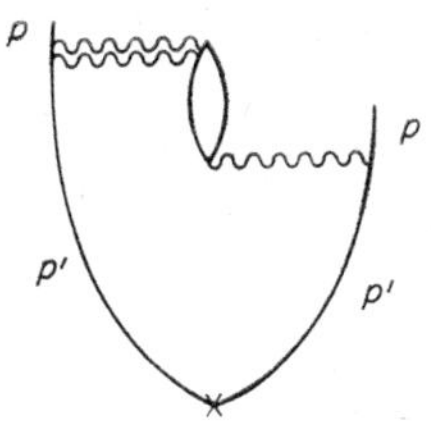

Fig. 6. $\Lambda^{(b)}(p)$.

Diagrammatically, $\Lambda^{(b)}(p)$ is given in Fig. 6. Now using Eq. (69) for the derivatives of $G(p)$, we may write

$$\begin{aligned}
[k \cdot \nabla(k \cdot \nabla G(p))]G(p) &= \frac{k^2}{m}\left(-G^3 + \frac{4}{3}\varepsilon_p G^4\right) \\
&= \frac{k^2}{m}\left(-\frac{1}{2}\frac{\partial^2 G}{\partial \mu^2} + \frac{2}{9}\frac{\partial^3 G}{\partial \mu^3}\right) \\
[k \cdot \nabla G(p)]^2 &= \frac{2}{3}\frac{k^2}{m}\varepsilon_p G^4 \\
&= \frac{1}{9} \cdot \frac{k^2}{m}\varepsilon_p \frac{\partial^3 G}{\partial \mu^3}
\end{aligned} \quad (76)$$

Using these results, we find for the k^2 coefficient

$$b_c^{(b)} = \frac{1}{m}\int dp \left(\frac{1}{2}\frac{\partial^2 G}{\partial \mu^2} - \frac{1}{9}\varepsilon_p \frac{\partial^3 G}{\partial \mu^3}\right)\Lambda^{(b)} \quad (77)$$

C. Diagram (c). The analysis parallels that for diagram b. The diagrams representing the expansion in k are shown in Fig. 7. The numerical factors again arise from the same counting as in diagram (b). The first two diagrams in Fig. 7 have the same structure as in Fig. 5, except that the vertex function $\Lambda^{(b)}$ is replaced by the vertex function $\Lambda^{(c)}$ defined by

$$\Lambda^{(c)}(p) = -4\int dq\,dp'\,\tilde{V}(q)^2 G(p-q)G(p'+q)G(p')G(p') \qquad (78)$$

and given by Fig. 8. The first two diagrams in Fig. 7 then give, an analogy with Eq. (77), a contribution to $b_c^{(c)}$ equal to Eq. (77), with $\Lambda^{(c)}$ replacing $\Lambda^{(b)}$. The remaining two diagrams in Fig. 7 give

$$-2\int dq\Big[\int dp\,G(p)G(p)G(p+q)\Big]^2[\tilde{V}(q)k\cdot\nabla_q(k\cdot\nabla_q\tilde{V}(q)) - (k\cdot\nabla_q\tilde{V}(q)^2\Big] \qquad (79)$$

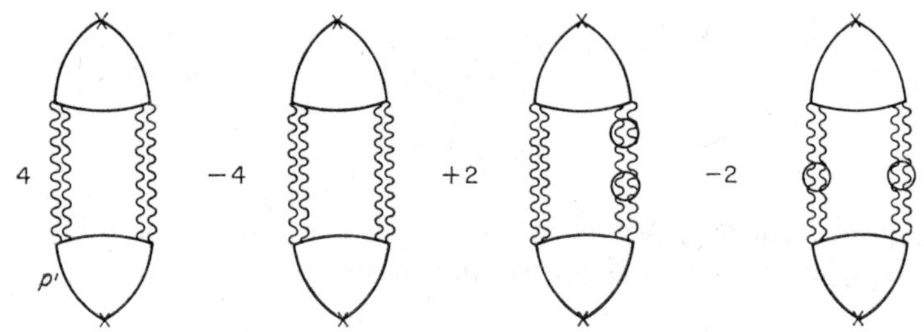

Fig. 7. k expansion diagram (c) of Fig. 4.

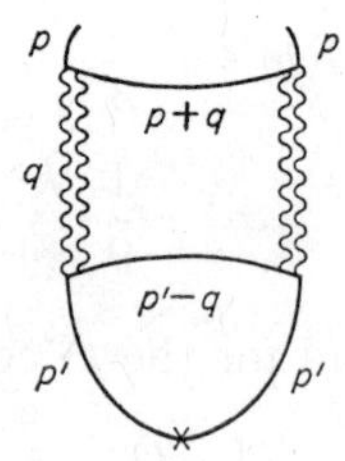

Fig. 8. Vertex function $\Lambda^{(c)}(p)$.

This may be rewritten using the relation

$$\int dp G(p)G(p)G(p+q) = -\frac{1}{2}\frac{\partial}{\partial\mu}\int dp G(p)G(p+q)$$
$$= -\frac{1}{4}\frac{\partial}{\partial\mu}F_0(q) \tag{80}$$

and averaging over the directions of k, as

$$-\frac{k^2}{24}\int dq\left(\frac{\partial F_0(\varepsilon)}{\partial\mu}\right)^2[\tilde{V}(q)\nabla_q^2\tilde{V}(q) - (\nabla q\tilde{V}(q))^2] \tag{81}$$

The coefficient b_c is the sum of the contributions from diagrams (a), (b), and (c). Collecting results, we find

$$b_c = \frac{1}{m}\int dp\left\{\left[-\frac{\partial^3 G}{\partial\mu^3} + \frac{1}{9}\varepsilon_p\frac{\partial^4 G}{\partial\mu^4}\right]\Sigma(p) + \left(\frac{1}{2}\frac{\partial^2 G}{\partial\mu^2} - \right.\right.$$
$$\left.\frac{1}{9}\varepsilon_p\frac{\partial^3 G}{\partial\mu^3}\right)(\Lambda^{(a)} + \Lambda^{(b)}) + \frac{m}{24\pi^2\mu}\Sigma(p_0, 0) -$$
$$\frac{1}{24}\int dq\left(\frac{\partial F_0(q)}{\partial\mu}\right)^2[\tilde{V}\nabla_q^2\tilde{V} - (\nabla_q\tilde{V})^2] \tag{82}$$

This result may be further simplified by using an identity connecting the vertex functions $\Lambda^{(a)}$ and $\Lambda^{(b)}$ with Σ. The diagram for $\Sigma(p)$ is shown in Fig. 9. Differentiating $\Sigma(p)$ with respect to μ

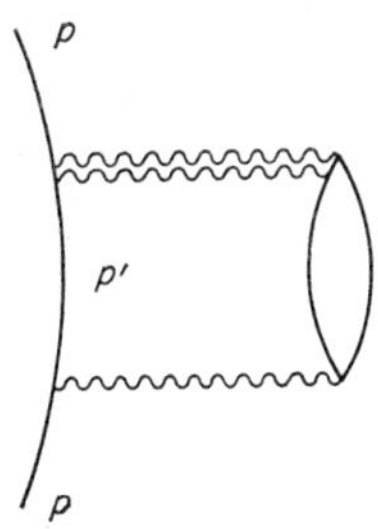

Fig. 9. Mass operator $\Sigma(p)$.

is equivalent to putting a zero momentum vertex on all particle or hole propagators, excluding those contained in the modified interaction $\tilde{V}(q)$. The result of the vertex introduced on the propagator $G(p')$ together with the derivative of $\tilde{V}$ is to give the negative of the vertex function $\Lambda^{(b)}$. The derivative of the closed loop in Fig. 9 together with the derivative of $\tilde{V}$ give the vertex function. $\Lambda^{(c)}$. Thus, we have the identity

$$\Lambda^{(b)} + \Lambda^{(c)} = -\frac{\partial \Sigma}{\partial \mu} \tag{83}$$

This result allows us to rewrite Eq. (82) as

$$b_c = -\frac{1}{m}\frac{\partial}{\partial \mu}\int dp\Sigma(p)\left(\frac{1}{2}\frac{\partial^2 G}{\partial \mu^2} - \frac{1}{9}\frac{\partial^3 G}{\partial \mu^3}\varepsilon_p\right) + \frac{1}{m}\int dp\Sigma(p)\,\frac{1}{6}\frac{\partial^3 G}{\partial \mu^3}$$
$$-\frac{1}{24}\int dq\left(\frac{\partial F_0(q)}{\partial \mu}\right)^2[\tilde{V}\nabla_q^2\tilde{V} - (\nabla_q\tilde{V})^2] + \frac{m}{24\pi^2\mu}\Sigma(p_0, 0) \tag{84}$$

The evaluation of these integrals is most easily carried out if Eq. (58) for $\Sigma(p)$ is introduced and the integrals over p first evaluated. In carrying out the final integral over q, only the small momentum part must be evaluated exactly. The analysis is lengthy but fairly straight forward and is given in detail in Ma and Brueckner (1967). The result cannot be obtained analytically but requires the numerical evaluation of two one-dimensional integrals. It may also be noted that the leading order in e^2, the last term in Eq. (84) may be dropped as of order $e^4 \ln e^2$.

The result is

$$b_c = 1.9756\,\frac{e^2m^2}{p_0^2(2\pi)^3}\,(1 + \theta(e^2)) \tag{85}$$

Now returning to Eq. (55), we have to leading order in the high density expansion

$$B(\rho) = 1.9756\,\frac{e^2m^2}{p_0^2(2\pi)^3}\,\frac{1}{2}\left(\frac{\pi^2}{p_0 m}\right)^2 \tag{86}$$

Expressing p_0 in terms of the density, we finally have for the gradient correction to the correlation energy

$$\Delta E_c = 8.470 \times 10^{-3}\int d\mathbf{r}\rho^{-4/3}|\nabla\rho|^2$$
$$\times\,[1 + \theta(\rho^{-1/3}\ln\rho) + \theta(\rho^{-1/3})]\ \text{Rydbergs} \tag{87}$$

with lengths now measured in atomic units $(1/e^2m)$. Before applying this result to atoms, we give an elementary semiquantitative derivation of the above results. The correlation energy of an electron gas is roughly the correction due to the effect of screening. If ρ' is the density fluctuation, the correlation energy is roughly

$$E_c \approx \frac{1}{2}\int d\mathbf{r}d\mathbf{x}\,\frac{e^2}{r}\,[\langle\rho'(\mathbf{x})\rho'(\mathbf{r}+\mathbf{x})\rangle_{\text{screened}} - \langle\rho'(\mathbf{x})\rho'(\mathbf{r}+\mathbf{x})\rangle_{\text{unscreened}}] \qquad (88)$$

The unscreened density fluctuation correlation function due to the statistics is

$$-\frac{9}{2}\rho^2\left[\frac{\sin p_0r - p_0r\cos p_0r}{p_0^3r^3}\right]^2 \qquad (89)$$

Thus the screened correlation function for $r \gg p_0^{-1}$ is

$$-\frac{9}{4}(3\pi^2)^{-4/3}\rho^{2/3}r^{-4}e^{-r/L_s} \qquad (90)$$

where we have substituted $(3\pi^2\rho)^{1/3}$ for p_0 and $\frac{1}{2}$ for $\cos^2 p_0r$. We need not consider the small r behaviour of the correlation function for small gradients since the effect of the gradient is appreciable only for large r.

Equation (90) is valid only for a uniform gas. However, to obtain an estimate, we regard the screening length $L_s(\rho)$ and ρ as slowly varying functions of r and evaluate the gradient-dependent term

$$\Delta(\mathbf{x},\mathbf{r}) = \left(\rho(\mathbf{x}+\frac{1}{2}\mathbf{r})\right)^{2/3}\left[\exp\left\{-\frac{r}{L_s(\rho(\mathbf{x}+\frac{1}{2}\mathbf{r}))}\right\}-1\right] - (\rho(\mathbf{x}))^{2/3}\left[\exp\left\{-\frac{r}{L_s(\rho(\mathbf{x}))}\right\}-1\right] \qquad (91)$$

The result to the second order is

$$\Delta(\mathbf{x},\mathbf{r}) = \left\{-\frac{r^3}{12}\nabla\rho^{2/3}\cdot\nabla L_s^{-1} + \frac{r^4}{24}\rho^{2/3}(\nabla L_s^{-1})^2\right\}e^{-r/L_s\mathrm{x}} \qquad (92)$$

Thus the gradient term in the correlation energy is

$$\Delta E_c \sim \frac{1}{2}\int d\mathbf{x}d\mathbf{r}\,\frac{e^2}{r}\left(-\frac{9(3\pi^2)^{-4/3}}{4r^4}\right)\Delta(\mathbf{x},\mathbf{r}) \qquad (93)$$

The $\mathbf{r}$ integrals may be easily carried out, giving the result

$$\Delta E_c = \frac{e^2}{8 \times 3^{1/3}\pi^{5/3}} \int d\mathbf{x} \left\{ L_s \nabla \rho^{2/3} \cdot \nabla L_s^{-1} - \frac{1}{2} \rho^{2/3} (L_s \nabla L_s^{-1})^2 \right\} \tag{94}$$

Using the fact that $L_s \sim (e^2 m \rho_0)^{-1/2}$ is proportional to $\rho^{-1/6}$ Eq. (94) reduces to (in atomic units)

$$\Delta E_c = 5.4 \times 10^{-3} \int d\mathbf{x} \rho^{-4/3} |\nabla \rho(\mathbf{x})|^2 \text{ Rydbergs}$$

This qualitative result confirms the form already found (which actually follows from dimensional considerations) and gives a coefficient surprisingly close to the detailed evaluation.

The preceding simple analysis shows clearly the origin of the gradient-dependent correlation energy and also the requirement for validity of the expansion used.

To summarize, the analysis of this section requires:

[1]. slow density change over the screening distance
[2]. high density
[3]. $k/\rho_0 \ll 1$

The first requirement justifies the treatment of the external field as weak, the second the ring diagram expansion, and the third the gradient expansion.

4. APPLICATION TO ATOMS

The estimate of correlation energy using the uniform electron gas result, as noted in the Introduction, overestimates the correlation energy by nearly a factor of two. The correction due to the density gradient may now be determined from Eq. (87). The integral has been evaluated for several atoms, using the Hartree–Fock densities. The result is, however, much too large, overcorrecting the electron gas result by about a factor of five. These overestimates can be easily understood as follows.

If the Fermi–Thomas density were used to evaluate (87), the integral would diverge logarithmically near the origin, since in the Fermi–Thomas approximation for small r

$$\rho \sim r^{-3/2} \tag{95}$$

so that

$$\Delta E_c \sim \int d\mathbf{r}(r^{-3/2})^{-4/3}(r^{-5/2})^2 \sim \frac{dr}{r} \tag{96}$$

This divergence is weak and can be cut off at very small r to match the result given by the Hartree–Fock density which is very high (although finite) at $r = 0$. This suggests that the density gradient is too high for the gradient expansion to be useful. In retrospect, this result is already apparent since the gradient expansion is useful only if the density is slowly varying over the screening length, which is approximately equal to

$$\begin{aligned}\lambda_s &\approx (e^2 m p_0)^{-1/2} \\ &\approx a_0 r_s^{1/2}\end{aligned} \tag{97}$$

The scaling length in the Fermi–Thomas approximation also has the same dependence on Z since r_s scales as $Z^{-2/3}$. The consequence is that for the electrons within the Fermi–Thomas scaling radius, the electron density changes rapidly over the screening length, as usually defined.

The preceding discussion and results leads to several important conclusions:

[1]. The density gradient expansion does not converge for atoms.

[2]. The local correlation structure of atoms is not closely related to electrons in a uniform medium.

[3]. The use of the uniform electron gas result to estimate correlation energy of atoms is entirely misleading, the rough agreement with experiment being fortuitous.

A further consequence of these results is that the full dependence of the correlation energy functional on the density and its derivatives is required to determine atomic properties.

Although the higher order terms in the density expansion must be summed to give a result not subject to the above criticisms, a crude estimate of the higher order terms can be made from the results so far obtained. We look for a function $f(\rho, |\nabla\rho|^2)$ which is well behaved for infinite $|\nabla\rho|^2$ but which reduces to the integrand of Eq. (34) for small $|\nabla\rho|^2$. A reasonable form is

$$A(n)\left[1 - \frac{B(n)|\nabla\rho|^2}{A(n)y}\right]^{-y} \tag{98}$$

with y a phenomenological parameter. This parameter has been matched to the experimental correlation energies to determine the best value for y. For $y = 0.32$, the correlation energies calculated from oxygen to potassium agree with experiment within 4%. A feature of the formula of Eq. (98) which is that the correlation energy of the inner electrons, where the density gradient is large, is far lower than given by the uniform gas estimate, is also confirmed by numerical calculations of pair correlation energies.

Reference

1. Gell-Mann, M., and Brueckner, K. A., *Phys. Rev.*, **106,** 364 (1957); Sawada, K., *Phys. Rev.*, **106,** 372 (1957).

ELECTRON CORRELATION IN ATOMS AND MOLECULES

OKTAY SINANOĞLU, *Sterling Chemistry Laboratory, Yale University, New Haven, Conn., U.S.A. and Teorik Kimya Bölümü Orta Doğu Teknik Üniversitesi, Ankara, Turkey*

CONTENTS

1. INTRODUCTION

Though Coulomb repulsions are of long range and would be expected to add to the complications introduced by the finiteness of an N-body system, a number of theoretical developments have indicated that correlations in atoms and molecules may have simplifying features which allow their evaluation by a combination of non-empirical, and theoretically based semiempirical methods. A crucial question in these developments is how and if the complicated behaviour of N electrons all coupled together may be simplified into that of a few sub-problems.[1] In the present article some of these developments are summarized with emphasis on recent results and applications to specific classes of atoms and molecules which indicate interesting trends and often "anomalous" correlation effects.

2. FORM OF THE EXACT N-ELECTRON WAVE FUNCTION OF A GROUND-STATE ATOM OR MOLECULE

The exact wave function of an N-electron closed shell or single determinantal state may be written with the "correlation normalization" as:

$$\psi(\mathbf{x}_1, \mathbf{x}_2 \ldots, \mathbf{x}_N) = \phi_0 + \chi \tag{1}$$

such that

$$\langle \phi_0 | \chi \rangle = 0, \langle \phi_0 | \phi_0 \rangle = 1$$

and hence $\langle \phi_0 | \psi \rangle = 1$. Then $\langle \chi | \chi \rangle / 1$ represents the "correlation probability" relative to the orbital theory wave function

$$\phi_0 = \mathscr{A}(123 \ldots N) \tag{2}$$

with $i \equiv i(\mathbf{x}_i)$ spin-orbitals and $\mathbf{x}_i = \mathbf{r}_i, \sigma_i$ space, spin coordinates.

The correlation wave function χ is analysed into $1, 2, 3, \ldots, n$-electron effects most conveniently by its successive Schmidt orthogonalization to the products of all $1, 2, \ldots, n$ spin-orbitals out of ϕ_0 $(n \leqslant N)$; ("method of successive partial orthogonalizations").[1,2] One then obtains

$$\chi = \sum_{i \geqslant 1}^{N} \{\hat{f}_i\} + \sum_{i>j}^{N} \{\hat{U}'_{ij}\} + \sum_{i>j>k}^{N} \{\hat{U}'_{ijk}\} + \ldots + \{\hat{U}'_{123 \ldots N}\} \tag{3}$$

where { } indicates antisymmetrization with the products of remaining $(N - n)$ spin-orbitals; e.g.

$$\{\hat{U}'_{ijk}\} \equiv \frac{1}{\sqrt{3!}} \mathscr{A}_N \left(\frac{123 \ldots N}{ijk} \hat{U}'_{ijk}\right) \tag{4}$$

The $(\hat{U}')$'s are closed form "cluster" functions with the properties

$$\hat{U}'_{ij}(\mathbf{x}_i, \mathbf{x}_j) = -\hat{U}'_{ij}(\mathbf{x}_j, \mathbf{x}_i) \tag{5}$$

and of "orbital" or "sea orthogonality",

$$\begin{aligned} \langle \hat{f}_i | k \rangle &= 0 \\ \langle \hat{U}'_{ij} | k \rangle &= 0 \\ \langle \hat{U}'_{ijk\ldots n} | k \rangle &= 0 \end{aligned} \tag{6}$$

for $k = 1, 2, \ldots N$. In Eq. (6), it is an $(n - 1)$ electron function that is rigorously zero, i.e.

$$\langle \hat{U}_{ij\ldots n} | k \rangle \equiv \int \hat{U}'^{*}_{ij\ldots n}(\mathbf{x}_i, \mathbf{x}_j, \ldots \mathbf{x}_n) k(\mathbf{x}_i) dx_i = 0 \tag{7}$$

The form of Eqs. (1)–(6) is general for any $\phi = \mathscr{A}(k_1 k_2 \ldots k_N)$ in Eq. (2) so long as these N spin-orbitals are orthonormal. They may be LCAO MO's, Hartree–Fock (HF) MO's, orthogonalized atomic orbitals (OAO), natural orbitals, or localized orbitals (LO's). A convenient and quite standard choice, however, is the HF ϕ_0 whereupon one has "correlation" in the strict sense.[3]

In Eq. (3), each $\hat{U}'_{ijk\ldots n}$ may in addition be split into true n-electron correlation terms $\hat{U}_{ijk\ldots n}$ (without the prime), and products of independent $\hat{U}$ functions of smaller n correlations ("unlinked clusters").[1,4,5]

Above formal procedure may be carried out also in practice on N-electron trial functions which go beyond ϕ_0, e.g. on multi-configurational SCF (MC–SCF), on functions with correlation factors,[3] etc., to display the types of correlation effects they include.

What happens if $\psi(\mathbf{x}_1, \mathbf{x}_2, \ldots, \mathbf{x}_N)$ refers to an excited state with a ϕ_0 that is not single determinantal (e.g. $1s^2 2s^2 2p^2 (^1S)$ of carbon)? The method of successive partial orthogonalizations (MSPO) may again be attempted, but the resulting χ is not sufficiently detailed. A different approach will be used for such states in Section 5.

3. FORM OF THE N-BODY DENSITY MATRIX AND ITS DECOMPOSITION INTO m-CORRELATION DENSITIES

In some problems, one may wish to deal with the density (matrix) operator D directly. Recently the MSPO-type process has been generalized to analyse the N-density D directly into a form which shows the $1, 2, 3, \ldots, m$-electron correlation effects explicitly.[6]

Let $D = |\chi\rangle\langle\chi|$ and $D_0 = |\phi\rangle\langle\phi_0|$ with the ϕ_0 of form Eq. (2). Then one obtains directly from D by taking successive subtraces (cf. Reference 6 for derivation):

$$D = D_0 + D_\chi; \;\; \mathrm{Tr}(D_0 D_\chi) = 0 \tag{8}$$

$$D_\chi = D_{\gamma(1)} + D_{\gamma(2)} + \ldots + D_{\gamma(N)} \tag{9}$$

with

$$D_{\gamma(m)} \equiv \frac{\binom{N}{m}}{m!N!} \mathscr{A}_N(\rho_1\rho_2 \ldots \rho_{(N-m)}\hat{\gamma}^{(m)}_{(N-m+1),\ldots,N)}\mathscr{A}_N^\dagger \tag{10}$$

The ρ_i is the "Fock–Dirac" density on the Hilbert space of the ith electron,

$$\rho_i = \sum_{k\geqslant 1}^{N} |k\rangle\langle k| \tag{11}$$

and $\hat{\gamma}^{(m)}$, the m-electron correlation density. These $\hat{\gamma}^{(m)}$ are again "sea" or "Fock–Dirac orthogonal", i.e.

$$\mathrm{Tr}_1(\rho_1\hat{\gamma}^{(M)}_{1,2,\ldots,m}) = 0 \tag{12}$$

The trace has been taken only over $\mathbf{x}_1$.

One advantage of the above formulation is that only "Fock–Dirac" operators and correlation operators $\hat{\gamma}^{(m)}$ appear. These operators are invariant under the unitary transformations of the "sea" spin-orbitals, $k = 1, 2, \ldots, N$ in ϕ_0. One can then start with the same formulation whether one wishes at the end to take a representation in delocalized SCF MO's, localized spin-orbitals (LO's), or other basis. The $\hat{\gamma}^{(m)}$ are related to the $\hat{U}'$ functions thus:

$$D_\chi = |\phi_0\rangle\langle\chi| + |\chi\rangle\langle\phi_0| + |\chi\rangle\langle\chi| \tag{13}$$

$$\hat{\gamma}^{(1)} = \sum_{k\geqslant 1}^{N} (|k\rangle\langle \hat{f}_k| + |\hat{f}_k\rangle\langle k| + |\hat{f}_k\rangle\langle \hat{f}_k|) \tag{14}$$

But $\hat{\gamma}^{(m)}$, $m \geqslant 2$, include the $\hat{U}'$'s of m as well as $n < m$; e.g.

$$\begin{aligned}\hat{\gamma}^{(2)} = 2! \sum_{l>k\geqslant 1}^{N} [|\mathscr{B}(kl)\rangle\langle \hat{U}'_{kl}| + |\hat{U}'_{kl}\rangle\langle \mathscr{B}(kl)| + |\hat{U}'_{kl}\rangle\langle \hat{U}'_{kl}| \\ + |\mathscr{B}(k\hat{f}_l)\rangle\langle \hat{U}'_{kl}| \\ + |\hat{U}'_{kl}\rangle\langle \mathscr{B}(k\hat{f}_l)| \\ + |\mathscr{B}(\hat{f}_k l)\rangle\langle \hat{U}'_{kl}| \\ + |\hat{U}'_{kl}\rangle\langle \mathscr{B}(\hat{f}_k l)| \\ + |\mathscr{B}(k\hat{f}_l)\rangle\langle \mathscr{B}(\hat{f}_k l)| + |\mathscr{B}(\hat{f}_k l)\rangle\langle \mathscr{B}(k\hat{f}_l)|] \end{aligned} \tag{15}$$

$\mathscr{B}$ is the two-electron antisymmetrizer, $\frac{1}{\sqrt{2}}(1 - P_{12})$.

The m densities $\hat{\gamma}^{(m)}$ deal with correlations of not more than m particles at a time explicitly and in covariant forms; they differ from other types of "reduced density matrices" used previously by Löwdin,[3] McWeeny,[7] Coleman,[8a] Ando,[8b] Primas,[9] and other authors.

4. SIGNIFICANT CORRELATIONS IN THE WAVE FUNCTIONS OF GROUND STATES

The different correlation effects in Eqs. (3) or (9) were investigated for the choice ϕ equal to the Hartree–Fock ϕ_0. The conclusions are quite equivalent if the ϕ is of natural spin-orbitals or if a part of $\phi_0 + \chi$ is evaluated by multi-configurational SCF (MCSCF).

If χ is solved formally only to first order[10] in perturbation theory, the resulting χ_1 contains only $\hat{U}_{ij}$'s. The "effect of correlation on orbitals", $\hat{f}_k$ in Eq. (3), and $n > 2$ electron correlations therefore contribute to the energy only in fourth and higher orders.[10,11] This of course does not guarantee the smallness of these effects. The $\hat{f}$, however, has been obtained[12,13] in closed form (to all orders) from N-electron trial functions and found to contribute to correlation energy negligibly except where ϕ_0 may be strongly near-degenerate with other determinants (e.g. in H_2 near dissociation).

The part of the Coulomb repulsion r_{12}^{-1} responsible for correlation between two electrons initially occupying the HF spin-orbitals $i(\mathbf{x}_i)$ and $j(\mathbf{x}_j)$ is the "fluctuation potential",*[4,10,11]

$$m_{ij} = \frac{1}{r_{ij}} - \bar{S}_i(\mathbf{x}_j) - \bar{S}_j(\mathbf{x}_i) + \bar{J}_{ij} \tag{16}$$

where

$$\bar{S}_i(x_j) \equiv \left\langle i \left| \frac{1}{r_{ij}} (1 - P_{ij}) \right| i \right\rangle_{\mathbf{x}_i} \tag{17}$$

the Coulomb and exchange potential of spin-orbital i acting on electron $\mathbf{x}_j$, and

$$\bar{J}_{ij} = \left\langle \mathscr{B}(ij) \left| \frac{1}{r_{ij}} \right| \mathscr{B}(ij) \right\rangle = J_{ij} - K_{ij} \tag{18}$$

Along certain contours, this potential is of much shorter range in atoms[1,4] than the full $1/r_{12}$. This aspect of m_{ij} along with the "exclusion effect" of the $(N - 2)$ electrons in the HF "sea" on a correlation (Eqs. (6)) leads one to expect the true $n = 3, 4, \ldots$ electron correlations ("linked" $\hat{U}$'s) to be considerably less important compared to the $\hat{U}_{ij}$'s.[1,4] This expectation may be tested

* The reader will note that this is the physical part of the fluctuation potential. In the total HF potential $V = V(\mathbf{x}_i)$ it is convenient to include the "self-potentials" $\bar{S}_i(\mathbf{x}_i)$ for formal purposes, so that

$$V(\mathbf{x}_i) = \sum_{j \geqslant 1}^{N} \bar{S}_j(\mathbf{x}_i)$$

j running $1, 2, \ldots, i, \ldots, N$. Since, however, $\bar{S}_i(\mathbf{x}_i) i(\mathbf{x}_i) = 0$, the action of V with or without self-potentials in the derivation of χ_1 is the same. As remarked, for example, in References 4 and 16 one may therefore take either Eq. (16) or one with self-potentials, i.e.

$$m_{ij}^s \equiv \frac{1}{r_{ij}} - \bar{S}_i(\mathbf{x}_j) - \bar{S}_j(\mathbf{x}_j) - \bar{S}_i(\mathbf{x}_i) - \bar{S}_j(\mathbf{x}_i) + \bar{J}_{ij}$$

with the same results. In the variational pair correlation energies $\tilde{\varepsilon}_{ij}$, m_{ij} occurs always in combination with $V_i + V_j$ (cf., for example, Reference 5 Eq. (86)). In an actual calculation it is therefore natural to take Eq. (16) and omit (cancel) the self-potentials both in the V's and in m_{ij}. (The statement appearing around Eq. (65) of Reference 14 is therefore erroneous).

most easily by successive configuration-interaction (CI) calculations to include up to 1, 2, 3, . . ., n excitations at a time at each Hamiltonian diagonalization. If true 4, 6, . . . correlations are small, the CI coefficients of 4, 6, . . . excitations of $\hat{U}_{1234\cdots}$ should be expressible in terms of products of bi-excitation coefficients ("unlinked clusters").[4] CI wave functions on Be, LiH, and benzene are quite in accord with these expectations,[4,15] though many more such CI studies would be most illuminating.*

On the basis of the expectations above, the simplified wave function obtained in the "many-electron theory" (MET)[1,4,5] is:

$$\chi_s' = \mathscr{A}\left\{(123\ldots N)\left[2^{-\frac{1}{2}}\sum_{i>j}^{N}\frac{\hat{u}_{ij}}{(ij)} + 2^{-1}\sum_{\substack{i>j\\(i,j\neq k,l)}}^{N}\sum_{k>l}^{N}\frac{\hat{u}_{ij}\hat{u}_{kl}}{(ijkl)} + 2^{-3/2}\sum_{\substack{i>j\\(i,j\neq k,l\neq m,n)}}^{N}\sum_{k>l}^{N}\sum_{m>n}^{N}\frac{\hat{u}_{ijk}\hat{u}_{l}\hat{u}_{mn}}{(ijklmn)} + \cdots\right]\right\} \tag{19}$$

and contains only the two-electron correlation functions $\hat{u}_{ij}$ (no $f_i f_j$ products if f_i are small) and their "unlinked" products.

This form so far appears satisfactory for the correlation energy of single determinantal states, the ground states of $1s^2 2s^n 2p^m$ atoms,[16,17] benzene,[15] and possibly also second-row atoms,[18] and some diatomic and larger molecules.†

The N-density operator corresponding to Eq. (19) is‡

$$D_{\chi_s}' = \sum_{s\geqslant 1}^{N/2} D_{\overbrace{\gamma(2)\gamma(2)\ldots\gamma(2)}^{s}} \tag{20}$$

* More detailed tests of various many-electron terms versus pair correlations may also be carried out by detailed "partial CI" procedures described in the sections of Eqs. (41), (42) and of Eq. (54) of Reference 4.

† Semiempirical methods based on Eq. (19) and other assumptions have been applied to various di- and polyatomic molecules.[28] No detailed, non-empirical tests of $n > 2$ correlation terms in molecules seem available, however, (other than Reference 15), although recent multi-configurational SCF studies (e.g. Reference 19) have features in common with Eq. (19) and could be analysed further by MSPO (successive partial orthogonalizations).[2]

‡ This equation has been derived by Tai-ichi Shibuya of the writer's laboratory.

where

$$D_{\overbrace{\gamma(2)\gamma(2)\ldots\gamma(2)}^{s}} = \frac{\binom{N}{2s}(2s)!}{(2!)^{2s}N!} \times$$

$$\mathscr{A}_N\left(\rho_1 \cdot \cdot \cdot \rho_{(N-2s)} \overbrace{\hat{\gamma}^{(2)}_{(N-2s+1),(N-2s+2)} \cdots \cdots \hat{\gamma}^{(2)}_{N-1,N}}^{s}\right)\mathscr{A}_N^\dagger$$

Thus $D_{\chi_s'}$ contains again only pair correlation densities $\hat{\gamma}^{(2)}$ and their products. Since for such states $\hat{f}_k$ are most likely negligible, the $\hat{\gamma}^{(2)}$ of Eq. (15) simplifies to:

$$\hat{\gamma}^{(2)} \cong 2! \sum_{i>j\geqslant 1}^{N} (|\mathscr{B}(ij)\rangle\langle\hat{u}_{ij}| + |\hat{u}_{ij}\rangle\langle\mathscr{B}(ij)| + |\hat{u}_{ij}\rangle\langle\hat{u}_{ij}|) \qquad (21)$$

5. SIGNIFICANT CORRELATION EFFECTS IN THE WAVE FUNCTIONS OF EXCITED STATES

For a general non-closed shell state, ϕ_0 is a combination of several determinants. There are also many variants of Hartree–Fock theory. In "non-closed shell many-electron theory" (NCMET),[20] a ϕ_0, again in

$$\psi = \phi_0 + \chi; \quad \langle\phi_0|\chi\rangle = 0$$

is chosen so that: (*a*) ϕ_0 has the same symmetry as ψ, hence as χ; (*b*) ϕ_0 has the lowest energy of all functions of the same form; (*c*) the orbitals of ϕ_0, i.e. of the partially filled "HF sea" and of the unfilled portions of the same "sea" are symmetry orbitals; and (*d*) ϕ_0 is sufficiently general so as to include strong near degeneracies (e.g. in dissociating diatomic molecules). The ϕ_0 that meets these requirements is the most general multi-configuration wave function that can be made by assigning N electrons into a set of M ($>N$) "sea" spin-orbitals in various ways, i.e. the complete multi-configurational SCF (MCSCF), or the general restricted Hartree–Fock (GRHF) wave function.* There are in general $\dfrac{M!}{N!(M-N)!}$ determinants Δ_K in ϕ_{GRHF},

* Variants of the MCSCF or GRHF have been developed and applied particularly by Yutsis and coworkers and many other authors. (For a list of references see, for example, "reference 9" of Reference 20*a*.)

$$\phi_{\mathrm{GRHF}} = \sum_{K \geqslant 1}^{\binom{M}{N}} C_K \Delta_K \tag{22}$$

$$\Delta_K = \mathscr{A}_N(k_1(\mathbf{x}_1)k_2(\mathbf{x}_2) \ldots k_N(\mathbf{x}_N))$$

The HF "sea" spin-orbitals are $1, 2, \ldots, M$ of which some N, $k_1, k_2, \ldots, k_N$ (in ascending order $k_1 < k_2 < \ldots < k_N$) are occupied in a Δ_K. Many of the C_K coefficients may be zero by symmetry. As an example, for the lowest ${}^3P$ of carbon, there are $1s, \overline{1s}, \ldots, \overline{2p_z}$ that is $M = 10$ spin-orbitals; the $\binom{M}{N}$ is $10!/6!4! = 210$, but due to ${}^3P$ symmetry, ϕ_{GRHF} contains only five determinants:

$$\begin{aligned}\phi_{\mathrm{GRHF}}({}^3P\text{-C}) = \mathscr{A}\{&C_1(1s^2 2s^2 2p_+ 2p_0)\\ &+ C_2(1s^2 2p_+^2 2p_- 2p_0)\\ &+ C_3(2s^2 2p_+^2 2p_- 2p_0)\\ &+ \frac{1}{\sqrt{2}} C_4[(1s\overline{2s} + 2s\overline{1s})2p_+^{a} 2p_- 2p_0]\}\end{aligned} \tag{23}$$

This example also demonstrates how, for a ground state, either the single-determinantal (det) type theory with

$$\phi_0 = \phi_{\mathrm{RHF}} = \mathscr{A}(1s^2 2s^2 2p_+ 2p_0) \tag{24}$$

and χ_s' given by Eq. (19), *or* the non-closed shell theory with ϕ_{GRHF} and the χ_d' to be given shortly may be used. The ϕ_{RHF} is the restricted HF theory wave function of the Roothaan-type.[21] The extra terms in Eq. (23), of which the second, i.e. the contribution to the $2s^2$ shell, is the largest effect, become in a broad sense the "near-degeneracy" or "non-dynamical" parts of pair correlations in Eq. (19). In that case C_1 is kept $C_1 = 1$ and C_2' evaluated by CI.[17,18] We have for such cases

$$\phi_{\mathrm{GRHF}} \cong \frac{\phi_{\mathrm{RHF}} + \chi^{\mathrm{CI}}}{\sqrt{1 + \langle \chi^{\mathrm{CI}} | \chi^{\mathrm{CI}} \rangle}} \tag{25}$$

Equation (25) differs from the MCSCF or fully optimized ϕ_{GRHF} mainly in the $\hat{f}_i$ terms (Eq. (3)), i.e. effects of non-dynamical correlations on RHF orbitals, which are known to be small[12] unless the C_K's in χ^{CI} are comparable to the C_K in ϕ_{RHF}.

Depending on the ϕ_0, a correlation function χ^{RHF} or χ^{GRHF} is defined. The NCMET function is obtained by solving[20a] the first-order Schrödinger equation formally with $\phi_0 = \phi^{\text{RHR}}$ for χ_1^{RHF} and then generalizing the form to include the indicated major correlations to all orders and their unlinked clusters. Thus one gets a variational theory[20b] (NCMET).

The NCMET wave functions are summarized:
If $\phi_0 \equiv \phi_{\text{RHF}}$, then

$$\psi \cong \phi_{\text{RHF}} + \chi_{\text{int}} + \chi_{\text{ext}} \tag{26}$$

where

$$\langle \phi_{\text{RHF}} | \chi_{\text{int}} \rangle = \langle \phi_0 | \chi_{\text{ext}} \rangle = \langle \chi_{\text{int}} | \chi_{\text{ext}} \rangle = 0 \tag{27}$$

χ_{int} includes only CI within the HF "sea", as in Eqs. (25) and (23) (with $C_1 \equiv 1$)—hence the name "*int*ernal correlation". The main part of χ_{int} is $\chi_{\text{non-dyn}}$, "non-dynamical correlations" as in the $2s^2$ correlation of first-row atoms[17]

$$\chi_{\text{int}} = \chi^{\text{CI}} \cong \chi_{\text{non-dyn}} \tag{28}$$

The χ_{ext} involves the rest of the correlations (the slowly convergent, remaining infinite CI) which contain at least some virtual excitations to the outside of the HF sea (hence "*ext*ernal" correlation).

If instead of ϕ_{RHF}, we start with $\phi_0 = \phi_{\text{GRHF}}$, Eq. (26) is replaced by

$$\psi \cong \phi_{\text{GRHF}} + \chi'_d \tag{29}$$

$$<\phi_{\text{GRHF}} | \chi'_d> = 0 \tag{30}$$

$$\phi_{\text{GRHF}} + \chi'_d = \sum_{k \geqslant 1}^{\binom{M}{N}} C_K(\Delta_K + \chi'_K) \tag{31}$$

where

$$\chi'_K = \mathscr{A}\Bigg[(k_1k_2 \ldots k_N)\Bigg(\sum_{a\geqslant 1}^{N} \frac{\hat{f}^K_{k_a}}{k_a} + \sum_{a>b}^{N} \sum_{\substack{l\geqslant 1 \\ (l\neq k_1,k_2,\ldots k_N)}}^{M} \frac{l\hat{f}^K_{k_ak_b;l}}{k_ak_b} + 2^{-\frac{1}{2}} \sum_{a>b}^{N} \frac{\hat{u}^K_{k_ak_b}}{k_ak_b} + \{\text{u.c.}\}\Bigg)\Bigg] \tag{32}$$

There is a χ'_K correlation function for each det Δ_K of GRHF. Each of these contain three types of correlation:

[1]. $k_a \rightarrow \hat{f}^K_{k_a}$: spin and symmetry polarization of spin-orbital k_a plus the effect of other correlations on k_a.

[2]. $k_a(\mathbf{x}_1)k_b(\mathbf{x}_2) \rightarrow l(\mathbf{x}_1)\hat{f}^K_{k_ak_b;l}(\mathbf{x}_2)$: a "semi-internal correlation". A contribution to the correlation between k_a and k_b (while in Δ_K) such that one is scattered into an empty spin-orbital of the HF "sea" ($l\in\{1, 2, \ldots M\}$ but not in Δ_K) while the other goes out of the "sea" into the function $\hat{f}^K_{k_ak_b;l}(\mathbf{x}_2)$.

[3]. $k_ak_b \rightarrow \hat{u}^K_{k_ak_b}$: a "dynamical" pair correlation all out of the "sea" and due to the short range, steep portions of the fluctuation potential.

The {u.c.} in Eq. (32) refers to the unlinked clusters of all these functions (cf. Eq. (8) of Reference 20*b*).

The $\hat{f}$'s and $\hat{u}$'s are external to the HF "sea" due to their full "sea-orthogonality":

$$\left.\begin{aligned} \langle \hat{f}^K_{k_a}|m\rangle &= 0 \\ \langle \hat{f}^K_{k_ak_b;l}|m\rangle &= 0 \\ \langle \hat{u}^K_{k_ak_b}|m\rangle &= 0 \end{aligned}\right\} m = 1, 2, \ldots, M \tag{33}$$

It may be noted that for the special case of a ϕ which is of a single determinantal form (e.g. all ground-state first-row atoms) either the single det MET or NCMET may be used. Depending on which is used, however, the definition of "dynamical" will differ. For instance in the ground-state $1s^22s^22p^m$ atoms calculated by MET, $\hat{u}(2s\bar{2}p)$, the $2s$–$2p$ correlation, is taken to be roughly dynamical.[17,28] If looked at in terms of NCMET, however, $\hat{u}(2s\bar{2}p)$ is separated into a "semi-internal" and a more dynamical part. In MET the two have been lumped together even though the semi-internal effect is noted (cf. p. 2693 of Reference 17). For such cases one has:

$$\hat{u}^{\text{dyn}}_{k_a k_b} \text{ (single det } \phi_0;\ \text{MET}) = \mathscr{B}(lf_{k_a k_b;l}) + \hat{u}_{k_a k_b} \text{ (NCMET)} \quad (34)$$

(Compare Eqs. (19) and (31), (32) for $\phi_{\text{GRHF}} \Rightarrow C_1\Delta_1 = \mathscr{A}(123\ldots N)$ only.)

Some theorems on the symmetry properties of the $\hat{f}$ and $\hat{u}$ functions of Eq. (32) and of Eq. (19) may be found in References 22*a*, 24, and 25.*

6. PAIR CORRELATIONS (IN GROUND STATES)

It was noted that if second-order energy is sufficient after taking ϕ_0 (Hartree–Fock) as the unperturbed state for a single determinantal case, the N-body correlation becomes the sum of $\frac{N(N-1)}{2}$ decoupled pair correlations and that the E_{Corr} of, for example, all ground-state first-row atoms and their ions may be constructed semi- and/or non-empirically from a set of such pair correlations.[11] A given such pair correlation ε_{ij} was noted to be influenced by the rest of the N electrons through their "exclusion effect"

$$(k = 1, 2, \ldots, N;\ k \neq i, j)$$

$$\langle \hat{u}_{ij} | k \rangle = 0$$

and through the full Hartree–Fock potential $V(\mathbf{x}_i)$ and $V(\mathbf{x}_j)$. Thus only when these "medium" or "environmental" effects on a given ε_{ij} are small can an ε_{ij} be transferred from an ion, to an atom, to a molecule[11,10] (like $\varepsilon\ (1s^2;\ Be^{2+}) \approx \varepsilon(1s^2;\ Be) \approx \varepsilon(1s^2;\ BeH)$). For the E_2 case

$$E_2 = \sum_{i>j}^{N} \varepsilon_{ij}^{(2)} = \sum_{i>j}^{N} \langle \mathscr{B}(ij) | g_{ij} | \hat{u}_{ij}^{(1)} \rangle \quad (35)$$

where ϕ is a single HF determinant.

* For the closed-shell case the properties given in References 10 and 23 apply. These properties were overlooked by some authors (see Reference 22*b*) as discussed on p. 858 of Reference 22*a*.

Many-electron theory (MET) was obtained by making these decoupled pair correlations exact to all orders ($\hat{u}_{ij}^{(1)} \rightarrow \hat{u}_{ij}$; $\varepsilon_{ij}^{(2)} \rightarrow \varepsilon_{ij}$) and by investigating the other effects ($\hat{f}_i$, $\hat{U}_{ijk}$, . . ., etc.) in the N-electron system which would affect the additivity and transferability of ε_{ij}'s.*[1] In this regard, the coupling terms in the energy, three and more electron correlations arriving there even with only pair functions in the wave function, are particularly important to examine in arriving at the MET notion of decoupled pair correlation energies.[4]

The correlation energy, E_{corr}, resulting from the MET wave function, Eq. (19), may be evaluated to different levels of approximation by a "generalized variation-perturbation approach" (VPA).[5] This approach which perhaps more appropriately may be

* The series of MET papers are as follows: MET. I compares $n > 2$ correlations with pair correlations in the energy and in the wave function.[4] New ways of doing CI to obtain few-electron correlations separately are suggested. MET. II Variational methods and procedures for the non-empirical calculation of ε_{ij}'s one at a time;[23] MET. III Effect of correlation on HF orbitals;[12] MET. IV Example of non-empirical calculation of E_{corr}, ε_{ij} at a time, environmental effects on $\varepsilon(1s^2)$ explicitly as $Be^{2+} \rightarrow Be^+ \rightarrow Be$;[16] MET. V Evaluation of ε_{ij}'s in ground state ($1s^2 2s^2 2p^n$) first-row atoms and ions by a combination of non-empirical and semiempirical methods.[17] The $\varepsilon(2s^2$ is found to decrease as $2p\alpha$ electrons are added along neutral atoms ($-\varepsilon(2s^2)$ in Ne^{6+} over 3 eV goes down to nearly zero in Ne). This decrease is compensated in the total E_{corr} by large $\varepsilon(2s2p)$ correlations.

NCMET I,[20a,b] II Generalization to arbitrary non-closed shells for excited states.

For the convenience of the reader some of the other applications or generalizations carried out in the writer's laboratory are summarized and include:

Non-transferability of some ε_{ij}'s in second-row atoms due to "exclusion effects" on "degeneracy effects".[18]

HF + corr. theory of intermolecular potential energy curves with application to the He . . . He curve over a wide range of interatomic separations, ε_{ij}'s in MO$\longleftrightarrow$LO forms.[27]

Semiempirical applications to the E_{corr} and ΔE_{corr} (contribution of correlation to binding energy) of various di- and polyatomic molecules.[28]

Non-empirical evaluation of Coulomb repulsion parameters γ_{pp}, γ_{pq} (for C, N, O) of the Pariser–Parr–Pople theory of π-electron systems. A valence-state Hartree–Fock (VHF) and correlation theory.[29] (The valence state HF orbital for $2p\pi$ already gives about 12.4 eV for the γ_{pp}).

Various types of pair correlations ("internal" parts, i.e. within finite CI) in the benzene π system.[30]

called the "varied-portions approach" since in general it is unrelated to perturbation theory, is outlined below:

Step-1 The full variational expression is,

$$\tilde{E} = \frac{\langle\tilde{\psi}|H|\tilde{\psi}\rangle}{\langle\tilde{\psi}|\tilde{\psi}\rangle} \tag{36}$$

or

$$C_{\text{corr}} \leqslant \tilde{E}_{\text{corr}} = \tilde{E} - E_{\text{HF}} =$$

$$\frac{2 < \phi_0|H - E_{\text{HF}}|\tilde{\chi}\rangle + \langle\tilde{\chi}|H - E_{\text{HF}}|\tilde{\chi}\rangle}{1 + \langle\tilde{\chi}|\tilde{\chi}\rangle} \tag{37}$$

A portion of $\tilde{E} \equiv \tilde{E}' + \Delta\tilde{E}'$, i.e. $\tilde{E}'$ such that

$$\delta\tilde{E}' = 0 \tag{38}$$

in principle has a solution, is minimized. One gets from

$$E' \leqslant \tilde{E}'[\tilde{\chi}'_{\text{VP}}] \tag{39}$$

at $\tilde{\chi}'_{\text{VP}}$.

Step-2 This $\tilde{\chi}'_{\text{VP}}$ is used in the full Eq. (37) and gives an upper bound to the total $\tilde{E}$,

$$E \leqslant \tilde{E}[\tilde{\chi}'_{\text{VP}}] = \tilde{E}' + \Delta\tilde{E}' \tag{40}$$

Different portions of $\tilde{E}_0[\phi + \tilde{\chi}'_s]$ with the χ'_s of the form of Eq. (19), when made stationary (Eq. (38)), lead as special cases to the perturbation theory results,[10] or differently to variational pair correlation equations*[1,5] (non-perturbative),

* The present article has been confined mainly to developments in MET since 1964. In these applications to excited states, for example, the dynamical pair correlations which are the more difficult ones to calculate are extracted from experimental data, then used in other states. Parallel to such applications, the "many-electron theory of atoms and molecules"[1,5] gives various non-empirical methods reviewed in 1963 for the calculation of each pair correlation by itself, together with ways of estimating the smaller effects such as the $n \geqslant 3$ correlations. That each pair correlation energy may be minimized even within closed shells is not at all obvious since all pair functions are coupled together (cf. Reference 4) in the total variational energy $<\tilde{\psi}|H|\,\tilde{\psi}>/<\tilde{\psi}|\tilde{\psi}>$. Moreover several pairs within the same atom (like ^1S, $(1s^22s^22p^6)$Ne), for example the $(2p^2,{}^1S)$, $(2S^2,{}^1S^1)$

and ($1s^2, {}^1S$) pairs have the same symmetry, yet the theory shows[1,5] each of these can still be minimized independently.

The pair correlation, ε_{ij}, methods given enable one to use the Hylleraas r_{ij}-coordinate method in an N-electron atom.[16,17] The reduction of the N-electron problem into such $N(N-1)/2$ electron pair problems, each a variation-perturbation one like the Hylleraas two-electron case, first appeared in the literature in a 1961 paper.[10]

In MET I,[4] separately, a non-empirical CI procedure has also been suggested which has become of considerable interest recently being perhaps the simplest way to estimate the $n \geqslant 3$-electron remainders, $\mathscr{R}$, in the energy. The procedure, as derived in detail with approximations involving the energy denominators, unlinked pair products, energy remainders, and exclusion effects and given on pp. 712, 715 (Eqs. (41), (42)), and 716 of MET I (Reference 4) is briefly the following:

[1]. A "partial CI" (or "pair CI") is made with ϕ_{RHF} plus cirtual excitations from *only one pair* of Hartree–Fock spin-orbitals, say i and j. This gives a single decoupled pair correlation energy, ε_{ij}

$$\tilde{\varepsilon}'_{ij} \equiv \frac{<\psi_{ij} \mid H \mid \psi_{ij}>}{<\psi_{ij} \mid \psi_{ij}>} - E_{\mathrm{HF}} \tag{41}$$

where
$$\psi_{ij} = \phi_{\mathrm{HF}} + \frac{\mathscr{A}}{\sqrt{2}}(123 \ldots \hat{u}_{ij} \ldots N)$$

and
$$\hat{u}_{ij} = \sum_{k>l>N} C^{ij}_{kl} \mathscr{B}(kl); \qquad (\text{if } \hat{f}_i, \hat{f}_j \approx 0)$$

[2]. One repeats such "pair CI's" for each ij, then adds up the resulting $\tilde{\varepsilon}'_{ij}$'s. However, the question remains: is $E_{\mathrm{corr}} \approx \sum_{i>j}^{N} \tilde{\varepsilon}'_{ij}$? The $\sum_{i>j}^{N} \tilde{\varepsilon}'_{ij}$ may have gone *below* the true E_{corr}. The $(E_{\mathrm{HF}} + \sum_{i>j}^{N} \tilde{\varepsilon}'_{ij})$ is not an upper bound[4] to E.

[3]. To get an upper bound (*Step*-2 in the VPA above), one must carry out *a larger* CI, *with all the pairs,* u_{ij}'s, *in.* The new energy $\tilde{E}_s - \sum_{i>j}^{N} \tilde{\varepsilon}'_{ij} = \mathscr{R}$ is now an estimate of $n \geqslant 3$-electron correlation energies.

[4]. However, this is not sufficient either. Even if $\mathscr{R}$ is small (Is it really in an atom like Neon? (Compare References 32*a* and 33 with 32*b* and 14 which contain conflicting statements on the magnitude of $\mathscr{R}$)) the $\sum_{i>j}^{N} \tilde{\varepsilon}'_{ij}$ expression is still not justified. *All unlinked products of pair functions must be added to the wave function,* to cancel out most of

$$D = (1 + \langle \tilde{\chi}_s \mid \tilde{\chi}_s \rangle)$$

energy denominator, to go from $(\Sigma \tilde{\varepsilon}_{ij}/D)$ to $(\sum_{i>j}^{N} \tilde{\varepsilon})'_{ij}$. (More detail will be

$$\varepsilon_{ij} \leqslant \tilde{\varepsilon}'_{ij} \equiv \frac{2\langle \mathscr{B}(ij)|g_{ij}|\hat{u}_{ij}\rangle + \langle \hat{u}_{ij}|e_i + e_j + m_{ij}|\hat{u}_{ij}\rangle}{1 + \langle \hat{u}_{ij}|\hat{u}_{ij}\rangle} \tag{41}$$

and to other approximate forms with or without remainders $\mathscr{R}$ containing $n > 2$ correlation energy effects (still only $\hat{u}_{ij}$'s). A non-variational method for the evaluation of the full energy obtainable with Eq. (19) has recently been given by Čížek.[26]

To varying degrees of accuracy in the uncoupling of different pairs in these methods, one gets (again for single-determinantal theory)[5,26]

$$E_{\text{corr}} \cong \sum_{i>j}^{N} \langle \mathscr{B}(ij)|g_{ij}|\hat{u}_{ij}\rangle = \sum_{i>j}^{N} \varepsilon_{ij} \tag{35'}$$

Even for a closed shell, near-degeneracies may exist leading to "non-dynamical" (finite, strong CI) parts of $\hat{u}_{ij}$'s. Though this is often an arbitrary division, in general one may write

$$\hat{u}_{ij} \equiv \hat{u}_{ij}(\text{non-dyn.}) + \hat{u}_{ij}(\text{dyn.}) \tag{42}$$

---

found in Reference 4). Thus to examine this decoupling, also to see the effect of "linked" $n \geqslant 3$ *correlations in the wave function*, still larger CI's with multiple excitations vs. with unlinked pair products only, are needed.

Various considerations in MET I[4] and later calculations (for example References 16 and 36) made it quite plausible that, in practice for an estimate of a ground state E_{corr}, steps [1] and [2] may suffice. When in doubt, steps [3] and [4] (especially [3] are needed).

For the Be and Ne atoms, recently Nesbet[33] has applied the above "pair CI" method, steps [1] and [2] only, of MET with possible extension with the VPA, *Step*-1, to larger clusters.[4,5] These procedures and decoupled pair notions have not been applied to finite nuclei. They are not contained in papers on Brueckner-type theories for infinite nuclear matter (e.g. Reference 31*c*) which could not be applied successfully to finite nuclei[31a,b] as discussed in Reference 23. Nesbet[33] has recently introduced the nomenclature "Bethe–Goldstone calculations", "Bethe–Goldstone increments", citing Reference 31*c*, for the *n*-electron clusters with VPA and partial CI, variational pair CI's methods of MET[1,4] mentioned above. In recent correspondence Professor H. A. Bethe has stated that his paper with Goldstone (Reference 31*c*) does not contain the approximations, procedures, and methods contained in the Sinanoğlu "Many electron theory of atoms and molecules" papers, (e.g. Reference 4), and that in fact his paper entitled "Effect of a repulsive core in the theory of complex nuclei" is on an unrelated problem of the nucleon–nucleon hard core which has no counterpart in atomic physics.

with

$$\langle \hat{u}_{ij}(\text{non-dyn.}) | \hat{u}_{ij}(\text{dyn.}) \rangle = 0$$

and neglecting the cross terms:

$$\varepsilon_{ij} \approx \varepsilon_{ij}(\text{non-dyn.}) + \varepsilon_{ij}(\text{dyn.}) \tag{43}$$

In first-row atoms, for example, ε_{1s^2}, $\varepsilon_{2p_z^2}$, are mainly $\varepsilon_{ij}(\text{dyn})$, $\varepsilon(2s^2)$ mainly $\varepsilon_{ij}(\text{non-dyn.})$.

Also, quite generally

$$\varepsilon_{ij}(\text{non-dyn.}) = \varepsilon_{ij}(\text{non-dyn.}; Z, N) \tag{44a}$$

strongly depending on "exclusion effect," and the "sea"; whereas

$$\varepsilon_{ij}(\text{dyn.}) \approx \text{const.} \tag{44b}$$

roughly independent of atomic number and N.

In Eq. (35′), there are $\dfrac{N(N-1)}{2}$ pairs, some with the same value due to symmetry (only $O\{N\}$ distinct ε values). In a molecule the ε_{ij} are MO-pair correlations[1,11] ("$\mathscr{B}(ij)$-type pairs") which may also be termed "reducible pairs" since $\mathscr{B}(ij)$ forms a reducible representation of the symmetry group, $\mathscr{G}$, (invariance group) of the Hamiltonian. Two other types of pair correlations were introduced:

[1]. Irreducible pairs[10,23], ε^{irr}
[2]. LO pairs,*[1,11] $\varepsilon_{\rho\nu}$.

The three types are related to each other by unitary transformations $\mathbf{T}$ which leave the E_{corr}, Eq. (35′), in value and form invariant (again single determinantal states).

$$\sum_{i>j}^{N} \varepsilon_{ij} = \sum_{\rho>\nu}^{N} \varepsilon_{\rho\nu} = \sum_{K\geqslant 1}^{N(N-1)/2} \varepsilon_K^{\text{irr}} \tag{45}$$

* Note that there are $N(N-1)/2$ localized spin-orbital (LO) pairs, $\varepsilon_{\rho\nu}$ obtained from Eq. (35′) by a unitary transformation, or calculated directly from the LO forms of Eq. (41).[5,27] As shown in Section 10, the inter-orbital correlation may contribute as much as the "localized pair" (two electrons in the same LO) correlations. Thus the LO pairs differ in concept from the $\dfrac{N}{2}$ Lewis–Langmuir type pairs formulated in "separate pair" or "geminal" theories[34] which assume the inter-orbital correlations to be negligible or at least much smaller.[35]

T acts on the two-electron space, e.g.

$$\psi_K^{\text{irr}} = \sum_{i>j}^{N} T_{K;ij}\mathscr{B}(ij) \tag{46a}$$

and with the same **T**,

$$\hat{u}_K^{\text{irr}} = \sum_{i>j}^{N} T_{K;ij}\hat{u}_{ij} \tag{46b}$$

The ψ_K^{irr} and $\hat{u}_K^{\text{irr}}$ belong to the irreducible representations of the "invariance group" ($\mathscr{G} = 0(3) \otimes SU(2)_{\text{spin}}$ in *LS*-coupling in atoms; $\mathscr{G} = G \otimes SU(2)_{\text{spin}}$ with point group *G* in molecules).

The **T** for LO pairs is the direct product

$$\mathbf{T} = \mathbf{t} \times \mathbf{t} \tag{47}$$

where **t** is the Lennard-Jones matrix transforming the HF spin-orbitals (say MO's) into "equivalent" or "localized" ones, η_ρ;

$$\{\eta\} \equiv \mathbf{t}\{i\} \tag{48}$$

Then,

$$\mathscr{B}(\eta_\rho\eta_\nu) = \sum_{i,j}^{N} t_{\rho i}t_{\nu j}\mathscr{B}(ij) \tag{49a}$$

and

$$\hat{u}_{\rho\nu} = \sum_{i,j}^{N} t_{\rho i}t_{\nu j}\hat{u}_{ij} \tag{49b}$$

With these transformed pair functions one has, in Eq. (45),

$$\varepsilon_K^{\text{irr}} = \langle\psi_K^{\text{irr}}|g_{ij}|\hat{u}_K^{\text{irr}}\rangle \tag{50}$$

$$\varepsilon_{\rho\nu} = \langle\mathscr{B}(\eta_\rho\eta_\nu)|g_{ij}|\hat{u}_{\rho\nu}\rangle \tag{51}$$

and still,

$$\langle\hat{u}_K^{\text{irr}}|k\rangle = 0 \qquad (k = 1, 2, \ldots, N) \tag{52a}$$

and

$$\langle\hat{u}_{\rho\nu}|k\rangle = 0 \qquad (k = 1, 2, \ldots, N) \tag{52b}$$

or

$$\langle\hat{u}_{\rho\nu}|\eta_\tau\rangle = 0 \qquad (\eta_\tau = \eta_1, \eta_2 \ldots, \eta_N) \tag{52c}$$

In terms of the invariant forms, Eqs. (12), (20), and the density operator form of Eq. (35′), one has

$$E_{\text{corr}} \cong \frac{1}{2!} \operatorname{Tr} g_{12}\rho^{(2)}\hat{\gamma}^{(2)} \tag{53}$$

where

$$\rho^{(2)} = \mathscr{B}(\rho_1\rho_2)\mathscr{B}^+ \tag{54}$$

$$\operatorname{Tr}\rho_1\hat{\gamma}^{(2)}_{1,2} = 0 \tag{55}$$

and $\hat{\gamma}^{(2)}$ is given by Eq. (21).*

Examples of Eq. (45) are:

Oxygen $(1s^2 2s^2 2p^4;\ ^3P)$:

In "reducible" or "$\mathscr{B}(ij)$ pairs":

$$\begin{aligned} E_{\text{corr}} = {} & \varepsilon_{1s}{}^2 + 2\varepsilon_{1s\overline{2s}} + 2\varepsilon_{1s2s} + \varepsilon_{2s}{}^2 + 4\varepsilon_{1s\overline{2p}} + 4\varepsilon_{1s2p} \\ & + 4\varepsilon_{2s\overline{2p}} + 4\varepsilon_{2s2p} + \varepsilon_{2p_-{}^2} + \varepsilon_{2p_-2p_+} \\ & + \varepsilon_{2p_-\overline{2p_0}} + \varepsilon_{2p_-\overline{2p_+}} + 2\varepsilon_{2p_+2p_0} \end{aligned} \tag{56}$$

In "irreducible pairs":

$$\begin{aligned} E_{\text{corr}} = {} & \varepsilon(1s^2) + \varepsilon(1s2s;{}^1S) + 3\varepsilon(1s2s;{}^3S) + \varepsilon(2s^2) \\ & + 2\varepsilon(1s2p;{}^1P) + 6\varepsilon(1s2p;{}^3P) \\ & + 2\varepsilon(2s2p;{}^1P) + 6\varepsilon(2s2p;{}^3P) + 4\varepsilon(2p^2;{}^3P) \\ & + \tfrac{5}{3}\varepsilon(2p^2;{}^1D) + \tfrac{1}{3}\varepsilon(2p^2;{}^1S) \end{aligned} \tag{57}$$

Neon $(1s^2 2s^2 2p^6;{}^1S)$:

In "reducible pairs":

$$\begin{aligned} E_{\text{corr}}(\text{Ne};\ L\ \text{shell}) = {} & \varepsilon_{2s^2} + 2\varepsilon_{2p_+{}^2} + \varepsilon_{2p_0{}^2} + 4\varepsilon_{2s2p_+} \\ & + 2\varepsilon_{2s2p_0} + 4\varepsilon_{2s\overline{2p_+}} + 2\varepsilon_{2s\overline{2p_0}} \\ & + 4\varepsilon_{2p_+2p_0} + 4\varepsilon_{2p_+\overline{2p_0}} \\ & + 2\varepsilon_{2p_+2p_-} + 2\varepsilon_{2p_+\overline{2p_-}} \end{aligned} \tag{58}$$

* The Eq. (45) invariance may be seen directly in Eq. (55):

$$\begin{aligned} E'_{\text{corr}} = \frac{1}{2!} \operatorname{Tr} \mathrm{T}\, g\rho^{(2)}\hat{\gamma}^{(2)}\, T^+ &= \frac{1}{2!} \operatorname{Tr} g\, \overbrace{T\rho^{(2)}\, T^+}\, \overbrace{T\hat{\gamma}^{(2)}\, T^+} \\ &= \frac{1}{2!} \operatorname{Tr} g\rho^{(2)}\hat{\gamma}^{(2)} \end{aligned}$$

for any T such that $TT^+ = T^+T = I$.

In "irreducible pairs":

$$E_{\text{corr}}(\text{Ne}; L \text{ shell}) = \varepsilon(2s^2) + 3\varepsilon(2s2p; {}^1P) + 9\varepsilon(2s2p; {}^3P) + 9\varepsilon(2p^2; {}^3P) + 5\varepsilon(2p^2; {}^1D) + \varepsilon(2p^2; {}^1S) \tag{59}$$

In "LO-pairs," e.g. tetrahedral hybrids ($2s2p^3 \to \eta^4$):

$$E_{\text{corr}}(\text{Ne}; L \text{ shell}) = 4\varepsilon_{\eta\bar{\eta}} + 12\varepsilon_{\eta\eta'} + 12\varepsilon_{\eta\bar{\eta}'} \tag{60}$$

The η and η' are two spatially differently oriented tetrahedral orbitals. The bar again denotes β spin, no bar α spin.

Benzene molecule (the π-system):

For the π-system alone,

$$\phi_0 = \mathscr{A}(0\bar{0}\; 1\bar{1}\; 5\bar{5})$$

where $0 \equiv a_{1u}\alpha$; $1, 5 \equiv e_{1g}\alpha$, the lowest π-MO's in complex form. Then:

In "reducible pairs":

$$E_{\text{corr}}(\pi) = \varepsilon_{0\bar{0}} + 2\varepsilon_{1\bar{1}} + 4\varepsilon_{01} + 2\varepsilon_{15} + 4\varepsilon_{0\bar{1}} + 2\varepsilon_{1\bar{5}} \tag{61}$$

In "irreducible pairs":

$$E_{\text{corr}}(\pi) = \varepsilon(a_{1u}^2; {}^1A_{1g}) + 2\varepsilon(e_{1g}^2; {}^1E_{2g}) + 6\varepsilon(a_{1u}e_{1g}; {}^3E_{1u}) + 3\varepsilon(e_{1g}^2; {}^3A_{2g}) + \varepsilon(e_{1g}^2; {}^1A_{1g}) + 2\varepsilon(a_{1u}e_{1g}; {}^1E_{1u}) \tag{62}$$

The above π-pair correlations have recently been evaluated in the "internal CI" approximation.[30] Values of atomic pair correlations will be found in References 16 and 17 and Sections 8 to 10 (cf. also Reference 36).*

* Equation (35) was also used but purely empirically in References 37*a* and 37b. However, these authors did not take into account the "exclusion effects" ($\langle \hat{u}_{ij} | k \rangle = 0$) and the N dependence of the "degeneracy" effects. Their conclusions ($\varepsilon(2s^2)$ was assumed unchanged for ($1s^22s^2$) ions as in neutral $1s^22s^22p^n$ atoms; $\varepsilon(2s\bar{2}p)$ came out ≈ 0) are in contradiction with theory (e.g. $-\varepsilon(2s^2) \approx 3.5$ eV in Ne versus the theoretical expectation,[1,4] $\varepsilon(2s^2; \text{Ne}; Z = N) \approx 0$) and other ε_{ij} values[16,17,33,36] (see also Section 8).

Quite generally, although the total number of pairs are $N(N-1)/2$, the number of distinct pairs, ε, to be evaluated are of the order of N.

7. PAIR, "CROSS PAIR", AND SEMI-INTERNAL CORRELATIONS IN THE ENERGIES OF EXCITED STATES

If ψ is the exact wave function of an excited state, ϕ its, in general, multi-determinantal "model" function (e.g. $\phi_{\text{RHF}}, \phi_{\text{GRHF}}$), and E the exact energy, the relations

$$\psi = \phi + \chi; \quad \langle\phi|\chi\rangle = 0; \quad \langle\phi|\phi\rangle = 1 \tag{63a}$$

$$\begin{aligned} E &= \langle\phi|\text{H}|\psi\rangle = \langle\phi|\text{H}|\phi\rangle + \langle\phi|\text{H}|\chi\rangle \\ &\equiv E_\phi + E_{\text{corr}}^\phi \end{aligned} \tag{63b}$$

still hold.

In Section 5, we gave the non-closed shell many-electron theory (NCMET) wave functions.

$$\phi = \phi_{\text{GRHF}} \approx \phi_{\text{RHF}} + \chi_{\text{int}} \tag{25'}$$

$$\psi \cong \phi + \chi_d' \tag{29'}$$

The χ_d' may be summarized as:

$$\chi_d' = \chi_f + \chi_u \tag{64}$$

χ_f contains only the f terms, the "orbital-average polarizations" $\hat{f}_{k_a}^K$ and the semi-internal correlations" $\hat{f}_{k_ak_b;l}^K$. The χ_u contains the $\hat{u}_{k_ak_b}^K$'s. It is the "dynamical" pair correlations part.

The variational energy expression[20b] resulting from Eqs. (29) and (32) is complicated. A simpler form is obtained if $\chi \approx \chi_d'$ is substituted in Eq. (63b). The result is no longer an upper bound to the exact E (as in the second-order perturbation energy form,[20a] $E_2 = \langle\phi|H|\chi_1\rangle$). These non-variational forms are convenient in sorting the different correlation effects in the energy.

From Eqs. (63b) and (64),

$$E \approx E_{\text{GRHF}} + E_f + E_u \tag{65}$$

or

$$E \approx E_{\text{RHF}} + E_{\text{int}} + E_f + E_u \tag{66}$$

Writing Eqs. (31) and (64) as

$$\chi_d' = \sum_{K \geqslant 1}^{\binom{M}{N}} C_K(\chi_K^f + \chi_K^u) \tag{67}$$

we have

$$E_f = \sum_{K,L} C_K^* C_L \langle \Delta_K | H | \chi_L^f \rangle \tag{68}$$

$$E_u = \sum_{K,L} C_K^* C_L \langle \Delta_K | H | \chi_L^u \rangle \tag{69}$$

Since $\hat{u}_{ij}^K$ in Eq. (32) are "dynamical" (hence not affected much by "environment"), it is reasonable to expect the $\hat{u}_{ij}^K$ to be quite independent of K, or to use some average $\hat{u}_{ij}$'s (anonymous parentage).[20b] Then:

$$E_u \cong \sum_{i>j}^{M} \rho_{ij} \varepsilon_{ij} + \sum_{\{i>j\} \neq \{k>l\}}^{M} \rho_{ijkl} \varepsilon_{ij;kl} \tag{70}$$

The

$$\rho_{ij} \equiv \sum_{\substack{K \\ i,j \ni K}} |C_K|^2 \tag{71a}$$

and

$$\rho_{ijkl} \equiv \sum_{\substack{K \neq L \\ i,j \ni K \\ k,l \ni L}} (-1)^{\sigma} C_K^* C_L \tag{71b}$$

sums over determinants containing the ij or kl spin-orbitals are occupation probabilities (e.g. of pair ij in ϕ). The σ is the number of interchanges needed to bring the $(N - 2)$ spin-orbitals in L into correspondence with the same ones in K. K and L differ in at most two spin-orbitals. Thus, Eq. (70) is a weighted sum of pair and "cross-pair" ($\varepsilon_{ij;kl}$) correlations (the last includes terms like $ij;il$).

$$\varepsilon_{ij} = \langle \mathscr{B}(ij) | g_{12} | \hat{u}_{ij} \rangle \tag{72}$$

$$\varepsilon_{ij;kl} = \langle \mathscr{B}(ij) | g_{12} | \hat{u}_{kl} \rangle$$

The off-diagonal "cross pairs" occur also in closed shells if, for example, an LO-pair $\varepsilon_{\rho\nu}$ is expressed in terms of the symmetry spin-orbitals i,j (cf. Eqs. (49) and (51)). There they cancel out in the total E_{corr}, Eq. (45), but remain even in the total E_{corr},

Eqs. (70) and (65) of non-closed shells. Their contribution can be comparable to that of ε_{ij}'s. Fortunately most of them are evaluated easily by transforming the $\{\mathscr{B}(ij)\}$ pair space basis into the irreducible pair basis, $\{\psi_I^{\text{irr}}\}$ as in Eq. (46).

ψ_I^{irr} is characterized by the quantum numbers of the invariance group $\mathscr{G}$. In atoms, $I = (LSM_LM_S, l_1s_1l_2s_2)$. Since $[g_{12}, \mathscr{G}] = 0$, in the $\{\psi_I^{\text{irr}}\}$ basis, cross terms vanish if $I \neq I'$.* Thus substituting the inverses of Eqs. (46) into Eqs. (70–72), we have, for example,

$$\varepsilon_{ij} = \sum_{I,I'} T_{ij;I} T^*_{ij;I'} \varepsilon^{\text{irr}}_{II'} \tag{73}$$

and if each I is distinct in its pair quantum numbers,

$$E_u = \sum_I \beta_I \varepsilon_I^{\text{irr}} \qquad \text{(all } I \neq I' \text{ only)} \tag{74a}$$

$$\beta_I = \sum_{i>j}^{M} \rho_{ij} |T_{ij;I}|^2 + \sum_{\substack{i>j \\ \{i,j\} \neq \{k,l\}}}^{M} \sum_{k>l}^{M} \rho_{ijkl} T_{ij;I} T^*_{kl;I} \tag{74b}$$

From Eq. (74a), it is straightforward to evaluate the $\varepsilon_I^{\text{irr}}$ semi-empirically if "experimental" E_u values are known for several states and ions of a system. The "experimental" E_u may be defined as:

$$\boxed{E_u^{(\text{``exp''})} \equiv E_{\text{exp}} - E_{\text{rel}} - E_{\text{GRHF}} - E_f} \tag{75a}$$

or as

$$\boxed{E_u^{(\text{``exp''})} \cong E_{\text{exp}} - E_{\text{rel}} - E_{\text{RHF}} - E_{\text{int}} - E_f} \tag{75b}$$

where E_{exp} is the experimental total energy, E_{rel} is equivalent to the relativistic correction to E. (Alternatively we shall refer to E_u as "all-external" correlation energy.)

These are the true "dynamical" correlation energies. For ground-state atoms where E_f may be lumped together with E_u, they were evaluated in References 17 and 18 (cf. Section 8) in combination with E_f.

* In $1s^22s^n2p^m$ atoms all irreducible cross terms vanish except $\varepsilon(2s^2; {}^1S|2p^2; {}^1S)$.

In more general excited states, E_f may vary considerably with state and therefore it may not be practical to lump parts of E_f with parts of E_u to get even approximately transferable pair correlations. Thus in general it is necessary to evaluate the E_f as well as the E_{int} (or E_{GRHF}) in Eq. (75) separately, to obtain the E_u which may then behave in a more additive and transferable way.*

The E_f, as well as E_{int}, may be evaluated by itself, again by the generalized variation-perturbation (or "varied-portions") approach[5] of MET. Equations (25′), (29′), and (64) substituted in the variational expression yield, $\tilde{E}$, an upper bound to E, but containing also cross terms between, χ_f and χ_u, χ_{int} and χ_f, etc. Parts of this $\tilde{E}$, however, may be varied to yield $\tilde{\chi}_f$, $\tilde{\chi}_u$, . . . separately. We have for each,

$$E - E_{\text{RHF}} < E_Q \leqslant \tilde{E}_Q \equiv \frac{2\langle\phi_{\text{RHF}}|H - E_{\text{RHF}}|\tilde{\chi}_Q\rangle + \langle\tilde{\chi}_Q|H - E_{\text{RHF}}|\tilde{\chi}_Q\rangle}{1 + \langle\tilde{\chi}_Q|\tilde{\chi}_Q\rangle} \tag{76}$$

where $Q =$ int. or f or u. If the E_{int} and E_f are evaluated by finite, limited CI, each Eq. (76) becomes a matrix diagonalization problem. (The minimum principle $E_Q \leqslant \tilde{E}_Q$, is of course only for the lowest state of each symmetry unless a form of the "variation-perturbation method for excited states"[38] is applied.) One has

$$E < E_Q + E_{\text{RHF}} \leqslant \tilde{E}_Q + E_{\text{RHF}} \quad \text{for } Q = \begin{cases} \text{int.} \\ f \\ u \end{cases} \tag{77a}$$

but $E_{\text{RHF}} + \sum\limits_Q \tilde{E}_Q$ is not an upper bound to E, only

$$(\langle\tilde{\psi}|H|\tilde{\psi}\rangle/\langle\tilde{\psi}|\tilde{\psi}\rangle) > E$$

with

$$\tilde{\psi} = \phi_{\text{RHF}} + \tilde{\chi}_{\text{int.}} + \tilde{\chi}_f + \tilde{\chi}_u$$

* It may be noted, however, that the magnitude and importance of $\hat{U}_{ijk}$, etc., higher correlations in excited states, are much less certain than in ground states. Tests of such quantities in excited states, for example by "partial CI" procedures (Reference 4; see also Reference 20*b* for NCMET), would be significant. The "triple virtual excitations" for example in such a test would refer to "excitations" from the excited state ϕ_{RHF}, rather than from the corresponding ground-state determinant ϕ_0 (cf. Reference 15*a*).

The theory outlined above, both MET and NCMET, may be used for purely non-empirical calculations (as in Reference 16) of both subsystem correlations and of total $\tilde{\psi}$ and $\tilde{E}$. In the next few sections, however, the emphasis will be on evaluation of correlation energies by a combination of non-empirical and semi-empirical methods. The non-dynamical effects E_{int} and E_f are calculated, the remaining dynamical ("all-external") correlations are transferred from ions to atoms to molecules.

8. APPLICATION: PAIR CORRELATIONS IN GROUND STATES OF FIRST-ROW ATOMS

A. The $2s^2$ Correlation and its Strong Decrease with $N = Z$ in Neutral Atoms

The $\varepsilon(1s^2)$, $\varepsilon(1s2s)$, $\varepsilon(2p^2)$ pair correlations in ground states looked at in terms of MET are mainly dynamical and quite transferable. The $\varepsilon(2s^2)$ is mainly non-dynamical. It has been found by Linderberg and Shull,[39a] Layzer,[39b] Watson,[39c] and others that the E_{corr} $(1s^22s^2)$ of the Be-like sequence due to a "near-degeneracy" with $(1s^22p^2)$ increases linearly in Z,

$$E_{\text{corr}}(1s^22s^2) \cong a + bZ \tag{77b}$$

A proportionality to Z is general along isoelectronic sequences of non-closed shell ground states $(1s^22s^22p^m)$. Along neutral $(N = Z)$ sequences $1s^22s^22p^n(n + 4 = N = Z)$, however, another term in E_{corr} is predicted[4,5] by MET due to the "exclusion effect":

$$\langle \hat{u}_{2s^2} | 2p \rangle = 0 \tag{78}$$

As $2p$ electrons are added the non-dynamical $\varepsilon(2s^2)$ decreases due to Eq. (78). The decrease is almost linear, and one gets

$$E_{\text{corr}}(1s^22s^2(2p\alpha)^{n_\alpha}(2p\beta)^{n_\beta};\ \text{ground states};\ Z = N = 4 + n_\alpha + n_\beta)$$
$$\cong a' + b'Z - c'n_\alpha \tag{79}$$

where a', b', $c' < 0$ constants and $n_\alpha \geqslant n_\beta$. One has also:

$$\varepsilon(2s^2)_{\text{non-dyn}} \cong E_{\text{GRHF}} - E_{\text{RHF}} \tag{80}$$

The $\varepsilon(2s^2)_{\text{non-dyn}}$ has been calculated[17] by CI, mixing $1s^22s^22p^m$ with $1s^2sp^{m+2}$ and with Roothaan HF orbitals. After the $2p$ shell has been half filled, i.e. from $N(1s^22s^22p^3;\ ^4S)$ to $\text{Ne}(1s^22s^22p^6;\ ^1S)$,

Eq. (78) excludes any virtual $2p^2$ with $2s^2$ mixing so that $\varepsilon(2s^2)_{\text{non-dyn}}$, e.g. in Ne, is approximately zero[4] (Table 1). The dynamical part of $\varepsilon(2s^2)$ in

$$\varepsilon(2s^2) \cong \varepsilon(2s^2)_{\text{non-dyn}} + \varepsilon(2s^2)_{\text{dyn}}$$

TABLE 1. Theoretical versus empirical $2s^2$ inner shell pair correlations $\varepsilon(2s^2)$ in neutral, ground-state atoms. Due to the "exclusion effect" of added $2p\alpha$ electrons the $|\varepsilon(2s^2)|$ decreases and remains very small in spite of residual dynamical correlations.

Neutral, ground-state atom ($1s^22s^22p^n$; $Z = N$)	Be	B	C	N	O	F	Ne
$-\varepsilon(2s^2)$; theoretical[a] (MET) (in eV)	1.13	0.81	0.46	~0 (0.17)[c]	~0	~0	(0.27)[c]
$-\varepsilon(2s^2)$; empirical[b] (in eV)	1.03	1.40	1.74	2.15	2.42	2.83	3.16

[a] McKoy, V., and Sinanoğlu, O., *J. Chem. Phys.*, **41**, 2689 (1964); Sinanoğlu, O., *Proc. Natl. Acad. Sci. U.S.*, **47**, 1217 (1961).

[b] Allen, L. C., Clementi, E., and Gladney, H. M., *Rev. Mod. Phys.*, **35**, 465 (1963).

[c] Dynamical correlation calculated using an r_{12} coordinate method in the variational pair equation.[a] The non-dynamical ("near-degeneracy") part is zero.

was also evaluated,[17] using an Hylleraas-type trial function $\psi_{34} = \mathscr{B}(2s^2)(1 + \bar{\gamma}r_{34})$ following the non-empirical MET procedure. The results indicate that Eq. (79) remains valid in spite of some dynamical $\varepsilon(2s^2)$ correlation which makes, for example, the $\varepsilon(2s^2)$ in Ne not exactly zero, but -0.27 eV.

These results are compared in Table 1 with those of Allen, Clementi, and Gladney[37] who have applied Eq. (35) in an empirical way and assumed the Eq. (77) and the $\varepsilon(2s^2)$ values in $1s^22s^2$ ions to be unchanged when $2p$ electrons are added.

In the total E_{corr} $(N = Z)$ of these atoms the assumption of unchanged $1s^22s^2$ cores also leads to very small values of $\varepsilon(2s\bar{2}p)$ or total dynamical correlations.

B. The Large Intershell Correlations

After the $\varepsilon(1s^2 2s^2)$ appropriate to neutral atoms are subtracted out from E_{corr} (cf. Eq. (79)) one finds the intershell s–p correlation $\varepsilon(2s\bar{2}p)$ to be large,[17] $\varepsilon(2s\bar{2}p) + \varepsilon(\bar{2}s\bar{2}p) \approx -1.13$ eV, and the $\varepsilon(2p^2)$ correlation -1.0 ± 0.3 eV. The $2p^n$ and $(2p^n \leftarrow 1s^2 2s^2)$ intershell correlation values obtained in Reference 17 are summarized in Table 2.

TABLE 2. Dynamical correlations in the L shell of $(1s^2 2s^2 2p^n)$ ground states. Once the non-transferable near-degeneracy effects (Table 1) have been removed from the correlation energy, these quite transferable dynamical correlations E_{corr} (L shell; dyn.) remain. $E_{\text{corr}}(\text{PPS}) \equiv \varepsilon(2p^n) + \varepsilon(2p^n \rightarrow 1s^2 2s^2)$ (in eV). The $\varepsilon(2p^n) \equiv$ correlation energy of n $2p$ electrons with each other; $\varepsilon(2p^n \rightarrow 1s^2 2s^2) \equiv$ correlation energy of n $2p$ electrons with the $1s^2 2s^2$ core.

n	Configuration	Z	$-E_{\text{corr}}(\text{PPS})^a$
1	$1s^2 2s^2 2p\ (^2P)$	5	1.14 eV
		6	1.23
		7	1.33
		8	1.41
		9	1.49
		10	1.60
2	$1s^2 2s^2 2p^2\ (^3P)$	5	—
		6	2.32
		7	2.48
		8	2.59
		9	2.74
		10	2.90
3	$1s^2 2s^2 2p^3\ (^4S)$	5	—
		6	—
		7	3.45
		8	3.53
		9	3.63
		10	3.75

[a] McKoy, V., and Sinanoğlu, O., *J. Chem. Phys.*, **41**, 2689 (1964).

These decreasing near-degeneracy effects have very recently been calculated also by Clementi and Veillard[40] who obtained the ϕ_{GRHF} (MCSCF) rather than $\phi_{\text{RHF}} + \chi_{\text{CI}}$ (of Eq. (25)). Their results are in agreement with those of Reference 17 to within a few hundredths of an eV indicating again the smallness of the $\hat{f}_i$ terms which makes ϕ_{GRHF} differ from $\phi_{\text{RHF}} + \chi_{\text{CI}}(\text{int.})$ very little.

Recent non-empirical calculations by Kelly[36] on oxygen (^3P) and Nesbet[33] on Ne(^1S) are in agreement with the small $\varepsilon(2s^2)$, and the large intershell values.

Similar conclusions appear to hold for $\varepsilon(3s^2)$ and $\varepsilon(3s\bar{3}p)$ in second-row atoms[18] where the $3s$–$3p$–$3d$ near-degeneracy effects play a role. Large "inter-shell" correlations between orbitals with the same radial quantum number but different l are a general phenomenon and are expected to be significant in the transition metals and the rare earths.

9. APPLICATION: THE SEMI-INTERNAL, PAIR, AND OTHER CORRELATION EFFECTS IN THE EXCITED STATES OF FIRST-ROW ATOMS

In excited states such as $1s^22s^22p^2(^1S)$, $1s^22s2p^2(^2D)$, the $\hat{u}_{ij}$ correlations of NCMET are orthogonal to all the GRHF "sea" spin-orbitals, m (cf. Section 5). The $\varepsilon(2s^2)$ in E_{int} exhibits a more complicated dependence on state than Eq. (79).

The "internal" and semi-internal correlations, E_{int} and E_f, have recently been calculated[41] for $1s^22s^n2p^m$ states by CI procedures (cf. Section 7). Some of these results and others on correlation effects on excited states are summarized below.

A finite CI is carried out using automatic programs. These CI include both the E_{int} and the E_f effects together. Separate and smaller CI's for E_{int} alone and for E_f alone were also made (Eq. (77)). The E_{int} and E_f *components in the larger* CI are very close to the separate E_{int} and E_f values. Tables 3 and 4 give the components from the larger CI.

The main contribution to the E_{int} comes from the mixing of $1s^22s^n2p^m \leftrightarrow 1s^22s^{n-2}2p^{m+2}$ using Eq. (76) and with the virtual orbitals the same as the occupied ones in their radial parts. Since differences from a two-configuration SCF calculation, where all the orbitals are reoptimized, and very small, this simpler procedure is preferred. The E_{int} for some states of C, N, and O are shown in Table 3.

TABLE 3. "Internal" correlation energy ($E^{\text{int}}_{\text{RHF}} \approx \varepsilon(2s^2)_{\text{non-dyn}}$) in some states of carbon, nitrogen, and oxygen[a] (in eV).

N	Multiplet	Carbon	Nitrogen	Oxygen
5	$(1s^2 2s^2 2p;\ {}^2P)$	(C^+) −1.10	(N^{2+}) −1.32	(O^{3+}) −1.52
6	$(1s^2 2s^2 2p^2;\ {}^3P)$	(C) −0.47	(N^+) −0.58	(O^{2+}) −0.68
	$(1s^2 2s^2 2p^2;\ {}^1D)$	−0.44	−0.56	−0.67
	$(1s^2 2s^2 2p^2;\ {}^1S)$	−1.63	−2.15	−2.57
7	$(1s^2 2s^2 2p^3;\ {}^2P)$	(C^-) −0.62	(N) −0.95	(O^+) −1.17
8	$(1s^2 2s^2 2p^4;\ {}^1S)$	—	(N^-) −1.10	(O) −1.50

[a] From the calculations of Reference 41. The values are the E_{int} components of a larger CI that yields $(E_{\text{int}} + E_f)$ together.

The E_f, i.e. the combination of the semi-internal correlations (due to $\hat{f}_{k_a k_b;l}$ in Eq. (32) and the orbital-average polarization terms (from $\hat{f}_{k_a}$)), are also obtained by a finite CI calculation and use of Eq. (76) as well as a component of a larger $(E_{\text{int}} + E_f)$ CI. The E_f from the larger CI are shown in Table 4.[41] The contribution of the $\hat{f}_{k_a}$ terms, i.e. $\hat{f}_{2s}$ polarization here, to the total E_f is small, ca. −0.02 eV. Thus the values in Table 4, though total E_f, give an indication of the semi-internal correlation effects.

In the lowest configurations, $1s^2 2s^2 2p^n$, the semi-internal correlations contribute mainly to the overall $(2s2p)$ correlations, being of the form

$$\mathscr{B}(2s\bar{2}p) \rightarrow \mathscr{B}(2p'\hat{f}_{2s\bar{2}p;2p'}) \tag{81}$$

In excited configurations like $1s^2 2s 2p^m$, semi-internal correlations contribute also to $\varepsilon(p^2)$ correlations, with terms of the type,

$$\mathscr{B}(2p^2) \rightarrow \mathscr{B}(2s\hat{f}_{2p^2;2s}) \tag{82}$$

The semi-internal effects are subject to "exclusion effects", the way the internal or non-dynamical effects were. For example in Ne, $1s^2 2s^2 2p^6$, the RHF "sea" is full and therefore virtual transitions

TABLE 4. The semi-internal and orbital polarization contribution, E_f, to correlation energy[a] (in eV).

Configuration	Multiplet	Carbon	Nitrogen	Oxygen
$1s^22s2p$	^3P	—[b]	−0.149	−0.155
	^1P	—	−1.12	−1.14
$1s^22p^2$	^3P	—	0	0
	^1D	—	−0.93	−0.88
	^1S	—	−0.21	−0.21
$1s^22s^22p$	^2P	−0.66	−0.71	−0.75
$1s^22s2p^2$	^4P	−0.14	−0.15	−0.15
	^2D	−1.12	−1.13	−1.10
	^2S	−0.32	−0.18	−0.14
	^2P	−1.51	−1.60	−1.64
$1s^22s^22p^2$	^3P	−1.01	−1.13	−1.17
	^1D	(−1.03)	(−1.15)	(−1.23)
	^1S	−0.14	−0.07	−0.05
$1s^22s2p^3$	^5S	0	0	0
$1s^22s^22p^3$	^4S	−1.02	−1.26	−1.37
	^2D	−1.07	−1.29	−1.43
	^2P	−0.54	−0.59	−0.64
$1s^22s^22p^4$	^3P	—	−0.93	−1.04
	^1D	—	(−0.82)	(−1.09)
	^1S	—	−0.10	−0.04
$1s^22s^22p^5$	^2P	—	—	−0.55

[a] Based on calculations of Reference 41. Values are E_f components of larger $(E_{\text{int}} + E_f)$ CI together.

[b] Dashes indicate no calculation.

TABLE 5. Dynamical, "all-external" parts[a] of pair correlations[b] according to "non-closed shell many-electron theory"[c] (in eV).

ε_{ij}(dyn.; NCMET)[a]	Carbon	Nitrogen	Oxygen
$\varepsilon(1s^2)$	−1.24	−1.27	−1.26
$\varepsilon(1s^2 \to 2s)$	−0.09	−0.07	−0.08
$\varepsilon(1s^2 \to 2p)$	−0.10	−0.10	−0.10
$\varepsilon(2s^2 \to 2p)^a$	−0.41	−0.43	−0.45
$\varepsilon(2p_0^2)$	−0.76	−0.78	−0.74
$\varepsilon(2p_+\bar{2}\bar{p}_+)$	−0.59	−0.59	−0.56
$\varepsilon(\bar{2}\bar{p}_+\bar{2}\bar{p}_0)$	−0.28	−0.24	−0.22
$\varepsilon(2p_+\bar{2}\bar{p}_0)$	−0.44	−0.42	−0.39
$\varepsilon(2p_+\bar{2}\bar{p}_-)$	−0.68	−0.65	−0.57
$\varepsilon(2p_+\bar{2}\bar{p}_0;\ 2p_0\bar{2}\bar{p}_+)$	−0.15	−0.17	−0.17
$\varepsilon(2p_+\bar{2}\bar{p}_-;\ 2p_-\bar{2}\bar{p}_+)$	−0.33	−0.36	−0.35
$\varepsilon(2p^2;\ ^3P)$	−0.28	−0.24	−0.22
$\varepsilon(2p^2;\ ^1D)$	−0.59	−0.59	−0.56
$\varepsilon(2p^2;\ ^1S)$	−1.11	−1.17	−1.11

[a] The E_u part in Eq. (66). To get the single determinantal state total pair correlations of MET[17], the semi-internal contribution is added where it exists to the values of this table. (cf. Text and Table 6).

[b] Averages over various states and ions.

[c] Based on Reference 41.

to $2s$, $2p$ orbitals in terms like Eqs. (81) and (82) are not possible. Thus $E_{\text{semi-int}} = 0$ as well as E_f. (The orbital-symmetry polarization too is zero due to spherical symmetry and the dynamical $\hat{f}_i$ are small.)[12]

Once the RHF,[42,37a] the E_{int},[41], and the E_f[41] are known, the dynamical ("all-external") E_u is obtained from the total correlation energies according to Eq. (75b). These NCMET dynamical parts are broken down into "reducible" or "irreducible" pair (and "cross-pair") energies. This procedure was followed previously in the context of the single determinantal MET using only the ground states of first-row atoms and ions. As discussed in Section 8, this data was sufficient to yield the $\varepsilon(2s^2)$, the $\varepsilon(2s^2 \rightarrow 2p)$, but only the average p^2 correlations.[17] The NCMET uses many more states including highly excited configurations $1s^2 2s^n 2p^m (0 \leqslant n \leqslant 2; 0 \leqslant m \leqslant 6)$ and treats them in more detail. Thus it can account for the variation of the $\varepsilon(2s^2 2p)$ values observed with Z and N,:[28] and also yields the various p^2 correlations, $\varepsilon(2p_x 2p_y)$, $\varepsilon(2p_x \bar{2}p_y)$, $\varepsilon(2p_z^2)$, and the cross terms separately. Table 5 summarizes some values of such dynamical, "all-external", terms.[41] The Table shows the pairs both in the "$\mathscr{B}(ij)$", i.e. "reducible pair" description ($\varepsilon(2p_+ \bar{2}p_-)$, cross terms, etc.) and in the "irreducible pairs" description ($\varepsilon(2p^2;\ LS)$, etc.).

The pair correlations in Table 5 as obtained from the NCMET, are found to be quite similar to the overall values of the single det, ground-state MET,[17] when combined with the semi-internal parts (Eqs. (81) and (82)). Table 6 compares the s–p intershell correlations as obtained by single det MET[17] and by NCMET.[41] The MET intershell values shown are obtained from the total $\varepsilon(2p^n) + \varepsilon(1s^2 2s^2 \rightarrow 2p^n)$, given for the high spin states,[17] by subtracting the now known $\varepsilon(\bar{2}p\,\bar{2}p)$ same spin correlations.

The results above indicate that *the correlation energy of a large number of atomic states may be expressed in terms of a few dynamical, "all-external" correlation parameters, once the Hartree-Fock and the internal and semi-internal effects have been calculated.* These values have been used to calculate the E_{corr}, electron affinity, negative ion excited state energy, and other properties of additional systems.[41]

10. APPLICATION: CORRELATION BETWEEN TETRAHEDRALLY LOCALIZED ORBITALS

The results above indicated that the $\varepsilon(2s\,\bar{2}p)$ correlations turn out to be very significant even though they are of an "inter-orbital" type. In the "separated pair" or "geminal" type theories[34] one assumes that "inter-orbital" effects will be small,[35] especially if one starts with "localized orbitals" (LO's). In the perturbation theory of electron correlation[10,11] or in MET[1], on the other hand, $N(N-1)/2$ "pair correlations", of which many more than $(N/2)$

TABLE 6. Comparison of overall $(s\text{–}p)$ intershell correlation, as obtained by single det. MET[a] and NCMET as the sum of semi-internal and all-external correlation effects[b] (in eV).

Multiplet	Carbon		Nitrogen		Oxygen	
	$\varepsilon^{s-p}_{\text{MET}}$ [a]	$\varepsilon^{s-p}_{\text{NCMET}}$ [b]	$\varepsilon^{s-p}_{\text{MET}}$ [a]	$\varepsilon^{s-p}_{\text{NCMET}}$ [b]	$\varepsilon^{s-p}_{\text{MET}}$ [a]	$\varepsilon^{s-p}_{\text{NCMET}}$ [b]
$(1s^22s^22p;\ ^2P)$	−1.23	−1.17	−1.33	−1.23	−1.41	−1.28
$(1s^22s^22p^2;\ ^3P)$	−0.98	−1.01	−1.10	−1.08	−1.17	−1.12
$(1s^22s^22p^3;\ ^4S)$	—	−0.85	−0.96	−0.93	−1.01	−1.01

[a] Table 3 in Reference 17 gives $E'_{sp} \equiv \varepsilon(sp^n_\alpha\alpha) + \varepsilon(1s^22s^2 \to 2p^n_\alpha\alpha)$ for these states. The values listed here are: $\varepsilon^{s-p}_{\text{MET}} \equiv [(E'_{sp}/n_\alpha) - (m\varepsilon(2p^2;\ ^3P)/n_\alpha)]$ where m = number of $(2p^2;\ ^3P)$ pairs contained in the multiplet. The $\varepsilon(2p^2;\ ^3P)$ values are from Reference 41.

[b] NCMET values of Reference 41 according to: $\varepsilon^{s-p}_{\text{NCMET}} \equiv \varepsilon(1s^2 \to 2p) + \varepsilon(2s^2 \to 2p)_{\text{all-ext}} + \left(\dfrac{\text{semi-int}}{n_\alpha}\right)$.

"geminal" ones may be significant, are indicated. Below we evaluate the $\varepsilon_{\eta_\rho\eta_\tau}$, localized orbital correlations between and within tetrahedral (sp^3) hybrids in Ne. The inter-orbital contribution to the total L-shell correlation is found to be at least as large as the "geminal" part. This system also forms a model for (CH_4) and indicates that the E_{corr} of (CH_4) is not made up of $\varepsilon(1s^2)$ and four bond correlations alone.*

Choosing the Z axis along one of tetrahedral orbitals, the "equivalent orbital" LO transformation is

$$\begin{pmatrix} \eta_1 \\ \eta_3 \\ \eta_5 \\ \eta_7 \end{pmatrix} = \frac{1}{2} \begin{pmatrix} 1 & \sqrt{3} & 0 & 0 \\ 1 & -\dfrac{1}{\sqrt{3}} & 2\sqrt{\dfrac{2}{3}} & 0 \\ 1 & -\dfrac{1}{\sqrt{3}} & -\sqrt{\dfrac{2}{3}} & \sqrt{2} \\ 1 & -\dfrac{1}{\sqrt{3}} & -\sqrt{\dfrac{2}{3}} & -\sqrt{2} \end{pmatrix} \begin{pmatrix} 2s\alpha \\ 2p_z\alpha \\ 2p_x\alpha \\ 2p_y\alpha \end{pmatrix} \tag{83}$$

for the α-spin set with a similar equation for the β-spin set, $\eta_2, \eta_4, \eta_6, \eta_8$.

The pair-functions $\hat{u}_{ij}$ are then transformed by the direct product matrix $\mathbf{T} = \mathbf{t} \times \mathbf{t}$ with the $\mathbf{t}$ in Eq. (83) (cf. Eq. (49b)). Each LO-pair correlation $\varepsilon_{\eta_i\eta_j}$ may then be expressed in terms of the "reducible" pairs and the "cross pairs" in the s, p_x, p_y, p_z description (Eq. (51), and similar to Eq. (73)). We have

$$\varepsilon_{\eta_1\eta_2} = \frac{1}{16}[\varepsilon(2s^2) + 6\varepsilon(2s\bar{2}p_z) + 9\varepsilon(2p_z^2) + 6\varepsilon(2s^2; 2p_z^2) + 6\varepsilon(2s\bar{2}p_z; 2p_z\bar{2}s)] \tag{84a}$$

$$\varepsilon_{\eta_1\eta_4} = \frac{1}{16}[\varepsilon(2s^2) + 6\varepsilon(2s\bar{2}p_z) + \varepsilon(2p_z^2) + 8\varepsilon(2p_z\bar{2}p_x) - 2\varepsilon(2s^2; 2p_z^2) - 2\varepsilon(2s\bar{2}p_z; 2p_z\bar{2}s)] \tag{84b}$$

$$\varepsilon_{\eta_1\eta_3} = \frac{1}{2}[\varepsilon(\bar{2}s\bar{2}p_z) + \varepsilon(\bar{2}p_z\bar{2}p_x)] \tag{84c}$$

* Only bond correlations are assumed also in some semiempirical work, for example References 43*a* and 43*b*.

These are the three types of pairs, $\varepsilon_{\eta_1\eta_2}$, a doubly occupied LO, $\varepsilon_{\eta_1\eta_4}$, two different LO's with opposite spins and $\varepsilon_{\eta_1\eta_3}$ with the same spin. The other $\varepsilon_{\eta_i\eta_j}$ have different expressions, but the same values as one of the above. In Eqs. (84), the notation means, for example,

$$\varepsilon(2s^2; 2p_z^2) \equiv \langle \mathscr{B}(2s^2)|g_{12}|\hat{u}_{2p_z^2}\rangle \tag{85}$$

and, for example, $\bar{2}s \equiv 2s\beta$. Also, one has

$$\varepsilon(2s^2; 2p_z^2) \simeq \varepsilon(2p_z^2; 2s^2) \tag{86}$$

i.e. symmetric cross terms have been assumed. This symmetry can be shown to be exactly valid in the second-order energy, i.e. with $\langle \mathscr{B}(ij)|g_{ij}|\hat{u}_{ij}^{(1)}\rangle$, but may be only approximately so, in higher orders.

All the terms above, except $\varepsilon(2s^2; 2p^2)$ are evaluated by group theory in terms of "irreducible pair correlations".

There are twelve inter-orbital terms of the type $\varepsilon_{\eta_1\eta_4}$ and four $\varepsilon_{\eta_1\eta_2}$'s. Thus the ratio of the inter-orbital part of the L-shell correlation, $E_{\text{corr}}^{\text{inter}}$, to the "geminal" part, $E_{\text{corr}}^{\text{intra}}$, is given by

$$\text{Rt} \equiv \frac{E_{\text{corr}}^{\text{inter}}}{E_{\text{corr}}^{\text{intra}}} = \frac{3(\varepsilon_{\eta_1\eta_4} + \varepsilon_{\eta_1\eta_3})}{\varepsilon_{\eta_1\eta_2}} \tag{87}$$

Then, the total L-shell E_{corr} (L shell) of Ne from Eq. (60) is

$$E_{\text{corr}}(L\ \text{shell}) = 4\varepsilon_{\eta_1\eta_2}(1 + \text{Rt}) \tag{88a}$$

The individual $\varepsilon_{\eta_i\eta_j}$ have been evaluated using pair correlations[41] in Eqs. (84) appropriate to Ne and similar to those in Table 5. The results are shown in Table 7, for two choices of $\varepsilon(2s^2; 2p^2)$.

If we neglect the small dynamical part of $\hat{u}(2s^2)$ in Ne, and also use Eq. (86), we have

$$\varepsilon(2p^2; 2s^2) \simeq \varepsilon(2p^2; 2s^2)_{\text{non-dyn}} = 0 \tag{89a}$$

since $\hat{u}(2s^2)_{\text{non-dyn}}$ is zero. On the other hand, an overestimate of $\varepsilon(2p^2; 2s^2)$ is obtained using two Cauchy–Schwartz inequalities and yields roughly

$$|\varepsilon(2s^2; 2p_z^2)| \approx 0.37\ \text{eV} \tag{89b}$$

Table 7 indicates that the $\varepsilon_{\eta_i\eta_j}$ values are quite insensitive to the value of $\varepsilon(2s^2; 2p_z^2)$, Eqs. (89); so is the conclusion

$$\text{Rt} > 1 \tag{88b}$$

*Thus inter-*LO *correlations in Ne, and very likely in CH_4 and other saturated molecules, contribute to the E_{corr} as much as or greater than the contribution of the "geminal", i.e. the "bond" correlations part alone.*

TABLE 7. Correlations between and within tetrahedral orbitals in the L shell of Ne[a] (in eV).

	A^b	B^b
$\varepsilon_{\eta_1\eta_2}$	−0.67	−0.78
$\varepsilon_{\eta_1\eta_4}$	−0.35	−0.31
$\varepsilon_{\eta_1\eta_3}$	−0.15	−0.15
$\sum_{i>j}^{N} \varepsilon_{\eta_i\eta_j} = E_{\text{corr}}^{(\text{Ne; L shell})}$	−8.64	−8.64
$\text{Rt} \equiv \dfrac{E_{\text{corr}}^{\text{inter}}}{E_{\text{corr}}^{\text{intra}}}$	2.2	1.8

[a] Using the pair correlation values for Ne of Reference 41 in Eqs. (84).

[b] Calculation A uses Eq. (89a); B uses Eq. (89b) in text.

11. APPLICATION: CORRELATION BINDING IN MOLECULES

It is well known that a large fraction of the binding energy of a molecule relative to ground-state atoms arises from the correlation energy difference. In the formation of a molecule (ABC . . .) from ground-state atoms A, B, C, . . ., i.e.

$$\text{A} + \text{B} + \text{C} + \ldots \rightarrow (\text{ABC} \ldots)$$

we have for the binding energy (B.E.)

$$\text{B.E.} = \Delta E = \Delta E_{\text{HF}} + \Delta E_{\text{corr}} \tag{90}$$

where

$$\Delta E_{\text{HF}} = \text{Hartree–Fock binding}$$

$$\Delta E_{\text{corr}} = \text{correlation binding}$$

For species in their ground states, the single determinant RHF is sufficient for each E_{HF} and the single determinant MET for the E_{corr}'s. As in atoms, it is also possible to extend the Hartree–Fock, to GRHF, or some version of multi-configurational HF, and the MET into NCMET based on some version of GRHF.

In this section, we shall outline two semiempirical methods,[28] both of which use the atomic pair correlation data similar to those of previous sections and give estimates of E_{corr} and ΔE_{corr} of a molecule. We consider first the following non-dynamical correlation effect which makes the $(2s^2)$ "inner shells" contribute indirectly to binding:

A. "Exclusion Binding"

In the ground state, the $2s^2$-"core" correlation of nitrogen atoms is zero (Section 8.A). Consider now what happens when an N_2 molecule forms. The $2p\alpha$ electrons are paired up forming the "triple bond" in the N_2 $(^1\Sigma_g^+)$. One result of this decrease in the $2p\alpha$ population on each atom is a partial elimination of the "exclusion effect", $\langle \hat{u}_{2s^2} | 2p\alpha \rangle = 0$. Thus the distorted $2s^2$ "cores" in N_2 have greater correlation energy magnitude than the free atom cores. The net result is a contribution to binding ("exclusion binding").[28]

This effect may be calculated by Hartree–Fock (RHF) plus CI or by GRHF (or a partial multi-configurational MCSCF). But it may also be estimated in a simple way, albeit roughly, from Eq. (79) and gross populations in the molecule. We have, for example,

$$\underbrace{\mathrm{N}}_{1s^2\underbrace{2s^2}_{2s'_a{}^2}2p^3(\uparrow\uparrow\uparrow)} + \mathrm{N} \rightarrow \underbrace{\mathrm{N}_2}_{1\sigma_g{}^2 1\sigma_u{}^2 \underbrace{2\sigma_g{}^2 2\sigma_u{}^2}_{``2s_a'^2 2s_b'^2"} 3\sigma_g{}^2 1\pi^4} \tag{91}$$

$$\mathrm{N}_2 \approx (1s^{2.0}2s^{1.8}2p^{3.2}_{\sigma,\pi})_2 \tag{92}$$

where the gross populations shown are based on a minimal basis LCAO MO SCF. Since the state is $^1\Sigma_g^+$, the $(2p\alpha)^3$ (i.e. $n_\alpha = 3$) of the atom has been replaced by

$$2p^3(\uparrow\ \uparrow\ \uparrow) \rightarrow (2p\alpha)^{1.6}(2p\beta)^{1.6} \tag{93}$$

Using Eq. (79) to interpolate for fractional $n_\alpha = 1.6$ to get the molecular "$(2s^2)$" correlation, and neglecting the effect of the distortion of the 2s orbitals, one obtains

$$\Delta E_{\text{exclusion}} \equiv \varepsilon(2\sigma_g^2 2\sigma_u^2) - 2\varepsilon(2s^2; \text{N}; {}^4S) \approx -1.51 \text{ eV} \quad (94)$$

a sizable contribution to the B.E., though a "core" effect.

The ΔE_{excl} of some homonuclear diatomics obtained by this semiempirical procedure are shown in Table 8. That the effect is

TABLE 8. Contribution of $2s^2$—"core" correlation to binding in molecules through the "exclusion effect"—Exclusion binding $\equiv \Delta E_{\text{exclusion}} = \varepsilon(2\sigma_g^2 2\sigma_u^2) - 2\varepsilon(2s^2; \text{A})$[a] (in eV).

A_2	$-\Delta E_{\text{excl}}$	$-\Delta E_{\text{corr}}^{(\text{pair pop})}$ [b]	$D_e^{(\text{exp})}$ [c]
Li_2			1.05
B_2	0.96	1.18	2.9
C_2	1.21	3.01	6.4
N_2	1.51	4.06	9.9
O_2	1.26	1.69	5.2
F_2	0.65	1.82	1.7

[a] Calculated semiempirically from Z and number of $2p\alpha$ electrons dependence and population analysis as described in text.

[b] Calculated by the approximate "pair populations" method (cf. text).

[c] $D_e = -$Binding energy. From Gaydon, A. G., *Dissociation Energies*, Chapman-Hall Ltd., London, 1953, (corrected for zero point energy) and other sources (cf. p. 17 of Reference 28).

appreciable, and the trend for different molecules is of interest, rather than the actual numbers which may be too approximate.

B. Two Semiempirical Methods for Estimating the Correlation Energy of Molecules

Both methods are developed and applied to a large number of molecules in Reference 28. Here they are summarized briefly.

(*a*) *Pair population method.* The gross populations as in Eq. (92) may be converted into approximate pair populations. In a singlet ground state, half of each gross population gives roughly the number of α–β opposite spin electrons occupying the same AO. The number of inter-orbital correlations of a given type, e.g. $2s$–$2p$, is obtained using the product of the random spin populations of each orbital. In N_2, Eq. (92), the average $2s$–$2p$ random spin correlation becomes

$$\varepsilon_{\text{corr}}^{(2s-2p)}(\text{N}_2) \cong 1.8 \times 3.2 \times \frac{\varepsilon[2s^2 \rightarrow (2p\alpha)^3;\ \text{N}(^4S)]}{2 \times 3} \tag{95}$$

For $E_{\text{corr}}(\text{N}_2)$ we have:

$$\begin{aligned} E_{\text{corr}}(\text{N}_2) \cong 2 \Bigg[& \frac{2.0}{2.0} \times \varepsilon(1s^2) + \frac{1.8}{2} \times \varepsilon(2s^2;\ \text{for } n_\alpha = 3.2/2) \\ & + \frac{3.2}{2}\,\varepsilon(2p^2) + \frac{2.0 \times 1.8}{2 \times 2} \times \varepsilon(1s^2 \rightarrow 2s^2) \\ & + \frac{2.0 \times 3.2}{2} \times \varepsilon(1s^2 \rightarrow 2p) \\ & + \frac{1.8 \times 3.2}{6} \times \varepsilon(2s^2 \rightarrow (2p\alpha)^3) \Bigg] \end{aligned} \tag{96}$$

The values thus obtained and using the average MET atomic ε_{ij}'s of Section 8 are shown in Table 9 for first-row homonuclear diatomics, and in Table 10 for hydrides AH (for larger molecules cf. Reference 28).

This simplified method has recently been extended and formalized and in its general form should yield more reliable values on quite large molecules.

(*b*) *"Shrunk-core" method.* For most molecules the ΔE_{HF} and $\Delta E_{\text{corr}}^{(\text{"exp"})}$ are still unknown. Therefore, especially in semi-empirical work, it is useful to have two or more rather different ways of estimating the same ΔE_{corr}. If the various methods agree, the values obtained may be fairly realistic.

For small interatomic separations, the correlation energy of a diatomic molecule appears quite insensitive to distance. A limiting case which also incorporates by its nature the important exclusion effects is provided by the "shrunk core" model for small molecules. For molecules like C_2, NO, HCN, CF_4, etc. the united atom is not useful for estimates of either MO-energy levels or for correlation. Because the $1s^2$ cores when forced together change character altogether, the united AO levels follow a different sequence even qualitatively than the MO levels. A closer analogue is the shrunk core atom where the core of the fictitious, composite atom that results is radially similar to the molecular one, in regard

TABLE 9. Comparison of correlation estimates obtained by the "pair population" method ($\Delta E_{\text{corr}}^{(\text{pair pop})}$, by the "shrunk core" method ($\Delta E_{\text{corr}}^{(\text{shr core})}$) and the $\Delta E_{\text{corr}}^{\text{"exp"}} \equiv \Delta E^{\text{exp}} - \Delta E_{\text{HF}}$; (assuming $\Delta E_{\text{rel}} \approx 0$) (in eV).

A_2	$-\Delta E_{\text{corr}}^{(\text{shr core})}$	$-\Delta E_{\text{corr}}^{(\text{pair pop})}$	$-\Delta E_{\text{corr}}^{\text{"exp"}\,b}$
$Li_2(^1\Sigma_g^+)$	—[a]	—	0.885
$B_2(^3\Sigma_g^-)$	1.42	1.18	$2.01 \pm 0.24_{\text{exp}}$ [c]
$C_2(^1\Sigma_g^+)$	3.31	3.01	$5.58 \pm 0.22_{\text{exp}}$
$N_2(^1\Sigma_g^+)$	3.20	4.06	4.63
$O_2(^3\Sigma_g^-)$	1.39	1.69	3.9
$F_2(^1\Sigma_g^+)$	2.01	1.82	3.05

[a] The shrunk core method is not properly applicable to Li_2 since binding arises from 2*s* electrons whose correlation is very sensitive to Z; hence it is omitted.

[b] ($\Delta E^{\text{"exp"}} - \Delta E_{\text{HF}}$). The ΔE_{HF} calculations are from the Chicago group, Cade, P. E., Huo, W., Roothaan, C. C. J., and coworkers (see p. 57, Reference 28 and later references).

[c] The error estimates shown are on the E_{exp} only and do not include any on the ΔE_{HF}.

to its nodal behaviour. Consider, for example, the $C_2(^1\Sigma_g^+)$. The shrunk core replaces the $(1\sigma_g^2 1\sigma_u^2)$ with two $C^{4+}(1s^2)$ cores. The shrunk valence electrons, on the other hand, move in the field of a single $1s^2$ core, but of roughly twice the effective charge. Thus the valence shells of C_2 are shrunk into the L shell of Ne. We have

$$C_2(^1\Sigma_g^+;\ 1\sigma_g^2 1\sigma_u^2 2\sigma_g^2 2\sigma_u^2 1\pi_u^4) \tag{97a}$$
$$\approx K_a(C^{4+};\ 1s_a^2) K_b(C^{4+};\ 1s_b^2) \text{Ne}(2s^2 2p^6;\ ^1S)$$

and

$$E_{corr}(C_2) \approx 2E_{corr}(C^{4+}) + E_{corr}(Ne) - E_{corr}(Ne^{8+}) \qquad (97b)$$

It may be noted that, the $2\sigma_g^2 2\sigma_u^2$ have also been shrunk bringing in at least partly the exclusion effect.

The shrunk core results are compared with the previous method for A_2 in Table 9. The "exp." ΔE_{corr} values (B.E. $- \Delta E_{HF}$; assuming $\Delta E_{rel} \approx 0$) are also shown. The methods agree fairly

TABLE 10. Comparison of correlation binding estimates obtained by the pair population method ($\Delta E_{corr}^{(pair\ pop)}$), by the shrunk core method, ($\Delta E_{corr}^{(shr\ core)}$) (for hydrides reduces to united atom, see text) and the $\Delta E_{corr}^{\text{"exp"}}$. (All in eV.)

AH	$\Delta E_{corr}^{(shr\ core)}$	$\Delta E_{corr}^{(pair\ pop)}$	$\Delta E_{corr}^{\text{"exp"}}$ [b]
$LiH(^1\Sigma^+)$	1.88[a]	0.93	1.03
$BH(^1\Sigma^+)$	1.46	0.80	0.80
$CH(^2\pi)$	1.68	1.21	1.13
$NH(^3\Sigma^-)$	1.90	1.35	1.70(?)
$OH(^2\pi)$	1.79	1.45	1.60
$HF(^1\Sigma^+)$	1.87	1.55	1.74

[a] See footnote *a*, Table 9.

[b] Cade, P. E., and Huo, W., Laboratory for Molecular Structure and Spectra, University of Chicago, report, 1967.

well with each other, but both differ considerably from $\Delta E_{corr}^{\text{"exp"}}$ for C_2, O_2, and F_2. The anomalous ΔE_{corr} of F_2 has also been noted in non-empirical (MCSCF) calculations.[44] The C_2 case is particularly puzzling and makes one wonder if the HF limit has been reached in some of these molecules.

For hydrides AH, the shrunk core model reduces to the united atom model, which was first used by Stanton[45] for estimating the ΔE_{corr} of several diatomic hydrides. In Table 10 recent shrunk core (or here united atom) values[28] on (AH) are compared with ΔE_{corr} (pair pop.) and $\Delta E_{corr}^{\text{"exp"}}$. Here all the values agree quite well, but ΔE_{corr} (pair pop.) on the whole appears better than the united atom case.*

The shrunk core method can be applied to a few small polyatomics; the pair population method is more general and has been used on, for example, benzene, borazine, and other molecules.[28] Reliable estimate of E_{corr} on molecules where HF is not yet available, provides also "experimental Hartree-Fock values" ($E_{\text{HF}}^{\text{“exp”}} \cong E^{\text{exp}} - \text{E}_{\text{corr}}^{(\text{theo})} - E_{\text{rel}}$ (atoms)) which may be useful as lower bounds to be approached in HF calculations ($E_{\text{HF}}^{\text{“exp”}} \leqslant E_{\text{HF}}^{(\text{calc})}$ if $\Delta E_{\text{rel}} \approx 0$).

12. CONCLUSION

This article has presented very recent results on: (*a*) correlation effects in excited states, the form of the E_{corr} in such non-closed shell systems; (*b*) the numerical values of three types of correlation effect in atoms of configurations $1s^22s^n2p^m$ $(0 \leqslant n \leqslant 2; 0 \leqslant m \leqslant 6)$; (*c*) correlations between localized orbitals, in particular the tetrahedral ones; and (*d*) the form of the N-density (matrix) operator in terms of $1, 2, 3, \ldots, n(n \leqslant N)$-electron "correlation densities" defined in Section 3. Some of the earlier results, primarily on ground state (and single determinantal) atoms and molecules, are also summarized including the simplification of an N-body correlation into $N(N-1)/2$ pair correlations, the introduction of three types of pair correlations defined first by perturbation theory,[10,11] later by a more general variational theory,[1] the numerical values of near-degeneracy effects and pair correlations in ground-state atoms, and the application of these atomic values in estimating the correlation energy and correlation binding of some molecules. A number of avenues such as the non-empirical application of the theory to molecules to simplify configuration-interaction (CI) or other ways of calculation which constitute another aspect of the MET have been left outside the scope of this article.

Some of the results discussed above indicate that:

[1]. *Correlations between electrons occupying different spatial orbitals are often just as important as correlations in doubly occupied*

* These values on hydrides have recently been compared and discussed also by Cade and Huo.[46]

orbitals. The effect first observed in the large $\varepsilon(2s\overline{2p})$ type correlations in atoms[17,18] is a quite general one including the correlations between the different molecular orbitals of a molecule. A rather drastic example of the latter is found in

$$He_2(\varepsilon(1\sigma_g\overline{1\sigma}_u) \approx \varepsilon(1\sigma_g^2))\ ^{47}$$

[2]. Going over to the localized orbital (LO) description does not change the situation of the above paragraph appreciably. *The correlations between tetrahedral hybrids are found in Section* 10 *to contribute to* E_{corr} *more than the "lone pair" (doubly occupied* LO*) correlations*. The effect is expected to contribute appreciably to the binding energies and heats of formation of saturated molecules.

The importance of inter-orbital and inter-bond correlations is in contradiction with the notion of significant correlations in only Lewis–Langmuir type electron pairs and the "geminal" or "separated-pair" type methods incorporating only these.

[3]. *The correlation energy of some "inner shells" like* $K2s^2$, $KL3s^2$, *sigma cores in conjugated molecules, . . . may change by as much as several electron volts depending on the number and state of the outer electrons*. Thus, not all "inner shell" correlations cancel as implied in valency or in molecular-orbital theory; they contribute to electron affinity, ionization potential, molecular binding energy, and excitation energy.

Correlation effects in sizable molecules, especially in their excited states, in transient species during chemical reaction or dissociation, have hardly been investigated. Calculations of molecular pair correlations, by MO "partial pair CI" 's, or possibly by the r_{12}-coordinate methods which require further development, may shed light on the validity and accuracy of semiempirical results. In this context a closer comparison of the notions and results based on the "many electron theory of atoms and molecules" above and those using multi-configurational SCF and other forms of partial CI may be needed. The relative importance of three or more electron correlations in excited states are difficult to ascertain and require extensive computations.

It is hoped that more use of abstract group theory in connection with theories of correlation will also add to an understanding of these phenomena.

Acknowledgements

Support of this work by the U.S. National Science Foundation and by the Alfred P. Sloan Foundation is gratefully acknowledged. It is a pleasure to thank Bolesh Skutnik and Önder Pamuk for many helpful discussions and Iskender Oksüz, Tai-ichi Shibuya, and Dr. Paul Westhaus for critical readings of the manuscript and checking of results.

References

1. Sinanoğlu, O., *Proc. Natl. Acad. Sci. U.S.A.*, **47**, 1217 (1961).
2. Sinanoğlu, O., *Rev. Mod. Phys.*, **35**, 517 (1963).
3. Löwdin, P.-O., *Advances in Chemical Physics, Vol.* 2, Interscience Publishers, New York, 1959), p. 207.
4. Sinanoğlu, O., *J. Chem. Phys.*, **36**, 706 (1962).
5. Sinanoğlu, O., *Advances in Chemical Physics, Vol.* 6, Interscience Publishers, New York, 1964, pp. 315ff.
6. Shibuya, T., and Sinanoğlu, O., *J. Math. Phys.*, to be published.
7. McWeeny, R., *Rev. Mod. Phys.*, **32**, 335 (1960).

8*a*. Coleman, A. J., *Rev. Mod. Phys.*, **35**, 668 (1963).

8*b*. Ando, T., *Rev. Mod. Phys.*, **35**, 690 (1963).

9. Primas, H., in *Modern Quantum Chemistry, Vol. II, Interactions*, Ed. O. Sinanoğlu, Academic Press, New York, 1965.
10. Sinanoğlu, O., *Proc. Roy. Soc.* (*London*), *Ser. A*, **260**, 379 (1961).
11. Sinanoğlu, O., *J. Chem. Phys.*, **33**, 1212 (1960).
12. Sinanoğlu, O., and Tuan, D. F., *J. Chem. Phys.*, **38**, 1740 (1963).
13. Nazaroff, G. V., and Hirschfelder, J. O., *J. Chem. Phys.*, **39**, 715, 3115 (1963).
14. Nesbet, R. K., *Advances in Chemical Physics, Vol. 9*, Interscience Publishers, New York, 1965, pp. 321ff.

15*a*. Koutecký, J., Čižek, J., Dubský, J., and Hlavatý, K., *Theoret. Chim. Acta*, **2**, 462 (1964).

15*b*. Čižek, J., and Paldus, J., to be published.

16. Tuan, D. F., and Sinanoğlu, O., *J. Chem. Phys.*, **41**, 2677 (1964).
17. McKoy, V., and Sinanoğlu, O., *J. Chem. Phys.*, **41**, 2689 (1964).
18. McKoy, V., and Sinanoğlu, O., in *Modern Quantum Chemistry, Istanbul Lectures,* Ed. O. Sinanoğlu, Vol. II, Academic Press, New York, 1965, p. 23.
19. Das, G., and Wahl, A. C., *J. Chem. Phys.*, **44**, 87 (1966).

20*a*. Silverstone, H. J., and Sinanoğlu, O., *J. Chem. Phys.*, **44**, 1899 (1966).

20*b*. Silverstone, H. J., and Sinanoğlu, O., *J. Chem. Phys.*, **44**, 3608 (1966).

21. Roothaan, C. C., *Rev. Mod. Phys.*, **32**, 179 (1960).

22*a*. Silverstone, H. J., and Sinanoğlu, O., *J. Chem. Phys.*, **46,** 854 (1967).
22*b*. Geller, M., Taylor, H. S., and Levine, H. B., *J. Chem. Phys.*, **43,** 1727 (1965); McKoy, V., *J. Chem. Phys.*, **43,** 1605 (1965).
23. Sinanoğlu, O., *J. Chem. Phys.*, **36,** 3198 (1962).
24. Steiner, E., *J. Chem. Phys.*, **45,** 328; **46,** 1727 (1967).
25. Paulus, K. F. G., *J. Chem. Phys.*, **46,** 3078 (1967).
26. Čižek, J., *J. Chem. Phys.*, **45,** 4256 (1966).
27. Kestner, N. R., and Sinanoğlu, O., *J. Chem. Phys.*, **94,** 194 (1966).
28. Hollister, C., and Sinanoğlu, O., *J. Am. Chem. Soc.*, **88,** 13 (1966).
29. Sinanoğlu, O., and Orloff, M. K., in *Modern Quantum Chemistry, Istanbul Lectures*, Ed. O. Sinanoğlu, Vol. I, Academic Press, New York, 1965, p. 221; for other aspects cf. also Orloff, M. K., and Sinanoğlu, O., *J. Chem. Phys.*, **43,** 49 (1965).
30. Sinanoğlu, O., and Čižek, J., *Chem. Phys. Letters*, **1,** 337 (1967).
31*a*. Eden, R. J., *Nuclear Reactions, Vol.* 1, North Holland Publishing Co., Amsterdam, 1958.
31*b*. Rodberg, L. S., *Ann. Phys.*, **2,** 199 (1957).
31*c*. Bethe, H. A., and Goldstone, J., *Proc. Roy. Soc. (London), Ser. A*, **238,** 551 (1957).
32*a*. Kestner, N. R., and Lowrey, A., *Bull. Ann. Phys. Soc., Ser. II*, **12,** 337 (1967).
32*b*. Geller, M., Taylor, H. S., and Levine, H. B., *J. Chem. Phys.*, **43,** 1727 (1965).
33. Nesbet, R. K., *Phys. Rev.*, **155,** 56 (1967).
34. Hurley, A. C., Lennard-Jones, J. E., and Pople, J. A., *Proc. Roy. Soc. (London), Ser. A.*, **220,** 446 (1953); Parks, J. M., and Parr, R. G., *J. Chem. Phys.*, **28,** 335 (1958); McWeeny, R., *Rev. Mod. Phys.*, **32,** 335 (1960); Coleman, A. J., *Rev. Mod. Phys.*, **35,** 668 (1963); Allen, T. L., and Shull, H., *J. Chem. Phys.*, **35,** 1644 (1961); Edmiston, C., and Ruedenberg, K., *Rev. Mod. Phys.*, **35,** 457 (1963).
35. As remarked by Kapuy, E., *J. Chem. Phys.*, **44,** 956 (1966), and Edmiston, C., *J. Chem. Phys.*, **39,** 2394 (1963).
36. Kelly, H. P., this Book and *Phys. Rev.*, **131,** 684 (1963); **144,** 39 (1966).
37*a*. Clementi, E., *J. Chem. Phys.*, **38,** 2248 (1963).
37*b*. Allen, L. C., Clementi, E., and Gladney, H. M., *Rev. Mod. Phys.*, **35,** 465 (1963).
38. Sinanoğlu, O., *Phys. Rev.*, **122,** 491 (1961).
39*a*. Linderberg, J., and Shull, H., *J. Mol. Spectry.*, **5,** 1 (1960).
39*b*. Layzer, D., *Ann. Phys.*, **8,** 271 (1959).
39*c*. Watson, R. E., *Phys. Rev.*, **119,** 170 (1960).
40. Clementi, E., and Veillard, A., *J. Chem. Phys.*, **44,** 3050 (1966).
41. Sinanoğlu, O., Skutnik, B., and Öksüz, I., to be published.
42. Roothaan, C. C. J., and Kelly, P. S., *Phys. Rev.*, **131,** 1177 (1963).
43. For example (*a*) Clementi, E., *J. Chem. Phys.*, **39,** 487 (1963); (*b*) Allen, L. C., and coworkers' papers on "Pair Correlations in Molecules", *J. Chem. Phys.*, in press.
44. Gilbert, T., and Wahl, A. C., private communication.

45. Stanton, R. E., *J. Chem. Phys.*, **36**, 1298 (1962).
46. Cade, P. E., and Huo, W., Laboratory for Molecular Structure and Spectra, Univ. of Chicago report, 1967.
47. Sinanoğlu, O., "*Modern Quantum Chemistry*"—*Istanbul Lectures Vol. II- "Interactions"*, Academic Press, New York, 1965, p. 231.

SOME ASPECTS ON THE CORRELATION PROBLEM AND POSSIBLE EXTENSIONS OF THE INDEPENDENT-PARTICLE MODEL*

PER-OLOV LÖWDIN, *Quantum Theory Project, University of Florida, Gainesville, Florida and Department of Quantum Chemistry, University of Uppsala, Uppsala, Sweden*

CONTENTS

* This work was supported in part by the National Science Foundation and the National Aeronautics and Space Administration under research grants NSF-512 and GP-1695 with the University of Florida, and in part by the Office of Scientific Research, United States Air Force, through the European Office of Aerospace Research under a contract with Uppsala University.

1. CORRELATION PROBLEM: INTRODUCTION

In the independent-particle model, one assumes that each particle moves in an "average field" of all the other particles.[1] This means that one neglects the correlation between the motions of the individual particles involved, their "collisions", etc. Electrons repel each other due to the Coulomb potential e^2/r_{12}, and each electron is surrounded by a "Coulomb hole" with respect to the other electrons. For electrons with parallel spins, the "Fermi hole" associated with the Pauli exclusion principle takes partly the place of the Coulomb hole, and the essential part of the correlation error is hence associated with pairs of electrons having antiparallel spins.

One may get an idea of the order of magnitude of the correlation error in the independent-particle model by studying the *correlation energy*,[2] which is defined as the difference between the exact energy of the Hamiltonian $\mathscr{H}$ associated with the system and the energy of the conventional Hartree–Fock (HF) scheme:

$$E_{\text{corr}} = E_{\text{exact}} - E_{\text{HF}} \tag{1}$$

For the ground state $(1s)^2$ of the He-like ions, H^-, He, Li^+, . . ., the correlation energy is approximately constant -1.1eV, of which amount +1.1eV refers to the kinetic energy and −2.2eV to the potential energy in accordance with the virial theorem. Similar figures are valid also for the ground state of the hydrogen molecule, H_2. It should be observed, however, that the correlation energy per electron pair in general is not a constant, and that, for the ground state of atoms, it goes up approximately linearly with increasing atomic number.[3] For atoms and small molecules, the correlation energy seems to be approximately one per cent of the total energy.

It is clear that the correlation effects are of essential importance in any accurate treatment of a many-particle system. For the periodic table of the atoms, the independent-particle model works very well in explaining the shell structure of the electronic clouds, and one usually assumed that this was connected with the fact that the interaction between the electrons was comparatively weak. It was hence rather surprising that the "shell-model" turned out to work so well also for atomic nuclei, since the interactions between the nucleons are certainly very strong, and one had then reasons to reinvestigate the entire basis of the independent-particle model. In this paper, we will briefly review some of the recent attempts to extend and generalize the original Hartree–Fock scheme.

2. BRIEF SURVEY OF THE HARTREE–FOCK SCHEME

A. Slater Determinant

Since the correlation energy is essentially a measure of the error in the Hartree–Fock scheme, it may be convenient to start our study with a brief discussion of the underlying principles of this approximation. The total Hamiltonian for a system of N identical particles has the form

$$\mathscr{H} = \mathscr{H}_c + \sum_i \mathscr{H}_i + \sum_{i<j} \mathscr{H}_{ij} \tag{2}$$

where $\mathscr{H}_c$ is a "constant", $\mathscr{H}_i$ is a term referring to particle i, and $\mathscr{H}_{ij}$ describes the interaction between the particles i and j. The fundamental problem is to solve the many-particle Schrödinger equation

$$\mathscr{H}\Psi = E\Psi \tag{3}$$

in order to find the stationary states of the system.

In the classical approach to the independent-particle model outlined independently by Slater[4] and Fock,[5] one would utilize the *variation principle:*

$$\delta\langle\mathscr{H}\rangle = 0 \tag{4}$$

where

$$\langle\mathscr{H}\rangle = \frac{\langle\Phi|\mathscr{H}|\Phi\rangle}{\langle\Phi|\Phi\rangle} \tag{5}$$

is the expectation value of the Hamiltonian $\mathscr{H}$ with respect to the approximate wave function Φ. Starting out from a trial function in the form of a "Hartree-product":

$$\Phi = \psi_1(1)\psi_2(2) \cdot \cdot \cdot \psi_N(N) \tag{6}$$

Slater and Fock could now derive the equations previously intuitively proposed by Hartree.[1] By introducing the combined space–spin coordinate $x_i = (\mathbf{r}_i, \zeta_i)$ and the convention that, in forming the binary product, one should integrate over the space coordinate and sum over all spin coordinates, Slater could further simplify the theory. For fermions, the Pauli principle could now be automatically taken into account by considering wave functions *antisymmetric* in the combined coordinates $x_1, x_2, \ldots x_N$. In the independent-particle model, the Slater determinant

$$D = \frac{1}{\sqrt{N!}}\begin{vmatrix} \psi_1(x_1) & \psi_2(x_1) & \cdots & \psi_N(x_1) \\ \psi_1(x_2) & \psi_2(x_2) & \cdots & \psi_N(x_2) \\ \cdots & \cdots & \cdots & \cdots \\ \psi_1(x_N) & \psi_2(x_N) & \cdots & \psi_N(x_N) \end{vmatrix} = (N!)^{-\frac{1}{2}}|\psi_k(x_i)| \tag{7}$$

becomes now the fundamental trial function, and we note that, except for a normalization factor, the Slater determinant is essentially the *antisymmetric component* of the Hartree product:

$$D = (N!)^{+\frac{1}{2}}O_{\mathrm{AS}}\Phi \tag{8}$$

where O_{AS} is the antisymmetric projection operator:

$$O_{\mathrm{AS}} = (N!)^{-1}\sum_P (-1)^p P \tag{9}$$

associated with the symmetric group S_N formed by all permutations over the particle coordinates. This operator satisfies the relations

$$O_{AS}^2 = O_{AS},\ O_{AS}^\dagger = O_{AS},\ PO_{AS} = (-1)^p O_{AS} \tag{10}$$

The functions $\psi_k(x)$ are *one-particle functions* or spin-orbitals of the general form

$$\psi_k(x) = \psi_k^+(\mathbf{r})\alpha(\zeta) + \psi_k^-(\mathbf{r})\beta(\zeta) \tag{11}$$

so each spin-orbital may contain two space functions. These functions belong to the one-particle Hilbert space, and binary products in this space will, in the following, also be denoted by brackets of the type $\langle | \rangle$, if there is no risk of a misunderstanding.

B. Fock–Dirac Density Matrix

The one-particle functions $\{\psi_k\}$ may be arranged in a row matrix:

$$\mathbf{\Psi} = \{\psi_1, \psi_2, \ldots \psi_N\} \tag{12}$$

and they span a subspace of the Hilbert space which will be denoted by M_N. This subspace is characterized by a projection operator

$$\rho = |\mathbf{\Psi}\rangle\langle\mathbf{\Psi}|\mathbf{\Psi}\rangle^{-1}\langle\mathbf{\Psi}| \tag{13}$$

where $\mathbf{d} = \langle\mathbf{\Psi}|\mathbf{\Psi}\rangle$ is the metric matrix having the elements $d_{kl} = \langle\psi_k|\psi_l\rangle$. It satisfies the relations

$$\rho^2 = \rho, \qquad \rho^\dagger = \rho, \qquad \mathrm{Tr}(\rho) = N \tag{14}$$

If one has a non-singular linear transformation $\mathbf{\Psi}' = \mathbf{\Psi}a$ or

$$\psi_k' = \sum \psi_l a_{lk} \tag{15}$$

one obtains for the associated determinants

$$\begin{aligned} D' &= (N!)^{-\frac{1}{2}}|\psi_k'(x_i)| \\ &= (N!)^{-\frac{1}{2}}\Big|\sum_l \psi_l(x_i)a_{lk}\Big| \\ &= (N!)^{-\frac{1}{2}}|\psi_l(x_i)| \,.\, |a_{lk}| \\ &= D \,.\, |a_{lk}| \end{aligned} \tag{16}$$

i.e. except for a normalization factor, the total wave function stays the same. This means that a Slater determinant D is essentially characterized by the linear manifold $\mathscr{M}_N$ spanned by the set of one-electron functions $\mathbf{\Psi} = \{\psi_1, \psi_2, \ldots \psi_N\}$, i.e. by the projection operator ρ for this subspace. It is also easily shown[5] that the expectation values of arbitrary physical quantities with respect to D are expressible explicitly in terms of ρ.

The projection operator ρ has a kernel $\rho(x_1, x_2)$ which may be written in the form

$$\begin{aligned}\rho(x_1, x_2) &= |\mathbf{\Psi}(1)\rangle \mathbf{d}^{-1}\langle\mathbf{\Psi}(2)| \\ &= \sum_{k,l=1}^{N} \psi_k(x_1)(\mathbf{d}^{-1})_{kl}\psi_l^*(x_2) \end{aligned} \tag{17}$$

Slater has emphasized that, without loss of generality, one may choose the set $\mathbf{\Psi}$ orthonormal, so that

$$\mathbf{d} = \langle\mathbf{\Psi}|\mathbf{\Psi}\rangle = 1 \tag{18}$$

In this case, the kernel $\rho(x_1, x_2)$ takes the simplified form

$$\rho(x_1, x_2) = \sum_{k=1}^{N} \psi_k(x_1)\psi_k^*(x_2) \tag{19}$$

This is the well-known quantity introduced by Fock[6] and by Dirac,[7] and, in the following, it will be referred to as the Fock–Dirac density matrix. This quantity will be discussed in greater detail below.

C. Hartree–Fock Equations

In order to evaluate the expectation value (5) for the energy, it is convenient to assume that the basic set $\mathbf{\Psi}$ is orthonormal and satisfies (18). If F is an operator which is symmetric in all coordinates, so that $PF = FP$, one obtains according to (8) and (10):

$$\begin{aligned}\langle D|F|D\rangle &= (N!)\langle O_{\mathrm{AS}}\Phi|F|O_{\mathrm{AS}}\Phi\rangle \\ &= (N!)\langle\Phi|O_{\mathrm{AS}}^{\dagger}FO_{\mathrm{AS}}|\Phi\rangle \\ &= (N!)\langle\Phi|FO_{\mathrm{AS}}|\Phi\rangle \\ &= \sum_P (-1)^p\langle\Phi|F|P\Phi\rangle \end{aligned} \tag{20}$$

where Φ is the Hartree product (6). This gives directly $\langle D|D\rangle = 1$. Introducing the abbreviation

$$\overline{\mathscr{H}}_{12} = \mathscr{H}_{12}(1 - P_{12}) \tag{21}$$

where P_{12} is the exchange operator changing x_1 into x_2, and vice versa, one obtains further for the energy

$$\begin{aligned}\langle\mathscr{H}\rangle &= \langle D|\mathscr{H}|D\rangle \\ &= \mathscr{H}_c + \sum_i \langle\psi_i(1)|\mathscr{H}_1|\psi_i(1)\rangle \\ &\qquad + \sum_{i<j} \langle\psi_i(1)\psi_j(2)|\overline{\mathscr{H}}_{12}|\psi_i(1)\psi_j(2)\rangle \qquad (22) \\ &= \mathscr{H}_c + \sum_i \langle\psi_i(1)|\mathscr{H}_1|\psi_i(1)\rangle \\ &\qquad + \tfrac{1}{2}\sum_{i,j} \langle\psi_i(1)\psi_j(2)|\overline{\mathscr{H}}_{12}|\psi_i(1)\psi_j(2)\rangle \qquad (23)\end{aligned}$$

Using the Fock–Dirac density matrix (19), the last expression may be written in the form:

$$\begin{aligned}\langle\mathscr{H}\rangle = \mathscr{H}_c &+ \int dx_1 \mathscr{H}_1 \rho(x_1, x_1')_{/x_1' = x_1} \\ &+ \tfrac{1}{2}\int dx_1 dx_2 \overline{\mathscr{H}}_{12} \rho(x_1, x_1')\rho(x_2, x_2')_{/x_i' = x_i} \qquad (24)\end{aligned}$$

where one has also introduced the convention that the operators $\mathscr{H}_1$ and $\overline{\mathscr{H}}_{12}$ work on the unprimed coordinates x_1 and x_2 only, and that one puts $x_1' = x_1$ and $x_2' = x_2$ after the operations are performed but before the integrations are carried out. In applying the variation principle (4), it may be convenient to use the form (23). Varying the one-electron functions ψ_i and introducing the notation

$$\begin{aligned}H_{\text{eff}}(1) &= \mathscr{H}_1 + \sum_j \langle\psi_j(2)|\overline{\mathscr{H}}_{12}|\psi_j(2)\rangle \\ &= \mathscr{H}_1 + \int d\chi_2 \overline{\mathscr{H}}_{12}\rho(x_2, x_2') \qquad (25)\end{aligned}$$

one obtains

$$\begin{aligned}\delta\langle\mathscr{H}\rangle = \sum_i &\langle\delta\psi_i(1)|H_{\text{eff}}(1)|\psi_i(1)\rangle \\ &+ \sum_i \langle\psi_i(1)|H_{\text{eff}}(1)|\delta\psi_i(1)\rangle \qquad (26)\end{aligned}$$

It is easily shown that H_{eff} (1) is self-adjoint, $H_{\text{eff}}^\dagger = H_{\text{eff}}$, and hence one gets further

$$\delta\langle\mathscr{H}\rangle = 2\sum_i \text{Re}\langle\delta\psi_i(1)|H_{\text{eff}}(1)\psi_i(1)\rangle \tag{27}$$

The one-electron functions ψ_i are varied subject to the orthonormalization condition (18), i.e. $\langle\psi_i|\psi_j\rangle = \delta_{ij}$. Varying $\delta\psi_i$ in this condition, introducing the Lagrangian multipliers λ_{ji} and combining with (27), one obtains finally

$$H_{\text{eff}}(1)\psi_i(x_1) = \sum_j \psi_j(x_1)\lambda_{ji} \tag{28}$$

The Lagrangian multipliers form a matrix $\boldsymbol{\lambda} = \{\lambda_{ji}\}$ which may be considered as the matrix representation of the operator $H_{\text{eff}}(1)$ with respect to the basis $\boldsymbol{\Psi} = \{\psi_i\}$. Since H_{eff} is a self-adjoint operator, this implies further that $\boldsymbol{\lambda}$ is a self-adjoint matrix:

$$\boldsymbol{\lambda}^\dagger = \boldsymbol{\lambda} \tag{29}$$

i.e. $\lambda_{ij} = \lambda_{ji}^*$. There exists hence a unitary matrix $\mathbf{U}$ which brings $\boldsymbol{\lambda}$ to the diagonal form $\boldsymbol{\epsilon}$:

$$\mathbf{U}^\dagger\boldsymbol{\lambda}\mathbf{U} = \boldsymbol{\epsilon} \tag{30}$$

Introducing the linear transformation $\boldsymbol{\Psi}' = \boldsymbol{\Psi}\mathbf{U}$, one has that $\rho = |\boldsymbol{\Psi}\rangle\langle\boldsymbol{\Psi}|$ as well as H_{eff} (1) are left invariant. Writing (28) in matrix form

$$H_{\text{eff}}(1)\boldsymbol{\Psi}(1) = \boldsymbol{\Psi}(1)\boldsymbol{\lambda} \tag{31}$$

substituting $\boldsymbol{\Psi} = \boldsymbol{\Psi}'\mathbf{U}^\dagger$, multiplying to the right by $\mathbf{U}$, and using (30), one obtains

$$H_{\text{eff}}(1)\boldsymbol{\Psi}'(1) = \boldsymbol{\Psi}'(1)\boldsymbol{\epsilon} \tag{32}$$

or

$$H_{\text{eff}}(1)\psi_k'(x_1) = \epsilon_k\psi_k'(x_1) \tag{33}$$

where the diagonal element ϵ_k is referred to as the *one-electron energy* of the spin-orbital ψ_k'. The physical importance of these quantities is expressed by Koopmans's theorem,[8] which relates them to the ionization potentials of the system.

The Hartree–Fock equations are usually treated in the eigenvalue form (33), and the solutions are often referred to as "Hartree–Fock functions" in a restricted sense; they are characterized by the fact that they are *delocalized* over the entire system. It should be observed, however, that the total wave function D defined by the Slater determinant (7) depends only on the manifold $\mathscr{M}_N$ spanned by the functions $\boldsymbol{\Psi}' = \{\psi_k'\}$ and its projection operator ρ and not on the individual functions. The essential feature of the Hartree–Fock manifold $\mathscr{M}_N$ is that it should be *stable* with respect to the effective Hamiltonian H_{eff}. There are other solutions $\{\psi_k\}$ to the general equations (28) which may be valuable from other points of view, i.e. strongly *localized* one-particle functions in the form of equivalent orbitals[9] or Wannier functions.[10]

It should be observed that, if there are no further restrictions on the total wave function D, one can always diagonalize the matrix $\boldsymbol{\lambda}$ of the Lagrangian multipliers and bring the conditions corresponding to the variation principle (4) to the eigenvalue form (33). However, if there are certain constraints on D, this may not always be the case, and one obtains modifications of the scheme which will be further discussed below.

D. Hartree–Fock Potentials

The effective Hamiltonian (25) may be written in the form

$$H_{\text{eff}}(1) = \mathscr{H}_1 + u_1 \tag{34}$$

where the second term

$$u_1 = \int d\chi_2 \overline{\mathscr{H}}_{12}\rho(x_2, x_2') \tag{35}$$

is often referred to as the *Hartree–Fock potential* associated with the interaction term $\mathscr{H}_{12}$ in the original Hamiltonian. For the Coulomb repulsion $\mathscr{H}_{12} = e^2/r_{12}$, one obtains particularly

$$u_1 = e^2 \int d\chi_2 \left\{\frac{\rho(x_2, x_2)}{r_{12}} - \frac{\rho(x_1, x_2)}{r_{12}} P_{12}\right\} \tag{36}$$

where the first term is the ordinary average Coulomb potential and the second is the famous "exchange potential", which contains the permutation operator P_{12} and is of non-local nature.

In the original *self-consistent-field* (SCF) scheme,[1] Hartree started out from an estimate of the average field, solved the eigenvalue problem for the effective Hamiltonian, used the eigenfunctions to calculate the electron distribution and the associated average field. Starting out from the new average field, the process was repeated until it became self-consistent, i.e. until a repetition of the cycle did not lead to any change in the average field.

In modern SCF technique, one is usually less concerned about the "fields", and the cycle is conveniently described by the diagram

$$\rho \rightarrow H_{\text{eff}} \rightarrow \{\psi_k\} \rightarrow \rho \tag{37}$$

The calculations are repeated until the Fock–Dirac density matrix ρ becomes "self-consistent". The convergence problems associated with this loop are still open for discussion.

E. Brillouin Theorem

Using the Hartree–Fock potentials (35), one may rewrite the original many-body Hamiltonian (2) in the form:

$$\mathscr{H} = \mathscr{H}_0 + V \tag{38}$$

where

$$\begin{cases} \mathscr{H}_0 = \mathscr{H}_c + \sum_i (\mathscr{H}_i + u_i) & (39) \\ V = -\sum_i u_i + \sum_{i<j} \mathscr{H}_{ij} & (40) \end{cases}$$

This form is convenient for discussing the improvement of the Hartree–Fock scheme by means of perturbation theory. It is also the key for understanding some fundamental features of the Hartree–Fock scheme itself.

Let $\boldsymbol{\Psi} = \{\psi_1, \psi_2, \ldots \psi_N\}$ be the set of one-electron functions associated with the Slater determinant D, and let $\bar{\psi}_m$ be a function in the orthogonal complement, so that $\langle \bar{\psi}_m | \psi_k \rangle = 0$ for $k = 1, 2, \ldots N$. If one replaces a specific orbital ψ_k in D by $\bar{\psi}_m$, one obtains a new Slater determinant:

$$D_k^m = (N!)^{-\frac{1}{2}} \det\{\psi_1, \psi_2, \ldots \psi_{k-1}, \bar{\psi}_m, \psi_{k+1}, \ldots \psi_N\} \tag{41}$$

which is often said to be associated with a "singly excited configuration". Using (2), (25), and (33), one obtains

$$\begin{aligned} \langle D_k^m | \mathscr{H} | D \rangle &= \langle D_k^m | \sum_i \mathscr{H}_i + \sum_{i<j} \mathscr{H}_{ij} | D \rangle \\ &= \langle \bar{\psi}_m(1) | H_{\text{eff}}(1) | \psi_k(1) \rangle = \epsilon_k \langle \bar{\psi}_m | \psi_k \rangle = 0 \end{aligned} \tag{42}$$

i.e. the matrix elements of $\mathscr{H}$ between D and all the singly excited (s.e.) determinants are vanishing:

$$\langle D_{\text{s.e.}}|\mathscr{H}|D\rangle = 0 \tag{43}$$

This is the Brillouin theorem.[11] It implies that, if one tries to build up the exact solution Ψ to the eigenvalue problem (3) from the Slater determinant D and its orthogonal complement, there are no contributions from the singly excited configurations through direct coupling with respect to $\mathscr{H}$, even if there may be small contributions of this type through indirect coupling over doubly and more highly excited configurations.

Using (39) and (33), one finds further that the Slater determinant D is an eigenfunction to the unperturbed Hamiltonian $\mathscr{H}_0$:

$$\mathscr{H}_0 D = E_0 D \tag{44}$$

where

$$E_0 = \mathscr{H}_c + \sum_i \epsilon_i \tag{45}$$

Substituting $\mathscr{H} = \mathscr{H}_0 + V$ into (43) and using (44), one obtains

$$\langle D_{\text{s.e.}}|V|D\rangle = 0 \tag{46}$$

or

$$\langle D_{\text{s.e.}}|-\sum_i u_i + \sum_{i<j} \mathscr{H}_{ij}|D\rangle = 0 \tag{47}$$

The last form of the Brillouin theorem may also be used as an alternative definition of the Hartree–Fock potentials, and Professor Lefebvre at this summer school has shown in a fine series of papers that the entire Hartree–Fock scheme including the numerical applications may be built up on this basis.

F. Expansion Methods

In order to solve the eigenvalue problem (33) of the effective Hamiltonian, it may be convenient to introduce a basis $\boldsymbol{\phi} = \{\phi_\mu\}$ consisting of M linearly independent one-electron functions with $M \geq N$. In order to obtain exact solutions, the basis should be complete with $M = \infty$. The idea is to express the Hartree–

Fock functions in the set $\mathbf{\Psi} = \{\psi_1, \psi_2, \ldots \psi_N\}$ as linear expansions of the form

$$\psi_k = \sum_{\mu=1}^{M} \phi_\mu c_{\mu k} \tag{48}$$

or

$$\mathbf{\Psi} = \mathbf{\phi c} \tag{49}$$

where $\mathbf{c} = \{c_{\mu k}\}$ is a rectangular matrix of order $M \times N$. In many practical applications, it has turned out to be practical to use a non-orthonormal basis $\mathbf{\phi}$ having a metric matrix $\mathbf{\Delta}$ of the form:

$$\mathbf{\Delta} = \langle \mathbf{\phi} | \mathbf{\phi} \rangle = \mathbf{1} + \mathbf{S} \tag{50}$$

where the overlap matrix $\mathbf{S}$ is a measure of the non-orthonormality. Since the effective Hamiltonian is self-adjoint, the eigenfunctions to (33) are orthonormal, and one has the condition

$$\langle \mathbf{\Psi} | \mathbf{\Psi} \rangle = \mathbf{c}^\dagger \mathbf{\Delta} \mathbf{c} = \mathbf{1} \tag{51}$$

According to (19), the Fock–Dirac density matrix ρ may now be written in the form

$$\rho = |\mathbf{\Psi}\rangle\langle\mathbf{\Psi}| = |\mathbf{\phi}\rangle \mathbf{cc}^\dagger \langle\mathbf{\phi}| = |\mathbf{\phi}\rangle \mathbf{R} \langle\mathbf{\phi}| \tag{52}$$

where

$$\mathbf{R} = \mathbf{cc}^\dagger \tag{53}$$

The quantity ρ is essentially the projection operator for the Hartree–Fock manifold in the one-electron Hilbert space, and it is here expressed in the form

$$\rho(x_1, x_2) = \sum_{\mu, \nu=1}^{M} \phi_\mu(x_1) R_{\mu\nu} \phi_\nu^*(x_2) \tag{54}$$

where

$$R_{\mu\nu} = \sum_{k=1}^{N} c_{\mu k} c_{\nu k}^* \tag{55}$$

In the MO–LCAO theory of conjugated systems, Coulson and Longuet-Higgins[12] have called the quantities $R_{\mu\mu}$ and $R_{\mu\nu}$ charge-orders and bond-orders, respectively, and the general matrix $\mathbf{R} = \{R_{\mu\nu}\}$ is often referred to as the charge- and bond-order matrix. According to (14), the Fock–Dirac density matrix

ρ satisfies the relations $\rho^2 = \rho$, $\rho^\dagger = \rho$, $\mathrm{Tr}(\rho) = N$, and it is then easily shown[5] that $\mathbf{R}$ satisfies the corresponding relations

$$\mathbf{R\Delta R} = \mathbf{R}, \mathbf{R}^\dagger = \mathbf{R}, \mathrm{Tr}(\mathbf{\Delta R}) = N \tag{56}$$

In the same way as ρ, it may be considered as the fundamental quantity in the Hartree–Fock scheme.

If the effective Hamiltonian is considered as fixed, one may solve the eigenvalue problem (33) by means of the variation principle $\delta\langle H_{\text{eff}}\rangle = 0$, which leads to a system of linear equations

$$\sum_\nu (H_{\mu\nu} - \epsilon\Delta_{\mu\nu})c_\nu = 0 \tag{57}$$

where $H_{\mu\nu} = \langle\phi_\mu|H_{\text{eff}}|\phi_\nu\rangle$ are the matrix elements of the effective Hamiltonian with respect to the basis $\boldsymbol{\phi}$. The system (57) has a non-trivial solution only if the secular equation

$$|H_{\mu\nu} - \epsilon\Delta_{\mu\nu}| = 0 \tag{58}$$

is satisfied. In general, one has M solutions which are divided into two groups: the N "occupied" solutions associated with the N lowest one-electron energies ϵ_k, and the $(M - N)$ "virtual" solutions associated with the higher one-electron energies which lack physical reality, since they correspond only to the behaviour of "test charges" having no influence on the system itself.

Using (25), one obtains directly for the matrix elements of the effective Hamiltonian:[5]

$$\begin{aligned}
H_{\mu\nu} &= \langle\phi_\mu(1)|H_{\text{eff}}(1)|\phi_\nu(1)\rangle \\
&= \langle\phi_\mu(1)|\mathscr{H}_1 + \textstyle\int dx_2\mathscr{H}_{12}(1 - P_{12})\rho(2, 2')|\phi_\nu(1)\rangle \\
&= \langle\phi_\mu(1)|\mathscr{H}_1 + \textstyle\int dx_2\mathscr{H}_{12}[\rho(2, 2')\phi_\nu(1) - \rho(1, 2')\phi_\nu(2)\rangle \\
&= \langle\phi_\mu(1)|\mathscr{H}_1 + \textstyle\int dx_2\mathscr{H}_{12}[\boldsymbol{\phi}(2)\mathbf{R}\boldsymbol{\phi}^\dagger(2')\phi_\nu(1) \\
&\qquad\qquad -\boldsymbol{\phi}(1)\mathbf{R}\boldsymbol{\phi}^\dagger(2')\phi_\nu(2)]\rangle \\
&= \int\phi_\mu^*(1)\mathscr{H}_1\phi_\nu(1)dx_1 \\
&\qquad + \sum_{\kappa,\lambda} R_{\kappa\lambda}\iint dx_1dx_2[\phi_\mu^*(1)\mathscr{H}_{12}\phi_\lambda(2)\phi_\kappa^*(2')\phi_\nu(1) \\
&\qquad\qquad -\phi_\mu^*(1)\mathscr{H}_{12}\phi_\lambda(1)\phi_\kappa^*(2')\phi_\nu(2)]
\end{aligned} \tag{59}$$

In this expression, $\mathscr{H}_{12}$ is not supposed to work on the primed coordinate x_2, but, before the integration, one should put $x_2' = x_2$.

In the case of Coulomb repulsion, $\mathscr{H}_{12} = e^2/r_{12}$, one obtains particularly

$$H_{\mu\nu} = \int \phi_\mu^*(1)\mathscr{H}_1\phi_\nu(1)dx_1 + e^2 \sum_{\kappa,\lambda} R_{\lambda\kappa} \int\int \frac{\phi_\mu^*(1)\phi_\nu(1)\phi_\kappa^*(2)\phi_\lambda(2) - \phi_\mu^*(1)\phi_\lambda(1)\phi_\kappa^*(2)\phi_\nu(2)}{r_{12}} \tag{60}$$

This result shows that the matrix $\{H_{\mu\nu}\}$ is linear in the matrix $\{R_{\lambda\kappa}\}$, and this dependence of the effective Hamiltonian of $\mathbf{R}$ reflects the non-linear character of the entire Hartree–Fock scheme.

The expansion method was first introduced into the SCF scheme by Coulson[13] in connection with a study of the H_2 molecule, and it was later systematized by Roothaan[14] and by Hall.[15] Roothaan emphasized the importance of the coefficients $\{c_{\mu k}\}$ and considered the iterative cycle:

$$\{c_{\mu k}\} \rightarrow H_{\text{eff}} \rightarrow \{H_{\mu\nu}\} \rightarrow \{c_{\mu k}\} \tag{61}$$

Today many important calculations on atomic and molecular systems have been carried out in this so-called Roothaan scheme.

The author[5] has tried to focus interest on the charge- and bond-order matrix $\mathbf{R}$ as a convenient representative of the Fock–Dirac density matrix ρ and on the iterative cycle:

$$\mathbf{R} \rightarrow H_{\text{eff}} \rightarrow \{H_{\mu\nu}\} \rightarrow \{c_{\mu k}\} \rightarrow \mathbf{R} \tag{62}$$

in which the coefficients $\{c_{\mu k}\}$ are used only for the evaluation of $\mathbf{R}$. Somewhat later McWeeny[16] has stressed the existence of an even simpler iterative cycle:

$$\mathbf{R} \rightarrow H_{\text{eff}} \rightarrow \{H_{\mu\nu}\} \rightarrow \mathbf{R} \tag{63}$$

in which one avoids the calculation of the coefficients $\{c_{\mu k}\}$ entirely. However, since the solution of high-order secular equations of type (57) are standard on most electronic computers, it seems as if the cycle (62) would be a happy compromise between (61) and (63). So far, the Roothaan scheme has dominated the applications, but it seems to me as if a closer numerical study of the behaviour and general properties of **R** would also be of essential interest.

In many of the semiempirical schemes, as, for instance, the Hückel scheme for treating conjugated systems, the matrix elements of the effective Hamiltonian have been considered as adjustable parameters which have been fitted to experimentally known data. These schemes have been most successful within a limited field of applications, but it is also clear that one cannot expect good agreement between these semiempirical parameters and values of the matrix elements obtained by using formulas of the *ab initio* type (59). The reason is, of course, that the semiempirical parameters have been forced to adjust themselves to data far beyond the accuracy of the Hartree–Fock scheme itself. These questions will be further discussed below.

3. FUNDAMENTALS OF PERTURBATION THEORY: WAVE AND REACTION OPERATORS

A. Transition Formula for the Energy

Let us start by considering the Schrödinger equation

$$\mathscr{H}\Psi = E\Psi \tag{64}$$

for a discrete state or a state in the continuum. Forming the binary product with an arbitrary function Φ in the Hilbert space, one obtains $\langle\Phi|\mathscr{H}\Psi\rangle = \langle\Phi|E\Psi\rangle = E\langle\Phi|\Psi\rangle$, i.e.

$$E = \frac{\langle\Phi|\mathscr{H}\Psi\rangle}{\langle\Phi|\Psi\rangle} \tag{65}$$

As a generalization of this expression, we will now consider the functional

$$\langle\mathscr{H}\rangle_{12} = \frac{\langle\Phi_1|\mathscr{H}|\Phi_2\rangle}{\langle\Phi_1|\Phi_2\rangle} \tag{66}$$

for arbitrary wave functions Φ_1 and Φ_2. The functional exists, if both functions belong to the Hilbert space, and the value is then independent of the normalization of the individual functions. We note further that, if one is using the standard realization of the binary product, i.e.

$$\langle\Phi_1|\Phi_2\rangle = \int\Phi_1^*\Phi_2(dx) \tag{67}$$

the functional (66) exists even if only one of the functions belongs to the Hilbert space and the other represents a scattering state, whereas the functional (66) does not exist if both functions are associated with scattering states.

In the diagonal case, $\Phi_1 = \Phi_2 = \Phi$, the functional (66) goes over into the *expectation value*

$$\langle\mathscr{H}\rangle = \frac{\langle\Phi|\mathscr{H}|\Phi\rangle}{\langle\Phi|\Phi\rangle} \tag{68}$$

and we note that this value is always an *upper bound* to the lowest eigenvalue of $\mathscr{H}$. The expression (68) forms the basis for the famous *variation principle*, which says that the eigenvalue problem (64) for the discrete states is equivalent to the condition

$$\delta\langle\mathscr{H}\rangle = 0 \tag{69}$$

In the non-diagonal case, $\Phi_1 \neq \Phi_2$, the functional (66) is said to give a *transition value* of the Hamiltonian with respect to the two states involved, but it should be observed that this quantity—in contrast to the expectation value—does not have any immediate physical interpretation. For $\Phi_2 = \Psi$, one obtains the special case (65), i.e. the transition value equals the eigenvalue E independent of the choice of the other function Φ_1. For a pair of orthogonal functions, Φ_1 and Φ_2, the functional (66) blows up, and it is evident that it is completely unbounded. It should finally be observed that, even if the diagonal form (68) is always real, the non-diagonal form (66) is usually complex. All in all, one would expect that the form (66) has too general properties to be really useful, and it is hence rather surprising that it forms the basis for an important part of modern perturbation theory.

B. Wave and Reaction Operators

Let us assume that the Hamiltonian has the form

$$\mathscr{H} = \mathscr{H}_0 + V \tag{70}$$

where V is a perturbation which is not necessarily small, and that φ_0 is an eigenfunction to the unperturbed Hamiltonian associated with the eigenvalue E_0, so that

$$\mathscr{H}_0\varphi_0 = E_0\varphi_0 \tag{71}$$

Putting $\Phi = \varphi_0$ in transition formula (65), one obtains

$$E = \frac{\langle\varphi_0|\mathscr{H}|\Psi\rangle}{\langle\varphi_0|\Psi\rangle} = \frac{\langle\varphi_0|\mathscr{H}_0 + V|\Psi\rangle}{\langle\varphi_0|\Psi\rangle} = E_0 + \langle\varphi_0|V|\Psi\rangle/\langle\varphi_0|\Psi\rangle \tag{72}$$

This formula goes back to the starting days of perturbation theory, and it forms the basis for the modern development of this theory during the last two decades. In molecular theory, Parr[17] and his group have recently used this formula under the name "integrated Hellman–Feynman theorem" in conformational studies, noting that the energy shift in connection with a conformational change depends only on the first-order transition-density matrix associated with the two states involved.

For the following, it is convenient to introduce the so-called *intermediate normalization*:

$$\langle\varphi_0|\Psi\rangle = 1 \tag{73}$$

This is a mathematical relation without immediate physical interpretation, but it is still very useful. We note that, if the unperturbed level E_0 is discrete, the eigenfunction φ_0 belongs to the Hilbert space, and one may choose the normalization

$$\langle\varphi_0|\varphi_0\rangle = 1 \tag{74}$$

Since the scalar product $\langle\varphi_0|\Psi\rangle$ exists for eigenfunctions Ψ associated with both the discrete and continuous part of the eigenvalue spectrum, the intermediate normalization (73) makes it possible to treat closed states and scattering states on the same basis. For the closed states, it is later possible to consider the ordinary normalization integral $\langle\Psi|\Psi\rangle$ separately. Putting (73) into (72), one obtains the simplified formula

$$E = E_0 + \langle\varphi_0|V|\Psi\rangle \tag{75}$$

for the energy shift.

The *wave operator* W is any operator which maps the unperturbed eigenfunction φ_0 into the perturbed eigenfunction Ψ, so that

$$\Psi = W\varphi_0 \tag{76}$$

For the energy shift, one obtains $\langle\varphi_0|V|\Psi\rangle = \langle\varphi_0|VW|\varphi_0\rangle$, and it is then convenient to introduce the *reaction operator* through the relation

$$t = VW \tag{77}$$

which gives

$$E = E_0 + \langle\varphi_0|t|\varphi_0\rangle \tag{78}$$

The intermediate normalization (73) now takes the form $\langle\varphi_0|W|\varphi_0\rangle = 1$, and the expectation value of an arbitrary physical quantity F with respect to a closed state Ψ is given by the relation

$$\langle F\rangle_{\text{AV}} = \frac{\langle\Psi|F|\Psi\rangle}{\langle\Psi|\Psi\rangle} = \frac{\langle\varphi_0|W^\dagger FW|\varphi_0\rangle}{\langle\varphi_0|W^\dagger W|\varphi_0\rangle} \tag{79}$$

The wave operator formalism is hence a convenient tool for transforming expectation values over the perturbed state Ψ back to expectation values over the unperturbed state φ_0.

Before proceeding, it should be observed that there is a great deal of ambiguity left in this approach. An operator is generally a *mapping* which maps a "domain" of elements onto a "range" of other elements. Since it is here neither necessary nor convenient to restrict the domain of the wave operator W to a single element, φ_0, its domain may be a subspace D_W of the Hilbert space which contains this specific element. Since the definition (76) tells how the element φ_0 is going to be mapped, $\varphi_0 \to \Psi$, but there is still no description how all the other elements of the domain D_W are going to be mapped, the wave operator W is only partly defined. If there are two wave operators, W_1 and W_2, satisfying (76):

$$\Psi = W_1\varphi_0, \qquad \Psi = W_2\varphi_0 \tag{80}$$

one has

$$(W_2 - W_1)\varphi_0 = 0 \tag{81}$$

i.e.

$$W_2 - W_1 = A\{1 - |\varphi_0\rangle\langle\varphi_0|\}$$

where A is an arbitrary linear operator expressing the difference between W_2 and W_1 with respect to the orthogonal complement to φ_0 in D_W. This ambiguity in the wave operator W is, of course, also reflected in the properties of the reaction operator t, and there is a certain confusion in the field depending on the fact that these ambiguities are not sufficiently well-known and understood. Below we will briefly discuss some of the most important types of wave operators which have been studied in the literature.

C. Different Types of Wave and Reaction Operators

(*a*) *Correlation factor*. In its absolutely simplest form, the wave operator is a multiplicative operator defined by the quotient

$$W = \frac{\Psi}{\varphi_0} \tag{82}$$

and the operator W is hence essentially a "function". Off hand, one would probably guess that such a "wave operator" is too simple to be of any real value, but it has still turned out to be quite useful in the literature. This depends on the fact that even an approximate solution Ψ may be written in the product form

$$\Psi = W \, . \, \varphi_0 \tag{83}$$

and the best form of W may then be determined by means of the variation principle.

This approach was first used by Hylleraas[18] in his study of the helium atom, and it has later been extended to other simple systems.[19] In studies of improvements of the Hartree–Fock scheme starting out from $\varphi_0 = D$, the quantity W in (83) is often described as a "correlation factor".[20]

In their studies of the collective motions of the electrons in solids and particularly the so-called plasma oscillations, Bohm and Pines[21] have used wavefunctions of type (83) involving a correlation factor. Krisement[22] has also used this idea in his analysis of Wigner's[23] classical treatment of the correlation energy in a solid.

In the last decade, there has been a renewed interest in this approach as a consequence of the new ideas in perturbation theory developed by Lewis and Dalgarno,[24] according to which one puts

$$W = 1 + F \tag{84}$$

and determines the function F variationally. The idea that the wave operator W should be a "function" has recently been further analysed by Hirschfelder[25] and his group. In conclusion, it should be observed that, if one has succeeded in deriving any form W of the wave operator, the corresponding "local" operator W_{loc} is, according to (82), defined by the relation

$$W_{loc} = \frac{W\varphi_0}{\varphi_0} \tag{85}$$

in close analogy with Bartlett's idea[26] of the "local energy".

(*b*) *Adiabatic wave operator.* An entirely different type of wave operator is obtained from time-dependent perturbation theory. The time-dependent Schrödinger equation has the form

$$-\frac{h}{2\pi i}\frac{\partial \Psi}{\partial t} = H\Psi \tag{86}$$

and one should find the solution $\Psi = \Psi(t)$ which corresponds to the *initial condition* $\Psi = \Psi_0$ for $t = t_0$. Using the theory of differential equations, it may be shown that each initial condition Ψ_0 corresponds uniquely to a solution $\Psi = \Psi(t)$, and that one has a unique mapping $\Psi_0 \rightarrow \Psi(t)$:

$$\Psi(t) = U(t, t_0)\Psi_0 \tag{87}$$

where $U = U(t, t_0)$ is the *evolution operator.* It is easily shown[27] that the operator U is unitary and linear, and its properties have been carefully studied in connection with field theory. The diagram technique is a special type of book-keeping procedure to account for the various terms in the evolution operator which has been discussed at this summer school.

In order to use the time-dependent theory to solve the eigenvalue problem in the time-independent case, one is using the *adiabatic theorem*[28] which says that an eigenstate φ_0 of the unperturbed Hamiltonian $\mathscr{H}_0$ at time t_0 goes "adiabatically" over into the perturbed eigenstate Ψ of the Hamiltonian $\mathscr{H}$ at time t, if the perturbation $V = \mathscr{H} - \mathscr{H}_0$ is switched on slowly enough, i.e. one has

$$\Psi = U\varphi_0 \tag{88}$$

In order to satisfy the conditions of the adiabatic theorem, it is convenient to choose $t_0 = -\infty$ and $t = 0$, but unfortunately the fundamental formulas may become divergent when t_0 goes towards $-\infty$. For this purpose, one introduces a new Hamiltonian

$$\mathscr{H}_\alpha = \mathscr{H}_0 + Ve^{-\alpha|t} \tag{89}$$

where α is a positive constant. The limiting process $t_0 \to -\infty$ now becomes convergent, and one can later let $\alpha \to +0$.

If $U_\alpha = U_\alpha(0, -\infty)$ is the evolution operator for the time interval between $t_0 = -\infty$ and $t = 0$, one obtains from (88) by introducing the intermediate normalization:

$$\Psi_\alpha = \frac{U_\alpha \varphi_0}{\langle \varphi_0 | U_\alpha | \varphi_0 \rangle} \tag{90}$$

For the wave and reaction operators, one hence obtains

$$W = \lim_{\alpha \to +0} \frac{U_\alpha}{\langle \varphi_0 | U_\alpha | \varphi_0 \rangle} \tag{91}$$

$$t = VW = \lim_{\alpha \to +0} \frac{VU_\alpha}{\langle \varphi_0 | U_\alpha | \varphi_0 \rangle} \tag{92}$$

The limiting procedure in this "adiabatic approximation" is by no means of an elementary nature, but it seems now clear that it leads to the same result as the ordinary time-independent perturbation theory. This approach hence gives one more aspect on the fundamental eigenvalue problem which is of essential importance. Since several lecturers at this summer school have spoken of the adiabatic method, it is perhaps of some interest to consider it from this somewhat broader point of view.

(c) *Schrödinger wave operator.* It is evident that an alternative form of the wave operator is provided by Schrödinger's perturbation theory.[29] The structure of the higher terms in the expansions have been carefully studied by Brueckner.[30] Denoting the eigenvalues of $\mathscr{H}_0$ by E_k^0 and the properly normalized eigenfunctions by φ_k^0, one obtains the familiar results:

$$\Psi = \varphi_0 + \sum_{k \neq 0} \frac{\varphi_k^0 \langle \varphi_k^0 | V | \varphi_0^0 \rangle}{E_0 - E_k^0} + \cdots \tag{93}$$

$$E = E_0 + \langle \varphi_0 | V | \varphi_0 \rangle + \sum_{k \neq 0} \frac{\langle \varphi_0 | V | \varphi_k^0 \rangle \langle \varphi_k^0 | V | \varphi_0 \rangle}{E_0 - E_k^0} + \cdots \tag{94}$$

where one should sum over all discrete energy levels and integrate over the continuum but leave out the "reference level" E_0, which may be the ground state or one of the excited states.

Introducing the set of projection operators O_k^0 associated with the unperturbed Hamiltonian $\mathscr{H}_0$ through the relation

$$O_k^0 = |\varphi_k^0\rangle\langle\varphi_k^0| \tag{95}$$

one obtains a resolution of the identity of the form

$$1 = \sum_k O_k^0 \tag{96}$$

and a spectral resolution of $\mathscr{H}_0$ and of any function $f(\mathscr{H}_0)$ of the type:

$$\mathscr{H}_0 = \sum_k E_k^0 O_k^0 \tag{97}$$

$$f(\mathscr{H}_0) = \sum_k f(E_k^0) O_k^0 \tag{98}$$

For the resolvent $(z - \mathscr{H}_0)^{-1}$, one has in particular

$$(z - \mathscr{H}_0)^{-1} = \sum_k \frac{O_k^0}{z - E_k^0} \tag{99}$$

In order to exclude the "reference level" E_0 from the summation, it is convenient to introduce the projection operator P for the orthogonal complement to the state φ_0, i.e.

$$P = 1 - |\varphi_0\rangle\langle\varphi_0| \tag{100}$$

From Brueckner's study,[30] it follows now that all the results in Schrödinger's perturbation theory may be expressed in terms of the *reduced resolvent*

$$R_0 = \frac{P}{E_0 - \mathscr{H}_0} = \sum_{k \neq 0} \frac{O_k^0}{E_0 - E_k^0} = \sum_{k \neq 0} \frac{|\varphi_k^0\rangle\langle\varphi_k^0|}{E_0 - E_k^0} \tag{101}$$

and the perturbation V. From (93) one obtains the Schrödinger wave and reaction operators

$$W = 1 + R_0 V + R_0(V - \epsilon_1) R_0 V + \ldots \tag{102}$$

$$t = VW = V + VR_0V + VR_0(V - \epsilon_1)R_0V + \ldots \tag{103}$$

where $\epsilon_1 = \langle\varphi_0|V|\varphi_0\rangle$, etc. It is not necessary to describe these operators in more detail at this point, since they will be derived in closed form in Section 6 by means of the partitioning technique.

4. CORRELATION ENERGY AS A PERTURBATION EFFECT

A. Pair Correlations in Transition Formula

In order to study the correlation problem from the point of view of perturbation theory, it is convenient to start from Eqs. (38)–(40). Let us further introduce a basis $\{\psi_k\}$ in the one-electron Hilbert space consisting of the eigenfunctions to the effective Hamiltonian (25), i.e. of the occupied and virtual Hartree–Fock functions. Any selection of N functions $\psi_{k1}, \psi_{k2}, \ldots \psi_{kN}$ with $k_1 < k_2 < \ldots < k_N$ is called an "ordered configuration" K, and it is easily shown[30] that any antisymmetric wave function Ψ may be expanded in terms of Slater determinants D_K over such configurations:

$$\Psi = \sum_K D_K C_K \tag{104}$$

For the eigenfunctions to $\mathscr{H}$, the coefficients C_K are determined by the secular equation

$$\sum_L (\mathscr{H}_{KL} - E\delta_{KL})C_L = 0 \tag{105}$$

where $\mathscr{H}_{KL} = \langle D_K|\mathscr{H}|D_L\rangle$. In comparison to the special Slater determinant $\varphi_0 = D$ over the occupied one-electron functions $\psi_1, \psi_2, \ldots \psi_N$, the various configurations may be described as "singly excited", "doubly excited", etc. Taking together all terms of the same type, one may in analogy to (41) introduce the notations

$$\varphi_0^{\text{s.e.}} = \sum_{k,\,m} D_k^m C_k^m \tag{106}$$

$$\varphi_0^{\text{d.e.}} = \sum_{k,\,l,\,m,\,n} D_{kl}^{mn} C_{kl}^{mn} \ldots \text{etc.} \tag{107}$$

The expansion (104) may now be written in the form of a sum of terms referring to the various degrees of "excitation" with respect to φ_0:

$$\Psi = \varphi_0 + \varphi_0^{\text{s.e}} + \varphi_0^{\text{d.e.}} + \varphi_0^{\text{t.e.}} + \ldots \varphi_0^{(N)} \tag{108}$$

It should be observed that this function satisfies the intermediate normalization $\langle\varphi_0|\Psi\rangle = 1$.

Using formula (75) for the energy shift and observing that the matrix elements of V between φ_0 and the triply and more highly excited states of φ_0 vanish identically because of the orthogonality integrals involved, one obtains

$$\begin{aligned} E &= E_0 + \langle\varphi_0|V|\Psi\rangle \\ &= E_0 + \langle\varphi_0|V|\varphi_0\rangle + \langle\varphi_0|V|\varphi_0^{\text{s.e.}}\rangle + \langle\varphi_0|V|\varphi_0^{\text{d.e.}}\rangle \end{aligned} \tag{109}$$

In the right-hand member, the third term vanishes because of the Brillouin theorem (46), and one obtains finally

$$\begin{aligned} E &= \langle\varphi_0|\mathscr{H}|\varphi_0\rangle + \langle\varphi_0|V|\varphi_0^{\text{d.e.}}\rangle \\ &= E_{\text{HF}} + \langle\varphi_0|-\sum_i u_i + \sum_{i<j}\mathscr{H}_{ij}|\varphi_0^{\text{d.e.}}\rangle \\ &= E_{\text{HF}} + \langle\varphi_0|\sum_{i<j}\mathscr{H}_{ij}|\varphi_0^{\text{d.e.}}\rangle \end{aligned} \tag{110}$$

i.e.

$$E_{\text{corr}} = \langle\varphi_0|\sum_{i<j}\mathscr{H}_{ij}|\varphi_0^{\text{d.e.}}\rangle \tag{111}$$

This formula forms essentially the basis for the discussion of the correlation energy given at this summer school by Nesbet and by Sinanoğlu. We note that it is essentially a transition formula referring to a transition between the original Slater determinant φ_0 and the sum $\varphi_0^{\text{d.e.}}$ of all doubly excited terms in (104). Since the latter sum may be described in terms of "pair functions", formula (111) leads to a mathematical interpretation of the correlation energy in terms of "pair correlations". For further details, we will refer to the original discussions by Nesbet and by Sinanoğlu. The physical interpretation of the correlation energy will be further studied below.

B. Correlation Potential

The transition formula for the energy may also be used to derive a local potential V_{corr} which is such that its expectation value for the unperturbed state φ_0 gives the correlation energy, provided that φ_0 is the Slater determinant associated with the Hartree–Fock scheme. According to (78), one has

$$\begin{aligned} E &= E_0 + \langle\varphi_0|t|\varphi_0\rangle \\ &= \langle\varphi_0|\mathscr{H}_0 + V|\varphi_0\rangle + \langle\varphi_0|t - V|\varphi_0\rangle \\ &= E_{\mathrm{HF}} + \langle\varphi_0|t - V|\varphi_0\rangle \end{aligned} \tag{112}$$

i.e.

$$E_{\mathrm{corr}} = \langle\varphi_0|t - V|\varphi_0\rangle = \langle\varphi_0|V_{\mathrm{corr}}|\varphi_0\rangle \tag{113}$$

where

$$V_{\mathrm{corr}} = t_{\mathrm{loc}} - V = \frac{t\varphi_0}{\varphi_0} - V \tag{114}$$

It still remains to investigate whether quantities of this type are useful in studying the general properties of the correlation energy, or whether there are other "local potentials" which are better suited for this purpose. It should also be remembered that, since the reaction operator is subject to a great deal of ambiguity, the same is also true for any general correlation potential $(t - V)$ and its "local" approximation.

C. Physical Interpretation of Correlation Energy in Terms of Expectation Values

It is evident that there is an essential difference between the interpretation of the correlation energy based on the virial theorem in Section 1 and the one based on the pair correlations in transition formula (111). In order to study this difference in greater detail, we will consider the Hamiltonian for an electronic system with fixed nuclei g having the atomic numbers Z_g:

$$\mathscr{H} = \mathscr{H}_c + T + L + C \tag{115}$$

Here $\mathscr{H}_c$ is a constant term representing the Coulomb repulsion between the nuclei, T is the kinetic energy of the electrons, L

is the Coulomb attraction between the electrons and the nuclei, and C is the Coulomb repulsion between the electrons:

$$\mathscr{H}_{\mathrm{c}} = e^2 \sum_{g<h} \frac{Z_g Z_h}{r_{gh}} \tag{116}$$

$$T = \sum_i \frac{p_i^2}{2m}, \qquad L = -e^2 \sum_i \sum_g \frac{Z_g}{r_{ig}} \tag{117}$$

$$C = \sum_{i<j} \frac{e^2}{r_{ij}} \tag{118}$$

Substituting (115) into the transition formula (65) for the energy, one obtains

$$E = \mathscr{H}_{\mathrm{c}} + \frac{\langle \Phi | T | \Psi \rangle}{\langle \Phi | \Psi \rangle} + \frac{\langle \Phi | L | \Psi \rangle}{\langle \Phi | \Psi \rangle} + \frac{\langle \Phi | C | \Psi \rangle}{\langle \Phi | \Psi \rangle} \tag{119}$$

where the sum in the right-hand member is independent of Φ, whereas the individual terms usually depend strongly on the choice of Φ. In evaluating the correlation energy in (111), one has chosen $\Phi = \varphi_0$, and neither T nor L then give any contribution to this quantity.

The physical interpretations of quantum mechanics are ordinarily associated with *expectation values* and, in order to evaluate the contributions of T, L, and C to the total energy, one should choose $\Phi = \Psi$ in (119), which gives

$$\begin{aligned} E &= \mathscr{H}_{\mathrm{c}} + \frac{\langle \Psi | T | \Psi \rangle}{\langle \Psi | \Psi | \rangle} + \frac{\langle \Psi | L | \Psi \rangle}{\langle \Psi | \Psi \rangle} + \frac{\langle \Psi | C | \Psi \rangle}{\langle \Psi | \Psi \rangle} \\ &= \mathscr{H}_{\mathrm{c}} + \langle T \rangle + \langle L \rangle + \langle C \rangle \end{aligned} \tag{120}$$

It is clear that this formula is going to give an entirely different interpretation of the contributions to the correlation energy than the transition formula (111) dealing only with "pair correlations". Because of the virial theorem, one is going to have the relation

$$\langle T \rangle_{\mathrm{corr}} = -\tfrac{1}{2} \langle L + C \rangle_{\mathrm{corr}} \tag{121}$$

One is apparently going to have one-electron contributions to the correlation energy from the kinetic energy T and the nuclear attraction L, whereas one is going to have electron-pair contributions from the Coulomb repulsion C. It is clear that $\langle T \rangle_{\mathrm{corr}}$ is going to be positive, it is further very likely that $\langle C \rangle_{\text{corr}}$ is going to be negative, whereas it may be hard to predict the sign of $\langle L \rangle_{\text{corr}}$ in general. It seems certainly worthwhile to investigate these contributions in greater detail, and work along these lines is now in progress in our group.

5. BRUECKNER SCHEME AND THE EXACT SCF METHOD

A. Brueckner's Generalization of the Hartree–Fock Method

As pointed out in Section 1, one believed that the success of the independent-particle model in explaining the shell structure of the electronic clouds of the atoms and molecules depended essentially on the fact that the interaction between electrons is comparatively weak, and that the perturbation V in (40) is small. It was hence very surprising that the independent-particle model turned out to work so well in the shell model of the atomic nuclei, since the interaction between the nucleons is certainly very strong.

In this situation, Brueckner[31] and his coworkers asked the question whether there existed any natural extension of the Hartree–Fock scheme which would incorporate also strong two-body interactions, so that it would become exact in the case of two particles. This leads to an interesting development of the many-body theory, and we will here briefly review only the fundamental ideas.

According to (4) and (5), the Hartree–Fock scheme is based on the variation principle starting out from the energy expectation value:

$$\langle \mathscr{H} \rangle = \langle \varphi_0 | \mathscr{H}_0 + V | \varphi_0 \rangle \tag{122}$$

where φ_0 is a single Slater determinant. In the Hamiltonian $\mathscr{H} = \mathscr{H}_0 + V$, the one-electron potential u_1 is still at our disposal according to (39) and (40). In treating the ground state of a system, the quantity $\langle \mathscr{H} \rangle$ gives always an *upper bound* to the true energy E, and the error in the optimum case is essentially the "correlation energy". On the other hand, if one starts out from formula (78), one obtains for the true energy

$$E = \langle \varphi_0 | \mathscr{H}_0 + f | \varphi_0 \rangle \tag{123}$$

i.e. an expression which is analogous to (122), if one replaces the perturbation V by the reaction t:

$$V \rightarrow t \tag{124}$$

Brueckner emphasized that, if one could expand t in the form

$$t = \sum_i t_i + \sum_{i<j} t_{ij} + \dots \tag{125}$$

and, according to (124), made the replacement

$$-\sum_i u_i + \sum_{i<j} \mathscr{H}_{ij} \rightarrow \sum_i t_i + \sum_{i<j} t_{ij} \tag{126}$$

i.e. $\mathscr{H}_{ij} \rightarrow t_{ij}$, one obtains an approximate scheme which includes all the two-body reactions so that it becomes exact in the limiting case of two particles. In this case, one can in analogy to (35) construct a self-consistent-field potential

$$u_1 = \int dx_2 t_{12}(1 - P_{12})\rho(x_2, x_2') \tag{127}$$

which leads to the formation of an effective Hamiltonian $H_{\text{eff}}(1) = \mathscr{H}_1 + u_1$ and an associated Slater determinant φ_0. For further details of this scheme, we will refer to the original papers.[31]

B. Exact SCF Theory

The question is now whether one can generalize Brueckner's idea in order to make the theory exact. It is a remarkable feature of the Hartree–Fock scheme itself that it may be extended[5] also to systems having many-body interactions and Hamiltonians of the type:

$$\mathscr{H} = \mathscr{H}_{\text{c}} + \sum_i \mathscr{H}_i + \sum_{i<j} \mathscr{H}_{ij} + \sum_{i<j<k} \mathscr{H}_{ijk} + \dots \tag{128}$$

Starting out from the assumption that the trial wave function φ_0 is a single Slater determinant (7) and applying the variation principle (4), one obtains for the one-particle functions ψ_k the optimum condition

$$H_{\text{eff}}(1)\psi_k(x_1) = \epsilon_k \psi_k(x_1) \tag{129}$$

where

$$\begin{aligned} H_{\text{eff}}(1) &= \mathscr{H}_1 + u_1 \\ &= \mathscr{H}_1 + u_1^{(2)} + u_1^{(3)} + \dots + u_1^{(N)} \end{aligned} \tag{130}$$

and the Hartree–Fock potential u_1 is a sum of contributions corresponding to the various terms in the Hamiltonian (128):

$$u_1^{(2)} = \int dx_2 \mathscr{H}_{12}(1 - P_{12})\rho(x_2, x_2')$$

$$u_1^{(3)} = \frac{1}{2}\int dx_2 dx_3 \mathscr{H}_{123}(1 - P_{12} - P_{13}) \begin{vmatrix} \rho(x_2, x_2') & \rho(x_2, x_3') \\ \rho(x_3, x_2') & \rho(x_3, x_3') \end{vmatrix} \tag{131}$$

. . . .

and analogously for the higher terms. For further details, we will refer to Reference 5.

One can now pursue the analogy between (122) and (123) completely, if one expands the reaction operator t in the form

$$t = -\sum_i u_i + \sum_{i<j} t_{ij} + \sum_{i<j<k} t_{ijk} + \ldots \tag{132}$$

and writes the replacement $V \to t$ accordingly:

$$\mathscr{H}_{ij} \to t_{ij}, \qquad \mathscr{H}_{ijk} \to t_{ijk}, \ldots \tag{133}$$

It is evident that, even if one has only two-body terms $\mathscr{H}_{ij}$ in the Hamiltonian, one may have many-body terms t_{ijk}, t_{ijkl}, . . . in the reaction operator. In analogy with (131), it may now be convenient to introduce the *exact* SCF *potential* u_1 by a sum of the type:

$$u_1 = u_1^{(2)} + u_1^{(3)} + u_1^{(4)} + \ldots + u_1^{(N)} \tag{134}$$

where

$$u_1^{(2)} = \int dx_2 t_{12}(1 - P_{12})\rho(x_2, x_2')$$

$$u_1^{(3)} = \frac{1}{2}\int dx_2 dx_3 t_{123}(1 - P_{12} - P_{13}) \begin{vmatrix} \rho(x_2, x_2') & \rho(x_2, x_3') \\ \rho(x_3, x_2') & \rho(x_3, x_3') \end{vmatrix} \tag{135}$$

. . . . . .

It is now possible to construct the effective Hamiltonian $H_{\text{eff}}(1) = \mathscr{H}_1 + u_1$, and the associated Slater determinant $\varphi_0 = D$. The energy (123) comes out exactly, and there is no variation principle involved. However, it should be observed that, in analogy to (46), one obtains the relation

$$\langle D_{\text{s.e.}}|t|D\rangle = 0 \tag{136}$$

which is sometimes referred to as the Brillouin–Brueckner theorem.

This theorem implies that, if the exact eigenfunction Ψ is expanded in terms of configurations over the occupied and virtual eigenfunctions of H_{eff}, the term corresponding to the "singly excited states" will vanish identically:

$$\varphi_0^{\text{s.e.}} \equiv 0 \tag{137}$$

For further details as to the "exact SCF scheme," we will refer to another paper.[32]

The main feature of this approach is that it makes it possible to formulate the exact theory in the same form as the self-consistent-field method, and that it leads to the correct energy even if the independent-particle model wave function φ_0 may be a rather poor approximation to the exact eigenfunction Ψ. Our study shows that the SCF method based on the exact reaction operator t has a great deal of flexibility, but the results do not necessarily prove that it is justified to use the "independent-particle model" in many-body systems having strong interactions between the particles. We will return to this question after discussing the importance of the symmetry properties.

6. EXPLICIT FORMULAE FOR WAVE AND REACTION OPERATORS OBTAINED BY PARTITIONING TECHNIQUE

A. Partitioning Technique

It is clear from the preceding sections that perturbation theory plays an important role in discussing the correlation effects, and we will here try to derive some closed formulae for the wave and reaction operators which are fundamental in this approach.

In order to study the solution of the eigenvalue problem

$$\mathscr{H}\Psi = E\Psi \tag{138}$$

for a self-adjoint Hamiltonian $\mathscr{H}$ bounded from below, we will introduce a *reference function* φ in the Hilbert space, which is normalized so that $\langle\varphi|\varphi\rangle = 1$. The projection operator O associated with the one-dimensional reference space has the form

$$O = |\varphi\rangle\langle\varphi| \tag{139}$$

and the properties $O^2 = O$, $O^\dagger = O$, and $\mathrm{Tr}(O) = 1$. If Φ is an arbitrary function, one has $O\Phi = \varphi\langle\varphi|\Phi\rangle$, which defines the meaning of the symbol (139). Let us further introduce the projection operator P for the *orthogonal complement* to φ, i.e.

$$P = 1 - O \tag{140}$$

It has the properties $P^2 = P$, $P^\dagger = P$, $OP = PO = 0$, and one has also $P\varphi = 0$. We will further define an auxiliary Hamiltonian

$$\overline{\mathscr{H}} = P\mathscr{H}P \tag{141}$$

which will be called the "outer projection" of $\mathscr{H}$ with respect to the orthogonal complement P. It is easily shown[33] that, for the eigenvalues of $\overline{\mathscr{H}}$ and $\mathscr{H}$, one has the theorem

$$\overline{E}_k > E_k \tag{142}$$

in order.

In order to proceed, it is convenient to introduce a complex variable $\mathscr{E}$ and the resolvent $(\mathscr{E} - \overline{\mathscr{H}})^{-1}$ for the auxiliary Hamiltonian. The key quantity in the theory is the *reduced resolvent*

$$T = (\mathscr{E} - \overline{\mathscr{H}})^{-1}P \tag{143}$$

which may blow up for $\mathscr{E} = \overline{E}_k$ but usually stays regular for $\mathscr{E} = E_k$. In order to handle this operator, it is useful to mention some elementary theorems.

If A is an arbitrary operator which has an inverse A^{-1} and commutes with P, it follows from $AP = PA$ by multiplication to the left and right by A^{-1} that

$$PA^{-1} = A^{-1}P \tag{144}$$

If the theorem is applied to (143), one has $\overline{\mathscr{H}}P = P\overline{\mathscr{H}}$ and putting $A = \mathscr{E} - \overline{\mathscr{H}}$, one obtains immediately

$$T = (\mathscr{E} - \overline{\mathscr{H}})^{-1}P = P(\mathscr{E} - \overline{\mathscr{H}})^{-1} = P(\mathscr{E} - \overline{\mathscr{H}})^{-1}P \tag{145}$$

If further B is an arbitrary operator, one has

$$\begin{aligned} AP - PBP &= PA - PBP \\ (A - PB)P &= P(A - PBP) \\ P(A - PBP)^{-1} &= (A - PB)^{-1}P \end{aligned} \tag{146}$$

provided that the inverse operators involved do exist. Applying (146) to formula (145), one obtains particularly

$$T = P(\mathscr{E} - P\mathscr{H}P)^{-1} = (\mathscr{E} - P\mathscr{H})^{-1}P \tag{147}$$

According to (145), one has also the relations

$$TP = PT = T \tag{148}$$

The importance of the reduced resolvent T depends essentially on the fact that it satisfies the following fundamental relation

$$P(\mathscr{E} - \mathscr{H})T = P \tag{149}$$

which is proved in the following way: $P(\mathscr{E} - \mathscr{H})T = \mathscr{E}T - P\mathscr{H}PT = (\mathscr{E} - P\mathscr{H}P)T = (\mathscr{E} - \overline{\mathscr{H}})(\mathscr{E} - \overline{\mathscr{H}})^{-1}P = P.$

It is now time to introduce the *trial wave function* $\Psi_{\mathscr{E}}$ through the relation

$$\Psi_{\mathscr{E}} = \varphi + \phi_{\mathscr{E}} = \varphi + T_{\mathscr{E}}\mathscr{H}\varphi = (1 + T_{\mathscr{E}}\mathscr{H})\varphi \tag{150}$$

where we, by the symbol $T_{\mathscr{E}}$, have indicated that the reduced resolvent depends on the variable $\mathscr{E}$. This function has the remarkable property that it satisfies a Schrödinger equation in the subspace of the orthogonal complement, i.e.

$$P(\mathscr{H} - \mathscr{E})\Psi_{\mathscr{E}} = 0 \tag{151}$$

This follows from the fact that

$$\begin{aligned} P(\mathscr{H} - \mathscr{E})\Psi_{\mathscr{E}} = P(\mathscr{H} - \mathscr{E})\varphi + P(\mathscr{H} - \mathscr{E})T\mathscr{H}\varphi \\ = P\mathscr{H}\varphi - P\mathscr{H}\varphi = 0 \end{aligned}$$

This implies also that

$$\begin{aligned} (\mathscr{H} - \mathscr{E})\Psi_{\mathscr{E}} &= (O + P)(\mathscr{H} - \mathscr{E})\Psi_{\mathscr{E}} \\ &= O(\mathscr{H} - \mathscr{E})\Psi_{\mathscr{E}} \\ &= \varphi\langle\varphi|\mathscr{H} - \mathscr{E}|\Psi_{\mathscr{E}}\rangle \end{aligned} \tag{152}$$

where the right-hand member is proportional to the reference function φ. It should be observed that $\Psi_{\mathscr{E}}$ consists of a term φ and a term $\phi_{\mathscr{E}}$ from the orthogonal complement, and that the intermediate normalization

$$\langle\varphi|\Psi_{\mathscr{E}}\rangle = 1 \tag{153}$$

is automatically fulfilled for both closed states and scattering states. In considering (152), it is now convenient to introduce the new quantity

$$\begin{aligned}\mathscr{E}_1 &= \langle\varphi|\mathscr{H}|\Psi_{\mathscr{E}}\rangle \\ &= \langle\varphi|\mathscr{H} + \mathscr{H}T_{\mathscr{E}}\mathscr{H}|\varphi\rangle \equiv f(\mathscr{E})\end{aligned} \tag{154}$$

It is clear that, for every value of $\mathscr{E}$, the trial wave function $\Psi_{\mathscr{E}}$ satisfies the *inhomogeneous Schrödinger equation*:

$$(\mathscr{H} - \mathscr{E})\Psi_{\mathscr{E}} = \varphi(\mathscr{E}_1 - \mathscr{E}) \tag{155}$$

and that one gets a solution to the ordinary Schrödinger equation $(\mathscr{H} - E)\Psi_E = 0$ in the case when $\mathscr{E}_1 = \mathscr{E} = E$, i.e. when $E = f(E)$.

The function $\mathscr{E}_1 = f(E)$ defined by (154) has some additional interesting features. From (143), it follows that

$$\frac{\partial T_{\mathscr{E}}}{\partial \mathscr{E}} = -(\mathscr{E} - \bar{\mathscr{H}})^{-2}P = -T_{\mathscr{E}}^2 \tag{156}$$

We will say that variable $\mathscr{E}$ is in the "discrete part" of the energy spectrum on the real axis, if the associated wave function $\Psi_{\mathscr{E}}$ is normalizable, so that

$$\langle\Psi_{\mathscr{E}}|\Psi_{\mathscr{E}}\rangle = \langle\varphi|\varphi\rangle + \langle\phi_{\mathscr{E}}|\phi_{\mathscr{E}}\rangle = 1 + \langle\phi_{\mathscr{E}}|\phi_{\mathscr{E}}\rangle \tag{157}$$

exists. According to (154), one obtains now for the derivative of the function $\mathscr{E}_1 = f(\mathscr{E}) = \langle\varphi|\mathscr{H} + \mathscr{H}T_{\mathscr{E}}\mathscr{H}|\varphi\rangle$:

$$\begin{aligned}f'(\mathscr{E}) &= -\langle\varphi|\mathscr{H}T_{\mathscr{E}}^2\mathscr{H}|\varphi\rangle \\ &= -\langle T_{\mathscr{E}}\mathscr{H}\varphi|T_{\mathscr{E}}\mathscr{H}\varphi\rangle = -\langle\phi_{\mathscr{E}}|\phi_{\mathscr{E}}\rangle \leq 0\end{aligned} \tag{158}$$

i.e. the derivative is always *negative* in the discrete part of the spectrum.

This property leads immediately to a *bracketing theorem* which says that, for any value $\mathscr{E}$ in the discrete part of the spectrum, the numbers $\mathscr{E}$ and $\mathscr{E}_1 = f(\mathscr{E})$ bracket always at least one true eigenvalue E. If one puts $\mathscr{E} = E + \epsilon$ and $\mathscr{E}_1 = E + \epsilon_1$, one has immediately by means of Lagrange's mean-value theorem:

$$E + \epsilon_1 = f(E + \epsilon) = f(E) + \epsilon f'(E + \theta\epsilon) \tag{159}$$

where $0 \leq \theta \leq 1$. This gives directly

$$\epsilon_1 = \epsilon f'(E + \theta\epsilon) \tag{160}$$

i.e. the errors ϵ and ϵ_1 have different signs, which proves the bracketing theorem. The function $\mathscr{E}_1 = f(\mathscr{E})$ will in the following often be referred to as the "bracketing function".

Since the eigenvalue problem is now reduced to finding the zero points of the function

$$F(\mathscr{E}) = \mathscr{E} - f(\mathscr{E}) \tag{161}$$

one could apply the Newton–Raphson method. If $\mathscr{E}$ is an approximate zero point, one obtains an improved value $\mathscr{E}^*$ from the tangential construction

$$\mathscr{E}^* = \mathscr{E} - \frac{F(\mathscr{E})}{F'(\mathscr{E})} = \mathscr{E} - \frac{\mathscr{E} - \mathscr{E}_1}{1 - f'(\mathscr{E})} \tag{162}$$

and repetition of this approach leads to a second-order iteration procedure. Using (155) and (158), one obtains for the expectation value of $\mathscr{H}$ with respect to the trial wave function $\Psi_{\mathscr{E}}$:

$$\begin{aligned}\langle\mathscr{H}\rangle &= \frac{\langle\Psi_{\mathscr{E}}|\mathscr{H}|\Psi_{\mathscr{E}}\rangle}{\langle\Psi_{\mathscr{E}}|\Psi_{\mathscr{E}}\rangle} = \mathscr{E} + \frac{\langle\Psi_{\mathscr{E}}|\mathscr{H} - \mathscr{E}|\Psi_{\mathscr{E}}\rangle}{\langle\Psi_{\mathscr{E}}|\Psi_{\mathscr{E}}\rangle} \\ &= \mathscr{E} + \frac{\langle\Psi_{\mathscr{E}}|\mathscr{E}_1 - \mathscr{E}|\varphi\rangle}{1 + \langle\phi_{\mathscr{E}}|\phi_{\mathscr{E}}\rangle} = \mathscr{E} - \frac{\mathscr{E} - \mathscr{E}_1}{1 - f'(\mathscr{E})}\end{aligned} \tag{163}$$

i.e.

$$\langle\mathscr{H}\rangle = \mathscr{E}^* \tag{164}$$

There is hence a close connection between the variation principle and the Newton–Raphson procedure.

B. Wave and Reaction Operators

Let us now consider the case when $\mathscr{H} = \mathscr{H}_0 + V$ and the reference function φ is chosen to be an eigenfunction φ_0 to $\mathscr{H}_0$ associated with the eigenvalue E_0, so that

$$\mathscr{H}_0\varphi_0 = E_0\varphi_0 \tag{165}$$

Introducing the projection operator $O = |\varphi_0\rangle\langle\varphi_0|$, one has $\mathscr{H}_0 O = O\mathscr{H}_0 = E_0 O$. For $P = 1 - O$, one obtains similarly

$$\mathscr{H}_0 P = P\mathscr{H}_0 \tag{166}$$

From $P\varphi_0 = 0$, it follows further that $T\varphi_0 = 0$. For the trial wave function (150), one gets

$$\begin{aligned}\Psi_{\mathscr{E}} &= \varphi_0 + T(\mathscr{H}_0 + V)\varphi_0 = \varphi_0 + T(E_0 + V)\varphi_0 \\ &= \varphi_0 + TV\varphi_0 = (1 + TV)\varphi_0 = W\varphi_0 \end{aligned} \tag{167}$$

and the quantity

$$W = 1 + TV \tag{168}$$

is a generalization of the *wave operator* which transforms the unperturbed eigenfunction φ_0 over into the trial wave function $\Psi_{\mathscr{E}}$. For the bracketing function $\mathscr{E}_1 = f(\mathscr{E})$, one obtains similarly according to (154):

$$\begin{aligned}\mathscr{E}_1 &= \langle\varphi_0|(\mathscr{H}_0 + V) + (\mathscr{H}_0 + V)T(\mathscr{H}_0 + V)|\varphi_0\rangle \\ &= \langle\varphi_0|(E_0 + V) + (E_0 + V)T(E_0 + V)|\varphi_0\rangle \\ &= E_0 + \langle\varphi_0|V + VTV|\varphi_0\rangle \\ &= E_0 + \langle\varphi_0|t|\varphi_0\rangle \end{aligned} \tag{169}$$

where the quantity

$$t = V + VTV \tag{170}$$

is a generalization of the *reaction operator*, which gives the energy shift from E_0 to the value of the bracketing function. We note that the basic formula (77) is valid also in the generalized case.

In order to proceed, we will consider the reduced resolvent T defined by (143). For this purpose, it is convenient to introduce also the reduced resolvent T_0 associated with the unperturbed Hamiltonian $\mathscr{H}_0$:

$$T_0 = (\mathscr{E} - \mathscr{H}_0)^{-1}P \tag{171}$$

Using (166), (146), and the operator rule $(ab)^{-1} = b^{-1}a^{-1}$, one obtains

$$\begin{aligned}T &= (\mathscr{E} - P\mathscr{H}_0P - PVP)^{-1}P \\ &= (\mathscr{E} - \mathscr{H}_0P - PVP)^{-1}P \\ &= (\mathscr{E} - \mathscr{H}_0 - PVP)^{-1}P \\ &= (\mathscr{E} - \mathscr{H}_0 - PV)^{-1}P \\ &= [(\mathscr{E} - \mathscr{H}_0) - PV]^{-1}P \\ &= [1 - (\mathscr{E} - \mathscr{H}_0)^{-1}PV]^{-1}(\mathscr{E} - \mathscr{H}_0)^{-1}P \\ &= (1 - T_0V)^{-1}T_0 \end{aligned} \tag{172}$$

which relation shows that T, W, and t are expressible in closed form in terms of T_0 and V. According to (168), one gets further

$$W = 1 + TV = 1 + (1 - T_0V)^{-1}T_0V = (1 - T_0V)^{-1} \tag{173}$$

and

$$t = VW = V(1 - T_0V)^{-1} \tag{174}$$

If one expands the inverse $(1 - T_0V)^{-1}$ in an infinite geometric series, one obtains expressions for W and t which, for $\mathscr{E}_1 = \mathscr{E} = E$, are identical with those obtained in the ordinary Schrödinger–Brillouin perturbation theory.[33]

In order to get the corresponding quantities associated with the Schrödinger perturbation theory, it is convenient to introduce the reduced resolvent R_0 defined by (101):

$$R_0 = (E_0 - \mathscr{H}_0)^{-1}P \tag{175}$$

which is a special case of T_0 for $\mathscr{E} = E_0$. If a is an arbitrary number subject to the condition $a \neq E_0$, one has, according to (146) and (166), the following transformation:

$$\begin{aligned} T &= (\mathscr{E} - \overline{\mathscr{H}})^{-1}P \\ &= [a - \overline{\mathscr{H}} + (\mathscr{E} - a)]^{-1}P \\ &= [a - \overline{\mathscr{H}} + (\mathscr{E} - a)P]^{-1}P \\ &= [a - \mathscr{H}_0P - PVP + (\mathscr{E} - a)P]^{-1}P \\ &= [a - \mathscr{H}_0 - PV + (\mathscr{E} - a)P]^{-1}P \\ &= [(a - \mathscr{H}_0) - P(V - \mathscr{E} + a)]^{-1}P \\ &= [1 - (a - \mathscr{H}_0)^{-1}P(V - \mathscr{E} + a)]^{-1}(a - \mathscr{H}_0)^{-1}P \end{aligned} \tag{176}$$

Letting $a \to E_0$ and introducing the notation

$$V' = V - (\mathscr{E} - E_0) \tag{177}$$

one obtains finally

$$T = (1 - R_0V')^{-1}R_0 \tag{178}$$

and further

$$W = 1 + TV = 1 + (1 - R_0V')^{-1}R_0V \tag{179}$$

$$t = VW = V + V(1 - R_0V')^{-1}R_0V \tag{180}$$

which are the closed expressions desired. The use of these quantities leads to the construction of a trial wave function $\Psi_{\mathscr{E}} = W\varphi_0$ which satisfies an inhomogeneous Schrödinger equation (155) and of a bracketing function $\mathscr{E}_1 = E_0 + \langle\varphi_0|t|\varphi_0\rangle$ according to (169).

In order to see the connection with the conventional Schrödinger perturbation theory, one should consider the special case $\mathscr{E}_1 = \mathscr{E} = E$, expand the inverse $(1 - R_0V')^{-1}$ in an infinite geometrical series, use the special relation

$$V' = V - \langle\varphi_0|t|\varphi_0\rangle \tag{181}$$

and rearrange the expansions of W and $t = VW$ after powers of the perturbation V. Putting

$$W = 1 + \sum_{k=1}^{\infty} W_k, \qquad t = \sum_{k=1}^{\infty} t_k \tag{182}$$

and introducing the notation $\epsilon_k = \langle\varphi_0|t_k|\varphi_0\rangle$, one obtains the well-known result:

$$\begin{aligned}
t_1 &= V \\
t_2 &= VR_0V \\
t_3 &= VR_0(V - \epsilon_1)R_0V \\
t_4 &= VR_0(V - \epsilon_1)R_0(V - \epsilon_1)R_0V - \epsilon_2VR_0^2V \\
&\quad \cdot \quad \cdot \quad \cdot \quad \cdot \quad \cdot \quad \cdot \quad \cdot \quad \cdot \quad \cdot
\end{aligned} \tag{183}$$

whereas the terms in the wave operator are given by the formal relation $W_k = V^{-1}t_{k+1}$. For further details, we will refer to Reference 33.

C. Lower Bounds to the Reaction Operator

An interesting question is now whether one could use closed expressions of the type (173) and (174)— or (179) and (180)— as a substitute for infinite-order perturbation theory or as tools for calculating approximate eigenfunctions and upper and lower bounds to energy eigenvalues to any accuracy desired.

Let us assume that the perturbation V is such that it has an inverse V^{-1}. Using formula (174) and the operator relation

$b^{-1}a^{-1} = (ab)^{-1}$, one obtains $t = V(1 - T_0 V)^{-1} = (V^{-1} - T_0)^{-1}$, i.e.

$$t^{-1} = V^{-1} - T_0 \tag{184}$$

It is remarkable that, if $\mathscr{H}$ is additive in $\mathscr{H}_0$ and V, the inverse reaction operator t^{-1} is additive in $-T_0$ and V^{-1}:

$$\mathscr{H} = \mathscr{H}_0 + V$$
$$t^{-1} = -T_0 + V^{-1} \tag{185}$$

Looking at the applications, however, it seems very unlikely that one should have any explicit use of the simple two-term formula (189), since the inverse reaction operator occurs nowhere.

In order to proceed, we will now introduce the concept of *operator inequalities* for self-adjoint operators. We will write

$$A > B \tag{186}$$

provided that $\langle\psi|A|\psi\rangle > \langle\psi|B|\psi\rangle$ for all wave functions ψ for which the expectation values exist; this means that ψ is situated in the common part of the domain D_A and D_B of the operators involved. Putting $\psi = \Omega\psi'$, where Ω is an arbitrary linear transformation, and using the turn-over rule, one obtains further

$$\Omega^\dagger A\Omega > \Omega^\dagger B\Omega \tag{187}$$

If Q is a self-adjoint *projection operator* satisfying the relations $Q^2 = Q$ and $Q^\dagger = Q$, one has the inequality

$$0 \leq Q \leq 1 \tag{188}$$

i.e. the operator Q is situated between the zero operator and the identity operator. The proof follows from the fact that $\langle\psi|Q|\psi\rangle = \langle\psi|Q^\dagger Q|\psi\rangle = \langle Q\psi|Q\psi\rangle \geq 0$ which gives $Q \geq 0$. Since the same reasoning applies to the projection operator $P = 1 - Q$, one has $1 - Q \geq 0$, i.e. $1 \geq Q$.

Modifying the inequality (188) by means of a linear transformation according to (187), one gets the new inequality

$$0 \leq \Omega^\dagger Q\Omega \leq \Omega^\dagger\Omega \tag{189}$$

If $A > 0$ is a *positive definite operator* having the positive square root $A^{\frac{1}{2}}$, one may choose $\Omega^\dagger = \Omega = A^{\frac{1}{2}}$ which gives

$$0 \leq A^{\frac{1}{2}}QA^{\frac{1}{2}} \leq A \tag{190}$$

The operator

$$A' = A^{\frac{1}{2}}QA^{\frac{1}{2}} \tag{191}$$

may be described as the "inner projection" of the operator A with respect to Q; for further details, we will refer to Reference 34.

Let M_n be a linear manifold in the Hilbert space which is spanned by the linearly independent functions $\mathbf{f} = \{f_1, f_2, \ldots f_n\}$ having the metric matrix $\mathbf{\Delta} = \langle \mathbf{f}|\mathbf{f}\rangle$. The projection operator Q on this manifold is then given by the expression:[34]

$$Q = |\mathbf{f}\rangle\langle\mathbf{f}|\mathbf{f}\rangle^{-1}\langle\mathbf{f}| = \sum_{k,l=1}^{n} |f_k\rangle(\mathbf{\Delta}^{-1})_{kl}\langle f_l| \tag{192}$$

Substituting (192) into (191), one obtains the inner projection

$$A' = A^{\frac{1}{2}}|\mathbf{f}\rangle\langle\mathbf{f}|\mathbf{f}\rangle^{-1}\langle\mathbf{f}|A^{\frac{1}{2}} \tag{193}$$

which forms a lower bound to the operator A and which converges from below towards A, as the set $\mathbf{f}$ becomes complete and Q converges towards the identity operator. Introducing the substitution $\mathbf{f} = A^{-\frac{1}{2}}\mathbf{h}$, one obtains instead

$$A' = |\mathbf{h}\rangle\langle\mathbf{h}|A^{-1}|\mathbf{h}\rangle^{-1}\langle\mathbf{h}| \tag{194}$$

which is a lower bound containing the inverse operator, A^{-1}, instead of the square root.

Let us now consider the special case of a positive definite perturbation:

$$V > 0 \tag{195}$$

and let us choose the variable $\mathscr{E}$ in the interval $-\infty < \mathscr{E} < E_1^0$, where E_1^0 is the first excited level of the unperturbed Hamiltonian $\mathscr{H}_0$. If one chooses the ground state of $\mathscr{H}_0$ as reference level, the operator

$$T_0 = (\mathscr{E} - \mathscr{H}_0)^{-1}P = \sum_{k\neq 0} \frac{O_k^0}{\mathscr{E} - E_k^0} \tag{196}$$

is certainly *negative definite*. According to the two-term formula (189), the operator t^{-1} is then positive definite, and one has the inequality

$$t^{-1} > V^{-1} > 0 \tag{197}$$

which leads to the inverse relation

$$0 < t < V \tag{198}$$

In Reference 34, it has been shown that this rough inequality may be used to determine upper and lower bounds of higher and higher accuracy in the Schrödinger–Brillouin type of perturbation theory, provided that the power series for $(1 - T_0 V)^{-1}$ is convergent. Here we will focus our attention on the inner projection (194) which leads to a lower bound

$$\begin{aligned} t' &= |\mathbf{h}\rangle\langle\mathbf{h}|t^{-1}|\mathbf{h}\rangle^{-1}\langle\mathbf{h}| \\ &= |\mathbf{h}\rangle\langle\mathbf{h}|V^{-1} - T_0|\mathbf{h}\rangle^{-1}\langle\mathbf{h}| \end{aligned} \tag{199}$$

for the reaction t which may be calculated in a finite number of steps. According to (169), one has the relation

$$\begin{aligned} \mathscr{E}_1 &= E_0 + \langle\varphi_0|t|\varphi_0\rangle \\ &> E_0 + \langle\varphi_0|t'|\varphi_0\rangle \\ &= E_0 + \langle\varphi_0|\mathbf{h}\rangle\langle\mathbf{h}|V^{-1} - T_0|\mathbf{h}\rangle^{-1}\langle\mathbf{h}|\varphi_0\rangle \end{aligned} \tag{200}$$

which gives a lower bound to the bracketing function $\mathscr{E}_1 = f(\mathscr{E})$ in the interval $-\infty < \mathscr{E} < E_1^0$. If there is an upper bound to the ground-state energy of $\mathscr{H}$ in this interval, formula (200) may be used to evaluate a lower bound to the same energy. Further modifications of (200) are possible,[35] but for more details as to the calculation of lower bounds to energy eigenvalues, we will refer to the current literature.[36]

The purpose of this section is essentially to point out the existence of simple closed formulae for the wave and reaction operators which may be useful, for instance, in the exact SCF theory. So far, little use has been made of this formalism in many-body theory, and it still remains to calculate a *lower bound* to the ground-state energy corresponding to a Hartree–Fock scheme and to give an upper bound for the correlation energy. This depends partly on a turn of the interest towards other aspects of the independent-particle model related to the symmetry properties.

7. TREATMENT OF SYMMETRY PROPERTIES IN THE INDEPENDENT-PARTICLE MODEL

A. Constants of Motion

In classifying the energy eigenvalues E of the Schrödinger equation

$$\mathscr{H}\Psi = E\Psi \tag{201}$$

the constants of motion and particularly the symmetry properties of the Hamiltonian play an essential role in labeling the levels and in describing their degeneracy. If Λ is a normal constant of motion satisfying the relations

$$\mathscr{H}\Lambda = \Lambda\mathscr{H}, \qquad \Lambda\Lambda^\dagger = \Lambda^\dagger\Lambda \tag{202}$$

every eigenfunction Ψ of $\mathscr{H}$ is either automatically an eigenfunction also to Λ or—in the case of a degenerate level—may be chosen in that way:

$$\mathscr{H}\Psi = E\Psi, \qquad \Lambda\Psi = \lambda\Psi \tag{203}$$

The proof is based on the fact that, in addition to Ψ, the function $\Psi_1 = \Lambda\Psi$ is also an eigenfunction to $\mathscr{H}$:

$$\mathscr{H}\Psi_1 = \mathscr{H}(\Lambda\Psi) = \Lambda\mathscr{H}\Psi = \Lambda(E\Psi) = E\Psi_1 \tag{204}$$

For a non-degenerate level, one has immediately $\Psi_1 = \lambda\Psi$, which proves the relation $\Lambda\Psi = \lambda\Psi$.

For a degenerate energy level, the proof is somewhat more complicated. It will here be given in terms of the projection operator method, since we will anyway need this formalism for our later discussions. Let us assume that the eigenvalues of Λ are situated in a finite number of points $\lambda_1, \lambda_2, \ldots \lambda_n$ in the complex plane (each of which may be infinitely degenerate). The normal operator Λ satisfies in such a case a reduced Cayley–Hamilton equation of the type:

$$F(\Lambda) \equiv \prod_{l=1}^{n} (\Lambda - \lambda_l \,.\, 1) = 0 \tag{205}$$

Let us further introduce the operator:[36]

$$Q_k = \prod_{l \neq k} \frac{\Lambda - \lambda_l \,.\, 1}{\lambda_k - \lambda_l} = \prod_{l \neq k} \left\{1 + \frac{\Lambda - \lambda_k \,.\, 1}{\lambda_k - \lambda_l}\right\} \tag{206}$$

which obviously satisfies the relation $(\Lambda - \lambda_k \,.\, 1)Q_k = 0$, i.e.

$$\Lambda Q_k = Q_k\Lambda = \lambda_k Q_k \tag{207}$$

so that Q_k is an *eigenoperator* to Λ associated with the eigenvalue λ_k. Using (210), one finds further

$$Q_k^2 = Q_k, \qquad Q_kQ_l = 0 \qquad (k \neq l) \tag{208}$$

which means that the operators Q_k are idempotent and mutually exclusive. Introducing the complex variable Z instead of the operator Λ, one finds that the function

$$G(Z) \equiv 1 - \sum_{k=1}^{n} Q_k(Z) \tag{209}$$

is a polynomial of degree $(n - 1)$ having n different zero points, namely $Z = \lambda_1, \lambda_2, \ldots \lambda_n$. This implies the algebraic identity $G(Z) \equiv 1 - \Sigma Q_k(Z) = 0$. Introducing the operator Λ instead of the complex variable Z, one obtains the relation

$$1 = \sum_k Q_k(\Lambda) \tag{210}$$

which is described as a "resolution of the identity". Since Λ is a normal operator, one can further prove that the operators Q_k are self-adjoint.

The operators $Q_1, Q_2, \ldots Q_n$ represent a set of *projection operators* associated with a partitioning of the Hilbert space into subspaces characterized by the eigenvalues $\lambda_1, \lambda_2, \ldots \lambda_n$, respectively. They may be used to carry out a "component analysis" of an arbitrary element Φ of the Hilbert space. Using (210), one obtains

$$\Phi = 1 \,.\, \Phi = (\sum_k Q_k)\Phi = \sum_k \Phi_k \tag{211}$$

where the component $\Phi_k = Q_k\Phi$ according to (207) satisfies the eigenvalue relation

$$\Lambda\Phi_k = \lambda_k\Phi_k \tag{211'}$$

It is hence possible to write an arbitrary wave function Φ as a sum of eigenfunctions to Λ, and, by using the properties of the projection operators Q_k, it is easily shown that this component analysis is unique.

Let us now consider an exact eigenfunction Ψ of the Hamiltonian associated with a degenerate energy eigenvalue E. According to (211), one has the component analysis

$$\Psi = (\sum_k Q_k)\Psi = \sum_k \Psi_k \tag{212}$$

where $\Psi_k = Q_k\Psi$ is the projection of Ψ on the subspace characterized by the eigenvalue λ_k. Let us assume that a specific Ψ_k is non-vanishing, i.e. $\Psi_k \neq 0$. One has immediately $\Lambda\Psi_k = \lambda_k\Psi_k$.

Since Q_k is a polynomial in Λ, it follows from (202) that Q_k is also a constant of motion:

$$\mathcal{H} Q_k = Q_k \mathcal{H} \tag{213}$$

This gives further $\mathcal{H}\Psi_k = \mathcal{H} Q_k \Psi = Q_k \mathcal{H}\Psi = Q_k E \Psi = E\Psi_k$, i.e. Ψ_k is an eigenfunction to $\mathcal{H}$ associated with the eigenvalue E. In conclusion, one has consequently

$$\mathcal{H}\Psi_k = E\Psi_k, \qquad \Lambda\Psi_k = \lambda_k \Psi_k \tag{214}$$

which proves our basic theorem.

If the constants of motion form a group $G = \{g\}$, one may instead classify the energy levels by means of the irreducible representations Γ^α. In this case, one is using projection and shift operators of the type:

$$P^\alpha_{km} = f^\alpha \underset{g}{\mathrm{M}} \Gamma^\alpha_{km}(g) g^{-1} \tag{215}$$

where f^α is the order of the representation Γ^α and the symbol M stands for the "invariant mean" or average over the group. The component analysis is now based on the fact that the diagonal elements form a "resolution of the identity". For further details, we will refer to a recent survey paper.[37]

B. Symmetry Dilemma in the Conventional Hartree–Fock Scheme

In the original self-consistent-field calculations, it was always assumed that, if the system has certain symmetry properties contained in the total Hamiltonian $\mathcal{H}$, the solutions to the Hartree–Fock equations would always be symmetry-adapted. It has been shown by several authors[38] that this assumption is certainly *self-consistent*, i.e. if the assumption is once introduced into the iterative cycle (37), the solutions will always remain symmetry-adapted. There is hence a specific extreme value of the total energy $\langle \mathcal{H} \rangle$ associated with this symmetry assumption, and the only question is now whether this extreme value is associated with the *absolute minimum* of $\langle \mathcal{H} \rangle$ or not.

It should be observed that, for the exact eigenfunction Ψ, the eigenvalue relation $\Lambda\Psi = \lambda\Psi$ is essentially a consequence of the eigenvalue relation $\mathcal{H}\Psi = E\Psi$. However, if the latter relation is replaced by the variation principle $\delta\langle D|\mathscr{H}|D\rangle/\langle D|D\rangle = 0$ for an approximate eigenfunction D in the form of a Slater determinant, no one has so far proven that there follows a second equation

$$\Lambda D = \lambda D \tag{216}$$

and this relation must then be considered as a *constraint* which is necessarily going to increase the energy $\langle\mathscr{H}\rangle$. By means of a series of simple examples,[39] one may convince oneself that, for approximate wave functions of a certain fixed type, the basic symmetry properties do not follow automatically from the variation principle $\delta\langle\mathscr{H}\rangle = 0$.

On the basis of these observations, it may be convenient to reclassify some of the self-consistent-field schemes. In the *conventional Hartree–Fock scheme,* one starts apparently from two basic equations:

$$\delta\langle D|\mathscr{H}|D\rangle = 0, \qquad \Lambda D = \lambda D \tag{217}$$

with the normalization $\langle D|D\rangle = 1$, and the corresponding "relative minimum" is then said to be Λ adapted. It may be shown[40] that, if Λ is a fundamental symmetric function of the one-particle operators $\Theta_1, \Theta_2, \ldots \Theta_n$, i.e.

$$\begin{aligned} \Lambda &= \sum_i \Theta_i, \text{ or} \\ \Lambda &= \sum_{i<j} \Theta_i\Theta_j, \text{ etc.} \\ &\quad\cdot \quad \cdot \quad \cdot \end{aligned} \tag{218}$$

then the one-particle functions $\psi_k(1)$ in D are either automatically eigenfunctions to Θ_1 or may be chosen in that way.

If one drops the constraint $\Lambda D = \lambda D$ and considers only the variation principle

$$\delta\langle D|\mathscr{H}|D\rangle = 0 \tag{219}$$

one obtains the *unrestricted Hartree–Fock* (UHF) *scheme,*[41] and the solution D corresponding to the absolute minimum of $\langle\mathscr{H}\rangle$ has now usually lost its eigenvalue property with respect to Λ, i.e. the corresponding Hartree–Fock functions are no longer symmetry-adapted.

There is apparently a certain *symmetry dilemma*[42] in the Hartree–Fock scheme based on a single Slater determinant D. If one looks for the absolute minimum of the energy, one loses the symmetry properties, and, if one restores the symmetry properties, the energy goes up considerably. Since the *correlation energy* in (1) essentially refers to the difference between the exact energy and the energy in the *conventional* Hartree–Fock scheme, a very large part of the correlation energy depends apparently on the symmetry constraint. By going over to the unrestricted Hartree–Fock scheme and using one-particle functions of the general type

$$\Psi(\chi) = \Psi_+(\mathbf{r})\alpha(\zeta) + \psi_-(\mathbf{r})\beta(\zeta) \tag{220}$$

one can hence certainly remove a large fraction of the correlation error in the independent-particle model. The main difficulty is then that the resulting Slater determinant D turns out to be a "mixture" between different symmetry types. If one carries out a component analysis of this wave function according to (211), one obtains the unique resolution

$$D = (\sum_k Q_k)D = \sum_k Q_k D = \sum_k D_k \tag{221}$$

where $D_k = Q_k D$ is the projection of D on the λ_k space and $\Lambda D_k = \lambda_k D_k$. If $D_k \neq 0$, one can further introduce the quantities

$$w_k = \frac{\langle D_k | D_k \rangle}{\langle D | D \rangle} = \frac{\langle D | Q_k | D \rangle}{\langle D | D \rangle} \tag{222}$$

$$\mathscr{E}_k = \frac{\langle D_k | \mathscr{H} | D_k \rangle}{\langle D_k | D_k \rangle} \tag{223}$$

where w_k is the amount of λ_k character in the wave function, and $\mathscr{E}_k$ is the expectation value of $\mathscr{H}$ with respect to the component Ψ_k. One has $0 \leq w_k \leq 1$ and $\sum_k w_k = 1$. Since further

$$\langle D_k | \mathscr{H} | D_l \rangle = \langle Q_k D | \mathscr{H} | Q_l D \rangle = \langle D | \mathscr{H} Q_k Q_l | D \rangle = 0 \tag{224}$$

for $k \neq l$, one obtains

$$\langle \mathscr{H} \rangle = \frac{\langle D | \mathscr{H} | D \rangle}{\langle D | D \rangle} = \sum_{k,l} \frac{\langle D_k | \mathscr{H} | D_l \rangle}{\langle D | D \rangle} = \sum_k w_k \mathscr{E}_k \tag{225}$$

i.e. $\langle \mathscr{H} \rangle$ is a weighed average of the quantities $\mathscr{E}_k$. This result implies also that, unless all quantities $\mathscr{E}_k$ are the same, there is at least one of them which is lower than $\langle\mathscr{H}\rangle$. By selecting a proper component D_k of D, one can hence both restore the symmetry and lower the energy. The wave function $D_k = Q_k D$ is still associated with the *independent-particle model* in the sense that it is uniquely defined by a Hartree product (6), but it is no longer any single Slater determinant. It should also be observed that, since the energy $\langle\mathscr{H}\rangle$ is optimized with respect to D, a further lowering of the energy may be possible by studying the component D_k in greater detail itself.

C. Projected Hartree–Fock Scheme

The ideas outlined above lead over to a new scheme in which the total wave function Ψ is approximated by the projection Q of a single Slater determinant D:

$$\Psi \approx QD \tag{226}$$

where $Q = Q_k(\Lambda)$ for a single constant of motion Λ, and $Q = \{P^{\alpha}_{kk}\}$ for a group $G = \{g\}$. For certain constants of motion, e.g. translations and angular momenta, one may have a choice between the product form (206) and the expansion (215) for $m = k$ which may influence the calculations but not the final results.

In this *projected Hartree–Fock* (PHF) *scheme* first suggested by the author[43] in 1954, one is using general spin-orbitals of the type (225) in the Slater determinant D and applies the variation principle to the expression

$$\langle\mathscr{H}\rangle = \frac{\langle QD|\mathscr{H}|QD\rangle}{\langle QD|QD\rangle} = \frac{\langle D|Q^\dagger\mathscr{H}Q|D\rangle}{\langle D|Q^\dagger Q|D\rangle} = \frac{\langle D|\mathscr{H}Q|D\rangle}{\langle D|Q|D\rangle} \tag{227}$$

which is the same as if one considered a single Slater determinant D but replaced the operators $\mathscr{H}$ and 1 in the ordinary variational expression $\langle D|\mathscr{H}|D\rangle/\langle D|1|D\rangle$ by the operators $\mathscr{H}Q$ and Q, respectively.

Considering (217), one may say that the conventional Hartree–Fock scheme is based on the two relations:

$$\delta\langle\mathscr{H}\rangle = 0, \qquad QD = D \tag{228}$$

In both the conventional and the projected scheme, one can hence say that one is using wave functions of the type QD, but that one has removed the constraint $QD = D$ in the latter scheme. It is clear that the wave function QD associated with the *absolute minimum* of (227) is no longer going to be a single Slater determinant. It should still be remembered, however, that the function QD is still associated with the *independent-particle model* in the sense that it is uniquely defined by a Hartree product of the type (6).

In this connection, it may be helpful to remember that Slater[4] introduced the Pauli principle explicitly into the independent-particle model through the construction of the Slater determinant (7), which may be considered as the antisymmetric projection (8) of the Hartree product. In the projected Hartree–Fock scheme, one goes on and forms those additional projections of the Hartree product which are required by the symmetry properties of the system under consideration.

It is true that a Slater determinant D is the sum of $N!$ Hartree products and that the projection QD is a sum of Slater determinants, but, in both cases, one is much better off, if one considers both quantities D and QD as *conceptual entities* which are essentially defined by a set of N one-particle functions $\mathbf{\Psi} = \{\psi_1, \psi_2, \ldots \psi_N\}$. If one has a linear transformation $\mathbf{\Psi}' = \mathbf{\Psi}\mathbf{a}$ of the type (15), one obtains by applying the projection Q to the relation (16):

$$QD' = (QD) \cdot |\mathbf{a}| \tag{229}$$

i.e. the projection is changed only by a constant factor. This implies also that the wave function QD is essentially determined by the linear manifold $\mathscr{M}_N$ spanned by the set $\mathbf{\Psi}$ and its projection operator ρ defined by (13) and (19). In practical applications using expansion methods based on a complete set $\boldsymbol{\phi} = \{\phi_\mu\}$, one may further replace ρ by the discrete charge- and bond-order matrix $\mathbf{R} = \{R_{\mu\nu}\}$ defined by (53).

Putting $\langle \mathscr{H} \rangle = \overline{\mathscr{H}}$ and varying D in (227), one obtains the condition

$$\langle \delta D | (\mathscr{H} - \overline{\mathscr{H}}) Q | D \rangle = 0 \tag{230}$$

which means that the resulting equations will be of the same type as those obtained in the ordinary UHF scheme with the difference that the original many-body Hamiltonian $\mathscr{H}$ is now replaced by the *composite* Hamiltonian

$$(\mathscr{H} - \bar{\mathscr{H}})Q \tag{231}$$

which contains both the projection operator Q and the expectation value $\bar{\mathscr{H}}$ of the energy. Since this composite Hamiltonian is also of the general type (128), the condition (230) leads to extended Hartree–Fock equations of the type

$$H_{\text{eff}}(1)\psi_k(1) = \epsilon_k \psi_k(1) \tag{232}$$

which are completely analogous to (33) but have a more complicated effective Hamiltonian. For some further details as to this scheme, we will refer to a recent review article.[44]

D. Correlation Error in the Projected Hartree–Fock Scheme

In the classical formulation of the Pauli exclusion principle, one would permit a maximum of two electrons in each orbital, provided that they had antiparallel spins. This idea of "pairing" of electrons has been of fundamental importance in atomic and molecular theory. However, as pointed out in the introduction, the correlation error in the conventional independent-particle model seems also to be closely associated with such electron pairs having antiparallel spins.

Intuitively, one can hence remove a very large part of the correlation error simply by permitting "different orbitals for different spins" (DODS), so that electrons having α spin can avoid those having β spin. Since one is using one-electron functions of either α or β type, the result is more restricted than the general case based on (220), but the approach has still been useful as a first step towards the use of more general spin-orbitals. It is clear that the Slater determinant D in such a scheme is no longer a pure eigenfunction to the total spin S^2, and the symmetry properties have to be restored by projection. The DODS scheme is hence a special case of the PHF method in which one is using somewhat restricted spin-orbitals.

The *alternant-molecular-orbital* (AMO) method suggested in 1953 by the author[45] is essentially a special case of the DODS scheme. In following an idea outlined by Slater[46] in 1930, it is suggested that electrons having different spins in alternant systems may accumulate on different subsystems of atoms and that their separation may be regulated by a single parameter θ or by a set of parameters $\{\theta_k\}$ according to the variation principle. Applications to simple conjugated systems indicate that more than 85 per cent of the original correlation energy may be removed in this way, and the method has now been applied to a large number of such systems including an infinite chain.[47] For an excellent survey of this approach, we will refer to a recent book by Pauncz.[48]

In solid-state physics, the AMO-method has successfully been used by Calais[49] in a study of the cohesive properties of the alkali metals.

Recent research has made evident that, in treating correlation effects and particularly spin-density problems, it is essential that one uses general spin-orbitals of the type (220) in the PHF scheme. Applications to simple atomic and molecular systems have shown that one may remove about 95 per cent of the conventional correlation energy in this way. As a rule of thumb, the correlation energy is usually about 1 per cent of the total energy in such systems, and this means that the accuracy could now be improved to about 0.05 per cent. The conclusion seems to be that, at least for electrons, the main part of the correlation error can be removed by a proper treatment of the symmetry properties. Since the "shell-structures" of atoms and atomic nuclei are essentially manifestations of fundamental symmetry properties, one gets the feeling that it may be more appropriate to explain these structures on the basis of the PHF method, and that the reaction operator approach and the exact SCF method should be reserved for the removal of only the last part of the correlation error. In order to reach full accuracy, it is anyway necessary to make a synthesis of the two approaches, and this problem will be discussed in Section 8.

E. Multiconfigurational SCF Methods

The Hartree–Fock method is characterized by the fact that, in treating an N-particle system, one has N one-particle functions $\mathbf{\Psi}_N = \{\psi_1, \psi_2, \ldots \psi_N\}$ at one's disposal which are chosen so that one optimizes the energy $\langle \mathscr{H} \rangle$. If one instead had M one-particle functions

$$\mathbf{\Psi}_M = \{\psi_1, \psi_2, \ldots \psi_M\} \tag{233}$$

with $M > N$, one can construct $\binom{M}{N}$ Slater determinants D_K corresponding to ordered configurations $K = (k_1, k_2, \ldots k_N)$ having $k_1 < k_2 < \ldots < k_N$. Let us now consider an approximate eigenfunction Ψ_M to $\mathscr{H}$ of the form:

$$\Psi_M = \sum_K D_K C_K \tag{234}$$

i.e. a sum over the configurations in analogy to (104). If one keeps the set $\boldsymbol{\Psi}_M$ fixed, the best coefficients C_K are obtained from the variation principle $\delta\langle\mathscr{H}\rangle = 0$ which gives

$$\sum_L (\mathscr{H}_{KL} - E\delta_{KL})C_L = 0 \tag{234'}$$

If one further varies the set $\boldsymbol{\Psi}_M$ for fixed M, one obtains better and better energies until finally a certain minimum is reached. This idea leads to the multiconfigurational self-consistent-field (MC–SCF) method which was developed long ago,[50] but which has recently become very popular among the people carrying out numerical applications.

It should perhaps be observed that, for fixed finite M, the MC–SCF method is subject to the same *symmetry dilemma* as the Hartree–Fock scheme. Since the expansion (234) does not represent an exact eigenfunction to $\mathscr{H}$, the symmetry relation $\Lambda\Psi_M = \lambda\Psi_M$ does not follow automatically from the variation principle $\delta\langle\mathscr{H}\rangle_M = 0$, and it is then a constraint which necessarily raises the energy.

The *conventional* MC–SCF *scheme* is hence characterized by the two conditions

$$\delta\langle\mathscr{H}\rangle_M = 0, \qquad \Lambda\Psi_M = \lambda\Psi_M \tag{235}$$

whereas, in the *unrestricted* MC–SCF *scheme*, one has dropped the symmetry constraint. This implies that the absolute minimum of $\langle\mathscr{H}\rangle_M$ is usually associated with an approximate wave function Ψ_M which is a "mixture" of different symmetry types. This wave function may again be studied by means of "component analysis" according to (211), and the final wave function hence takes the form

$$\Psi \approx O\Psi_M \tag{236}$$

Further, the improvements are then possible by varying the parameters in this wave function until the energy $\langle\mathscr{H}\rangle$ reaches its absolute minimum, and one is in this way led to the *projected* MC–SCF *scheme.*

It is sometimes believed that the MC–SCF scheme would be a good substitute for the PHF-method, but the complications mentioned in connection with the conditions (235) indicate instead that the "projection idea" may become a natural element also in the multiconfigurational approach.

8. TREATMENT OF SYMMETRY PROPERTIES IN PERTURBATION THEORY

A. Treatment of Symmetry Properties in Partitioning Technique

In the theory of wave and reaction operators described in Section 6, we have not considered the simplifications which are possible if there are constants of motion Λ or $G = \{g\}$. If operator Λ has the eigenvalues $\lambda_1, \lambda_2, \ldots \lambda_n$ associated with the projection operators $Q_1, Q_2, \ldots Q_n$, it follows from the relations

$$Q_k^\dagger Q_l = 0, \qquad Q_k^\dagger \mathscr{H} Q_l = 0, \qquad (k \neq l) \tag{237}$$

that wave functions in subspaces associated with different eigenvalues are orthogonal and non-interacting with respect to H, which means that the subspaces are completely independent of each other. The author[51] has previously emphasized that this leads to a simplification of the partitioning technique: if one starts out from a reference point in one of the subspaces, one is going to stay within this subspace, and it is sufficient to consider the "orthogonal complement" with respect to this subspace only.

Let us consider a specific eigenvalue λ_k and the associated projection operator Q_k; in the following, we will omit the index k, if there is no risk for misunderstanding. Let us assume that we start out from an arbitrary function φ, which has a non-vanishing component

$$\bar{\varphi} = Q\varphi \neq 0 \tag{238}$$

in the subspace of Q. Choosing this function as the new reference function, one obtains for the associated projection operator in the partitioning technique:

$$O = \frac{|\bar{\varphi}\rangle\langle\bar{\varphi}|}{\langle\bar{\varphi}|\bar{\varphi}\rangle} = \frac{Q|\varphi\rangle\langle\varphi|Q}{\langle\varphi|Q|\varphi\rangle} = Q|\varphi\rangle\langle\varphi|Q|\varphi\rangle^{-1}\langle\varphi|Q \tag{239}$$

The operator $P = 1 - O$ is then the projection operator for the orthogonal complement with respect to the entire Hilbert space, whereas

$$\begin{aligned} P_1 &= QP = PQ = Q - O \\ &= Q - Q|\varphi\rangle\langle\varphi|Q|\varphi\rangle^{-1}\langle\varphi|Q \end{aligned} \tag{240}$$

is the projection operator for the orthogonal complement of O with respect to the Q space.

The reduced resolvent T is defined by formula (143). Since Q commutes with $\mathscr{H}$ as well as P, one gets immediately $TQ = QT$, and we will now study the operator

$$T_1 = TQ = QT \tag{241}$$

in greater detail. From the relation $P_1 = P_1Q$, one obtains $P_1(\mathscr{E} - P\mathscr{H}P) = (\mathscr{E} - P_1\mathscr{H}P_1)P_1$ and, after multiplication by the proper inverses:

$$(\mathscr{E} - P_1\mathscr{H}P_1)^{-1}P_1 = P_1(\mathscr{E} - P\mathscr{H}P)^{-1} = QT = T_1 \tag{242}$$

i.e.

$$T_1 = (\mathscr{E} - P_1\mathscr{H}P_1)^{-1}P_1 \tag{243}$$

The operator T_1 is apparently the reduced resolvent in the case when all considerations are limited to the Q space alone.

According to (150), one obtains now for the trial wave function in the partitioning technique:

$$\begin{aligned} \Psi_\varepsilon &= \bar{\varphi} + T\mathscr{H}\bar{\varphi} = Q\varphi + T\mathscr{H}Q\varphi \\ &= Q\varphi + T_1\mathscr{H}\varphi = Q(\varphi + T_1\mathscr{H}\varphi) \end{aligned} \tag{244}$$

showing that this function is also situated properly in the Q space. It has the intermediate normalization

$$\langle\bar{\varphi}|\Psi_\mathscr{E}\rangle = \langle\bar{\varphi}|\bar{\varphi}\rangle = \langle\varphi|Q|\varphi\rangle \tag{245}$$

and it satisfies the inhomogeneous Schrödinger equation

$$(\mathscr{H} - \mathscr{E})\Psi_\mathscr{E} = (\mathscr{E}_1 - \mathscr{E})Q\varphi \tag{246}$$

where the quantity $\mathscr{E}_1$ is now given by the relation

$$\mathscr{E}_1 = \frac{\langle\bar{\varphi}|H|\Psi_\mathscr{E}\rangle}{\langle\bar{\varphi}|\Psi_\mathscr{E}\rangle} = \frac{\langle\varphi|H|\Psi_\mathscr{E}\rangle}{\langle\varphi|Q|\varphi\rangle} = \frac{\langle\varphi|HQ + HT_1H|\varphi\rangle}{\langle\varphi|Q|\varphi\rangle} \tag{247}$$

which defines the bracketing function $\mathscr{E}_1 = f(\mathscr{E})$. The formulae obtained show that it is hence possible to carry out the entire discussion of the partitioning technique in the Q space alone, considering one eigenvalue λ at a time.

B. Treatment of Symmetry Properties in Wave and Reaction Operator Formalism

If the Hamiltonian has the form $\mathscr{H} = \mathscr{H}_0 + V$, the treatment of the symmetry properties of $\mathscr{H}$ becomes simple if the projection operator Q commutes with $\mathscr{H}_0$ as well as $\mathscr{H}$. In the general case, however, one has

$$Q\mathscr{H}_0 \neq \mathscr{H}_0 Q \tag{248}$$

and this implies usually that $Q\varphi_0 \neq \varphi_0$. This occurs, for instance, in the treatment of the correlation effects, if $\mathscr{H}_0$ corresponds to the unperturbed Hamiltonian associated with the *unrestricted Hartree–Fock scheme.* Even in this case, it is convenient to start from a specific unperturbed reference state φ_0, satisfying the relation

$$\mathscr{H}_0\varphi_0 = E_0\varphi_0 \tag{249}$$

In the partitioning technique, we will now make the choice $\varphi = \varphi_0$. According to (240), one has

$$\begin{aligned} P_1 &= Q - Q|\varphi_0\rangle\langle\varphi_0|Q|\varphi_0\rangle^{-1}\langle\varphi_0|Q \\ &= w^\dagger Q = Qw \end{aligned} \tag{250}$$

where we have introduced the new notations:

$$w = 1 - |\varphi_0\rangle\langle\varphi_0|Q|\varphi_0\rangle^{-1}\langle\varphi_0|Q \tag{251}$$

$$w^\dagger = 1 - Q|\varphi_0\rangle\langle\varphi_0|Q|\varphi_0\rangle^{-1}\langle\varphi_0| \tag{252}$$

It follows immediately that

$$P_1\varphi_0 = 0, \qquad w^\dagger\mathscr{H}_0 P_1 = \mathscr{H}_0 P_1 \tag{253}$$

In the reduced Hamiltonian (243), the essential operator $P_1\mathscr{H}P_1$ takes now the following form:

$$\begin{aligned} P_1\mathscr{H}P_1 &= w^\dagger Q\mathscr{H}P_1 = w^\dagger\mathscr{H}P_1 = w^\dagger(\mathscr{H}_0 + V)P_1 \\ &= \mathscr{H}_0 P_1 + w^\dagger V P_1 = (\mathscr{H}_0 + w^\dagger V)P_1 \end{aligned} \tag{254}$$

According to (146), one hence obtains

$$T_1 = (\mathscr{E} - P_1\mathscr{H}P_1)^{-1}P_1 = [\mathscr{E} - (\mathscr{H}_0 + w^\dagger V)P_1]^{-1}P_1$$
$$= (\mathscr{E} - \mathscr{H}_0 - w^\dagger V)^{-1}P_1 \tag{255}$$

The reduced resolvent for the unperturbed Hamiltonian $\mathscr{H}_0$ is given by the expression:

$$T_0 = (\mathscr{E} - \mathscr{H}_0)^{-1}\{1 - |\varphi_0\rangle\langle\varphi_0|\} \tag{256}$$

According to (252), one has further $\langle\varphi_0|w^\dagger = 0$, and multiplying (256) to the right by $w^\dagger$ and $P_1 = w^\dagger Q$, one obtains

$$T_0 w^\dagger = (\mathscr{E} - \mathscr{H}_0)^{-1}w^\dagger, \qquad T_0 P_1 = (\mathscr{E} - \mathscr{H}_0)^{-1}P_1 \tag{257}$$

From (255) and the operator rule $(ab)^{-1} = b^{-1}a^{-1}$, it then follows

$$T_1 = [1 - (\mathscr{E} - \mathscr{H}_0)^{-1}w^\dagger V]^{-1}(\mathscr{E} - \mathscr{H}_0)^{-1}P_1$$
$$= (1 - T_0 w^\dagger V)^{-1}T_0 P_1 \tag{258}$$

This relation is analogous to (172). In ordinary perturbation theory, it is possible to express the reduced resolvent T explicitly in terms of T_0 and V. It is remarkable that, even in the case (248), it is possible to express the corresponding operator T_1 in terms of T_0, V, $w^\dagger$, and $P_1 = w^\dagger Q$.

In the Schrödinger case, one obtains analogously

$$T_1 = (1 - R_0 w^\dagger V')^{-1}R_0 P_1 \tag{259}$$

which expression should be compared with (178). A detailed discussion of the expressions (258) and (259) will be given elsewhere.[52]

Here we will only emphasize that, even in the case (248), the wave and reaction operators take a very simple form. From $P_1\varphi_0 = 0$, it follows that $T_1\varphi_0 = 0$. According to (244), one hence obtains

$$\Psi_{\mathscr{E}} = Q\varphi_0 + T_1 V\varphi_0 = (Q + T_1 V)\varphi_0 \tag{260}$$

Taking the proper intermediate normalization into account, one gets the following expressions for W and t:

$$W = \frac{Q + T_1 V}{\langle \varphi_0 | Q | \varphi_0 \rangle} \tag{261}$$

$$t = VW = \frac{VQ + VT_1 V}{\langle \varphi_0 | Q | \varphi_0 \rangle} \tag{262}$$

It is easily checked that formula (262) is in accordance with (247), and that one obtains

$$\begin{aligned} \mathscr{E}_1 &= E_0 + \langle \varphi_0 | t | \varphi_0 \rangle \\ &= E_0 + \frac{\langle \varphi_0 | VQ + VT_1 V | \varphi_0 \rangle}{\langle \varphi_0 | Q | \varphi_0 \rangle} \end{aligned} \tag{263}$$

It is hence possible to incorporate the symmetry properties of the total Hamiltonian in the reaction operator formalism and in the exact SCF theory even in the general case (248).

This may be of importance for the ultimate studies of the correlation effects, since it may be simpler to use the unperturbed Hamiltonian $\mathscr{H}_0$ associated with the UHF scheme than the one associated with the PHF scheme. In all events, these questions are still rather far in the future and, for the moment, we believe that it is of particular interest to study the various aspects of the *projected Hartree–Fock scheme* in greater detail before trying to proceed. In this connection, one should also observe the interesting results reported by Lefebvre at this summer school.

References

1. Bohr, N., *Proc. Phys. Soc.* (*London*), **35**, 296 (1923); Hartree, D. R., *Proc. Cambridge Phil. Soc.*, **24**, 89 (1928).
2. Löwdin, P.-O., *Advan. Chem. Phys.*, *Vol.* 2, Ed. I. Prigogine, Interscience Publishers, New York, 1959, p. 207.
3. Linderberg, J., and Shull, H., *J. Mol. Spectry*, **4**, 30 (1960); see also E. Clementi, *IBM J. Res. Develop.*, **9**, No. 1, January, 1965.
4. Slater, J. C., *Phys. Rev.*, **34**, 1293 (1929); *Phys. Rev.*, **35**, 210 (1930).
5. Löwdin, P.-O., *Phys. Rev.*, **97**, 1490 (1955).
6. Fock, V., *Z. Physik*, **61**, 126 (1930).
7. Dirac, P. A. M., *Proc. Cambridge Phil. Soc.*, **26**, 376 (1930); **27**, 240 (1931).
8. Koopmans, T., *Physica*, **1**, 104 (1933).
9. Lennard-Jones, J., *Proc. Roy. Soc.* (*London*), *Ser. A*, **198**, 1, 14 (1949), and a long series of papers with G. G. Hall and J. A. Pople.
10. Koster, G. F., *Phys. Rev.*, **89**, 67 (1953).

11. Brillouin, L., *Actualités Sci. et Ind.* No. **71** (1933); No. **159** (1934); Møller, C., and Plesset, M. S., *Phys. Rev.*, **46,** 618 (1934).
12. Coulson, C. A., and Longuet-Higgins, H. C., *Proc. Roy. Soc.* (*London*), *Ser. A*, **191,** 39; **192,** 16 (1947); **193,** 447, 456; **195,** 188 (1948).
13. Coulson, C. A., *Proc. Cambridge Phil. Soc.*, **34,** 204 (1938).
14. Roothaan, C. C. J., *Rev. Mod. Phys.*, **23,** 69 (1951).
15. Hall, G. G., *Proc. Roy. Soc.* (*London*), *Ser. A*, **202,** 336 (1950); **205,** 541 (1951); **213,** 113 (1952).
16. McWeeny, R., *Proc. Roy. Soc.* (*London*), *Ser. A*, **235,** 496 (1956); **237,** 355 (1956).
17. Wyatt, R. E., and Parr, R. G., *J. Chem. Phys.*, **43,** S217 (1965), and a series of later papers.
18. Hylleraas, E. A., *Z. Physik*, **48,** 469 (1928); **54,** 347 (1929).
19. Frost, A. A., Braunstein, J., and Schwemer, W., *J. Am. Chem. Soc.*, **70,** 3292 (1948); Löwdin, P.-O., and Redei, L., *Phys. Rev.*, **114,** 752 (1959).
20. See Reference 2, particularly pp. 303–306 and pp. 316–317; compare further Löwdin, P.-O., *Rev. Mod. Phys.*, **32,** 328 (1960).
21. Bohm, D., and Pines, D., *Phys. Rev.*, **92,** 609 (1953).
22. Krisement, O., *Phil. Mag.*, **2,** 245 (1957).
23. Wigner, E., *Phys. Rev.*, **46,** 1002 (1933); *Trans. Faraday Soc.*, **34,** 678 (1938).
24. Dalgarno, A., and Lewis, J. T., *Proc. Roy. Soc.* (*London*), *Ser. A*, **233,** 70 (1955).
25. For a survey, see Hirschfelder, J. O., Byers Brown, W., and Epstein, S. T., *Advan. Quantum Chem.*, **1,** 255 (1964).
26. Bartlett, J. H., *Phys. Rev.*, **98,** 1067 (1955).
27. See e.g. the survey of the evolution operator technique in Löwdin, P. O., *Advan. Quantum Chem.*, **3,** 323 (1967).
28. Gell-Mann, M., and Low, F., *Phys. Rev.*, **84,** 350 (1951).
29. Schrödinger, E., *Ann. Physik*, (4) **80,** 437 (1926); Nissen, K. F., *Phys. Rev.*, **34,** 253 (1929).
30. Löwdin, P.-O., *Phys. Rev.*, **97,** 1474 (1955).
31. Brueckner, K. A., Levinson, C. A., and Mahmoud, H. M., *Phys. Rev.*, **95,** 217 (1954); Brueckner, K. A., *Phys. Rev.*, **96,** 508 (1954); **97,** 1353 (1955); **100,** 36 (1955); Brueckner, K. A., and Levinson, C. A., *Phys. Rev.*, **97,** 1344 (1955); Bethe, H. A., *Phys. Rev.*, **103,** 1353 (1956); Goldstone, J., *Proc. Roy. Soc.* (*London*), *Ser. A*, **239,** 267 (1957); Bethe, H. A., and Goldstone, J., *Proc. Roy. Soc.* (*London*), *Ser. A*, **238,** 511 (1957); Rodberg, L. S., *Ann. Phys.* (*N.Y.*), **2,** 199 (1957); to mention only a selection of the rich literature on this subject.
32. Löwdin, P.-O., *J. Math. Phys.*, **3,** 1171 (1962).
33. Löwdin, P.-O., *J. Math. Phys.*, **3,** 969 (1962).
34. Löwdin, P.-O., *Phys. Rev.*, **139,** A357 (1965).
35. Löwdin, P.-O., *J. Chem. Phys.*, **43,** S175 (1965); Reid, C. E., *J. Chem. Phys.*, **43,** S186 (1965); Choi, J. H., and Smith, D. W., *J. Chem.*

Phys., **43,** S189 (1965); Bunge, C. F., and Bunge, A., *J. Chem. Phys.*, **43,** S194 (1965); Löwdin, P.-O., *Perturbation Theory and its Applications in Quantum Mechanics,* Ed. C. H. Wilcox, John Wiley and Sons, New York, 1966, p. 255; Choi, J. H., and Smith, D. W., *J. Chem. Phys.*, **45,** 4425 (1966); Wilson, T. M., *J. Chem. Phys.*, **47,** 1200, 3912, 4706 (1967); Wilson, T. M., and Reid, C. E., *J. Chem. Phys.*, **47,** 3920 (1967); Goscinski, S. O., *Intern. J. Quantum Chem.*, **1,** 769 (1967); Wilson, T. M., *Intern. J. Quantum Chem.*, **S1,** 511 (1967).

36. Löwdin, P.-O., *Phys. Rev.*, **97,** 1509 (1955); *Rev. Mod. Phys.*, **32,** 328 (1960); **34,** 520 (1962).
37. Löwdin, P.-O., *Rev. Mod. Phys.*, **39,** 259 (1967).
38. Delbrück, M., *Proc. Roy. Soc. (London), Ser. A,* **129,** 686 (1930); Roothaan, C. C. J., *Rev. Mod. Phys.*, **32,** 179 (1960); Löwdin, P.-O., *J. Appl. Phys.* Suppl., **33,** 251 (1962).
39. See, for instance, the discussions in Slater, J. C., *Phys. Rev.*, **35,** 509 (1930); *Rev. Mod. Phys.*, **6,** 209 (1934); *Phys. Rev.*, **81,** 385 (1951); **82,** 538 (1951); *Rev. Mod. Phys.*, **25,** 199 (1953); Pauncz, R., de Heer, J., and Löwdin, P.-O., *J. Chem. Phys.*, **66,** 2247, 2257 (1962); Overhauser, A. W., *Phys. Rev. Letters,* **4,** 415, 462 (1960); Kohn, W., and Nettel, S. J., *Phys. Rev. Letters,* **5,** 8 (1960); Sawada, K., and Fukuda, N., *Progr. Theoret. Phys. (Kyoto),* **25,** 653 (1961); Henley, E. M., and Ruijgrok, Th. W., *Ann. Phys. (N.Y.),* **12,** 409 (1961); Henley, E. M., and Wilets, L., *Ann. Phys. (N.Y.),* **14,** 120 (1961); Arai, T., *Tech. Rep.* 1961, Argonne National Laboratories (unpublished).
40. Löwdin, P.-O., *J. Phys. Chem.*, **61,** 55 (1957).
41. Compare the discussion in Nesbet, R. K., *Proc. Roy. Soc. (London), Ser. A,* **230,** 312 (1955); Pratt, Jr., G. W., *Phys. Rev.*, **102,** 1303 (1956); Wood, J. H., and Pratt, Jr., G. W., *Phys. Rev.*, **107,** 995 (1957); Nesbet, R. K., and Watson, R. E., *Ann. Phys. (N.Y.),* **9,** 260 (1960); Sachs, L. M., *Phys. Rev.*, **117,** 1504 (1960); Watson, R. E., and Freeman, A. J., *Phys. Rev.*, **210,** 1125, 1134 (1960).
42. Löwdin, P.-O., *Rev. Mod. Phys.*, **35,** 496 (1963).
43. Löwdin, P.-O., *Quart. Progr. Rept. Solid-State Molecular Theory Group M.I.T.*, June 15, 1954; *Phys. Rev.*, **97,** 1509 (1955); *Ann. Acad. Reg. Sci. Upsalien,* **2,** 127 (1958); *J. Appl. Phys. Suppl.*, **33,** 251 (1962); *Rev. Mod. Phys.*, **34,** 520 (1962).
44. Löwdin, P.-O., *Quantum Theory of Atoms, Molecules, and the Solid State,* Slater dedicatory volume, Academic Press, New York, 1966, p. 601.
45. Löwdin, P.-O., *Nikko Symp. Mol. Phys.* (Symposium on molecular physics held at Nikko on the occasion of the Internat. Conf. on Theoret. Phys., Sept. 1953 in Tokyo and Kyoto; Maruzen, Tokyo 1954), p. 13; *Phys. Rev.*, **97,** 1509 (1955); Proc. 10th Solvay Conf., 1954, p. 71 (Inst. internat. de physique Solvay. 10e conseil de physique tenu à Bruxelles 1954: Les electrons dans le métaux, Rapports et discussions, Bruxelles 1955); *Rev. Mod. Phys.*, **32,** 328 (1960).
46. Slater, J. C., *Phys. Rev.*, **35,** 509 (1930).

47. Itoh, T., and Yoshizumi, H., *J. Phys. Soc. Japan*, **10**, 201 (1955); *J. Chem. Phys.*, **23**, 412 (1955); *Busseiron Kenkyu*, **83**, 13 (1955); Lefebvre, R., Dearman, H. H., and McConnell, H. M., *J. Chem. Phys.*, **32**, 176 (1960); Löwdin, P.-O., Pauncz, R., and de Heer, J., *J. Chem. Phys.*, **36**, 2247, 2257 (1962); de Heer, J., *J. Chem. Phys.*, **37**, 2080 (1962); Pauncz, R., *J. Chem. Phys.*, **37**, 2739 (1962); de Heer, J., *Rev. Mod. Phys.*, **35**, 631 (1963); Pauncz, R., in *Molecular Orbitals in Chemistry, Physics, and Biology*, Ed. P.-O. Löwdin, Academic Press, New York, 1964, p. 433; *Tetrahedron*, **19**, Suppl. 2, 43 (1963); *J. Chem. Phys.*, **43**, S69 (1965); Goscinski, O., and Calais, J. L., *Arkiv Fysik*, **29**, 135 (1965); de Heer, J., and Pauncz, R., *J. Chem. Phys.*, **39**, 2314 (1963).
48. Pauncz, R., *Alternant Molecular Orbital Method*, W. B. Saunders, Philadelphia, 1967.
49. Calais, J. L., *Arkiv Fysik*, **28**, 479, 511, 539 (1965); **29**, 255 (1965); Calais, J. L., *Intern. J. Quantum Chem.*, **13**, 661 (1967).
50. Frenkel, J., *Wave Mechanics, Advanced General Theory*, Clarendon Press, Oxford, 1934, pp. 460–462; Slater, J. C., *Phys. Rev.*, **91**, 528 (1953); Löwdin, P.-O., *Phys. Rev.*, **97**, 1474 (1955).
51. Löwdin, P.-O., *J. Math. Phys.*, **3**, 1171 (1962), particularly p. 1183; *Molecular Orbitals in Chemistry, Physics, and Biology* (Ed. P.-O. Löwdin), Academic Press, New York, 1964, p. 37.
52. Löwdin, P.-O., *Studies in Perturbation Theory. XIII. Treatment of Constants of Motion in Partitioning Technique and Perturbation Theory*, Technical Note from the Uppsala Quantum Chemistry Group, 1967.

CORRELATION EFFECTS IN DIATOMIC MOLECULES OBTAINED FROM CONFIGURATION INTERACTION USING HARTREE–FOCK ORBITAL. EFFECTS ON ENERGY AND MONOELECTRONIC OPERATORS

F. GRIMALDI, *Commissariat a l'Energie Atomique, 29, Rue de la Fédération, Paris, France*

CONTENTS

1. INTRODUCTION

Extensive configuration interaction calculations can now be made currently on diatomic molecules whose ground state is a closed-shell singlet. So far calculations have been made on N_2[1] and CO.[2]

The problem is how to obtain the best limited set of configurations to span the wave function subspace useful for the actual calculation.

In his calculation of Be, Watson[3] built this set by optimizing the second-order contribution to the energy of each basis orbital.

We choose the set determined once and for all by the "restricted molecular Hartree–Fock" program of Nesbet–Stevens.[4] It gives a total number of orbitals, occupied and unoccupied, equal to the number of corresponding basis functions (STO), for each symmetry type.

These orbitals form an orthonormal set. Starting then from the Hartree–Fock wave function ϕ_0 a large number (several thousands) of orthonormal configurations can be obtained by substituting in ϕ_0 occupied orbitals (holes) by unoccupied ones (particles).

We shall use the following notation: ϕ_a^α for a single substitution or monoexcitation; $\phi_{ab}^{\alpha\beta}$ for a double substitution or diexcitation, the lower indices corresponding to holes and the upper ones, to particles. In this notation the proper projection operators have been applied so that the configurations are symmetry adapted for space and spin.

Up to now we have restricted ourselves to single and double substitutions.

The Nesbet–Stevens program gives all one- and two-electron integrals involving any orbitals. The matrix representation of the Hamiltonian can then be obtained in any configuration subspace. The one- and two-electron operator matrix elements are computer determined.

As we have a good reference state, perturbation expansion to any order is used to find the lower eigenvalue and eigenstate.

Two kinds of calculations are made: (*a*) *Exact* within a limited subspace of configurations; so far we have used 200, this could be increased. (*b*) *Approximate* to second order, using every available doubly substituted configuration. Their number amounts to several thousands.

2. CONFIGURATION INTERACTION

A. Choice of Configurations

(*a*) *Diexcited configurations.* From results of previous calculations we have been led to use the second-order energy contribution to a Rayleigh–Schrödinger series:

$$E^{(2)} = |\langle\phi_0|V|\phi_{ab}^{\alpha\beta}\rangle|^2/[\langle\phi_0|H^0|\phi_0\rangle - \langle\phi_{ab}^{\alpha\beta}|H^0|\phi_{ab}^{\alpha\beta}\rangle] \qquad (1)$$

as a criterion for choosing diexcited configurations. The unperturbed Hamiltonian H^0 is chosen to be the sum of the one-electron Hartree–Fock Hamiltonians $\mathscr{H}(k)$

$$H^0 = \sum_{\text{electrons } k} \mathscr{H}(k) \tag{2}$$

The perturbation is defined by $V = H - \sum_{\text{electrons } k} \mathscr{H}(k)$ where H is the total electrostatic Hamiltonian. Molecular orbitals are eigenfunctions of $\mathscr{H}$:

$$\mathscr{H}\psi_i = \varepsilon_i\psi_i \tag{3}$$

ε_i is the one-electron Hartree–Fock energy.

Configurations $\phi_0 \ldots \phi_i$ are eigenfunctions of H^0 with eigenvalues

$$\sum_{\text{conf. } \phi_0} \varepsilon_j, \ldots \sum_{\text{conf. } \phi_i} \varepsilon_j$$

(summations are made over molecular spin orbitals involved in one determinant occurring in configurations $\phi_0 \ldots \phi_i$).

This choice has the advantage that $E^{(2)}$ is relatively simple to calculate. One can write it as follows:

$$E^{(2)} = |\langle\phi_0|H|\phi_{ab}^{\alpha\beta}\rangle|^2/(\varepsilon_a + \varepsilon_b - \varepsilon_\alpha - \varepsilon_\beta) \tag{4}$$

The list of configurations is set up in decreasing order of $E^{(2)}$.

In the Reference 1 calculation on the N_2 molecule, comparisons are made between $E^{(2)}$ and the "exact contribution" for the 20 most important configurations. The "exact contribution" is difficult to define. An exact definition would be the contribution obtained when one adds this configuration to a basis set which will be complete after this operation. This is rather difficult to achieve. The order of importance may be false for the very first configurations which may strongly interact. It is physically useless to separate contributions of such a group which must be considered as a whole. This is not of great practical importance as we go further than these first configurations: when the order in the list increases the difference between $E^{(2)}$ and the "exact contribution" will tend to stabilize (the subspace spanned by the first configurations does not increase in a significant manner with each addition of a new configuration).

(*b*) *Singly excited configurations.* For closed-shell systems there is no simple energy criterion to select monoexcitations as $\langle\phi^0|V|\phi_i^\alpha\rangle = 0$ (Brillouin's theorem).

Those particles which are energetically and spatially closest to the holes, i.e. the first unoccupied orbitals in each symmetry group, can be considered to be the most important.

Anyway their total number is relatively small (≈ 50) in our problems and the effect of including each of them can be tested easily. We shall come back to this point when dealing with the problem of calculating other monoelectronic operator mean values.

B. Perturbation Expansions

In the following H designates the total electrostatic Hamiltonian, H^0 the unperturbed Hamiltonian, and V the perturbation.

$$H = H^0 + V \tag{5}$$

H^0 is defined in such a way that the reference state ϕ_0 and substituted configurations $\phi_{ab}^{\alpha\beta}$, ϕ_d^γ are its eigenfunctions corresponding to eigenvalues E_0, $E_{ab}^{\alpha\beta}$, $E(^\gamma_d)$. These configurations form an orthonormalized subspace in which we seek the Ψ eigenfunction of H corresponding to the lowest eigenvalue E.

Let us define

$$\begin{aligned} \delta E &= E - E_0 \\ \delta\phi &= \Psi - \phi_0 \end{aligned} \tag{6}$$

We then have:

$$H^0|\phi_0\rangle = E_0|\phi_0\rangle \tag{7}$$

$$(H^0 + V)|\phi_0 + \delta\phi\rangle = (E_0 + \delta E)|\phi_0 + \delta\phi\rangle \tag{8}$$

to which one adds without restriction the condition that $\delta\phi$ is orthogonal to ϕ_0; this is the so-called "intermediate normalization" expressed by

$$\langle\phi_0|\Psi\rangle = 1 \tag{9}$$

The energy shift δE is obtained by projection of (8) on $|\phi_0\rangle$:

$$\delta E = \langle\phi_0|V|\phi_0 + \delta\phi\rangle = \langle\phi_0|V|\phi_0\rangle + \langle\phi_0|V|\delta\phi\rangle \tag{10}$$

To obtain $\delta\phi$ one can either solve directly Eq. (8) and obtain the Brillouin–Wigner (BW) expansion or try an expansion in powers of the interaction V and obtain the Rayleigh–Schrödinger (RS) expansion.

(*a*) *Brillouin–Wigner expansion.* Equation (8) can be written $(E - H^0)|\Psi\rangle = V|\Psi\rangle$ which suggests that:

$$|\Psi\rangle = |\phi_0\rangle + \frac{P}{E - H^0} V|\Psi\rangle \tag{11}$$

We have separated in Ψ the unperturbed component ϕ_0, using the projection operator

$$P = 1 - |\phi_0\rangle\langle\phi_0| = \sum_{i\neq 0} |\phi_i\rangle\langle\phi_i| \tag{12}$$

In what follows we shall write $|i\rangle$ for ϕ_i, $|0\rangle$ for ϕ_0.

We shall verify, using Eqs. (7), (8), and (10), that Ψ verifies the perturbed Schrödinger equation (Eq. (8)). It is sufficient to show that this is true after projection on the complete set of eigenfunctions of H_0.

Equation (11) yields:

$$(E - H^0 - PV)|\Psi\rangle = (E - H^0)|\phi_0\rangle = \delta E|\phi_0\rangle \tag{13}$$

Projection of (13) on $|\phi_0\rangle$ gives:

$$\langle\phi_0|E - H^0 - PV|\Psi\rangle = \delta E \tag{14}$$

As P projects out ϕ_0 one can suppress PV, and, replacing δE by its values (Eq. (10)), obtain:

$$\langle\phi_0|E - H^0 - V|\Psi\rangle = 0 \tag{15}$$

Projection of (13) on the other eigenstates $|\phi_i\rangle$ of H^0 gives:

$$\langle\phi_i|E - H^0 - V|\Psi\rangle = \delta E\langle\phi_i|\phi_0\rangle = 0 \tag{16}$$

we have suppressed the projection operator P which is useless since we project in the relevant space.

$$|\Psi\rangle \text{ verifies } (E - H)|\Psi\rangle = 0$$

as it is true after projection on all basis vectors.

Iterating Eq. (11) one obtains the Brillouin–Wigner expansion:

$$\left.\begin{aligned}\Big|\Psi> &= \left\{1+\sum_{k=1}^{\infty}\left[\frac{P}{E-H^0}V\right]^k\right\}\Big|\phi_0> \\ |\Psi\rangle &= |\phi_0> + \sum_{k=1}^{\infty}|\delta\phi_k>\end{aligned}\right\} \tag{17}$$

$\delta\phi_k$ is obtained from $\delta\phi_{k-1}$ by applying once the operator $\dfrac{P}{E-H^0}V$ which can be written using (10) as follows

$$\frac{P}{E-H^0}V = \sum_{i\neq 0}\frac{|i\rangle\langle i|}{E-\langle i|H^0|i\rangle}V$$

To solve a CI this recursive form of building the successive corrections to Ψ is used

$$|\delta\phi_k\rangle = \sum_{i\neq 0}|i\rangle\frac{\langle i|V|\delta\phi_{k-1}\rangle}{E-\langle i|H^0|i\rangle} \tag{18}$$

$\delta\phi_{k-1}$ is known and given by its expansion:

$$|\delta\phi_{k-1}\rangle = \sum_{j\neq 0}\alpha_{k-1}^j|j\rangle$$

which gives:

$$|\delta\phi_k\rangle = \sum_{i\neq 0}|i\rangle\frac{1}{E-\langle i|H^0|i\rangle}\Big(\sum_{j\neq 0}\alpha_{k-1}^j\langle i|V|j\rangle\Big) \tag{19}$$

Call N the number of configurations. The calculation of the ith component α_k^i of $|\delta\phi_k\rangle$ necessitates N operations and N^2 operations are needed to obtain all of them. In numerical calculation one has to fix a maximum value of k, say k_m. Then k_mN^2 operations are used to obtain $|\delta\phi\rangle$ to this order. $k_m = 12$ or 13 gives a precise evaluation of the series with any of the Hamiltonian partitioning which will be considered later.

The energy shift δE is calculated from Eq. (8):

$$\delta E = \Big\langle\phi_0|V|\left(1+\sum_{k=1}^{km}\frac{P}{E-H^0}V\right)|\phi_0\rangle \tag{20}$$

As we are usually interested in terms beyond first-order correction, we shall write the operator $(E - H^0)^{-1}$ in a different form.

Let us define

$$\delta E' = \delta E - \langle \phi_0 | V | \phi_0 \rangle$$

Then:

$$\begin{aligned}(E - H^0)^{-1} &= [\langle 0|H^0|0\rangle + \delta E - H^0]^{-1} \\ &= [\langle 0|H|0\rangle + \delta E' - H^0]^{-1}\end{aligned}$$

Equation (20) has now changed to an implicit equation for $\delta E'$

$$\delta E' = \left\langle \phi_0 \left| \sum_{k=1}^{km} \left(\frac{P}{\langle 0|H|0\rangle + \delta E' - H^0} \right)^k \right| \phi_0 \right\rangle \tag{21}$$

which is solved by iterations, starting on the right-hand side with $\delta E' = 0$.

We shall now consider, for example, two different choices for H^0.

[1]. H^0 is the diagonal part of H. The matrix representation of the unperturbed Hamiltonian and perturbations V are in the configurations subspace:

$$H^0 \rightarrow \begin{pmatrix} \langle 0|H|0\rangle & 0 & 0 \\ 0 & \langle 1|H|1\rangle & \\ | & & \\ 0 & & \langle N|H|N\rangle \end{pmatrix} \quad V \rightarrow \begin{pmatrix} 0 & & \\ & 0 & \langle j|H|i\rangle \\ \langle i|H|j\rangle & & \\ & & 0 \end{pmatrix}$$

Then $\delta E = \delta E'$ and as $\langle i|V|i\rangle = 0$ Eq. (19) is:

$$|\delta\phi_k\rangle = \sum_{i\neq 0} \Big(\sum_{\substack{j\neq i \\ \neq 0}} \alpha_{k-1}^{j} \langle i|H|j\rangle \Big) \frac{1}{\delta E + \langle 0|H|0\rangle - \langle i|H|i\rangle} |i\rangle \tag{22}$$

It is convenient to separate the quantities $\langle 0|H|0\rangle - \langle i|H|i\rangle$ which are those actually calculated.

[2]. H^0 is the sum of the one-electron Hartree–Fock Hamiltonian (already defined by Eq. (21)).

Then:

$$i \neq j \; \langle i|V|j\rangle = \langle i|H - H^0|j\rangle = \langle i|H|j\rangle$$

The diagonal matrix elements of H are in fact calculated taking $E = \langle 0|H|0\rangle$ for zero energy and has the form:

$$\langle \phi_{ab}^{\alpha\beta} | H | \phi_{ab}^{\alpha\beta} \rangle = \varepsilon_\alpha + \varepsilon_\beta - \varepsilon_a - \varepsilon_b + \langle \phi_{ab}^{\alpha\beta} | \mathscr{V} | \phi_{ab}^{\alpha\beta} \rangle \tag{23}$$

where

$$\mathscr{V} = V - \langle 0|V|0\rangle \tag{24}$$

We shall then consider for convenience this new splitting of the Hamiltonian into a perturbed and unperturbed part:

$$\left.\begin{aligned} H^0 &= \sum_{\text{electrons } k} \mathscr{H}(k) + \langle 0|V|0\rangle \\ \mathscr{V} &= H - \sum_{\text{electrons } k} \mathscr{H}(k) - \langle 0|V|0\rangle \end{aligned}\right\} \tag{25}$$

Then $\delta E = \delta E'$ and the operator $P(\delta E + \langle 0|H|0\rangle - H^0)^{-1}$ can now be written as

$$P(\delta E + \langle 0|\Sigma\mathscr{H}k|0\rangle - \Sigma\mathscr{H}(k))^{-1}$$

and when acting on configuration $\phi_{ab}^{\alpha\beta}$, for instance, gives

$$P(\delta E + \varepsilon_a + \varepsilon_b - \varepsilon_\alpha - \varepsilon_\beta)^{-1} \tag{26}$$

The perturbation matrix elements are

$$\left.\begin{aligned} &i \neq j, \langle i|\mathscr{V}|j\rangle = \langle i|H|j\rangle \\ &i = j, \langle i|\mathscr{V}|i\rangle = \langle i|H - \Sigma\mathscr{H}(k)|i\rangle - \langle 0|H - \Sigma\mathscr{H}(k)|0\rangle \end{aligned}\right\} \tag{27}$$

(*b*) *Rayleigh–Schrödinger expansion.* Let us expand the total Hamiltonian, exact wave functions, and energy in the Schrödinger equation in powers of an interaction parameter λ

$$[(H^0 + \lambda V) - (E_0 + \lambda E_1 + \ldots + \lambda^n E_n \ldots)] \\ |\psi_0 + \lambda\psi_1 + \ldots + \lambda^n\psi_n + \ldots\rangle = 0 \tag{28}$$

This is a development in powers of λ which is identically null. All orders have to be equated separately to zero. The zeroth order gives

$$(H^0 - E_0)|\psi_0\rangle = 0 \qquad \psi_0 = \phi_0 \tag{29}$$

The nth order

$$(H^0 - E_0)|\psi_n\rangle = \sum_{r=0}^{n-2} E_{n-r}\psi_r - (V - E_1)\psi_{n-1} \tag{30}$$

Projection of (30) on $|\phi_0\rangle$ gives:

$$E_n = \langle \phi_0|V|\phi_{n-1}\rangle \tag{31}$$

Projecting on the subspace orthogonal to ϕ_0 by application of P (Eq. 12) and using the fact that $[P, H^0] = 0$:

$$|\psi_n\rangle = P(H^0 - E_0)^{-1}[\sum_{r=0}^{n-2} E_{n-r}|\psi_r\rangle + (V - E_1)|\psi_{n-1}\rangle] \quad (32)$$

One has to proceed step by step using Eqs. (31) and (32) alternatively. The same splittings of the Hamiltonian as those considered for the BW expansion may be used. For the same choice of Hamiltonian partitioning, the convergence is slower than in the Brillouin–Wigner procedure.

C. Approximate Calculations

Since the second-order approximation (Eq. 4) follows within 10 per cent of the exact value in the case of N_2 calculations[1] we thought it of interest to sum contributions of all possible configurations which can be built with the available particles. Sums can be made for each pair of holes, this gives an estimate of pair correlation energies.

3. ENERGY CALCULATION RESULTS

Calculations made for N_2 and CO ground states show that at least 14 per cent of the correlation energy is due to the near degeneracy of the highest hole and lowest particle ($1\pi_u$, $1\pi_g$ for N_2; 1π, 2π for CO). This is shown very well in the natural orbital scheme. There appears in N_2 a π_g natural orbital of occupation number 0.1 and a π_u natural orbital of occupation number 3.9. The same thing happens in CO: the first π natural orbital has an occupation number 3.93 and the second 0.07. This results in a decrease of electron density in the bonding region as $1\pi_g(N_2)$ and $2\pi(CO)$ have an antibonding character.

An analysis of pair correlation estimated from $E^{(2)}$ (summation for any configurations which have the same pair of holes) is made in Reference 1 for N_2.

The monoexcitations do not give an important contribution to the energy. Few of them should be added to diexcitation in an exact calculation. More precisely the monoexcitation, characterized by the highest hole and lowest particle of the same symmetry, is significant as expected. In the case of N_2, a 201-configuration interaction gives $\delta E = 0.277$ a.u. (=50 per cent of the total correlation energy); the $\phi_{1\pi_u}^{2\pi_u}$ configuration gives a contribution of 0.001 a.u. In the case of CO, a 201-configuration interaction gives $\delta E = 0.25$ a.u. (=47 per cent of the total correlation energy) and the $\phi_{1\pi}^{2\pi}$ configuration gives an improvement of 0.002 a.u. Though not important for energy we shall see that these monoexcitations may produce charge transfer important for other operators.

As a consequence of the change of symmetry produced by the building of the molecule, the MO of symmetry type other than those occurring in the HF calculation may be significant. As the Nesbet–Stevens program can handle those kinds of orbitals, we investigated δ_u, δ_g, ϕ_u, ϕ_g orbitals in the N_2 case, and found that the two independent singlets $\phi_{1\delta_u^+ 1\delta_u^-}^{1\delta_g^+ 1\delta_g^-}$ together contribute 0.01 a.u., which is important.

We think that the total approximate second-order calculation (sum of $E^{(2)}$ over any configurations) gives a fair amount of outer-shell correlation energy. The argument given in Reference 1 is very rough (and incorrect as far as 2s electrons are concerned).[5] A good test is to calculate a potential curve and spectroscopic constants. Using Huo's Hartree–Fock calculations for a number of internuclear distances of CO,[6] we performed a total approximate calculation to second order. As this approximation seems to be 10 per cent optimistic, when compared to an exact calculation, we decreased the total sum by this amount. The results are given in Table 1. Fitting then a third-degree polynominal by a "rigorous

TABLE 1. Points used for the potential curve of CO. The CI starts from Huo's Hartree–Fock calculations. The second-order estimate has been decreased by 10 per cent.

R (Bohr)	1.898	2.015	2.081	2.249	2.366	2.483	2.6
E (a.u.) Hartree–Fock + $E^{(2)}$ estimate of CI	113.0243	113.0607	113.0694	113.0700	113.0646	113.0421	112.9880

least square method",[7] we found the results given in Table 2. According to the criterion of the method which fits the polynomial to all possible sets of points taken in the list, we select the two best: calculation (1) and calculation (2). Calculation (2), which uses the outer part of the potential curve (third to seventh points), gives of course the best value for the $\omega_e x_e$, calculation (1) does not give a good value for it. Both give results for ω_e and

TABLE 2. Spectroscopic constants for the ground state of CO. Calculation (1) fits a third-degree curve by a least square method[7] from third to seventh point of the potential curve; calculation (2) fits from first to fifth point (cf. Table 1).

	Huo Hartree–Fock	Experimental	Calculation (1)	Calculation (2)
ω_e (cm^{-1})	2431	2169,829	2163	2297
$\omega_e x_e$ (cm^{-1})	11,69	13,295	12,54	21,18
B_e (cm^{-1})	2.027	1.931	1.934	1.928
R_e (Bohr)	2.081	2.132	2.130	2.134

R_e which are too good. This has to be confirmed by other calculations. The same method of fitting a third-degree polynomial when applied to Huo's potential curve gives the same values of spectroscopic constant as those calculated by Huo.

4. MEAN VALUE OF MONOELECTRONIC OPERATORS

Up to now the reference values, namely Hartree–Fock values, were supposed to give a correct answer within a few per cent. The calculated mean value in the Born–Oppenheimer approximation is the sum of an electronic term and of a nuclear term of opposite sign. The electronic part seems to be obtained with the accuracy of a few per cent. More careful calculations are necessary and contributions from excited configurations must not be neglected when the two terms are of close absolute value. We shall examine in a wave function interaction of double and single substitutions the different kinds of terms and their relative contributions.

A. Expression of the Mean Value with a CI Wave Function

Let $\theta = \sum_{\text{electrons } k} \theta(k) + \theta(N)$ be a single-particle operator sum of monoelectronic operators $\theta(k)$ and of $\theta(N)$ which has the same form as the $\theta(k)$'s but which acts on the coordinates of the nuclei. In the Born–Oppenheimer approximation, the total wave function is the product of the electronic wave function ψ and the nuclei wave χ function. The mean value separates into two terms:

$$\langle\theta\rangle = \langle\psi|\Sigma\theta(k)|\psi\rangle + \langle\chi|\theta_N|\chi\rangle \tag{33}$$

The electric charges involved in each term are generally of opposite sign and $\langle\theta\rangle$ will be the difference of electronic and nuclear contribution. We shall note $\Sigma\theta(k) = \theta_e$.

We shall deal with the electronic part which is the more difficult to obtain with a given accuracy. Ψ will be taken as a CI wave function. The configurations are restricted to single and double excitations. The results of the discussion will show that higher excitations will not give an important contribution.

The normalized electronic wave function has the form:

$$\Psi = c_0\phi_0 + \sum_{\binom{\alpha\beta}{ab}} c_{ab}^{\alpha\beta}\phi_{ab}^{\alpha\beta} + \sum_{\binom{\gamma}{d}} c_d^{\gamma}\phi_d^{\gamma} \tag{34}$$

(the c's are supposed to be real). Then

$$\langle\Psi|\sum_k\theta(k)|\Psi\rangle = c_0^2\langle\phi_0|\theta_e|\phi_0\rangle$$

Contribution of diexcitations
$$\left\{\begin{aligned} &+\sum_{\binom{\alpha\beta}{ab}\neq\binom{\alpha'\beta'}{a'b'}} c_{ab}^{\alpha\beta}c_{a'b'}^{\alpha'\beta'}\langle\phi_{ab}^{\alpha\beta}|\theta_e|\phi_{a'b'}^{\alpha'\beta'}\rangle \\ &+\sum_{\binom{\alpha\beta}{ab}} (c_{ab}^{\alpha\beta})^2\langle\phi_{ab}^{\alpha\beta}|\theta_e|\phi_{ab}^{\alpha\beta}\rangle \end{aligned}\right.$$

Terms given by coupling di- and monoexcitations
$$\Big\}+\sum_{\binom{\alpha\beta}{ab}\binom{\alpha'}{a'}} c_{ab}^{\alpha\beta}c_{a'}^{\alpha'}\langle\phi_{ab}^{\alpha\beta}|\theta_e|\phi_{a'}^{\alpha'}\rangle \tag{35}$$

Terms given by monoexcitations

$$\left\{\begin{array}{l} + \sum_{\binom{\alpha}{a}\binom{\beta}{b}} c_a^\alpha c_b^\beta \langle \phi_a^\alpha | \theta_e | \phi_b^\beta \rangle \\ + \sum_{\substack{\alpha \\ a}} (c_a^\alpha)^2 \langle \phi_a^\alpha | \theta_e | \phi_a^\alpha \rangle \end{array}\right.$$

Terms given by coupling the reference state to monoexcitations

$$+2c_0 \sum_{\substack{\alpha \\ a}} c_a^\alpha \langle \phi_0 | \theta_e | \phi_a^\alpha \rangle$$

We have separate summation on diagonal and off-diagonal terms. $\phi_{ab}^{\alpha\beta}$ (resp ϕ_a^α) is different from $\phi_{a'b'}^{\alpha'\beta'}$ (resp $\phi_{a'}^{\alpha'}$) if at least one of the holes or particles in each configuration is different.

Notice the last term: it represents a coupling by a single-particle operator of the reference state and monoexcitations, when there could not be by the Hamiltonian in the case of a closed-shell system (Brillouin's theorem).

We shall now write Eq. (35) in a slightly different form so that the reference mean value $\langle \phi_0 | \theta_e | \phi_0 \rangle$ will be outlined. We express c_0^2 by its value given by the normalization condition:

$$c_0^2 = 1 - \sum_{\binom{\alpha\beta}{ab}} (c_{ab}^{\alpha\beta})^2 - \sum_{\binom{\alpha}{a}} (c_a^\alpha)^2 \tag{36}$$

and group factors of the same c's. This will replace the first, third, and sixth terms (diagonal terms) by the following:

$$\begin{aligned} \langle \Psi | \theta_e | \Psi \rangle = {} & \langle \phi_0 | \theta_e | \phi_0 \rangle \\ & + \cdots \\ & + \sum_{\binom{\alpha\beta}{ab}} (c_{ab}^{\alpha\beta})^2 [\langle \phi_{ab}^{\alpha\beta} | \theta_e | \phi_{ab}^{\alpha\beta} \rangle - \langle \phi_0 | \theta_e | \phi_0 \rangle] \\ & + \cdots \\ & + \sum_{\binom{\alpha}{a}} (c_a^\alpha)^2 [\langle \phi_a^\alpha | \theta_e | \phi_a^\alpha \rangle - \langle \phi_0 | \theta_e | \phi_0 \rangle] \end{aligned} \tag{37}$$

The other terms being unchanged. This gives the exact expression of the variation $\delta\langle \theta_e \rangle$ due to substituted configurations. In the first form the diagonal terms are relatively large: the normalization condition weights the reference state by the coefficient c_0^2, and so the excited states must take care, in some way of a part, $(1 - c_0^2)$, of the reference value.

Now call $\theta_a \ldots \theta_\alpha$ the mean value of $\theta(\mathbf{r})$ over the one-particle states $a \ldots \alpha$. We note $\theta(\mathbf{r})$ for one of the one-particle operators $\theta(k)$

$$\theta_a = \langle a|\theta(\mathbf{r})|a\rangle \tag{38}$$

The diagonal terms may also be written explicitly

$$\begin{aligned}\langle\Psi|\theta_e|\Psi\rangle = & \sum_{\text{holes } a} 2\theta_a \\ & + \sum_{\binom{\alpha\beta}{ab}} (c_{ab}^{\alpha\beta})^2[\theta_\alpha - \theta_a + \theta_\beta - \theta_b] \\ & + \sum_{\binom{\alpha}{a}} (c_a^\alpha)^2[\theta_\alpha - \theta_a]\end{aligned} \tag{39}$$

Thus in the diagonal terms a meaningful quantity is the shift for each single excitation, weighted by the probability of the configuration.

The off-diagonal terms can also be made explicit.

Those in which two doubly excited configurations

$$\langle\phi_{ab}^{\alpha\beta}|\theta_e|\phi_{a'b'}^{\alpha'\beta'}\rangle$$

occur are non-vanishing when at least three of the first configuration's holes and particles are equal to three of the second configuration. These terms are proportional to the elementary one-particle integral: of hole–hole type, say $\langle a|\theta(\mathbf{r})|a'\rangle$ if $b = b'$ $\alpha = \alpha'$ $\beta = \beta'$ or of particle–particle type, say $\langle\alpha|\theta(\mathbf{r})|\alpha'\rangle$ if $a = a'$ $b = b'$ $\beta = \beta'$.

Those in which singly and doubly excited configurations $\langle\phi_{ab}^{\alpha\beta}|\theta_e|\phi_{a'}^{\alpha'}\rangle$ occur are of hole–particle type and are non-vanishing only if one single excitation is common to both configurations

$$\langle\phi_{ab}^{\alpha\beta}|\theta_e|\phi_{a'}^{\alpha'}\rangle \simeq \langle b|\theta(\mathbf{r})|\beta\rangle$$

if $a = a'$ $\alpha = \alpha'$.

Those in which singly excited configurations occur are either of hole–hole type: $\langle\phi_a^\alpha|\theta_e|\phi_b^\beta\rangle \simeq \langle a|\theta(\mathbf{r})|b\rangle$ if $\alpha = \beta$ or particle–particle type $\langle\phi_a^\alpha|\theta_e|\phi_b^\beta\rangle \simeq \langle\alpha|\theta(r)|\beta\rangle$ if $a = b$.

The last term $\langle\phi_0|\theta_e|\phi_a^\alpha\rangle$ is never vanishing and is of hole–particle type:

$$\langle\phi_0|\theta_e|\phi_a^\alpha\rangle \simeq \langle a|\theta(\mathbf{r})|\alpha\rangle$$

One can then rewrite the electronic mean value

$$
\begin{aligned}
\langle\Psi|\theta_e|\Psi\rangle = \sum_{\text{holes } a} 2\theta_a \\
&\left\{\begin{array}{l} +\sum_{\substack{\alpha\beta \\ aba'}} c_{ab}^{\alpha\beta} c_{a'b}^{\alpha\beta}[\simeq\langle a|\theta(\mathbf{r})|a'\rangle] \\ +\sum_{\substack{\alpha\beta\alpha' \\ ab}} c_{ab}^{\alpha\beta} c_{ab}^{\alpha'\beta}[\simeq\langle\alpha|\theta(\mathbf{r})|\alpha'\rangle] \end{array}\right. \\
&+\sum_{\substack{\alpha\beta \\ ab}} (c_{ab}^{\alpha\beta})^2[\theta_\alpha - \theta_a + \theta_\beta - \theta_b] \\
&+2\sum_{\binom{\alpha\beta}{ab}} c_{ab}^{\alpha\beta} c_a^\alpha[\simeq\langle b|\theta(\mathbf{r})|\beta\rangle] \\
&\left\{\begin{array}{l} +\sum_{\binom{\alpha}{ab}} c_a^\alpha c_b^\alpha[\simeq\langle a|\theta(\mathbf{r})|b\rangle] \\ +\sum_{\binom{\alpha\beta}{a}} c_a^\alpha c_a^\beta[\simeq\langle\alpha|\theta(r)|\beta\rangle] \end{array}\right. \\
&+\sum_{\substack{\alpha \\ a}} (c_a^\alpha)^2[\theta_\alpha - \theta_a] \\
&+2\sum_{\substack{\alpha \\ a}} c_0 c_a^\alpha[\simeq\langle a|\theta(r)|\alpha\rangle]
\end{aligned} \tag{40}
$$

(The $\simeq$ sign indicates an unwritten numerical factor which is in general different in each case.)

B. Order of Magnitude of Different Terms

We can now investigate the relative magnitude of these different terms. We have to analyse two factors: the c's products, and the brackets representing effects of charge transfer weighted by the calculated operator.

[1]. The first point to look at is the order of magnitude of the coefficients $c_{ab}^{\alpha\beta}$, c_d^γ . . . and of the products involved in Eq. (40). The coefficients $c_{ab}^{\alpha\beta}$ of diexcitations are of first order which is obvious from their expression in the perturbation expansion. In a RS scheme, for example,

$$
c_{ab}^{\alpha\beta} \sim \langle\phi_0|V|\phi_{ab}^{\alpha\beta}\rangle/[\langle\phi_0|H^0|\phi_0\rangle - \langle\phi_{ab}^{\alpha\beta}|H^0|\phi_{ab}^{\alpha\beta}\rangle] \tag{41}
$$

The coefficients c_d^γ of a single excitation vanish to first order for closed shells and the series begin with a second-order term:

$$c_d^\gamma \simeq \sum_{\substack{\alpha\beta \\ ab}} \langle\phi_0|V|\phi_{ab}^{\alpha\beta}\rangle\langle\phi_{ab}^{\alpha\beta}|V|\phi_d^\gamma\rangle / \\ [(\langle\phi_0|H^0|\phi_0\rangle - \langle\phi_{ab}^{\alpha\beta}|H^0|\phi_{ab}^{\alpha\beta}\rangle)(\langle\phi_0|H^0|\phi^0\rangle - \langle\phi_d^\gamma|H^0|\phi_d^\gamma\rangle)] \tag{42}$$

Two factors occur in the numerator of (42): $\langle\phi_0|V|\phi_{ab}^{\alpha\beta}\rangle$ never vanishes; $\langle\phi_{ab}^{\alpha\beta}|V|\phi_d^\gamma\rangle$ will be different from zero if at least one of the following conditions is fulfilled: a or $b = d$, α or $\beta = \gamma$. This leaves a great number of possible interactions.

Considering then the c's products in Eq. (40): the second and third terms involving only diexcitations are of second order; the fourth involving di- and monoexcitations is of third order; the fifth involving monoexcitations only is of fourth order; the sixth one resulting from the coupling of the reference state and monoexcitation, is of second order as the reference state is not a bad approximation and $c_0 \simeq 1$.

The important conclusion is that through this last term monoexcitations must be treated at the same stage as diexcitations.

The last-mentioned considerations show that excitations of higher order, $\phi^{\alpha\beta\gamma}$ . . . etc., would give higher order contributions. The c's coefficients of triexcitations and quadriexcitations would be of second order. Their expression to lowest order would be similar to that of a single excitation coefficient, Eq. (42), with the restriction that the number of non-zero V matrix elements coupling these excitations to diexcitations would be smaller. Anyway there would be no coupling to the reference state by the single-particle operator. These excitations would enter in the expression of mean value only when coupled to diexcitations or monoexcitations. The corresponding c's products would be of third order.

We are left now in Eq. (40) with terms concerning diexcitations (second and third) and interactions of the reference state with monoexcitations (last term).

[2]. The second point to look at now is the relative magnitude of the single-particle matrix elements in these remaining terms. They are: hole–hole or particle–particle type in the second

term (off-diagonal, diexcitations); differences of particle and hole total contributions (diagonal, diexcitations); hole–particle type (interaction of reference state and monoexcitations).

In the calculations made up to now, this last term turned out to be the most important. The sum is given mainly by the configuration characterized by the highest hole and the lowest particle. Furthermore this $\langle a|\theta(r)|\alpha\rangle$ integral may be much larger than the corresponding hole contribution θ_a. It can be understood as this hole and particle (of same symmetry type) have very close energies and though orthogonal, are localized in the same area. To insure orthogonality, the wave functions are of opposite sign in at least one region of space, or are of different character: in CO 1π is bonding, 2π is antibonding. The product $a(r)\,.\,\alpha(r)$ is now multiplied by say Z_A, dipole moment, measured on nucleus A, which itself has a sign different in two regions of space. If the regions of constant sign for $a(r)\,.\,\alpha(r)$ and Z_A overlap enough, the product will be of the same sign in the greatest part of space and the result of integration may be large.

The other point which makes the highest hole–lowest particle term important is that the corresponding c_a^α is only apparently of second order and may be as large as the first-order coefficients $c_{ab}^{\alpha\beta}$. This is due to the summation over intermediate diexcited states (Eq. (42)), a great number of which may interact strongly with the relevant monoexcitation. Moreover, the most important diexcitations involve this highest hole and lowest particle. The situation is not the same in each symmetry group of orbitals. It may happen that for some of these typical pairs (highest hole–lowest particle) the energy difference is large. The effects decrease roughly with increasing energy difference and are at a maximum for the most energetic electron. This is the case for the calculation of the electric field gradient on the nucleus, $\langle q\rangle$, in the N_2 molecule. The ground-state configuration is $1\sigma_g^2 2\sigma_g^2 3\sigma_g^2 1\sigma_u^2 2\sigma_u^2 (1\pi_u)^4$; the higher energetic holes are $2\sigma_u(-0.77$ a.u.), $3\sigma_g(-0.63$ a.u.), $1\pi_u(-0.609$ a.u.), and the lowest particles $3\sigma_u(0.62$ a.u.), $4\sigma_g(0.53$ a.u.), $2\pi_u(0.94$ a.u.). The $\phi_{2\sigma_u}^{n\sigma_u}$ configurations do not contribute at all. The $\phi_{3\sigma_g}^{4\sigma_g}\ .\ .\ .\ \phi_{3\sigma_g}^{8\sigma_g}$ give roughly a 2 per cent variation of $\langle q\rangle$ (each of them produces a variation of different sign so we have summed the absolute variations). The effects rise strongly with the π-symmetry group:

$$\phi_{1\pi}^{2\pi} \text{ gives a 3.4 per cent change}$$

$$\phi_{1\pi}^{3\pi} \text{ gives a 1.9 per cent change}$$

$$\phi_{1\pi}^{4\pi} \text{ gives a 1.7 per cent change}$$

and the coefficient $c_{1\pi u}^{2\pi u}$ is of the same magnitude as those of diexcitations.

One can understand why the excitation from highest hole to lowest particle has a c coefficient which may be significant: the electron in this highest orbital contributing to a small extent to the total energy, is not as much concerned by the variational calculation as an inner electron, and is not so well described.

The excitations to higher particles will be less significant as the c coefficient will be small; in this case the $a(r)\alpha(r)$ product will also, in general, be small for hole and particle having different localizations and the particle wave function will oscillate with many secondary minima.

These considerations are qualitative and it may happen that for a particular operator which weights more or less different regions of space, an excitation to a high-energy particle gives a large value of the hole–particle integral. For instance, in the $\langle q \rangle$ calculation of N_2 the single-particle excitation $\phi_{3\sigma_g}^{8\sigma_g}$ ($8\sigma_g$ has a monoelectronic HF energy of 11.36 a.u.) is important, in spite of the small size of $c_{3\sigma_g}^{8\sigma_g}$, because $\langle 3\sigma_g | q | 8\sigma_g \rangle$ is very large.

All the other terms are in general smaller: the diagonal term, if $\theta_a - \theta_\alpha$ is large, corresponds in general to high excitations and small c coefficients; if the c's are not small the hole and particle are neighbours $\theta_a \simeq \theta_\alpha$.

The other off-diagonal terms involve only hole–hole or particle–particle elementary integrals. In general these $a(r)a'(r)$ or $\alpha(r)\alpha'(r)$ products are not very large compared to the highest hole–lowest particle product since they are multiplied by second order c's product and give small contributions.

C. Natural Orbitals Scheme

All this can be stated in a more compact way, using the natural orbitals. They are eigenfunctions of the first-order density matrix

$\Gamma(\mathbf{r}, \mathbf{r}')$. We shall express the density matrix and natural orbitals in terms of molecular orbitals.

$$\Gamma(\mathbf{r}, \mathbf{r}') = \sum_{\lambda} \sum_{ij} \Gamma_{ij}^{\lambda} \psi_i^{\lambda}(\mathbf{r}) \psi_j^{(\lambda)}(\mathbf{r}')$$

Only products of MO of same symmetry λ appear. i, j are the indices labelling the different orbitals of the symmetry group. The density matrix is obtained from Eq. (35) by ordering it in

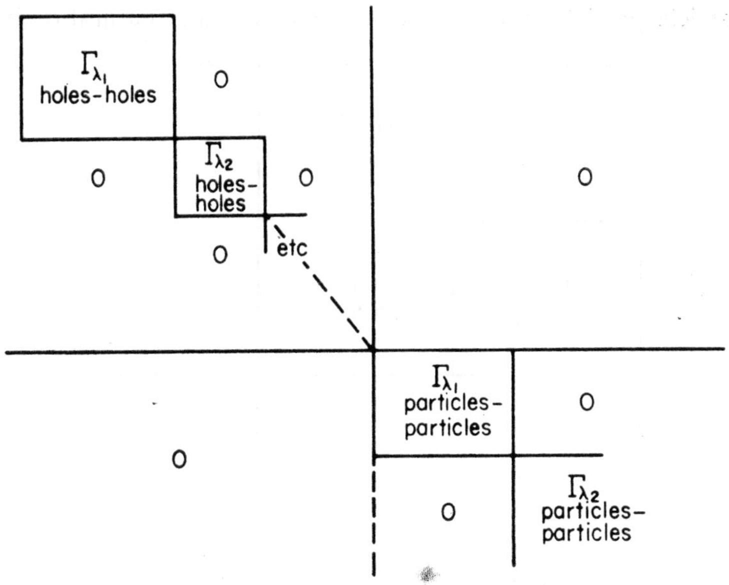

Fig. 1. Γ_{ij}^{λ}'s matrix (CI diexcitations only).

terms of elementary integrals $\langle \phi_i^{\lambda}(\mathbf{r}) | \theta(\mathbf{r}) | \phi_j^{\lambda}(\mathbf{r}) \rangle$ and replacing each matrix element by $\phi_i^{\lambda}(\mathbf{r}) \phi_j^{\lambda}(\mathbf{r}')$. The mean value of a monoelectronic operator θ_e is then:

$$\langle \theta_e \rangle = \mathrm{Tr}[\theta_e \Gamma(\mathbf{r}, \mathbf{r}')]$$

(θ_e operates on $\mathbf{r}'$ and then the trace is calculated).

When the CI involves only diexcitations, the Γ_{ij}^{λ}'s matrix is made of submatrices Γ_{λ_i} of hole–hole or particle–particle products corresponding to the symmetry type λ_i (Fig. 1). To each particular hole corresponds one natural orbital having its main component on this hole and small components on other holes. The corresponding occupation number is large and near to the main component occupation number in the reference state. The eigenvector combination of particles has a very small occupation number, except in the case of near degeneracy.

When the CI involve both mono- and diexcitations, the density matrix is not reduced as above (Fig. 2). In each λ-type submatrix there are now products of hole–particle as well as hole–hole or particle–particle matrix elements. Now the first natural orbitals

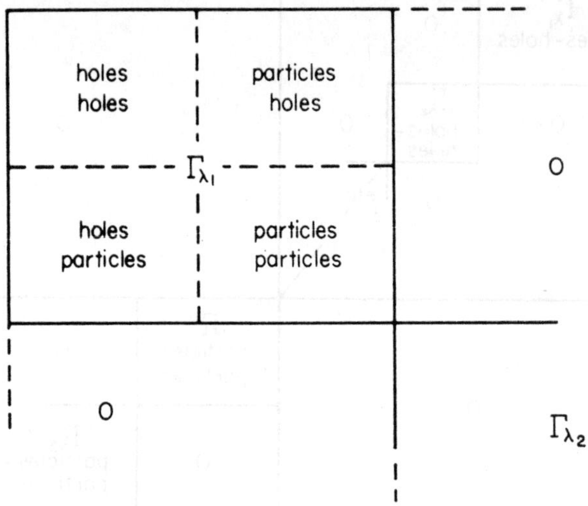

Fig. 2. Density matrix (CI of mono- and diexcitation).

with large occupation number have components on particles. When expressed in terms of natural orbitals Φ_i^λ, $\langle\theta_e\rangle$ is in diagonal form:

$$\langle\theta_e\rangle = \sum_\lambda \sum_i n_i^\lambda \langle\Phi_i^\lambda|\theta(\mathbf{r})|\Phi_i^\lambda\rangle$$

n_i^λ is the Φ_i^λ occupation number.

A diagonal term of this sum can be written:

$$n_i^\lambda \langle\sum_j c_{ij}^\lambda \varphi_j^\lambda|\theta(\mathbf{r})|\Sigma c_{ij}^\lambda \varphi_j^\lambda\rangle$$

c_{ij}^λ is the component on MO ϕ_j^λ of natural orbital Φ_i^λ.

One can see on this form that diexcitations will not change $\langle\theta_e\rangle$ much: the occupation numbers are not very different from those of the main components in the Hartree–Fock state and the elementary one-particle integrals involved are indeed the same, for the first terms. If a natural orbital composed only of holes has an appreciable occupation number, it corresponds to a near degeneracy of MO: for instance, the first π_g natural orbital of N_2, occupation number 0.1, has its main component on $1\pi_g$ nearly degenerate with $1\pi_u$. In accordance the first π_u natural orbital has occupation number 3.9 and main component on $1\pi_u$. It is obvious that as the $\langle\pi_g|\theta(\mathbf{r})|\pi_g\rangle$ and $\langle\pi_u|\theta(\mathbf{r})|\pi_u\rangle$ integrals are similar the result will not be very different from the reference value.

On the other hand, we see that the inclusion of monoexcitations will involve another type of elementary one-particle integral multiplied by a large occupation number. An inspection of the list of these integrals, if available, will show at once which single-excited configurations could change $\langle\theta_e\rangle$.

D. Results

We have applied CI involving monoexcitations to a number of diatomic molecules:

In the Reference 2 calculation, the dipole moment of CO is shown to be negative. This is a typical case where nuclear and electronic contributions have similar absolute values. In Table 3 we give the detailed results of dipole moment calculations for a number of diatomic molecules: CO, LiH, BH, AlH, FH (ground state and experimental R_e). The nuclear and electronic parts are indicated for Hartree–Fock, CI involving 200 diexcitations, CI involving diexcitations and a small number of monoexcitations selected by the indicated criterion.

BH is an exception: monoexcitations do not contribute.

In AlH the nuclear and electronic parts are identical up to the second decimal; we do not ascertain the result to be negative, we may say it is very small.

In all cases, the changes from HF values are significant, and agreement with those experimental values which are known (LiH, FH) is very good. The electric field's gradient on nucleus

$\langle q \rangle$ has been calculated in the N_2 ground state: the HF value is 1.34 a.u.; CI using 200 diexcited configurations gives 1.30 a.u.; CI using 190 diexcitations and ten significant monoexcitations $(\phi_{3\sigma_g}^{n\sigma_g}, \phi_{1\Pi_g}^{n\Pi_g})$ gives 1.22 a.u.

TABLE 3. Electric dipole moment $\langle Z \rangle$ a.u. Results given by: Hartree-Fock; CI involving 200 diexcitations; CI involving 200 diexcitations and significant monoexcitations. Electronic and nuclear part are indicated separately

Molecule	CO	LiH	BH	AlH	FH
Nuclear part $\langle Z_N \rangle$ a.u.	12.486	3.015	2.336	3.114	1.7328
Electronic part $\langle Z_e \rangle$ a.u. Hartree–Fock	−12.426	−5.376	−1.654	−3.04783	−0.9690
Interaction of 200 diexcitations	−12.460	−5.366	−1.6660	−3.04040	−0.9864
Interaction of 200 di- and monoexcitations	−15.560	−5.303	−1.6663	−3.11756	−1.0116
Total $\langle Z_N \rangle + \langle Z_e \rangle$ a.u. Hartree–Fock	+0.060	−2.361	+0.682	+0.06617	+0.764
Interaction of 200 diexcitations	+0.026	−2.351	+0.6701	+0.07360	+0.746
Interaction of 200 di- and monoexcitations	−0.074	−2.288	+0.6697	−0.00356	+0.721
Experimental values	−0.0472	−2.29			+0.716

E. Conclusion

We may say in other words that in configuration space there is a gradient $\nabla\langle\theta_e\rangle$ characterized by the elementary one-particle integrals. If this gradient is large, even small configuration coefficients will lead to a significant result. In the CI calculations performed in a closed-shell case we need, in order to take a full account of the monoexcitations, a large number of diexcitations strongly interacting with them. Which of the last-mentioned excitations are useful can be guessed easily by inspection of the one-electron elementary integrals of the relevant operator. It seems that in a Hartree–Fock calculation the high-lying occupied orbitals are not "Brueckner orbitals" and they must always be corrected to calculate with higher accuracy the mean value of a monoelectronic operator. This correction consists mainly in single excitation to the lowest unoccupied orbitals in the configurations interaction.

References

1. Grimaldi, F., *J. Chem. Phys.*, **43,** S 59 (1965).
2. Grimaldi, F., Lecourt, A., and Moser, C., *Intern. J. Quantum Chem.*, **1S,** 153 (1967).
3. Watson, R. E., *Phys. Rev.*, **119,** 170 (1960).
4. Nesbet, R. K., *Rev. Mod. Phys.*, **32,** 272 (1960). Program rewritten the 7094 by R. M. Stevens, I.B.M. Watson laboratory and Columbia University.
5. Hollister, C., and Sinanoğlu, O., *J. Am. Chem. Soc.*, **88,** 13 (1967); and different other publications of O. Sinanoğlu.
6. Huo, W., *J. Chem. Phys.*, **43,** 624 (1965).
7. Wentworth, W. E., *J. Chem. Educ.*, **42,** 2 (1965); Guerin, B., *J. Phys.* to be published; Fraga, S., and Ransil, B., *J. Chem. Phys.*, **35,** 2 (1961).

A LINKED DIAGRAM TREATMENT OF CONFIGURATION INTERACTION IN OPEN-SHELL ATOMS

P. G. H. SANDARS, *The Clarendon Laboratory, Oxford University, England*

CONTENTS

1. INTRODUCTION

A common and useful method of treating configuration interaction in open-shell atoms is to express its effects in terms of appropriate equivalent operators acting within the open shell. In this article we consider the feasibility of using the linked diagram perturbation methods of Brueckner[1] and Goldstone[2] in combination with the angular momentum techniques of Yutsis and coworkers[3] to obtain a perturbation theory of configuration interaction which yields results directly in terms of the equivalent operators.

Our main emphasis here will be on the various theorems and their derivation. Numerical calculations on atoms by means of diagrammatic techniques are discussed by Kelly,[4] and a more detailed discussion of equivalent operators is given by Judd.[5] Some applications of these methods to hyperfine structure have been discussed by the author elsewhere.[6]

Although our main interest lies in degenerate open-shell atoms, we consider first in some detail the Goldstone linked diagram perturbation theory for the non-degenerate closed-shell atom in order to develop the methods and notation which we shall need in the more complex open-shell problem. In our derivation of the linked diagram (LD) theorems we shall use time-independent methods since we feel that these are more familiar to workers in the field of atomic and molecular configuration interaction than the time-dependent methods more commonly used.[2,7] A rather different time-independent treatment of the linked diagram expansion for the degenerate situation has recently been given by Brandow.[8]

Having established our perturbation theory in equivalent operator form, we turn to the angular properties of the operators. To do this we make use of the topological equivalence[6,9] of the Feynman interaction diagrams used in the perturbation expansion and the angular momentum diagrams of Yutsis and coworkers.[3] We make extensive use of the powerful theorems given by Yutsis and coworkers[3] (YLV) to express our equivalent operators in their most convenient spherical tensor form.

Finally, we make some remarks concerning the usefulness of the diagrammatic approach in treating the radial part of the configuration interaction problem.

2. THE EQUIVALENT OPERATOR APPROACH

A. The Central Field Approximation

A familiar first approximation in atomic calculations is the replacement of the exact non-relativistic many electron Schrödinger equation (atomic units):

$$\left\{\sum_i \left(p_i^2/2 - \frac{Z}{r_i}\right) + \sum_{i \pm j} \frac{1}{2r_{ij}}\right\}\Psi = E\Psi \tag{1}$$

by the approximate equation:

$$H^0\psi = \sum_i (p_i^2/2 - Z/r_i + U_i(\mathbf{r}_i))\psi = E^0\psi \tag{2}$$

whose Hamiltonian H^0 is the sum of single-particle operators. The eigenstates of H^0 are of course just Slater determinants

$$\psi = \frac{1}{\sqrt{n!}} \operatorname{Det} \phi_1 \ldots \phi_n \tag{3}$$

constructed out of single-particle functions ϕ_i which satisfy

$$(p^2/2 - Z/r + U_i(\mathbf{r}))\phi_i = \varepsilon_i\phi_i \tag{4}$$

In open-shell atoms several Slater determinants may be degenerate, in which case the many-electron function ψ will in general be a linear combination of them.

The potential $U_i(\mathbf{r})$ is chosen as a compromise between the desire for ease of solution of the single-particle Schrödinger equation (4) and the necessity to minimize the effects of the residual perturbation

$$H^1 = \sum_{i \neq j} 1/2r_{ij} - \sum_i U_i(\mathbf{r}_i) \tag{5}$$

It is customary to require that $U_i(\mathbf{r}_i)$ be central and spin-independent so that the single-particle function ϕ_i can be written as the product of a radial function, a spherical harmonic and a spin function

$$\phi_{nlmm_s} = \frac{R_{nl}(r)}{r} Y^l_m(\theta, \phi)\chi_{m_s} \tag{6}$$

The radial function $R_{nl}(r)$ satisfies the Schrödinger equation

$$\left\{-\frac{1}{2}\frac{d^2}{dr^2} - Z/r + \frac{l(l+1)}{2r^2} + U_{nl}(r)\right\} R_{nl}(r) = \varepsilon_{nl} R_{nl}(r) \tag{7}$$

The choice of the particular form of the central potential $U_{nl}(r)$ depends on a compromise between convenience and the necessity to minimize the residual perturbation H^1 (5). In a certain variational sense, the Hartree–Fock potential constitutes a best choice. But as Kelly has pointed out,[4] the Hartree–Fock potential has many undesirable features when used for configuration interaction calculations. The best choice almost certainly depends on the particular problem under attack. We do not need to consider this interesting problem here because our analysis is independent of the precise form of $U_{nl}(r)$ provided that: (*a*) the single-particle functions ϕ_{nlmm_s} constitute a complete set, and (*b*) the residual perturbations (5) are sufficiently small that a perturbation theory approach converges rapidly enough to be useful.

B. The Equivalent Operator Approach

The central field approximation is used in two rather different ways. It forms the first approximation in a theoretical discussion of the structure and properties of atoms. Qualitatively, it performs this function very satisfactorily. The central field model adequately accounts for the main features of atomic structure. On a quantitative level, the situation particularly for many atomic properties is much less satisfactory. The configuration interaction effects which have been neglected in the model are not adequately small.

However, the central field approximation is still extremely useful because it can be used to parameterize the results of experiment. The data is then in its most convenient form for inter-relating different experiments and for the comparison with theory. To see how this comes about we consider two important examples.

(*a*) *Parameterization of the energy levels within a configuration.* On the central field model the relative energies of the different terms of a given configuration are given by the eigenvalues of the perturbation H^1 taken within the degenerate set of states belonging to the configuration. The central potential term produces no relative splitting and can be neglected here. The $\frac{1}{2}r_{ij}$ term is best treated by expanding it in spherical tensor form

$$\frac{1}{2}r_{ij} = \sum_{k=0}^{\infty}(-1)^k \frac{1}{2}\frac{r_<^k}{r_>^{k+1}}\sum_q(-1)^{k-q}C_q^k(i)C_{-q}^k(j) \tag{8}$$

where the $C_q^k(\theta, \phi)$ are the usual modified spherical harmonics.[10] For simplicity, we confine our discussion to an l^n configuration which has only a single open shell, although our conclusions have more general relevance. On taking the matrix elements of (8) within the l^n configuration we have the parity and angular momentum restrictions that k must be even and less than $2l + 1$. It follows that the relative term energies can be expressed in terms of the l radial integrals

$$F^k = \iint R_{nl}^2(1)R_{nl}^2(2)\,\frac{r_<^k}{r_>^{k+1}}\,dr_1dr_2 \tag{9}$$

(the $k = 0$ term does not contribute to energy differences within the configuration and so can be neglected).

In a straightforward application of the central field model, one would use solutions of the radial equations (7) to calculate the F^k integrals from (9). These integrals would then be used to calculate the term energies to be compared with experiment. In the majority of atoms the resulting agreement between theory and experiment would be poor.

The equivalent operator approach to the problem is to treat the F^k as experimental parameters and replace (8) by the effective Hamiltonian

$$H^1_{\text{equ}} = \sum_{i \neq j} \tfrac{1}{2} \sum_{k \text{ even}} (-1)^k F^k \sum_q (-1)^{k-q} C^k_q(i) C^k_{-q}(j) \tag{10}$$

The F^k are determined by least squares fitting to the experimental data. The resulting fit with experiment is normally very much better than from the straightforward application of the central field model; there is clearly some physical content in the use of (10).

The fit is not as good as one would like, and it has been known for several years[11] that it can be considerably improved if one adds to (10) operators $\sum_{\text{odd } k} F^k \mathbf{V}^k(i) \,.\, \mathbf{V}^k(j)$ with odd tensor rank, though with even parity. The improvement in the fit is better than one would expect simply on the grounds of the increased number of parameters and clearly corresponds to some physical effect.

Even with the odd k terms the fit is by no means exact and further extensions to (10) to include three-body operators have been proposed (Judd Reference 5).

(*b*) *Parameterization of the magnetic hyperfine interaction.* As a second example of the equivalent operator approach we consider the magnetic hyperfine interaction in the non-relativistic approximation. In the central field model the hyperfine interaction is given by the expectation value of the Hamiltonian

$$H^{\text{hfs}} = H^{\text{contact}} + H^{\text{spin dipolar}} + H^{\text{orbital}} \tag{11}$$

with (atomic units)

$$H^{\text{contact}} = \sum_i \frac{4\pi}{3} \alpha^2 \boldsymbol{\mu}_N \,.\, \mathbf{S}_i \delta(\mathbf{r}_i) \tag{12}$$

$$H^{\text{spin dipoler}} = \sum_i -\frac{\alpha^2}{2} \sqrt{10}\, \boldsymbol{\mu}_N \,.\, \left\{\frac{\mathbf{SC}^2}{r^3}\right\}^1 (i)$$

$$H^{\text{orbital}} = \sum_i \frac{\alpha^2}{2} \frac{\boldsymbol{\mu}_N \,.\, \mathbf{l}_i}{r_i^3}$$

Again restricting ourselves to the simple case of a simple l^n configuration ($l \neq 0$). we see that the hyperfine interaction depends on the single radial integral

$$\langle r^{-3}\rangle = \int R_{nl}(r)^2 \frac{1}{r^3}\, dr \tag{13}$$

since the contact interaction does not contribute for $l \neq 0$. Using a radial wave function from (7) one can calculate $\langle r^{-3}\rangle$ from (13) and compare the resulting magnetic hyperfine structure with experiment. The agreement is usually poor.[12]

In the simplest equivalent operator approach one treats $\langle r^{-3}\rangle$ as an experimental parameter and uses the effective Hamiltonian

$$H_{\text{equ}} = \sum_i \frac{\alpha^2}{2}\, \boldsymbol{\mu}_N \cdot \{\mathbf{l}_i - \sqrt{10}\, \{\mathbf{SC}^2\}^1_{j(i)}\}\langle r^{-3}\rangle \tag{14}$$

In most cases, the use of (14) improves the agreement with experiment.

Much better agreement can be obtained by regarding all the radial integrals appropriate to the hyperfine Hamiltonian (12) as parameters; this leads to the three-parameter Hamiltonian

$$H'_{\text{equ}} = \sum_i \frac{\alpha^2}{2}\, \boldsymbol{\mu}_N \cdot \{\mathbf{l}_i\langle r_l^{-3}\rangle + \mathbf{S}_i\langle r_c^{-3}\rangle - \sqrt{10}\, \{\mathbf{SC}^2\}^1_{j(i)}\langle r_{sd}^{-3}\rangle\} \tag{15}$$

Experimentally, one finds that $\langle r_{sd}^{-3}\rangle \neq \langle r_l^{-3}\rangle$ and that $\langle r_c^{-3}\rangle \neq 0$.

Since (15) is the most general single-particle hyperfine Hamiltonian any extension of the method would require equivalent operators which were two-body operators. As yet there is insufficient experimental data to justify this.

C. Equivalent Operators and Perturbation Theory

As we pointed out in the introduction, our aim here is to derive an LD perturbation theory for the open-shell atom which yields results directly in terms of equivalent operators in the open shell. There are three main reasons why we use the LD perturbation theory rather than the more familiar Rayleigh–Schrödinger or Brillouin–Wigner forms:

[1]. The linked diagram perturbation theory can be expressed directly in terms of equivalent operators. This is not so easy to do in the other forms of perturbation theory.

[2]. The results of the LD theory occur directly as a series of one-particle, two-particle, three-particle operators up to the number of particles in the open shell.

[3]. The topological equivalence of Feynman interaction diagrams and Yutsis angular momentum diagrams makes the extraction of the spherical tensor form of the equivalent operator extremely simple. As we saw above, the equivalent operators used to analyse experimental data are normally expressed in this form. In this respect, it is important to note that the disappearance of the exclusion principle which is an important feature of the LD method is vital for the successful use of the angular momentum diagrams.

3. NON-DEGENERATE LINKED DIAGRAM PERTURBATION THEORY

Before attacking the more complex degenerate problem, we shall outline here a simple time-independent treatment of the LD perturbation theory. This will help us to gain familiarity with the methods and notation to be used in the degenerate case.

A. Creation and Annihilation Operator Formalism

We make considerable use of a creation and annihilation formalism in which the zeroth-order Hamiltonian and the perturbation H^1 are expressed in the form

$$H^0 = \sum_p \varepsilon_p a_p^* a_p \tag{16}$$

and

$$H^1 = \sum_{pqrs} \langle pq|\tfrac{1}{2}r_{12}|rs\rangle a_p^* a_q^* a_s a_r + \sum_{pq} \langle p| - u(r)|q\rangle a_p^* a_q \tag{17}$$

where as usual we have the anticommutation rules

$$[a_p^*, a_q^*]_+ = [a_p, a_q]_+ = 0 \qquad [a_p^*, a_q]_+ = \delta_{pq} \tag{18}$$

It is important to note that the sum over single-particle states in (16) and (17) spans the complete set of solutions of (4). The reader who is unfamiliar with these methods is referred to the book by Judd.[5]

We also make considerable use of Feynman diagrams, an example being given in Fig. 1. As we shall see, each diagram represents an operator made up of four elements: creation and annihilation operators, phase, one- and two-particle matrix

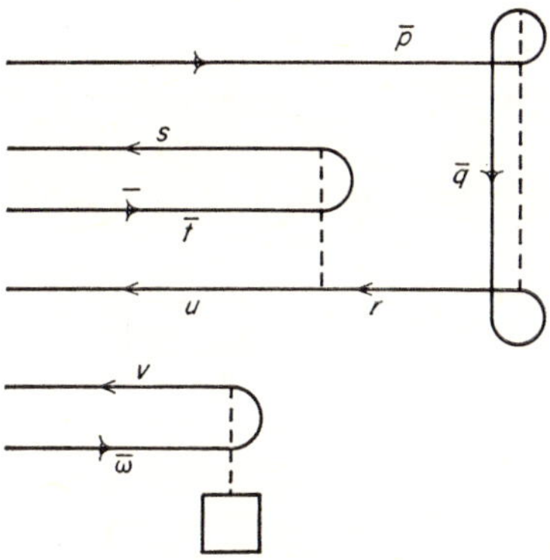

Fig. 1. A typical diagram.

elements, and energy denominators. The operator corresponding to Fig. 1 is

$$\frac{a_{\bar{p}} a_s^* a_{\bar{t}} a_u a_v^* a_{\bar{w}} \langle v| -u(r)|\bar{w}\rangle \left\langle su \left| \frac{1}{2r_{12}} \right| \bar{t}r \right\rangle \left\langle \bar{q}r \left| \frac{1}{2r_{12}} \right| \bar{p}\bar{q} \right\rangle}{(\varepsilon_{\bar{p}} + \varepsilon_{\bar{t}} + \varepsilon_{\bar{w}} - \varepsilon_s - \varepsilon_u - \varepsilon_v)(\varepsilon_{\bar{p}} + \varepsilon_{\bar{t}} - \varepsilon_s - \varepsilon_u)(\varepsilon_{\bar{p}} - \varepsilon_r)}$$

The general form of an operator associated with a diagram is
Product of creation and annihilation operators

$$(-1)^{n_c} \frac{1}{E_n} M_n \frac{1}{E_{n-1}} M_{n-1} \ldots M_2 \frac{1}{E_1} M_1 \tag{19}$$

The matrix elements $M_1, M_2, \ldots M_n$ are read off the diagram from right to left according to the rules given in Fig. 2. Directed horizontal lines represent states and vertical dotted lines represent operators. The intersection of the horizontal and vertical lines at a vertex represents a matrix element.

Our diagrams differ slightly from those used by Goldstone,[2] in that ours are horizontal where Goldstone's are vertical. The horizontal diagram makes the relationship between diagrams and operators much more transparent. The vertical form of Goldstone's diagrams is related to his use of time-dependent methods; the time axis being conventionally drawn vertically. Since we are

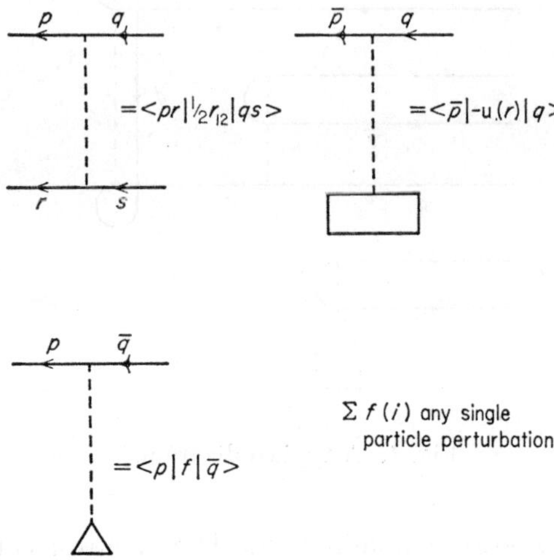

Fig. 2. Rules for relating diagrams and matrix elements.

using time-independent methods we are at liberty to draw them horizontally.

We find it convenient to distinguish those "unexcited" single-particle states which are present in the zeroth-order determinant $|\psi_0\rangle$ from those "excited" single-particle states which are not. We represent the unexcited states with a "closed" arrow and the excited states with an "open" arrow. We also distinguish the arguments of the states with a "bar", thus $\bar{p}$ represents an unexcited state and p an excited state.

The energy denominator E_m is found in the following way. One draws an imaginary line between the m and $m + 1$ interactions from the right and

$E_m =$ The sum of the single-particle energies of the unexcited-state lines minus the sum of the excited-state lines; both sums being taken over the state lines cut by our imaginary vertical line

The product of creation and annihilation operators is found by assigning a creation operator to each free end of a state line with an arrow pointing towards the open end and an annihilation operator to each free state line with an arrow pointing away from the open end. The ordering of the operators from left to right is that of the free ends on the diagram taken from top to bottom on the left-hand side followed by bottom to top on the right-hand side.

The phase is given by a very simple rule. The quantity n_c is simply the number of points at which two state lines cross. In order that this simple rule holds it is necessary that a state line which originates on one side of its interaction point and finishes up on the other side should go around and *underneath* this interaction point. Our diagrams differ in this respect from the usual one but the simple phase rule is the result.

One more piece of notation will be important in what follows. We represent a diagram and its associated operator having n interaction lines by $D_{n,i}$ the subscript i spanning the various different diagrams with n interactions. By different diagrams in this context we mean both diagrams which are different in a topological sense and also those that are different because the arguments of the states are different. A useful extension to this is to represent by $D'_{n,i}$ the operator which is identical to $D_{n,i}$ except that it is missing the final energy denominator $1/E_n$.

We now use the diagrams and their associated operators to define new "perturbed" wave functions by operating to the right on the unperturbed function $|\psi_0\rangle$. We call these functions perturbed because $D_{n,i}|\psi_0\rangle$ will have the form

$$(\text{perturbation})^n/(\text{energy denominator})^n|\psi_0\rangle$$

which is typical of the perturbed functions in ordinary perturbation theory.

A very important feature of the operation of creation and annihilation operators on $|\psi_0\rangle$ is that

$$a_p|\psi_0\rangle = a_{\bar{p}}^*|\psi_0\rangle = 0 \tag{20}$$

where of course $\bar{p}$ is an unexcited state and p an excited state. Equation (20) is physically obvious since both the creation of a particle in a state which is already occupied and the annihilation from a state which is unoccupied must give zero.

To make use of (20) we draw all our diagrams in a form in which the excited-state creation operators and the unexcited state annihilation operators are to the left of the excited shell annihilation operators and the closed shell creation operators. It follows from the association of operators with the free lines on the diagram together with (20) that any diagram with free lines at the right gives zero when operated on $|\psi_0\rangle$. Thus, we need only consider "standard diagrams" which have no free lines to the right.

B. The Linked Diagram Expansion

Our aim in this section is to see if we can find a solution of the Schrödinger equation (1) using diagrams and their associated perturbed functions which were introduced in the previous section. Specifically, we want to try to find a suitable set of diagrams such that if we define

$$S = \sum_{n=0}^{\infty} S_n \text{ and } S_n = \sum_i D_{n,i} \tag{21}$$

then $S|\psi_0\rangle$ satisfies the complete atomic Schrödinger equation

$$[E^0 - H^0 - H^1]S|\psi_0\rangle = -\delta E\, S|\psi_0\rangle \tag{22}$$

If we can do this and the series is convergent so that the wave function can be normalized then we have achieved a perturbation type solution.

(*a*) *Operation by H^1 on a standard diagram.* In order to see how to achieve this aim, we consider first the form of the diagrams which result when we operate with H^1 on one of our perturbed functions $D_{n-1,i}|\psi_0\rangle$. This is illustrated diagrammatically in Fig. 3 for a typical example. Now this diagram on the left-hand side of Fig. 3 is not in standard form. We need to move the associated closed shell creation operators and excited shell annihilation operators to the right. To do this we use the anti-commutation rules (18) to obtain

$$
\begin{aligned}
a_t^* a_{\bar{u}}^* a_s a_{\bar{r}} a_p^* a_{\bar{q}} \\
&= -a_t^* a_{\bar{q}} \delta_{\bar{r}\bar{u}} \delta_{ps} + a_t^* a_{\bar{r}} \delta_{\bar{u}\bar{q}} \delta_{ps} a_t^* a_{\bar{r}} a_{\bar{q}} a_{\bar{u}}^* \delta_{ps} \qquad (23) \\
&\quad - a_t^* a_{\bar{r}} a_p^* a_s \delta_{\bar{u}\bar{q}} - a_t^* a_p^* a_{\bar{q}} a_s \delta_{\bar{r}\bar{u}} + a_t^* a_{\bar{r}} a_p^* a_{\bar{q}} a_u^* a_s
\end{aligned}
$$

The appropriate diagrams are given on the right-hand side of Fig. 3. We see that the process of anti-commutation of the operators can be directly represented by the diagrams, the δ functions being represented by the joining of two state lines with

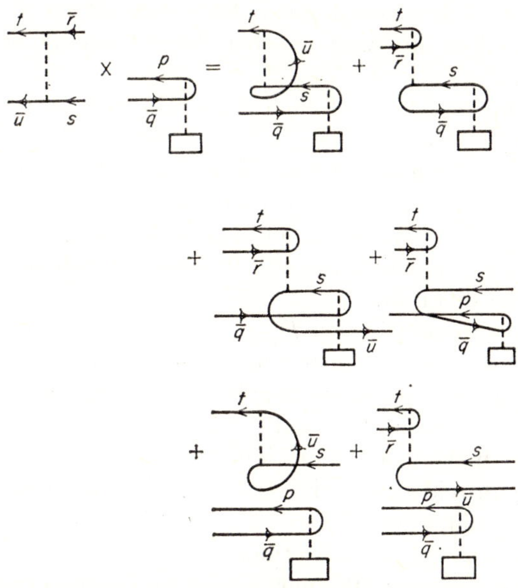

Fig. 3. Operation on a diagram with H^1.

compatible arrows, e.g. both unexcited or both excited states and pointing in the same direction.

Now the last four diagrams on the right-hand side of Fig. 3 have free lines to the right. They will give zero when operated on $|\psi_0\rangle$ and hence are of no interest. We are only interested in the standard diagrams which result from the operation of H^1 on a standard diagram. By considering the relationship between the diagrams and the commutation process illustrated above one can show that these standard diagrams can be constructed by the following very simple rules:

[1]. Draw the nth interaction to the left of the diagram $D_{n-1,i}$.

[2]. Take each excited state line which is free to the right, extend it in the same direction, and join it up with an excited state line on $D_{n-1,i}$ which is free to the left.

[3]. Take the unexcited state line on H^1 which is free to the left, bring it around and underneath its interaction point, and join it up with another unexcited state line to its right in such a way that the arrows are continuous. This rule allows an unexcited state line to join up with another line belonging to the same interaction.

[4]. Extend any excited shell lines still free to the left of the diagram.

[5]. Take any unexcited state lines which are still free to the right on H^1, bring them around and underneath their interaction point, and extend them to the left of the diagram. Any unexcited state lines in $D_{n-1,i}$ which are still free should also be extended to the left of the diagram.

The result of operating on $D_{n-1,i}$ with H^1 is given by all those diagrams which can be constructed using these rules. In accordance with the notation of the previous section these new diagrams will be primed since the final energy denominator is missing compared with the usual definition.

Note that provided we bring the state lines around and underneath their interaction points, the sign is automatically given correctly by our simple phase rule.

(*b*) *Linked diagrams.* We now introduce the important concept of a *linked* diagram. We do this by first defining an *unlinked* diagram as one which has a *closed part* which: (*a*) is not joined to another part of the diagram by any state or interaction lines; and (*b*) has no free state lines. A special case of an unlinked diagram is a diagram which has no free state lines at all. A *linked* diagram is a diagram which contains no closed parts. Examples of linked and unlinked diagrams are given in Fig. 4.

It is clear that all standard diagrams are either linked or unlinked so that we can write the effect of H^1 in the form

$$H^1 S_{n-1}|\psi_0\rangle = \{\sum_i D'_{n,i} \text{ (linked)} + \sum_j D'_{n,j} \text{ (unlinked)}\}|\psi_0\rangle \quad (24)$$

Note that the diagrams on the right of (24) are primed since they are missing the final energy denominator.

We can now make use of (24) to give an inductive definition of the set of diagrams which constitute S_n. We define S_n to be given by

$$S_n = \sum_i D_{n,i} \text{ (linked)}, \quad S_0 = 1 \tag{25}$$

where the sum over linked diagrams is the same as in (24). In other words, S_n consists of all the linked diagrams which result from the operation of H^1 on S_{n-1}.

To begin to see why we choose this form for S_n, we operate to the right on $S_n|\psi_0\rangle$ with $E^0 - H^0$. One can readily see that this

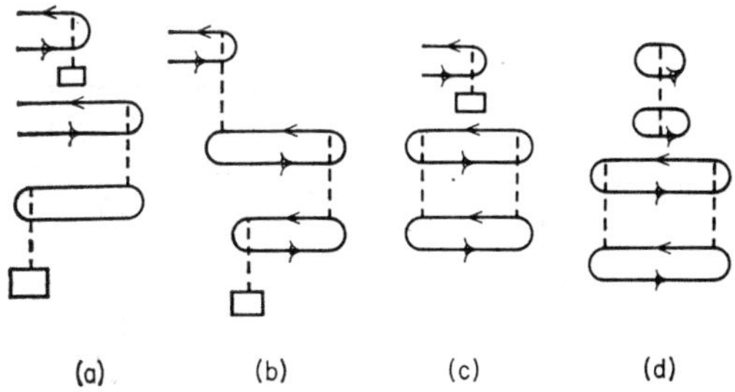

Fig. 4. Examples of linked and unlinked diagrams.

simply cancels out the final energy denominator so that with an obvious notation

$$(E^0 - H^0)S_n|\psi_0\rangle = \sum_i D'_{n,i} \text{ (linked)}|\psi_0\rangle = S'_n|\psi_0\rangle \tag{26}$$

But from (24) and (25) we also have

$$H^1S_{n-1}|\psi_0\rangle = S'_n|\psi_0\rangle + \sum_j D_{n,j} \text{ (unlinked)}|\psi_0\rangle \tag{27}$$

Subtracting (27) from (26) we obtain

$$\{(E^0 - H^0)S_n - H^1S_{n-1}\}|\psi_0\rangle = -\sum_j D_{n,j} \text{ (unlinked)}|\psi_0\rangle \tag{28}$$

This cancellation of the linked diagrams which leads to (28) is crucial to our argument.

(c) *Factorization of unlinked diagrams.* We now look in more detail at the form of the unlinked diagrams which constitute the right-hand side of (28). Each such diagram must have a part with no free state lines. Furthermore, the nth and final interaction line must belong to this closed part otherwise the diagram would have been unlinked in lower order and would not have been included in S_{n-1}. This is, however, the only restriction on the relative ordering of the interaction lines in the closed part and in the rest of the diagram.

From the method of construction one can show that the two parts are built up quite independently and that *all possible* relative

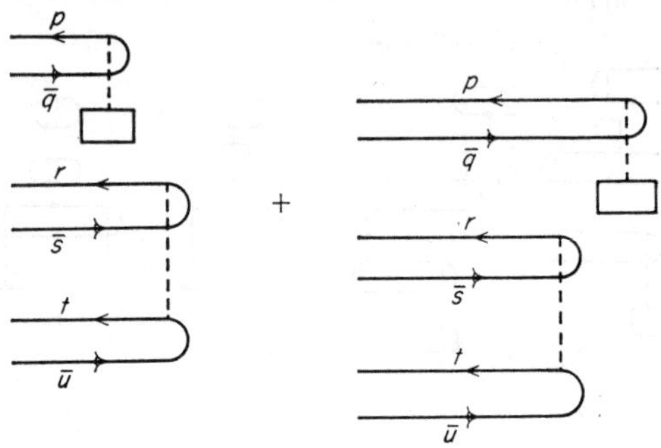

Fig. 5. Example of diagrams which differ in the relative ordering of their interaction lines.

orderings will be present. This result is crucial to the LD derivation. Its proof (see Appendix A) requires a little care, and the reader is advised to check that it is true in a simple example.

Each of the diagrams which differ only in the relative ordering of the interaction lines of the closed part and the rest has the same creation and annihilation operators, the same sign and the same matrix elements; they only differ in their energy denominators. However, there is a very general theorem which we prove in Appendix B which states that the sum of all the energy denominators corresponding to all possible relative orderings is just equal to the product of the energy denominators for each part taken separately. A simple example is given by the diagrams in Fig. 5 whose energy denominators sum up

$$\frac{1}{(\varepsilon_{\bar{q}}+\varepsilon_{\bar{s}}+\varepsilon_{\bar{u}}-\varepsilon_p-\varepsilon_r-\varepsilon_t)}\left\{\frac{1}{(\varepsilon_{\bar{s}}+\varepsilon_{\bar{u}}-\varepsilon_r-\varepsilon_t)}+\frac{1}{(\varepsilon_{\bar{q}}-\varepsilon_p)}\right\}$$

$$=\frac{1}{(\varepsilon_{\bar{s}}+\varepsilon_{\bar{u}}-\varepsilon_r-\varepsilon_t)}\times\frac{1}{(\varepsilon_{\bar{q}}-\varepsilon_p)} \qquad (29)$$

to give the product of the denominators for the two parts taken separately.

If we now collect together all the sets of diagrams in (28) which just differ in their ordering of the closed part with respect to the rest, we can use the result above to factorize their energy denominators and rewrite the right-hand side of (28) in the form

$$\sum_j D'_{n,j}\text{ (unlinked)} = \sum_{m,k} D'_{n-m,i}\text{ (closed) } D_{m,k}\text{ (linked)} \qquad (30)$$

m in (28) ranges from zero to $n - 1$ because the final interaction line must be in the closed part of the diagram.

From the way that the diagrams are constructed we can also show that for each closed diagram with $n - m$ interaction lines in (30) one will have all the possible linked diagrams with m interaction lines. But the sum of all possible linked diagrams with m interaction lines is just S_m. It follows that we can factorize (30) and write it in the form

$$\sum_j D'_{n,j}\text{ (unlinked)} = \sum_{m=0}^{n-1} \sum_i D'_{n-m,i}\text{ (closed)}S_m \qquad (31)$$

where the sum over i spans all possible closed diagrams of order $n - m$ which can be constructed from a linked diagram of order $n - m - 1$. One can readily show that

$$\sum_i D'_{n-m,i}\text{ (closed)} = \langle\psi_0|H^1 S_{n-m-1}|\psi_0\rangle \qquad (32)$$

Expressions (30), (31), and (32) can now be substituted into (28) to yield

$$\{(E^0 - H^0)S_n - H^1 S_{n-1}\}|\psi_0\rangle \qquad (33)$$

$$= -\sum_m \langle\psi_0|H^1 S_{n-m-1}|\psi_0\rangle S_m|\psi_0\rangle$$

We now sum (33) over all orders from $n = 1$ to ∞ and obtain

$$\{E^0 - H^0 - H^1\}S|\psi_0\rangle = -\sum_{n=1}^{\infty}\sum_{m=0}^{n-1}\langle\psi_0|H^1S_{n-m-1}|\psi_0\rangle S_m|\psi_0\rangle \tag{34}$$

where we have used $\{E^0 - H^0\}S_0|\psi_0\rangle = 0$.

We now assume that our perturbation approach is justified and that the series in (34) are absolutely convergent. If so, we can interchange the order of summation and rewrite (34) in the form

$$\{E^0 - H^0 - H^1\}S|\psi_0\rangle = -\sum_{p=0}^{\infty}\langle\psi_0|H^1S_p|\psi_0\rangle\sum_{m=0}^{\infty}S_m|\psi_0\rangle \tag{35}$$

which is equivalent to

$$\{E^0 - H^0 - H^1\}S|\psi_0\rangle = -\langle\psi_0|H^1S|\psi_0\rangle S|\psi_0\rangle \tag{36}$$

This has just the desired form of the complete Schrödinger equation (22) so that $S|\psi_0\rangle$ is a solution with eigenvalue

$$E = E^0 + \langle\psi_0|H^1S|\psi_0\rangle \tag{37}$$

From our analysis, we see that the perturbed part of the energy $\langle\psi_0|H^1S|\psi_0\rangle$ is given by the sum of all diagrams which were linked to order $n - 1$ but are closed by the final interaction.

C. Expectation Value of a Perturbation

Very often one does not want an expression for the energy but for the expectation value of some additional perturbation which is almost always a one- or two-particle operator.

The simplest and most direct way of treating such an additional perturbation is to note that we have not made use of any particular properties of H^1 in our derivation of the LD expansion. Thus, no change is nccessary if we redefine H^1 to be the sum of our original perturbation and the additional perturbation H^{p}. We still have exactly the same form of linked diagram expansion.

Usually, H^{p} is a small perturbation so that any expansion in powers of H^{p} converges rapidly, and diagrams which contain more than a small number of H^{p} interactions will be very small. If we collect the diagrams into groups depending on the number of H^{p} interactions that they have, then the first-order expectation value of H^{p} will be just the sum of all diagrams with a single H^{p} interaction, the second-order effect of H^{p} the sum of all diagrams with two H^{p} interactions and so on.

D. Connection with Ordinary Perturbation Theory

In order to understand the connection between the LD expansion and the more familiar perturbation theories, it is useful to express the ordinary perturbation theory in terms of diagrams.

The first-order energy is given by

$$\delta E^1 = \langle \psi_0 | H^1 | \psi_0 \rangle \tag{38}$$

Using the usual rules for evaluating the matrix elements of one- and two-particle operators between determinantal states, we can expand (38) in terms of matrix elements between the single-particle states in $|\psi_0\rangle$. But these matrix elements can be expressed

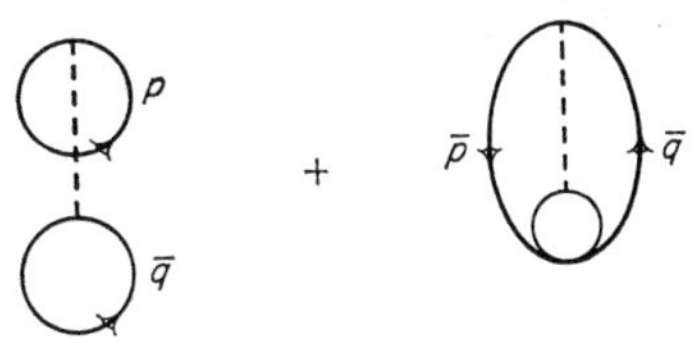

Fig. 6. First-order diagrams which cancel if $p = q$.

in terms of diagrams by reverse use of the rules of the previous section. We leave it to the reader to satisfy himself that the $1/2r_{12}$ diagrams that he gets are those illustrated in Fig. 6. The LD expansion also gives just these diagrams but there is an important difference. In the LD expansion the sum over $\bar{p}$ and $\bar{q}$ spans all possible unexcited states including $\bar{p} = \bar{q}$ while the antisymmetry of the determinant $|\psi_0\rangle$ rules this term out in (38). The two approaches do of course give the same answer because in the case $\bar{p} = \bar{q}$ the two diagrams in Fig. 6 have identical matrix elements but opposite signs because of the extra cross-over in the right-hand diagram.

This example illustrates a very important difference between the LD and ordinary perturbation theories. In the LD expansion the single-particle states are summed over all possibilities, the exclusion principle being completely neglected. In effect, the LD method makes use of the fact that if $|\psi_0\rangle$ is antisymmetric and H^1 is a symmetric operator, then it does not matter if one includes in one's intermediate states $|\psi_i\rangle$ not only the usual antisymmetric states but also states which are not completely antisymmetric. This neglect of the exclusion principle is a major advantage of the LD expansion.

The relation between ordinary second-order perturbation theory and the LD expansion is much as for the first order with the exclusion principle violating diagrams in the LD expansion cancelling out in pairs to give the same result as ordinary perturbation theory.

It is in third order that a really important difference between the theories first appears. The usual Rayleigh–Schrödinger expression for the energy is

$$\delta E^3 = \sum_i{}' \sum_j{}'' \left[\frac{\langle\psi_0|H^1|\psi_i\rangle\langle\psi_i|H^1|\psi_j\rangle\langle\psi_j|H^1|\psi_0\rangle}{(E^0 - E_i)(E^0 - E_j)} - \frac{\langle\psi_0|H^1|\psi_i\rangle\langle\psi_0|H^1\psi_0\rangle}{(E^0 - E_i)} \frac{\langle\psi_i|H^1|\psi_0\rangle}{(E^0 - E_i)} \right] \tag{39}$$

We can again express (39) in terms of diagrams and compare with the LD expansion. We find an additional difference between them. Some diagrams from (39) contain closed parts whereas the LD expansion has no such diagrams. These closed parts are present in both terms of (39). One can show that if one adds and subtracts suitable exclusion principle violating diagrams to the diagrams from (39) then the two sets of diagrams with closed parts cancel by virtue of the minus sign in (39) leaving just the LD result. It is important to note that the exclusion principle violating terms in the LD expansion no longer cancel out in pairs as some of them have been used to cancel out the closed part of the diagram.

If one had sufficient patience one could extend the derivation of the LD expansion from Rayleigh–Schrödinger perturbation theory to any order. One would find that if in each order one added and subtracted suitable exclusion principle violating diagrams then all those diagrams with closed parts would cancel leaving the LD expansion.

4. THE LINKED DIAGRAM EXPANSION FOR AN OPEN-SHELL ATOM

We now turn our attention to the more difficult problem of an open-shell atom. The most important way in which the open-shell problem differs from the closed shell treated in the previous section is that there are normally several degenerate zeroth-order determinantal states $|\psi_0^\alpha\rangle$. Their existence invalidates the arguments used in the previous section which assumed a single zeroth-order determinant. In this section we extend our arguments to cover this more general case. We follow the same basic method as in the non-degenerate case. We attempt to find a perturbed function

$$|\Psi_\alpha\rangle = S|\psi_0^\alpha\rangle \tag{40}$$

which satisfies the Schrödinger equation

$$\{E^0 - H^0 - H^1\}S|\psi_0^\alpha\rangle = -\delta E^\alpha S|\psi_0^\alpha\rangle \tag{41}$$

where

$$S = \sum_{n=0}^{\infty} S_n \tag{42}$$

and S_n is defined in terms of a suitable sum over diagrams. There are two complications over the non-degenerate case. The suitable choice of diagrams is rather more complicated, and the unperturbed function $|\psi_0^\alpha\rangle$ has to be correctly chosen.

We find it convenient in our analysis to make a slight extension to our diagrammatic notation. Whereas previously we divided our single-particle states into two classes depending on their presence or absence in the zeroth-order determinant, we now need three classes:

[1]. Unexcited states. Those single-particle states which are present in all the zeroth-order degenerate determinants.

[2]. Excited states. Those single-particle states which are not present in any of the determinants.

[3]. Open-shell states. Those single-particle states which are present in some but not all the determinants.

We represent the unexcited and excited states as before with closed and open arrows. The open-shell states we distinguish with a double arrow.

In this analysis, we restrict ourselves to the situation where all the open-shell states have the same energy ε_0. This is sufficiently general for many purposes, and the extension to a wider case does not appear to be simple. The energy denominator E_m now takes the slightly more complicated form

$$E_m = \Sigma(\varepsilon_{\bar{u}} - \varepsilon_e) + n\varepsilon_0$$

where ū and e span the unexcited and excited state lines cut by our imaginary vertical line drawn between the mth and $(m + 1)$th interactions. n is the difference between the number of unexcited and excited state lines which are cut.

A. The Linked Diagram Expansion

We note first that in analogy with (26) the operation of $E^0 - H^0$ has the effect of removing the final energy denominator so that we have

$$(E^0 - H^0)S_n|\psi_0^\alpha\rangle = S_n'|\psi_0^\alpha\rangle \tag{43}$$

(*a*) *Operation of* H^1. We now turn our attention to the operation of H^1 on our perturbed states $D_{n-1,i}|\psi_0^\alpha\rangle$. As before we can represent this in diagrammatic form. We first commute to the right the excited state annihilation operators and the unexcited state creation operators since these give zero when operated to the right on $|\psi_0^\alpha\rangle$.

The open shell creation and annihilation operators do not in general give zero when operated on $|\psi_0^\alpha\rangle$. For reasons which should become clear as we go along, we move the open shell creation operators to the left and the annihilation operators to the right; we treat them in this respect as if they were excited states. There is, however, a big difference because the open-shell annihilation operators do not give zero when operated on $|\psi_0^\alpha\rangle$. Thus, when we represent the process of commutation of the creation and annihilation operators in terms of our diagrams, we must retain diagrams with open-shell lines free at the right. An example is given in Fig. 7a.

As before, we introduce the concepts of *linked* and *unlinked* diagrams as diagrams which do not or do have a *closed part.* However, the definition of a closed part is now slightly more complex. It is defined to be a part of a diagram which: (*a*) has no free excited or unexcited state lines, and (*b*) is not joined to another part of the diagram by any state or interaction line. In other words, when considering whether a part is closed or not we neglect the presence of free open shell lines. An example of an unlinked diagram with a closed part is given in Fig. 7.

It is clear that all diagrams are either linked or unlinked so that we must have

$$H^1 S_{n-1}|\psi_0^\alpha\rangle = \{\sum_i D'_{n,i} \text{ (linked)} + \sum_j D'_{n,j} \text{ (unlinked)}\}|\psi_0^\alpha\rangle \quad (44)$$

(*b*) *Backwards diagrams.* At this point we come to the crucial difference between the degenerate and the non-degenerate problems. In the non-degenerate case we defined our perturbed

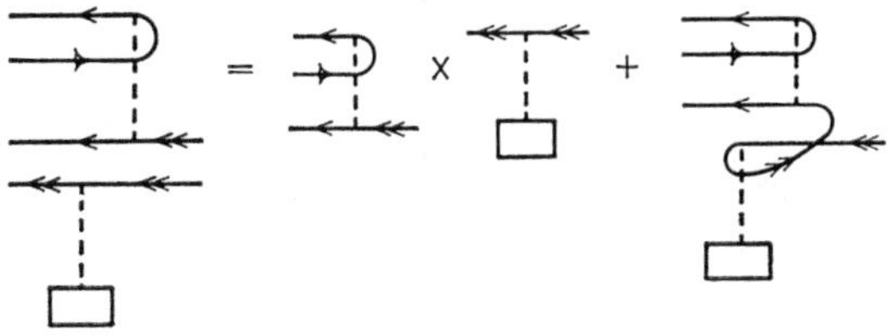

Fig. 7. Illustrating treatment of closed part of diagram leading to backwards diagrams.

function $S_n|\psi_0\rangle$ in terms of the sum over linked diagrams. This turned out to be a solution of Schrödinger's equation because the sum over all the unlinked diagrams could be factorized and written as the sum of terms of the form (30).

In the degenerate case we can still sum over all possible orderings of the interaction lines in the closed part with respect to the rest of the diagram. This allows us to write the matrix elements and energy denominators in a factorized form as we did in the non-degenerate case. But we cannot yet write down the equivalent of (30) in which the unlinked diagram is written as the product of a closed diagram and a linked one. The reason is that both parts of the diagram will in general contain open shell creation and annihilation operators. These are intermixed and we cannot simply separate them since we have to take into account the anticommutation rules (18). We express this difficulty formally by writing the equivalent of (30) in the form

$$\sum_{\text{all orderings}} D'_{n,j}\ (\text{unlinked}) = \frac{D_{m,k}\ (\text{linked})}{D'_{n-m,i}\ (\text{closed})} \tag{45}$$

where the vertical juxtaposition implies that the creation and annihilation operators are still intermixed. The dotted line implies that the sum over relative orderings has been taken so that the matrix elements and energy denominators are in factorized form.

We now use the anticommutation rules (18) to move all the open shell creation and annihilation operators belonging to the closed part to the right of those belonging to the open part. This process introduces some new diagrams which are the difference between the right-hand side of (45) and the separated product

$D_{m,k}$ (linked) $D_{n-m,i}$ (unlinked):

$$\sum_b D'_{n,b}\ (\text{backwards}) = \frac{D_{m,k}\ (\text{linked})}{D'_{n-m,i}\ (\text{closed})} - D_{m,k}\ (\text{linked})\ D'_{n-m,i}\ (\text{closed}) \tag{46}$$

We refer to these new diagrams as "backwards diagrams" as they are distinguished by having open-shell lines with arrows pointing to the right unlike the normal open-shell line which points to the left. An example of a backwards diagram is given in Fig. 7. Note that any diagram which consists simply of a single closed part can be considered to be already in a separated product form and gives rise to no backwards diagrams.

The rule for constructing these backwards diagrams is very simple:

Take one or more free open-shell lines at the left-hand side of the closed part of the diagram, bring them around and underneath their interaction point and join them up to free open shell lines at the right of the linked part of the diagram. All possible different ways of doing this lead to allowed backwards diagrams. Provided the open-shell line passes underneath both its interaction points the phase is correctly given by our crossing rule.

This process is illustrated schematically in Fig. 7. Note that in general the dotted line is still present implying that the matrix elements and energy denominators are still factorized. It is sometimes convenient to reverse the factorization process and replace the dotted line by the sum over all orderings of the two parts.

(*c*) *Definition of* S_n. We define S_n to be the sum of all the linked diagrams and backwards diagrams which result when H^1 operates on S_{n-1}, the definition being completed by $S_0 = 1$. Formally, we can write this

$$S_n = \sum_{\substack{\text{linked}\\+\\\text{backwards}}} \frac{1}{(E^0 - H^0)} H^1 S_{n-1} \tag{47}$$

The rules for obtaining the diagrams which make up S_n are collected together for reference in Appendix C.

(*d*) *Factorization.* From this definition together with (46) we obtain

$$H^1 S_{n-1}|\psi_0^\alpha\rangle = S'_n|\psi_0^\alpha\rangle + \sum_{m,k,i} D_{m,k}\,(\text{linked})\, D'_{n-m,i}\,(\text{closed})\,|\psi_0^\alpha\rangle \tag{48}$$

On comparing this with (43), we find

$$\{(E^0 - H^0)S_n - H^1 S_{n-1}\}|\psi_0^\alpha\rangle = -\sum_{m,k,i} D_{m,k}\,(\text{linked})\, D'_{n-m,i}\,(\text{closed})\,|\psi_0^\alpha\rangle \tag{49}$$

From the way that the diagrams are built up one can see that for each closed diagram $D'_{n-m,i}$ (closed) one will get all possible open diagrams. The sum of these is just S_m so that (49) becomes

$$\{(E^0 - H^0)S_n - H^1 S_{n-1}\}|\psi_0^\alpha\rangle = -\sum_{m=0}^{n-1} S_m \sum_i D'_{n-m,i}\,(\text{closed})\,|\psi_0^\alpha\rangle \tag{50}$$

We now insert a complete set of determinantal states between S_m and $D_{n-m,i}$(closed) on the right-hand side of (50). Since the latter has no free excited or unexcited state lines it is diagonal in the degenerate set $|\psi_0^\beta\rangle$. Experiment (50) then becomes

$$\{(E^0 - H^0)S_n - H^1 S_{n-1}\}|\psi_0^\alpha\rangle = -\sum_{m=0}^{n-1} S_m \sum_\beta |\psi_0^\beta\rangle\langle\psi_0^\beta| \sum_i{}' D_{n-m,i}\,(\text{closed})|\psi_0^\alpha\rangle \tag{51}$$

It is convenient to note at this point that one can write

$$\langle\psi_0^\beta|\sum_i D'_{n-m,i}\ (\text{closed})|\psi_0^\alpha\rangle = \langle\psi_0^\beta|H^1S_{n-m-1}|\psi_0^\alpha\rangle \tag{52}$$

giving (51) the simpler form

$$\{(E^0 - H^0)S_n - H^1S_{n-1}\}|\psi_0^\alpha\rangle = -\sum_{m=0}^{n-1} S_m \sum_\beta |\psi_0^\beta\rangle\langle\psi_0^\beta|H^1S_{n-m-1}|\psi_0^\alpha\rangle \tag{53}$$

(*e*) *Solution to the Schrödinger equation.* Summing over n as in the previous section, we obtain

$$\{E^0 - H^0 - H^1\}S|\psi_0^\alpha\rangle = \sum_n\sum_m\sum_\beta - S_m|\psi_0^\beta\rangle\langle\psi_0^\beta|H^1S_{n-m-1}|\psi_0^\alpha\rangle \tag{54}$$

Once again assuming the requisite convergence properties we rearrange the order of summation to obtain

$$\{E^0 - H^0 - H^1\}S|\psi_0^\alpha\rangle = -\sum_\beta S|\psi_0^\beta\rangle\langle\psi_0^\beta|H^1S|\psi_0^\alpha\rangle \tag{55}$$

We deduce from (55) that if we can choose $|\psi_0^\alpha\rangle$ such that it is an eigenstate of H^1S

$$H^1S|\psi_0^\alpha\rangle = \delta E^\alpha|\psi_0^\alpha\rangle \tag{56}$$

then we have

$$\{E^0 - H^0 - H^1\}S|\psi_0^\alpha\rangle = -\delta E^\alpha S|\psi_0^\alpha\rangle \tag{57}$$

This is just the complete Schrödinger equation with solution

$$|\Psi^\alpha\rangle = S|\psi_0^\alpha\rangle \tag{58}$$

and energy

$$E^\alpha = E^0 + \langle\psi_0^\alpha|H^1S|\psi_0^\alpha\rangle \tag{59}$$

5. THE EQUIVALENT HAMILTONIAN

In the previous section we expressed the problem of finding the energy and wave function of the atom in terms of the reduced problem of finding the eigenvalues and eigenstates of the equivalent operator

$$H^{\text{equ}} = H^1S = \sum_{n=0}^{\infty} H^1S_n \tag{60}$$

acting within the open shell. In this section we discuss some of the properties of E^{equ}.

A. The Perturbation Expansion

The equivalent operator H^{equ} is defined in terms of an infinite series in the perturbation expansion $H^1/\Delta E$. It is important to note that neither the atomic wave function $|\Psi^\alpha\rangle$ nor the energy E^α has a simple expansion of this form. The reason is that the powers of the perturbation H^1 are also involved in the diagonalization process which leads to the correct choice of $|\psi_0^\alpha\rangle$. In this respect the LD expansion is more akin to Brillouin–Wigner than to Rayleigh–Schrödinger perturbation theory.

An exception to this situation occurs in those cases where the same choice of $|\psi_0^\alpha\rangle$ diagonalizes each order of H^1S separately. This normally only happens for symmetry reasons, an example being a state which is the only one of a given L and S in the degenerate set. In this special case, Eq. (53) takes the simpler form

$$\{(E^0 - H^0)S_n - H^1S_{n-1}\}|\psi_0^\alpha\rangle = \sum_{m=0}^{n-1} S_m|\psi_0^\alpha\rangle\langle\psi_0^\alpha|H^1S_{n-m-1}|\psi_0^\alpha\rangle \quad (61)$$

It is clearly possible in this case to make an explicit expansion of both the energy and the wave function in powers of $H^1/\Delta E$.

We call this situation "quasi-degenerate". It differs from the problem considered in Section 3 in that the zeroth-order function cannot in general be expressed in terms of a single determinant. On the other hand, it is simpler than the general degenerate problem considered in Section 4 in that the diagonalization of H^{equ} within the open shell is taken care of automatically. Kelly[4] has argued that it is possible to derive an LD expansion for the quasi-degenerate situation without having to use the full degenerate treatment on Section 4. However, the diagrams that one obtains depend in detail on the particular state $|\psi_0^\alpha\rangle$ involved. From our point of view, Kelly's method has the disadvantage that the results cannot be expressed in equivalent operator form.

B. Non-Hermiticity of H^{equ}

One of the most important features of H^{equ} is that it is not in general Hermitian. One can readily demonstrate that the Hermitian conjugate of a diagram is obtained by reflecting it in a vertical plane but leaving the directions of the arrows unchanged. An example is given in Fig. 8a and Fig. 8b which are Hermitian conjugates of each other.

The majority of diagrams which make up H^{equ} are either unchanged by this reflection process or lead to other diagrams already included in H^{equ}. Thus, the major part of H^{equ} is Hermitian. The exceptions are some, though not all, of the backwards diagrams introduced in the previous section. Because of

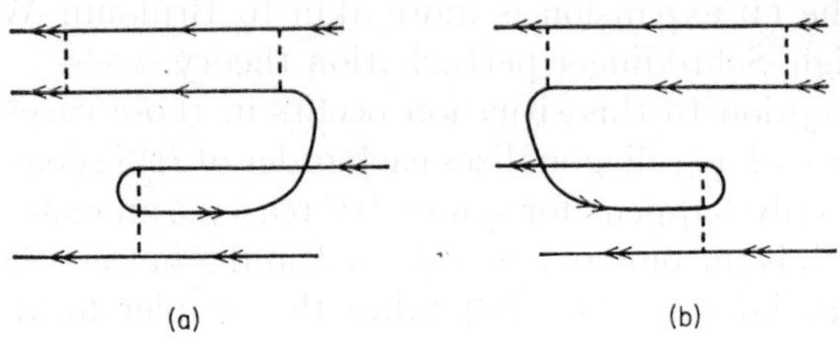

Fig. 8. Diagrams which are Hermitian conjugates of each other.

their method of construction, they do not have reflection symmetry, and their Hermitian conjugate diagrams are not allowed by the rules. An example of this is Fig. 8b which is the Hermitian conjugate of the allowed diagram Fig. 8a but is not itself an allowed diagram.

The presence of a non-Hermitian equivalent Hamiltonian should not come as a surprise since it is well known (e.g. Messiah,[13] p. 718) that this generally occurs when one carries out degenerate perturbation theory in terms of equivalent operators.

The most obvious result of the non-Hermiticity of H^{equ} is that the finding of the eigenvalues and eigenstates of H^{equ} in Eq. (56) is not the same as diagonalizing H^{equ}. Thus, the eigenstates $|\psi_0^\alpha\rangle$ and $|\psi_0^\beta\rangle$ belonging to different eigenvalues will not in general be orthogonal. The true wave functions $|\Psi^\alpha\rangle$ and $|\Psi^\beta\rangle$ must be orthogonal, but this does not require that $|\psi_0^\beta\rangle$ and $|\psi_0^\alpha\rangle$ are.

In actual practice, the non-Hermiticity of H^{equ} causes little trouble for the following reason. One can readily see from the construction of the backwards diagrams that the lowest order non-Hermitian diagram will occur in third order so that it is of order $H^1 \times (H^1/\Delta E)^2$. It is convenient to separate out the Hermitian and non-Hermitian parts of H^{equ}.

$$H_+^{\text{equ}} = \tfrac{1}{2}\{H^{\text{equ}} + (H^{\text{equ}})^\dagger\}$$
$$H_-^{\text{equ}} = \tfrac{1}{2}\{H^{\text{equ}} - (H^{\text{equ}})^\dagger\} \tag{62}$$

If our perturbation treatment is reasonably convergent then H_-^{equ} should be appreciably smaller than H_+^{equ}, and a perturbation treatment of H_-^{equ} should make sense.

It is simple to show from the reality of the matrix elements in the defining diagrams that both H_-^{equ} and H_+^{equ} have real matrix elements in the representation that diagonalizes H_+^{equ}. It follows that H_-^{equ} must be antisymmetric: $(H_-^{\text{equ}})_{\text{AB}} = -(H_-^{\text{equ}})_{\text{BA}}$. From this one can see that the first-order contribution from H_-^{equ} vanishes and that it first contributes to the energy in second order. This means that the lowest order contribution to the energy from the non-Hermiticity of H^{equ} is of order $H^1 \times (H^1/E)^4$. It seems hopeful that terms of this order will not be of major importance.

C. Connection with Rayleigh–Schrödinger Perturbation Theory

It may be useful at this point to consider briefly the relationship between the method introduced in the previous section and the familiar Rayleigh–Schrödinger perturbation theory for the degenerate problem.

We restrict ourselves to the situation in which the degeneracy is completely raised by the first-order perturbation H^1. The first- and second-order perturbations do not raise any new points of interest; the relationship between the two theories is essentially the same as in the non-degenerate case. In third order the comparison becomes more significant. The Rayleigh–Schrödinger energy is given by

$$\begin{aligned}\delta E^3 = &\sum_{iji} \frac{\langle\psi_0^\alpha|H^1\psi_i\rangle\langle\psi_i|H^1|\psi_j\rangle\langle\psi_j|H^1|\psi_0^\alpha\rangle}{(E^0 - E^i)(E^0 - E^j)}\\ &-\sum_i \frac{\langle\psi_0^\alpha|H^1|\psi_i\rangle\langle\psi_0^\alpha|H^1|\psi_0^\alpha\rangle\langle\psi_i|H^1|\psi_0^\alpha\rangle}{(E^0 - E^i)(E^0 - E^i)}\\ &+\sum_{ij}\sum_{\beta\neq\alpha} \frac{\langle\psi_0^\alpha|H^1|\psi_i\rangle\langle\psi_i|H^1\psi_0^\beta\rangle\langle\psi_0^\beta|H^1|\psi_j\rangle\langle\psi_j|H^1|\psi_0^\alpha\rangle}{(E^0 - E^i)\ \ (\delta E_1^\alpha - \delta E_1^\beta)\ \ (E^0 - E^i)}\end{aligned} \tag{63}$$

Here the sum over i and j spans the complete set of eigenstates of the zeroth-order Hamiltonian (2) with the exception of the degenerate set $|\psi_0^\beta\rangle$. $|\psi_0^\alpha\rangle$ and $|\psi_0^\beta\rangle$ are states from this set which diagonalize the first-order perturbation H^1. δE_1^α and δE_1^β are just the first-order expectation values of H^1.

The third term on the right-hand side of (63) can be readily recognized as the effect of the equivalent Hamiltonian H^1S taken to second order. The first term on the right-hand side looks very much like the expectation value of H^1S_2. It differs from it in three respects: (*a*) the absence of exclusion principle violating terms, (*b*) the presence of unlinked diagrams, and (*c*) the absence of the backwards diagrams. In the non-degenerate case, we added and subtracted exclusion principle violating diagrams and then found that the unlinked diagrams cancelled out with the renormalization term. In the present degenerate problem taking care of the exclusion principle violating terms is not sufficient to ensure the cancellation of the unlinked diagrams with the renormalization term. One has to add some rearrangement of diagrams with free open-shell lines which yields the desired cancellation but introduces some new diagrams which are just the "backwards" diagrams of the previous section.

Presumably this identification with ordinary perturbation theory can be carried out to higher orders, but the complexity of high order degenerate Rayleigh–Schrödinger perturbation theory rules this out as a useful method of derivation. Degenerate Brillouin–Wigner perturbation theory has a somewhat simpler form, and Brandow[8] has derived a linked diagram expansion from this starting point.

D. Expansion of H^{equ} in Terms of n-Body Operators

Thus far we have considered the equivalent Hamiltonian H^{equ} primarily in terms of an expansion in the perturbation parameter $(H^1/\Delta E)$. One of the advantages of the formulation in terms of diagrams and creation and annihilation operators is that we can immediately expand H^{equ} in terms of its operator form in the open shell. It follows immediately from the association of creation and annihilation operators with the diagrams that all diagrams with n open-shell lines at left and right contribute to an equivalent

n-particle operator within the open shell. As we shall see in the next section we can make some general statements about all one-particle operators, all two-particle operators and so on.

We shall now consider some of the more interesting contributions to the equivalent Hamiltonian. We confine ourselves mainly to the equivalent two-particle operator since the constant term and the one-particle operator do not lead to any relative splitting of the levels, and the three-particle operator is discussed by Judd.[5]

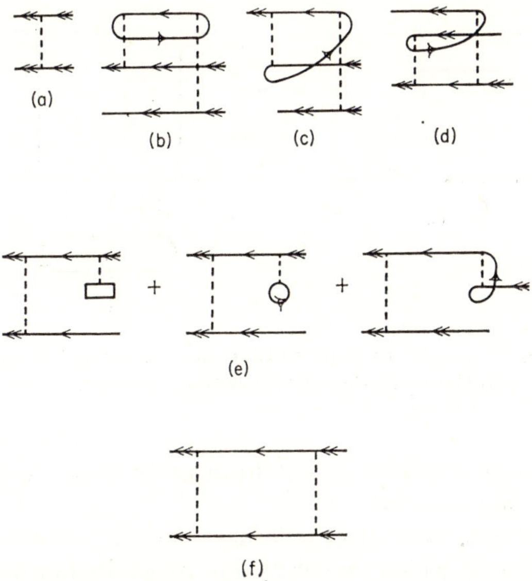

Fig. 9. Examples of diagrams which contribute in second order to the equivalent two-particle Hamiltonian.

In lowest order, the only contribution to the equivalent two-particle interaction comes from Fig. 9a. No energy denominators are involved since this is just H^1S_0 and $S_0 = 1$. The lowest order contribution is thus $\sum_{i \neq j} \frac{1}{2} r_{ij}$ as one would expect.

In second order, there are many contributions, and it is convenient to divide them into three groups:

(*a*) *Core polarization.* Examples of this type are given in Figs. 9b, c, and d. As the name suggests, these diagrams represent the polarization of the core by the open-shell electrons. This polarization then interacts back on the open shell to give an effective two-particle interaction. We can distinguish three different forms of core polarization depending on the number of exchange matrix elements. These different types which are illustrated in Figs. 9b, c, and d, respectively, have important differences in their properties.

(*b*) *Open-shell wave function modification.* This important class of contributions, illustrated in Fig. 9e arises from the use of a central potential $U(r)$ which was not the best from this point of

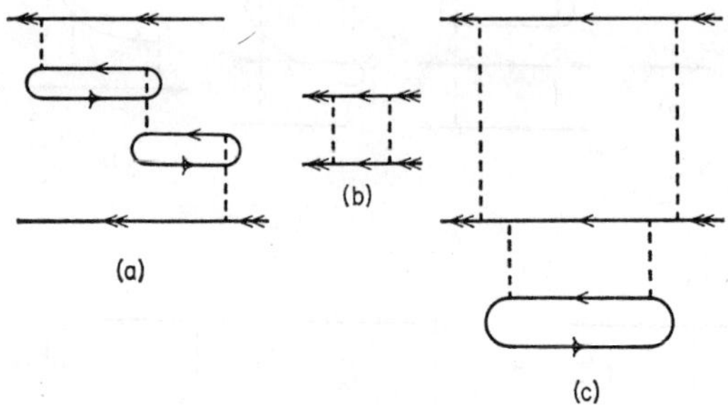

Fig. 10. Examples of higher order diagrams which contribute to the equivalent two-particle operator.

view. For a good choice of potential these terms tend to cancel and can be unimportant.

(*c*) *Correlation.* We collect together under this heading diagrams such as Fig. 9f in which two electrons are excited or de-excited at the same time. This double jump makes the calculation of these correlation effects much harder than the processes such as core polarization which involve only a single excitation.

When one comes to consider diagrams of higher order, there are so many that space does not permit one to give even a set of representative illustrations. However, before we move on, we want to point out a very useful method of classifying diagrams. This is in terms of the maximum number of excited state energies that occur in any of the energy denominators corresponding to the diagram. Thus, Fig. 10a is a single excitation diagram because there is never more than a single excited state energy in the energy denominators. Likewise, Figs. 10b and c are double and triple excitation diagrams. This method of classification is useful because it is closely connected with the complexity of calculation of the contribution from the diagram. The main problem in calculating the contributions from the diagrams comes from the infinite sum over excited states. This difficulty obviously increases very rapidly with the number of excited states that have to be summed simultaneously.

It is interesting to note that single and double excitations occur in all orders from second up, that triple excitations first occur in fourth order, and in general an n-tuple excitation first occurs in order $2n - 2$. The proof of this requires a certain amount of diagrammatic manoeuvering and we shall leave it to the reader. In particular, the argument relies on the fact that diagrams which are separable either to the right or to the left can have their energy denominators partially factorized by summing over all possible orderings. This factorization clearly reduces the number of excited states present in any given energy denominator. This use of the factorization theorem is an important advantage of the LD expansion.

E. One- and Two-Particle Perturbations

A problem that often arises in atomic problems is the calculation of the effect of an additional perturbation on an open-shell atom; examples are the fine and hyperfine structures. Provided this additional perturbation H^{p} can be written in terms of one- and two-particle operators, it can be immediately incorporated into our derivation of the LD expansion.

According to the rules derived in the previous section the energy is given by the eigenvalue of H^{equ} which operates in the open shell. H^{equ} will now include both H^1, the configuration interaction perturbation, and H^{p}. We confine ourselves to the situation where H^{p} is much smaller than H^1 so that it makes sense to expand H^{equ} in powers of H^{p}:

$$H^{\mathrm{equ}} = H_0^{\mathrm{equ}} + H_p^{\mathrm{equ}} + H_{pp}^{\mathrm{equ}} + \cdots \tag{64}$$

In most cases of interest terms higher than H^{equ} are negligible and can be neglected.

We are not now interested in the total eigenvalue of H^{equ} but only that part which is due to H_p^{equ}. In normal perturbation theory this would be given by the expectation value of H_p^{equ} taken over the eigenstate of H_0^{equ}. Unfortunately the derivation of this result assumes that the zeroth-order Hamiltonian H_0^{equ} is Hermitian which as we know is incorrect.

To get around this problem, we assume as in the earlier part of this section that the non-Hermitian part of H_0^{equ} is small compared to the Hermitian part. We then treat this together with H_p^{equ} as our perturbation. If $|\psi_0^\alpha\rangle$, $|\psi_0^\beta\rangle$, etc. are the orthonormal eigenstates of H_{0+}^{equ}, the Hermitian part of H_0^{equ}, then the perturbation to second order is given by

$$\delta E^p = \langle\psi_0^\alpha|H_p^{\text{equ}}|\psi_0^\alpha\rangle + \sum_\beta[\langle\psi_0^\alpha|H_p^{\text{equ}}|\psi_0^\beta\rangle\langle\psi_0^\beta|H_{0-}^{\text{equ}}|\psi_0^\alpha\rangle + \langle\psi_0^\alpha|H_{0-}^{\text{equ}}|\psi_0^\beta\rangle\langle\psi_0^\beta|H_p^{\text{equ}}|\psi_0^\alpha\rangle]\,\frac{1}{\delta E^\alpha - \delta E^\beta} \tag{65}$$

This expression can be considerably simplified if we separate out the Hermitian and anti-Hermitian parts of H_p^{equ}

$$H_p^{\text{equ}} = H_{p+}^{\text{equ}} + H_{p-}^{\text{equ}} \tag{66}$$

Then, as earlier, we can deduce from the reality of the defining matrix elements that H_{p+}^{equ} is symmetric and H_{p-}^{equ} is antisymmetric. It follows that (65) reduces to

$$\delta E^p = \langle\psi_0^\alpha|H_{p+}^{\text{equ}}|\psi_0^\alpha\rangle + 2\sum_\beta \frac{\langle\psi_0^\alpha|H_{p-}^{\text{equ}}|\psi_0^\beta\rangle\langle\psi_0^\beta|H_{0-}^{\text{equ}}|\psi_0^\alpha\rangle}{(\delta E^\alpha - \delta E^\beta)} \tag{67}$$

Note that the cross terms between the Hermitian part of H_p^{equ} and the anti-Hermitian part of H_0^{equ} have disappeared. The second term on the right-hand side of (67) should be very much smaller than the first term since it is of relative order $H^p \times (H^1/\Delta E)^4$. For most purposes it can be completely neglected, and we can use the simple form

$$\delta E^p = \langle\psi_0^\alpha|H_{p+}^{\text{equ}}|\psi_0^\alpha\rangle \tag{68}$$

where as we mentioned before $|\psi_0^\alpha\rangle$ is an eigenstate of the Hermitian part of H_0^{equ}.

The most important examples of H^p are single-particle operators The equivalent Hamiltonian H_p^{equ} of course can be expanded as a constant, a single-particle operator, a two-particle operator, and so on. The big difference between this expansion and that of H_0^{equ}

is that the single-particle part of the equivalent Hamiltonian is normally of primary importance.

In Fig. 11 we illustrate some of the more interesting diagrams which contribute to H_p^{equ}. Figs. 11a and b represent core polarization effects by direct and exchange interactions, respectively. Figs. 11c, d, and e represent various open-shell wave function modifications. Fig. 11f is an example of a higher order effect

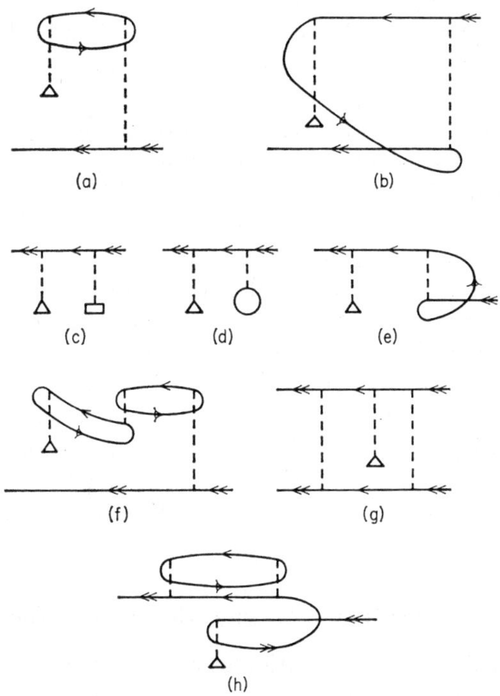

Fig. 11. Some interesting single-particle diagrams.

involving only single excitation. Fig. 11g is one of the lowest order correlation effects. Fig. 11h is an example of a backwards diagram contribution.

6. SUMMATION OVER MAGNETIC QUANTUM NUMBERS

In the previous section we showed that the contribution to each type of equivalent operator was given by the sum of the contributions from all allowed diagrams having the correct number of open valence shell lines.

This sum over allowed diagrams first of all spans the set of geometrically different diagrams and then for each of these we must sum over all possible single-particle states involved. Obviously, unexcited state lines are only summed over unexcited states and so on, but with these restrictions the sum must be carried over a complete set of single-particle states. In particular we can distinguish: (*a*) the sum over principal and orbital angular momentum quantum numbers, (*b*) over orbital magnetic quantum numbers, and (*c*) over spin quantum numbers.

The sum (*a*) over principal and orbital angular momentum quantum numbers is of course the most important part of the summation since it is this part that determines the magnitude of the equivalent operator. However, useful information can often be obtained by carrying out the much easier sum over magnetic quantum numbers. In particular one can often make useful statements about the tensor form of H^{equ} at this stage.

As we shall see in this section, this summation over magnetic quantum numbers can be carried out in a very powerful and elegant manner using the very close resemblance between the interaction diagrams used in the previous sections and the diagrams introduced by Yutsis, Levinson, and Vanagas (YLV) to describe quantities which appear in the quantum theory of angular momentum.

A. Application of Wigner–Eckart Theorem

In order to obtain the separation between the magnetic quantum numbers and the rest, we must factorize the contribution from each interaction diagram into a part which depends on the magnetic quantum numbers and a part which is independent of them.

This is completely trivial for the matrix element of the central potential which can be written in the form

$$\langle a|u(r)|b\rangle = \delta_{m_a m_b}\delta_{m_{s_a} m_{s_b}}\delta_{l_a l_b}\int R_{n_a l_a} R_{n_b l_b} U(r)\, dr \tag{69}$$

and the magnetic dependence just consists of the Dirac delta functions.

The separation is more complicated for the $1/2r_{12}$ interaction since this is not in a convenient spherical tensor form. We have first to write it in the expansion over spherical harmonics

$$\frac{1}{2r_{12}} = \frac{1}{2}\sum_K (-1)^K \frac{r_<^K}{r_>^{K+1}} \sum_Q (-1)^{K-Q} C_Q^K(1) C_{-Q}^K(2) \qquad (70)$$

It is rather convenient for the analysis that follows to adopt a notation in which the tensor argument K is treated in a manner analogous to the angular momentum l. In other words, when we want to represent a particular value of K we put this explicitly on the diagram, and the absence of any explicit value implies a sum over all allowed values.

We can now apply the Wigner–Eckart theorem to the matrix element of each term of (70) and write it in the form

$$\frac{(-1)^K}{2}\left\langle pq \left| \frac{r_<^K}{r_>^{K+1}} \sum_Q (-1)^{K-Q} C_Q^K(1) C_{-Q}^K(2) \right| rs \right\rangle$$

$$= \sum_Q (-1)^{l\ -m_p}(-1)^{K-Q}\begin{pmatrix} l_p & l_r & K \\ m_p & -m_r & Q \end{pmatrix}(-1)^{l_q-m_q}\begin{pmatrix} l_q & l_s & K \\ m_q & -m_s & Q \end{pmatrix}$$

$$\times \tfrac{1}{2}(-1)^K \langle l_p || C^K || l_r \rangle \langle l_q || C^K || l_s \rangle$$

$$\times \int R_{n_pl_p}^{(1)} R_{n_ql_q}^{(2)} \frac{r_<^K}{r_>^{K+1}} R_{n_rl_r}^{(1)} R_{n_sl_s}^{(2)} \mathrm{d}r_1 \mathrm{d}r_2 \qquad (71)$$

The Wigner–Eckart theorem can obviously be applied to all the interaction matrix elements in a particular diagram. The result can then be written in the factorized form

$$D_\gamma = S_\gamma T_\gamma R_\gamma \qquad (72)$$

where

$$S_\gamma = \sum_{\text{all } a} \text{the product of } (-1)^{K-Q}(-1)^{l_p-m_p}\begin{pmatrix} l_p & l_r & K \\ m_p & -m_n & -Q \end{pmatrix} \qquad (73)$$

$$\times (-1)^{l_q-m_q}\begin{pmatrix} lq & l_r & K \\ m_q & -m_r & Q \end{pmatrix}$$

for each $1/2r_{12}$ interaction and $\delta_{m_am_b}$ for each central potential matrix element.

$T_\gamma =$ The product of $\frac{1}{2}(-1)^K\langle l_p||C^K||l_r\rangle\langle l_q||C^K||l_s\rangle$ for each $\dfrac{1}{2r_{12}}$ interaction and $\delta_{l_a l_b}$ for each central potential matrix element (74)

and

$R_\gamma = (-1)^{n_c} \times$ The product of the appropriate radial integrals and energy denominators (75)

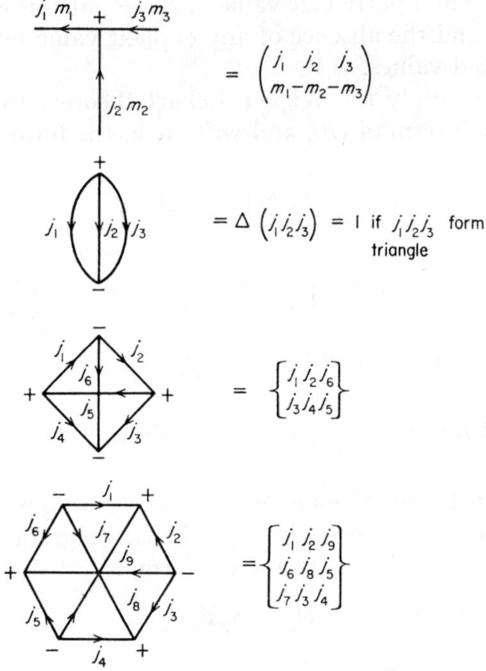

Fig. 12. Diagrammatic representation of $3j$, $6j$, and $9j$ symbols.

B. Angular Momentum Diagrams

In their book *The Mathematical Apparatus of the Theory of Angular Momentum* Yutsis, Levinson, and Vanagas have introduced an extremely powerful and useful graphical method of treating angular momentum problems. In this scheme, the Wignor $3j$ symbol $\begin{pmatrix} j_1 & j_2 & j_3 \\ m_1 & m_2 & m_3 \end{pmatrix}$ is represented by an oriented node which links three directed lines (Fig. 12). These lines correspond to j_1, j_2, and j_3 and their free ends to the projections m_1, m_2, and m_3. The direction of the arrow indicates the sign of the magnetic quantum number in the $3j$ symbol. A line directed away from the node corresponds to the positive sign of the magnetic quantum number and a line directed towards the node to the negative sign. A node in which the clockwise orientation of the lines is the same as the order from left to right of the arguments in the $3j$ symbol is indicated by a minus sign, one in which the order corresponds to an anti-clockwise orientation by a minus sign.

Summation over magnetic quantum numbers is graphically represented by joining the free ends of the corresponding lines with the requirement that the arrows be continuous and that the $3j$ symbols be multiplied by $(-1)^{j-m}$ where m is the magnetic quantum number to be summed over. Thus, we have the diagrammatic equivalence

$$\sum_{m_1}(-1)^{j_1-m_1}\begin{pmatrix} j_1 & j_2 & j_3 \\ m_1 m_2 m_3 \end{pmatrix}\begin{pmatrix} j_1 & j_4 & j_5 \\ m_1 m_4 m_5 \end{pmatrix} = $$

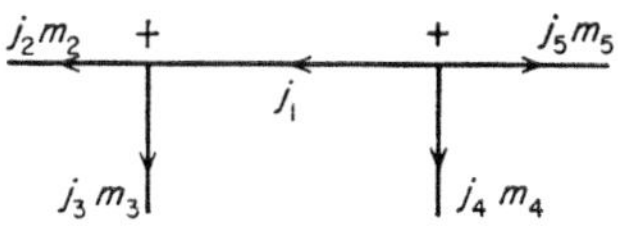

(76)

Since the usual angular momentum quantities such as the triangular delta function, the $6j$ symbol, and the $9j$ symbol can be expressed as sums over $3j$ symbols, they can be represented by diagrams. These are illustrated in Fig. 12.

The importance of the angular momentum diagrams lies in the ease with which angular momentum manipulations can be carried out. In particular it is possible to prove a series of very basic theorems which are illustrated in Fig. 13. For proofs and a discussion of the detailed use of the theorems, the reader is referred to YLV. As we shall see, the theorems of Fig. 13 are very useful in putting an angular momentum diagram with free lines into a standard form and then in reducing it to its most simple form.

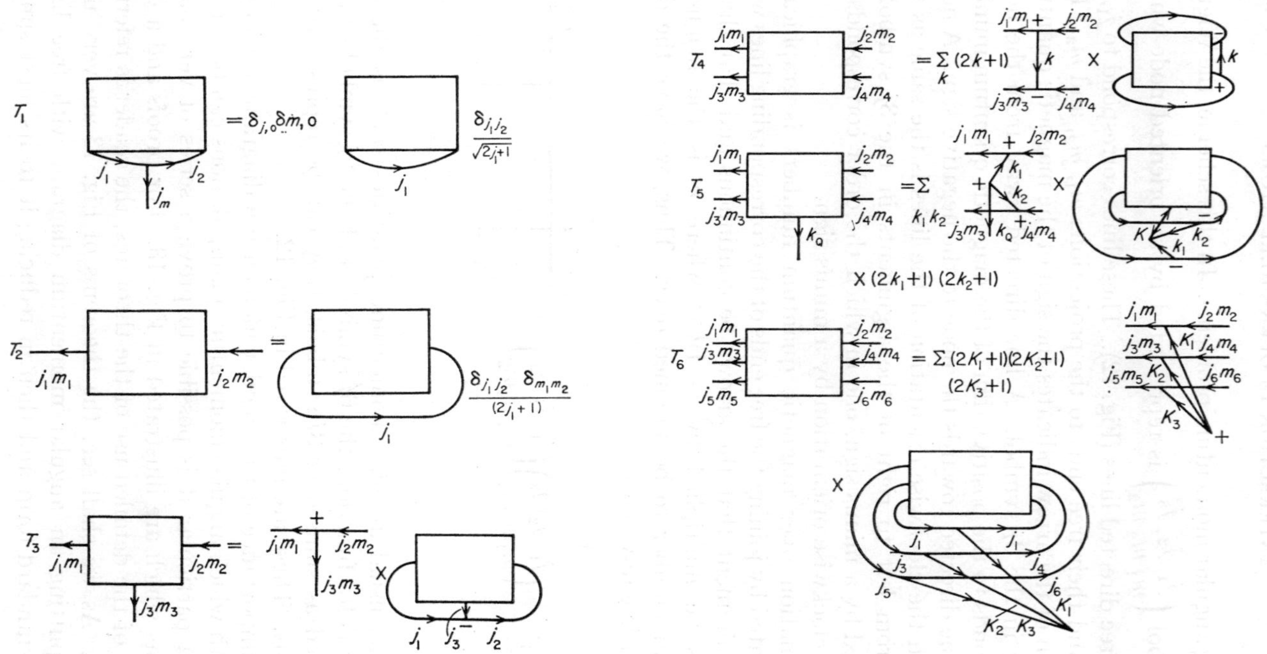

Figs. 13a and 13b. Some theorems for the simplification of angular momentum diagrams.

C. Interaction and Angular Momentum Diagrams

The usefulness of the angular momentum diagrams follows from a fascinating similarity between the interaction diagrams and the associated angular momentum diagrams. Thus a typical interaction vertex

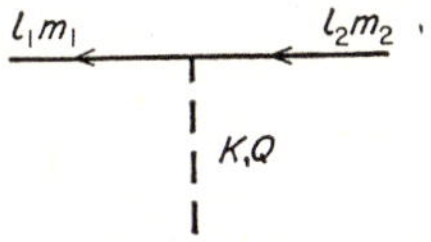

has associated with it according to our rules the $3j$ symbol

$$\begin{pmatrix} l_1 & l_2 & K \\ +m_1 & -m_2 & -Q \end{pmatrix}$$

which is represented by the angular momentum diagram

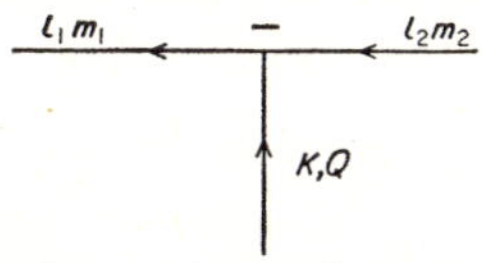

Furthermore, the equivalence extends to the situation when we join up the lines to make the complete diagram. A feature of our diagrammatic notation is that the intermediate state lines are summed over all magnetic quantum numbers. Thus it is clear that the interaction diagram and the angular momentum diagram which represents the magnetic part of the matrix elements are closely related. It turns out to be convenient to define a new quantity

$$Y_\gamma = (-1)^{\sum_i (l_i - m_i)} S_\gamma \tag{77}$$

where the sum is over all the open valence shell lines at the left of the diagram. Then one finds that the diagram which represents Y_γ can be obtained from the interaction diagram D_γ by the following rules:

[1]. Make each interaction line into a full line.

[2]. Remove any central potential lines.

[3]. Remove the arrows on all the lines except the open valence shell lines.

The angular momentum diagrams which result will appear slightly non-standard because they are missing any plus or minus signs on the interaction vertices and arrows on internal lines.

These simplifications are possible for two reasons: first, $\langle l_1 || C^k || l_2 \rangle$ vanishes for parity reasons unless $l_1 + l_2 + k$ is even. But changing a plus into a minus sign on a vertex is equivalent to permuting the arguments in the $3j$. This is well known to have the result of multiplying the $3j$ symbol by $(-1)^{l_1+l_2+k}$ which is unity by the parity condition. Second, the effect of changing the direction of the arrow on an internal line is to multiply the result by $(-1)^{2j}$. But in our problem all the angular momenta are orbital and hence integers, so that this factor is always unity, and the direction of the arrow on an internal line is irrelevant.

D. Tensor Form of H^{equ}

We now illustrate the usefulness of the angular momentum diagrams by examining the spherical tensor form of H^{equ}.

We consider first the two-particle part $\sum_{i \neq j} g^{\text{equ}}(i, j)$. Each diagram, γ, in the expansion of $g^{\text{equ}}(i, j)$ will be of the form

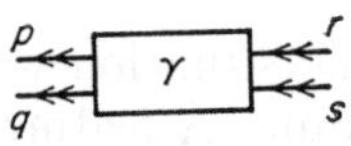

The associated angular momentum diagram will look almost the same

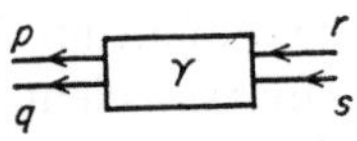

But by theorem T4, (Fig. 13), any such diagram can be rewritten in the form

Y_γ 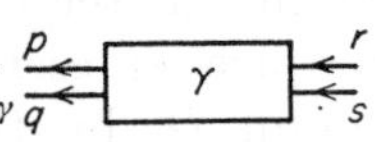

$\sum_k (2k + 1)$ 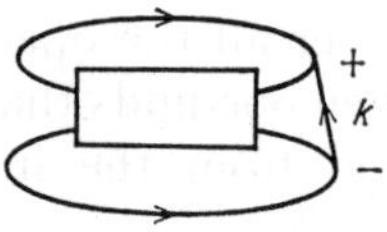

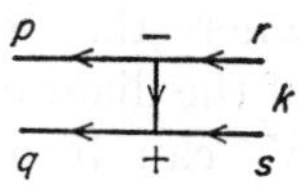

(78)

The part with open lines on the right of (78) when multiplied by the phase factor (77) is just the matrix element of the unit double tensor $\mathbf{U}^k(1) \,.\, \mathbf{U}^k(2)$ with $\langle l || U^k || l \rangle = 1$. Equation (78) therefore, can be rewritten

$$S_\gamma = \sum_k \overline{Y}_\gamma(k) \langle pq | \mathbf{U}^k(1) \,.\, \mathbf{U}^k(2) | rs \rangle \tag{79}$$

On substituting this into Eq. (72) we can write the equivalent two-particle operator in the form of a tensor operator

$$\sum_{i\neq j} g^{\text{equ}}(i, j) = \sum_k d^k \sum_{i\neq j} \mathbf{U}^k(i) \cdot \mathbf{U}^k(j) \tag{80}$$

where the parameters d^k are given by

$$d^k = \sum \overline{Y}_\gamma(k) T_\gamma R_\gamma \tag{81}$$

The sum in (81) is taken over all allowed diagrams with two open lines at right and left, bearing in mind that the sum over orbital magnetic quantum numbers has already been carried out.

An important point to note here is that the sum over k spans both odd and even values. Whereas the original Hamiltonian only had even k matrix elements in the open shell for parity reasons the equivalent Hamiltonian H^{equ} has contributions from both odd and even k. This is of course in agreement with the observations mentioned in the introduction.

We now show that the anti-Hermitian part of H^{equ} cannot contribute to the two-particle operator $g^{\text{equ}}(1, 2)$. To see this we note that the contribution from the Hermitian conjugate of a diagram is the same as that from the original diagram except that: (*a*) all the reduced matrix elements have their angular momenta reversed, and (*b*) the diagram representing the new $\overline{Y}$ is the mirror image of the diagram representing the original. It is straightforward to show that the reversal of the arguments in the reduced matrix elements leaves their product unchanged. One can also show that $\overline{Y}$ is identically equal to the mirror image diagram because taking a mirror image of a diagram is equivalent to changing plus signs to minus signs on the nodes. This has no effect as we pointed out earlier. Since the diagram and its Hermitian conjugate contribute equally to the two-particle operator the anti-Hermitian part $H^{\text{equ}} = \frac{1}{2}\{H^{\text{equ}} - (H^{\text{equ}})^+\}$ cannot contribute at all.

Applying the same general arguments to the three-particle part of H^{equ} we see using theorem T6 that it can be written in the form

$$g^{\text{equ}}_{(1,2,3)} = \sum_{k_1,k_2,k_3} \{\mathbf{u}^{k_1}_{(1)}\ \mathbf{u}^{k_2}_{(2)}\ \mathbf{u}^{k_3}_{(3)}\} d^{k_1k_2k_3} \tag{82}$$

The parameters $d^{k_1k_2k_3}$ are given by

$$d^{k_1k_2k_3} = \sum_{\gamma} \overline{Y}_{\gamma} T_{\gamma} R_{\gamma} \tag{83}$$

where the reduced diagram $\overline{Y}_{\gamma}$ is obtained from Y_{γ} in the manner illustrated in theorem T6. By means of the same arguments that we used previously, we can readily demonstrate that the Hermitian part of H^{equ} only contributes to the operators with $k_1 + k_2 + k_3$ even and the anti-Hermitian part to the operators with $k_1 + k + k_3$ odd. Thus, to a rather good approximation, we can restrict ourselves to the even case. A more detailed discussion of the use of these three-particle equivalent operators is given by Judd in this volume.[5]

E. Single-Particle Perturbations

In this section we extend the angular momentum analysis of H^{equ} to the perturbation operator $H_{\text{P}}^{\text{equ}}$. We shall restrict ourselves to the most interesting case where the perturbation is a single-particle operator which can be written as a sum of uncoupled spherical tensors of the form

$$H^{p} = d^{\kappa}(r)^{K} s_0^{\kappa} t_0^{K} \tag{84}$$

This covers all the usual perturbations such as hyperfine structure, spin–orbit coupling, and interaction with external electric fields. On applying the Wigner–Eckart theorem to the orbital single-particle matrix element of (84) we find that the angular momentum diagram appropriate to

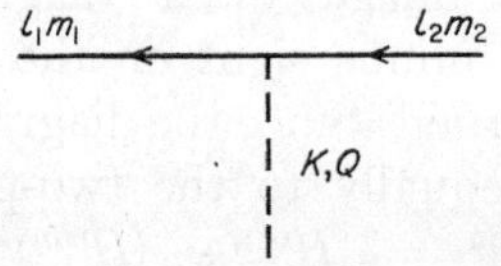

is

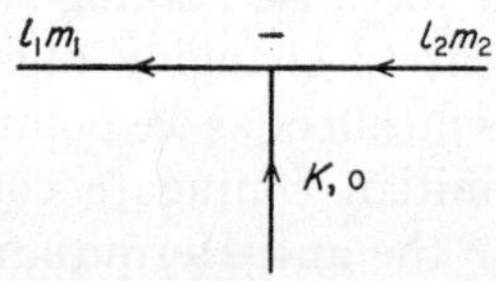

Thus the rules for finding the appropriate angular momentum diagrams from the interaction diagrams must be modified by the addition of the rule above. A very important point to note here is that the + and − signs are in general necessary for the perturbation vertex.

Having found the equivalent angular momentum diagram we can immediately apply the YLV theorems. Thus from T1 we see that any diagram with no open valence shell lines must be zero unless $K = 0$.

From T3 we see that all single-particle diagrams contribute to the single-particle equivalent operator

$$\sum_i d^{\kappa K} \mathbf{u}^K(i) \tag{85}$$

where the radial parameter is given by

$$d^{\kappa K} = \sum \overline{Y}_\gamma T_\gamma R_\gamma \tag{86}$$

To find an expression for the single-particle interaction (85) we have also to consider the spin which we do in the next section. Similar arguments can be applied to the two and higher particle parts of $H_{\mathrm{p}}^{\mathrm{equ}}$.

F. Spin Summation

The two-particle electrostatic interaction and the central potential are diagonal in the spin quantum numbers. It follows that the spin quantum numbers of the two single-particle states that meet at an interaction vertex must be equal. We exploit this by breaking all the $1/2r_{12}$ interaction lines, thus splitting the diagram into a number of parts each of which consists of a "string" of connected states. The spin quantum numbers of all the states in a string must be equal but there is no relation between the quantum numbers in different strings. From this one finds that the spin sum reduces to the simple rule for H^{equ}:

The spin sum multiplies the contribution from a given diagram by a factor two for each string which does not contain any free valence shell line.

When one has a single-particle perturbation present the situation becomes more complicated. For $\kappa = 0$, for example, a spin independent single-particle operator rule above still holds. But when $\kappa = 1$ one has to add the rule:

The spin summation gives zero unless the perturbation vertex lies on the same string as one of the open valence shell lines. If it lies on the string containing the ith open valence shell line then the form of the equivalent operator will be

$$\Sigma S_{i} g(1, 2, 3 \ldots i \ldots q)$$

where $g(1, 2, 3, i, q)$ is the orbital part.

7. EXAMPLES

In this section we consider some examples of the application of the angular momentum methods to the treatment of linked

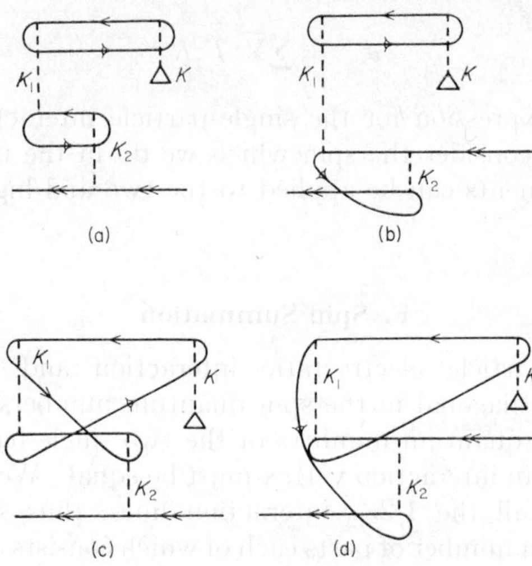

Fig. 14. Examples of diagrams.

diagrams. The diagrams we shall discuss are some typical third-order contributions to the single-particle effective hfs operator. They are illustrated in Fig. 14. All four diagrams have the same set of interaction lines but they differ in the relative number of direct and exchange matrix elements. As we shall see these differences enormously effect the contributions of the diagrams.

A. Spin Sum

Applying the rules of Section 6 to the diagrams we see immediately that in diagrams 14a, b, and c the hfs operator lies on a different string from the open-shell lines. Thus the contribution to these diagrams from the spin dependent hfs operator must be zero. Diagram d on the other hand need not be zero.

B. Diagram a

Using the rules of Section 6 we can immediately write $\overline{Y}$ the angular momentum diagram corresponding to (a), it is Fig. 15a.

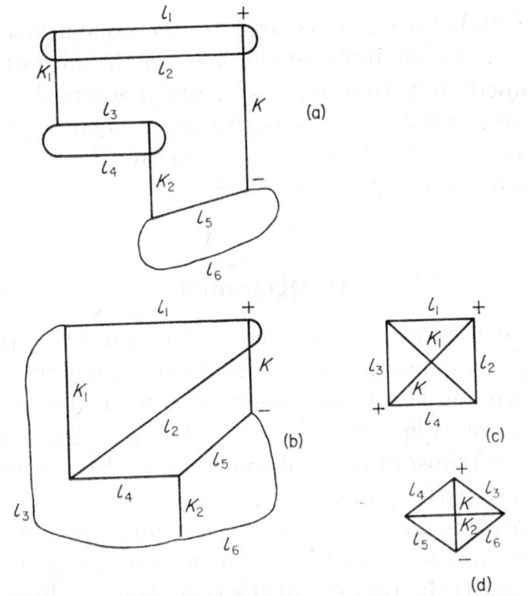

Fig. 15. Reduction of angular momentum diagrams corresponding to Fig. 14.

On comparing it with our theorems we see that it contains pairs of closed loops which have the form of theorem T2. We can therefore immediately deduce that we must have $K_1 = K$ and $K_2 = K$. The reduced diagram becomes

$$\overline{Y} = \delta_{K_1K}\delta_{K_2K}\Delta(l_1l_2K)\Delta(l_3l_4K)\Delta(l_5l_6K) \times \frac{1}{(2K+1)}$$

where we have used Fig. 13b. But these restrictions mean that the contribution from this diagram must vanish. For from the usual parity rules both K_1 and K_2 must be even. On the other hand, we have above that K_1 and K_2 must equal K. The only spin-independent contribution to the hfs interaction has $K = 1$, and the contribution from the diagram therefore vanishes.

C. Diagrams b and c

Exactly similar arguments lead to the conclusion that these diagrams do not contribute to the hfs. In the case of diagram b we find immediately that $K_1 = K$ from theorem T2 but $K = 1$ which is in contradiction to the parity requirement that K be even. Likewise in c we have $K_2 = K$ by theorem T2 so that K_2 is odd but it must be even for parity reasons.

D. Diagram d

Diagram d is more complicated and requires the full angular momentum treatment. The associated angular momentum diagram can readily be written down and then factorized to get the reduced diagram (Fig. 15b). Note that we have omitted all arrows on the lines and most of the signs but that we do need to retain the signs on the hyperfine interaction vertex.

This diagram is not yet in its simplest form since it is divisible on the three lines K, l_4, and l_3 so that we can apply to it theorem T3. The result is the product of the two diagrams Figs. 15c and d which after some slight rearrangement are seen to be $6j$ symbol diagrams. $\overline{Y}$ then reduces to

$$\overline{Y} = (-1)^{l_5+l_6+K} \begin{Bmatrix} l_1 & l_3 & K_1 \\ l_4 & l_2 & K \end{Bmatrix} \begin{Bmatrix} l_4 & l_3 & K \\ l_6 & l_5 & K_2 \end{Bmatrix} \tag{87}$$

There are no parity restrictions on K_1 and K_2 in this case except the usual ones provided by the wave functions. The only restrictions on K_1, K_2, and K are contained in the triangular relation implicit in the $6j$ symbols. Thus, in general, we should expect contributions to all the hfs operators from all values of K_1 and K_2.

8. CONCLUSION

In this paper we have investigated the usefulness of the linked cluster expansion in treating configuration interaction in open-shell atoms. While it is clear that the linked cluster method does not allow one to do anything that cannot be done by other methods, we do feel that it has certain advantages:

[1]. The many particle perturbation theory is given in a well-defined form in which the appropriate matrix elements, energy denominators, and sign can be determined in a straightforward manner.

[2]. The results appear directly in the form of effective operators, which is particularly advantageous because of the tendency to analyse experimental data in terms of standard operators multiplied by parameters which are to be determined.

[3]. The exclusion principle is automatically taken care of. This is vitally important when one comes to sum over magnetic quantum numbers.

[4]. The diagrams give a particularly simple way of characterizing and distinguishing the various types of configuration interaction and most important allow one to readily distinguish those contributions which can readily be calculated from those which are appreciably more difficult.

[5]. The topological equivalence between Feynman and Yutsis diagrams allows one to find the spherical tensor form of the equivalent interaction Hamiltonian in a simple and powerful manner.

In conclusion, we wish to reiterate that the real problem in atomic calculations is to get reliable numerical results on atoms beyond the first row. While linked diagrams and angular momentum methods may help a little, the real problem remains.

APPENDIX A

Proof of Theorem on All Possible Orderings of Disconnected Parts

Our task in this Appendix is to show that if $D^{AB}_{N_A+N_B,i}$ is an allowed diagram which contains two disconnected parts A and B neither of which is closed, then all those diagrams which differ from it in the relative ordering of the interactions in the two parts are also allowed. In other words the set of allowed diagrams is made up of subsets of the form

$$G^{AB}_{N_A+N_B} = \sum_{\substack{\text{all possible}\\ \text{orderings}}} D^{AB}_{N_A+N_B,i} \tag{A1}$$

The different diagrams $D^{AB}_{N_A+N_B,i}$ can conveniently be distinguished by the relative ordering of the interaction lines in the two parts A and B. We introduce a notation of the form

$$D^{AB}\begin{pmatrix} N_A \cdot \cdot \cdot 2 \quad 1 & \\ & N_B \;.\; 321 \end{pmatrix}$$

where the upper row refers to the interactions in part A and the lower row to those in B. The ordering of the interactions on the diagram is the same as that of the numbers in the brackets. We can also make use of this notation to refer to a diagram which consists of the first n_A interactions of A and the first n_B interactions of B, in some specified order, e.g. $D^{AB}\begin{pmatrix} n_A \cdot \;.\; .2 \quad 1 & \\ & n_B \cdot \;.\; 321 \end{pmatrix}$

It is convenient to extend the notation to cover the sum of all diagrams which differ only in the relative orderings of the first n'_A interactions of A and the first n'_B interactions of B. We call this sum $D^{AB}\begin{pmatrix} n_A \cdot \cdot \cdot n'_A \\ \hdashline n_B \cdot \;.\; n'_B \end{pmatrix}$ where the dashed line implies all possible orderings.

In this notation our task is to prove that

$$G^{AB}_{N_A+N_B} = D^{AB}\begin{pmatrix} N_A \\ \hdashline N_B \end{pmatrix} \tag{A2}$$

We proceed in induction. Let us define G^{AB}_n to be the set of all allowed diagrams of order $n = n_A + n_B$ which consist of the first n_A interactions of A and the first n_B interactions of B. We then assume that this set of diagrams has the form

$$\begin{aligned} G^{AB}_n = D^{AB}\begin{pmatrix} n \\ \hdashline 0 \end{pmatrix} + D^{AB}\begin{pmatrix} n-1 \\ \hdashline 1 \end{pmatrix} \ldots D^{AB}\begin{pmatrix} m+1 \\ \hdashline n-m-1 \end{pmatrix} \\ + \frac{1}{2}\left\{ D^{AB}\begin{pmatrix} m \\ \hdashline n-m \end{pmatrix} + \ldots D^{AB}\begin{pmatrix} n-m \\ \hdashline m \end{pmatrix} \right\} \\ + D^{AB}\begin{pmatrix} n-m-1 \\ \hdashline m+1 \end{pmatrix} + \ldots D^{AB}\begin{pmatrix} 0 \\ \hdashline n \end{pmatrix} \end{aligned} \tag{A3}$$

which is a sum over all possible values of n_A and n_B and all possible orderings of the two parts. The factor $\frac{1}{2}$ allows for the fact that the first m interactions of A and B will in general be identical. In that case $D^{AB}\begin{pmatrix} q \\ \hdashline p \end{pmatrix} = D^{AB}\begin{pmatrix} p \\ \hdashline q \end{pmatrix}$, and we must not count the same diagram twice. If the first diagram in A is different from the first diagram in B then $m = 0$, and we do not need the terms in brackets. In (A3) we assumed that $n < N_A$ and $n < N_B$. If this is not the case the series at the beginning and end must be cut off at $D^{AB}\begin{pmatrix} N_A \\ \hdashline n - N_A \end{pmatrix}$ and $D^{AB}\begin{pmatrix} n - N_B \\ \hdashline N_B \end{pmatrix}$, respectively.

Turning now to the diagrams with $n + 1$ interactions we see that the $n + 1$th interaction can either be in the part A or B. This gives rise to pairs of sets of diagrams of the form

$$D^{AB}\left(p + 1 \frac{p}{n - p}\right) + D^{AB}\left(n - \frac{p + 1}{p\, n - p - 1}\right) = D^{AB}\begin{pmatrix} p + 1 \\ \hdashline n - p \end{pmatrix} \tag{A4}$$

With results of this sort one can readily demonstrate that if G_n^{AB} is of the form specified in (A3) then so is G_{n+1}^{AB}.

But we know that

$$G_1^{AB} = D^{AB}\begin{pmatrix} 1 \\ \hdashline 0 \end{pmatrix} + D^{AB}\begin{pmatrix} 0 \\ \hdashline 1 \end{pmatrix} \tag{A5}$$

provided the first interaction in A is different from the first one in B or

$$G_1^{AB} = \frac{1}{2}\left\{D^{AB}\begin{pmatrix} 1 \\ \hdashline 0 \end{pmatrix} + D^{AB}\begin{pmatrix} 0 \\ \hdashline 1 \end{pmatrix}\right\} \tag{A6}$$

if the interactions are the same. (A5) and (A6) are just special cases of (A3) so that by induction we know that (A3) is correct for all n. The special case (A2) when $n_A = N_A$ and $n_B = N_B$ is therefore proven.

APPENDIX B

Proof of Rearrangement Theorem for Energy Denominators

We start by defining two parts of a diagram to be disconnected to order n and m if the first interaction line or state line joining them occurs after n interaction lines in A and m interaction lines in B. Next, we define the sum of the energy factors of all possible rearrangements of the relative order of the n interactions in A with respect to the m interactions in B to be R^n_m. We also define the energy denominator appropriate to A taken alone to be P_n and that appropriate to B taken alone to be Q_m. In terms of these quantities our theorem states that

$$R^n_m = P_n Q_m \tag{B1}$$

In words, this means that the sum of the energy factors from all possible relative orderings of the interactions in A and B is equal to the product of the energy factor appropriate to A taken alone with that appropriate to B taken alone.

We shall prove the theorem by induction, but first it is convenient to write P_n and Q_m in terms of the individual energy denominators. From the way the perturbation expansion was defined it is clear that they may be written in the form

$$P_n = \frac{1}{p_n} P_{n-1}, \quad P_0 = 1$$

$$Q_m = \frac{1}{q_m} Q_{m-1}, \quad Q_0 = 1 \tag{B2}$$

where p_n has the form

$$p_n = \sum_A (\varepsilon_u - \varepsilon_e) + n_A \varepsilon_{\text{open}} \tag{B3}$$

The sum is as described in Sections 3 and 4 except that only part A is considered. A similar expression exists for q_m.

Returning to our diagrams which are disconnected to order n and m, we consider the $n + m$th interaction line in the combined diagram. This interaction line will have an associated energy denominator given by

$$\frac{1}{r_{n+m}} = \frac{1}{\sum_A \{(\varepsilon_u - \varepsilon_e) + n_A \varepsilon + \sum_B (\varepsilon_u - \varepsilon_e) + n_B \varepsilon_0} \tag{B4}$$

Splitting up the sum into those lines which belong to part A and those that belong to part B, we see that

$$\frac{1}{r_{n+m}} = \frac{1}{p_n + q_m} \tag{B5}$$

Now the $n + m$th interaction line either belongs to diagram A or B depending on the relative ordering to the two parts. When it belongs to A we have associated with it all possible orderings of $n - 1$ interactions of A with respect to the m interactions of B. The sum of the energy factors for all these orderings is just by definition R^{n-1}_m so that the energy factor for the $n + m$ interactions when the $n + m$th is in A is just $\frac{1}{P_n + q_m} R^n_{m-1}$. Similarly the sum of the energy factors when the $n + m$th interaction is in B is $\frac{1}{p_n + q_m} R^{n-1}_m$. Adding these together we have

$$R^n_m = \frac{1}{p_n + q_m} [R^{n-1}_m + R^n_{m-1}] \tag{B6}$$

But if

$$R^{n-1}_m = P_{n-1}Q_m \tag{B7}$$

and by symmetry

$$R^n_{m-1} = P_nQ_{m-1} \tag{B8}$$

then it follows immediately from (B2), (B6), (B7), and (B8) that

$$R^n_m = P_nQ_m \tag{B9}$$

Carrying this induction process to its limit we see that the equation (B9) follows provided

$$R^1_1 = P_1Q_1 \tag{B10}$$

But

$$R^1_1 = \frac{1}{p_1 + q_1}\left[\frac{1}{p_1} + \frac{1}{q_1}\right] = P_1Q_1 \tag{B11}$$

and the theorem is proved.

APPENDIX C

Rules for Constructing Linked Diagrams

A linked diagram D_n in S_n is a diagram which can be obtained from a linked diagram D_{n-1} according to the following rules:

[1]. Draw the nth interaction to the left of the linked diagram D_{n-1}.

[2]. Take each excited state (single arrow) line which is free to the right, extend it in the same direction, and join it up to an excited state line in D_{n-1} which is free to the left.

[3]. Take each unexcited state (closed arrow) line which is free to the left, bring it around and underneath its interaction point, and join it up with another unexcited state line to its right in such a way that the direction of the arrows on the line is continuous. This rule allows an unexcited state line to join up with another line belonging to the same interaction.

[4]. Extend any excited state lines still free to the left until they reach the left-hand edge of the diagram.

[5]. Take any unexcited lines still free to the right, bring them around and underneath their interaction points and extend them to the left of the diagram. Any unexcited state lines in D_{n-1} which are still open should be extended to the left of the diagram.

[6]. The open shell (double arrow) lines are treated the same as excited state-lines (Rule 2, 4) except that it is not essential for the line in the nth interaction to be joined to one in D_{n-1}; instead it can be extended to the right-hand side of the diagram.

Our rules are now complete provided the resultant diagram has no closed disconnected parts. Two parts of the diagram are said to be disconnected if they have no state or interaction lines in common. A part of a diagram is said to be closed if it contains no free excited or unexcited lines; e.g. the only free lines on the part are open-shell lines. For such a diagram one must add the following rule:

[7]. If the diagram contains a disconnected closed part, then one must join it to the open part of the diagram by joining one or more open-shell lines on the left of the closed disconnected part to the open-shell lines on the right of the open part. The various ways of doing this lead to different allowed diagrams.

The linked diagrams with a single interaction line are obtained by applying these rules to a null diagram having no state or interaction lines.

References

1. Brueckner, K. A., *Phys. Rev.*, **100**, 36 (1955).
2. Goldstone, J., *Proc. Roy. Soc.* (*London*), *Ser. A*, **239**, 269 (1957).
3. Yutsis, A. P., Levinson, I. B., and Vanagas, V. V., *Mathematical*

Apparatus of the Theory of Angular Momentum, Translation published by Israel Programme for Scientific Translations (1962).

4. Kelly, H. P., *Perturbation Theory and its Applications in Quantum Mechanics*, Ed. C. H. Wilcox, John Wiley and Sons, 1966, p. 215 and references therein. Also article in this volume p. 215.
5. Judd, B. R., *Second Quantisation and Atomic Spectroscopy*, Johns Hopkins Press, 1967. Also article in this volume p. 207.
6. Sandars, P. G. H., *La Structure Hyperfine Magnetique des Atomes et Des Molecules*, Ed. R. Lefebvre and C. Moser, Editions du C.N.R.S., Paris, 1967, p. 111.
7. Morita, T., *Progr. Theoret. Phys.*, **29**, 351 (1963).
8. Brandow, B. H., *Many Body Description of Nuclear Structure and Reactions*, Ed. C. Bloch, Academic Press, New York, 1966.
9. Bolotin, A., Levinson, Y., and Tolmachev, V., *Lietuvos Fizikos Rinkinys IV*, **1**, 25.
10. Edmonds, A. R., *Angular Momentum in Quantum Mechanics*, Princeton University Press, Princeton, 1957.
11. Wybourne, B. G., *Spectroscopic Properties of Rare-Earths*, Interscience Publishers, New York, 1965.
12. Bleaney, B., *La Structure Magnetique Des Atomes et Des Molecules*, Eds. R. Lefebvre and C. Moser, Editions du C.N.R.S., Paris (1967), p. 13.
13. Messiah, A., *Quantum Mechanics*, North Holland Publishing Co., Amsterdam, 1961.

THE FIELD-THEORETIC FORM OF THE PERTURBATION THEORY FOR MANY-ELECTRON ATOMS

I. ABSTRACT THEORY

V. V. TOLMACHEV, *Mathematics-Mechanics Faculty, University of Moscow, Moscow, U.S.S.R. and Theoretical Department, Institute of Electrochemistry, Academy of Sciences, Moscow, U.S.S.R.*

CONTENTS

1. INTRODUCTION

The methods of quantum-field theory (with second quantization, Wick's pairing, and Feynman diagram techniques, as well as Green's function methods) are now widely used in the theory of atoms.[1–11]

In the first part of the present work we are dealing with the abstract theory. We consider the general system of interacting Fermions with the two-body interaction in the case of the perturbation theory for degenerate levels. For this system we obtain the rules of construction for the appropriate Feynman diagrams and their contributions to the calculation of energy shifts ΔE. In the second part the rules of construction for atomic Feynman diagrams, which are applied to the evaluation of ΔE, are obtained.

The field theoretic form of the perturbation theory for energy eigenvalues of the interacting Fermion system, and particularly the Feynman diagram technique for the case when the energy eigenvalues of non-interacting Fermions are non-degenerate, was developed by Goldstone[12] (see also References 13 and 14).

(The general structure of perturbation theory, i.e. that the contributions from unlinked diagrams drop out, was stated by Brueckner[15] by means of a detailed study of the third and fourth orders.)

The field theoretic form of the perturbation theory as well as the Feynman diagram technique, when the appropriate energy eigenvalues of the system of non-interacting Fermions are degenerate, has been developed in References 16, 17, and 18. In this case all the ΔE related to the particular degenerate level of the system without interaction should be examined simultaneously. In order to obtain the ΔE a finite-dimensional algebraic system of secular equations is set up.

In References 17 and 18 the algebraic set of secular equations is given in the following form:

$$T\varphi = \Delta E\varphi$$

and Feynman diagrams for the matrix elements of secular operator T are constructed. In Reference 16 another set of secular equations is dealt with:

$$M(\Delta E)\varphi = \Delta E\varphi$$

where the form of the secular operator M depends on ΔE: the Feynman diagrams for matrix elements of secular operator $M(\Delta E)$ are also constructed.

The purpose of the present work is to develop a more convenient formulation than that in References 17 and 18 for the field theoretic perturbation theory for degenerate states. Although in general our development (particularly the use of the Gell–Mann and Low adiabatic formula, the dynamic operator $S_\alpha(0, -\infty/g)$, etc.) is rather similar to the one in References 17 and 18, we construct the diagram technique for matrix elements of a simpler secular operator $M(\Delta E)$ than the one used in References 17 and 18. The results of the present work are close to those of Reference 16, though there are substantial distinctions in the method. In fact the present work is merely an extension of Reference 16 for the case of general degenerate states, while Reference 16 deals only with the ground state: the thermodynamic perturbation theory is examined and the ground-state energy is obtained as a zero-temperature free-energy limit.

In Section 2 and in Appendix A we shall consider for the case of a degenerate level the formulae for energy shifts ΔE in lower orders of the perturbation theory. These formulae essentially differ from those for a non-degenerate level, because starting from the third order singular denominators with interaction terms appear, and these denominators cannot be represented as sums of one-particle energies.

Since the expressions denoted with Feynman diagrams should contain only those denominators which are sums of one-particle energies, the particular terms in the perturbation theory expansions for ΔE cannot correspond to Feynman diagrams. Otherwise, it is not clear how to introduce Feynman time propagators for the particular particle lines of the diagram.

In Section 2 and Appendix A it will be shown that the perturbation theory expansions for matrix elements of the secular operator T do not contain these singular denominators with interaction terms contrary to the expansions for ΔE. That is why it is quite reasonable to develop the field theoretic perturbation theory in the case of a degenerate level not for the energy shifts ΔE themselves, but rather for the matrix elements of the secular operator T. Energy shifts ΔE can be easily found from the known secular operators.

The analogous situation arises in scattering problems, where it would be more convenient to introduce Feynman diagrams directly for the phase shifts δ (the analog of energy shifts ΔE). But they should be constructed not for the scattering phases, but for the matrix elements of the S matrix (the analog of the secular operator T). Phase shifts δ can be easily obtained from the known S matrix (as well as ΔE from the known T).

In Section 2 we obtain exact adiabatic formulae for the secular operator T, which express this secular operator in terms of a dynamical operator $S_\alpha(0, -\infty/g)$; these are the main starting formulae for the development of field theoretic perturbation theory for the degenerate level and construction of Feynman diagrams.

The construction of the secular operator is not a unique procedure; generally, we may find any number of different operators, acting in the space of the considered degenerate level and having true energy shifts ΔE_γ as eigenvalues. In fact the operator is fully specified only when both its eigenvalues and eigenfunctions are given.

In Section 2 the following fact is stated; if the projections of true complete eigenfunctions Ψ_γ on the degenerate level E_0 space are chosen as eigenfunctions φ_γ of the secular operator T, then the latter is naturally related to the dynamical operator $S_\alpha(0, -\infty/g)$. The adiabatic formulae obtained in Section 2 for the secular operator T enable one to develop nearly automatically the diagram technique for the matrix elements of this operator: the consideration is quite similar to the appropriate consideration for the non-degenerate case.

In order to construct the field theoretic perturbation theory in the non-degenerate case the following operations are carried out. First, the exact adiabatic formula of Gell–Mann and Low for ΔE is taken, which expresses ΔE through the logarithm of the matrix element $(\Phi_0, S_\alpha(0, -\infty/g)\Phi_0)$; here Φ_0 is the wave function of the non-interacting system, $S_\alpha(0, -\infty/g)$ is the dynamical operator. Then Dyson's formula for the perturbation theory expansion with the so-called product is applied to the operator $S_\alpha(0, -\infty/g)$. Next, the averages appearing from the Fermi-operator product are expanded by means of Wick's pairing technique; the particular systems of pairing are interpreted in terms of Feynman diagrams. Then ΔE is proved to be represented by a sum of contributions from topologically linked diagrams; the unlinked diagrams drop out, when we go over to the logarithm of $(\Phi_0, S_\alpha(0, -\infty/g)\,\Phi_0)$.

The construction of the field theoretic perturbation theory for the degenerate case in Section 3 is carried out closely to the presentation considered above. Only instead of proceeding from the adiabatic formula expressing ΔE through $(\Phi_0, S_\alpha(0, -\infty/g)\Phi_0)$ we turn to the adiabatic of T through the operator $S_\alpha(0, -\infty/g)$. Applying to this operator Dyson's perturbation theory expansion and Wick's pairing technique, we construct Feynman diagrams for the matrix elements of $S_\alpha(0, -\infty/g)$.

Inserting the diagrams, which have been obtained, into the adiabatic formula for the secular operator T we construct Feynman diagrams for the matrix elements of T and actually go to the limit $\alpha \to +0$, just as in the non-degenerate case diagrams for ΔE are obtained from diagrams for $(\Phi_0, S_\alpha(0, -\infty/g), \Phi_0)$.

In Section 3 it is shown that the convenient form of Feynman diagram technique cannot be found for T itself. Going to the limit $\alpha \to +0$ we obtain a rather complicated equation for T, which nevertheless can be easily formulated by means of Feynman diagram technique in terms of the new and simpler operator $M(\Delta E)$ introduced in Section 3. Using the equation for T considered above enables us to transform the previously mentioned secular equations with T into new ones with $M(\Delta E)$.

In Section 4 some transformations of the new secular equations with $M(\Delta E)$ are carried out as well as the formulation of final results.

2. THE SECULAR OPERATOR AND ITS ADIABATIC FORMULAE*

Let us take the total Hamiltonian $H = H_0 + gH_{\text{int}}$ and the free Hamiltonian H_0 and consider some degenerate level E_0 of the free Hamiltonian H_0; we suppose that the interaction completely removes the degeneracy of the degenerate level E_0 under consideration.

Consider the space which is spanned by those eigenstates of H_0 which belong to the degenerate level E_0 and take the correct linear combinations of these eigenstates, i.e. those states Φ_γ which are themselves eigenfunctions belonging to E_0. This space,

* Compare with References 16 and 17 and also References 19 to 22; for the details about the adiabatic formulae see Reference 3.

which is spanned in the described way by the eigenstates of the free Hamiltonian H_0, which belong to the degenerate level E_0, may also be considered to be spanned by the eigenstates Φ_γ. We shall call this space the space of the degenerate level E_0, the projection operator on this space P, and the projection operator on the complementary space $Q = I - P$.

The correct linear combinations of eigenstates Φ_γ are those eigenstates of H_0, $H_0\Phi_\gamma = E_0\Phi_\gamma$, which belong to the degenerate level E_0 and have the following property: when the interaction is switched off adiabatically the appropriate eigenfunctions Ψ_γ of the total Hamiltonian $H = H_0 + gH_{\text{int}}$ transform continuously to the eigenfunctions of H_0, Φ_γ. Besides the states Φ_γ related to Ψ_γ, we shall consider also the states φ_γ, which are projections of states Ψ_γ on the space of degenerate level E_0, i.e. $\varphi_\gamma = P\Psi_\gamma$.

The states Φ_γ are orthogonal and normalized; they form the basis for the space of the degenerate level E_0. The states φ_γ are neither normalized nor orthogonal to each other, although they also form a basis for the space of the degenerate level E_0.

Let us project the total Schrödinger equation on the space of the degenerate level E_0 which belongs to the Hamiltonian H_0. We begin with the Schrödinger equation for states Ψ_γ, which has the following form:

$$(H_0 - E_0 + gH_{\text{int}} - \Delta E_\gamma)\Psi_\gamma = 0 \tag{1}$$

where ΔE_γ are the energy shifts for the states.

Operating with the projection operator P on both sides of (1), we obtain:

$$\begin{aligned} O = P(H_0 - E_0 + gH_{\text{int}} - \Delta E_\gamma)\Psi_\gamma \\ = (H_0 - E_0)P\Psi_\gamma + PgH_{\text{int}}\Psi_\gamma - \Delta E_\gamma P\Psi_\gamma \end{aligned}$$

because P commutes with H_0. We then use the relation

$$(H_0 - E_0)P\Psi_\gamma = 0$$

this follows immediately from the fact that the state $P\Psi_\gamma$ belongs to the subspace of E_0. Therefore we have

$$PgH_{\text{int}}\Psi_\gamma = \Delta E_\gamma\varphi_\gamma \tag{2}$$

The states $PgH_{\text{int}}\Psi_\gamma$ belong to the space of E_0 and they may be considered as the result of applying some operator T to the states φ_γ which belong to the same space; i.e. we can write

$$PgH_{\text{int}}\Psi_\gamma \equiv T\varphi_\gamma \tag{3}$$

This equation is in fact the definition of the operator T; we shall call T the secular operator. Taking into account Eqs. (2) and (3) it may be written in the following form

$$T\varphi_\gamma = \Delta E_\gamma \varphi_\gamma \tag{4}$$

which we shall call the secular equations. Evidently, they have to be considered as resulting from the projection of the total Schrödinger equation on the space of the degenerate level E_0 belonging to the free Hamiltonian H_0. The set of equations (4) tells us that the states φ_γ are eigenstates of the operator T with real eigenvalues ΔE_γ. The states φ_γ do not form an orthogonal system, and therefore the operator T is not Hermitian, i.e.

$$T^\dagger \neq T$$

The secular operator T acts in the space of E_0; let us extend its domain of definition to include the whole space and introduce for this purpose an operator F, for which (5) holds in the whole space

$$\begin{cases} F\varphi_\gamma = gH_{\text{int}}\Psi_\gamma \\ F\psi \;= 0 \end{cases} \tag{5}$$

where ψ is any state belonging to the complementary space of E_0. According to the definition of operators F and T, we have,

$$PF = T \tag{6}$$

Operator F will be called the energy operator.

Now let us define (in the whole space again) the wave operator Ω which, applied to the functions φ_γ gives the Ψ_γ's, but applied to any state ψ, which belongs to the complementary space of E_0, gives zero, i.e.

$$\begin{cases} \Omega\varphi_\gamma = \Psi_\gamma \\ \Omega\psi \;= 0 \end{cases} \tag{7}$$

In accordance with the definition above, we have:

$$F = gH_{\text{int}}\Omega \tag{8}$$

$$P\Omega = P \tag{9}$$

$$\Omega P = \Omega \tag{10}$$

Let us set up an equation for the wave operator Ω. First using the total Schrödinger equation we get

$$(E_0 - H_0 - gH_{\text{int}} + \Delta E_\gamma)\Omega\varphi_\gamma = 0 \tag{11}$$

Then it follows from (11), that

$$\begin{aligned}(E_0 - H_0)\Omega\varphi_\gamma &= gH_{\text{int}}\Omega\varphi_\gamma - \Delta E_\gamma\Omega\varphi_\gamma \\ &= F\varphi_\gamma - \Omega F\varphi_\gamma = QF\varphi_\gamma - Q\Omega F\varphi_\gamma \end{aligned} \tag{12}$$

where we used the fact that $\Omega F\varphi_\gamma = \Omega PF\varphi_\gamma = \Omega T\varphi_\gamma = \Delta E_\gamma\Omega\varphi_\gamma$ (because $\Omega = \Omega P$), and also $PF\varphi_\gamma - P\Omega F\varphi_\gamma = T\varphi_\gamma - T\varphi_\gamma = 0$ (because $P\Omega = P$).

Due to the definitions of F and Ω (see Eqs. (5) and (7)), we then get from (12):

$$(E_0 - H_0)\Omega = QF - Q\Omega F \tag{13}$$

Acting on both sides of Eq. (13) with operator $\dfrac{Q}{E_0 - H_0}$, we obtain finally the equation

$$Q\Omega = \frac{Q}{E_0 - H_0} F - \frac{Q}{E_0 - H_0} \Omega F$$

The solution of this equation is given by an iterative expansion

$$Q\Omega = \sum_{n=1}^{+\infty} (-1)^{n-1} \frac{Q}{(E_0 - H_0)^n} F^n$$

By virtue of $P\Omega = P$ we obtain from the above relation

$$\Omega = P + \sum_{n=1}^{+\infty} (-1)^{n-1} \frac{Q}{(E_0 - H_0)^n} F^n \tag{14}$$

It is easily seen, that the operator (14) satisfies the following equation (we make use of (8)):

$$\Omega = P + X(E_0)gH_{\text{int}}\Omega \tag{15}$$

where

$$X(E_0) = \sum_{n=0}^{+\infty} (-1)^n \frac{Q}{(E_0 - H_0)^{n+1}} F^n \tag{15a}$$

(it is quite evident that from (15a) it follows that

$$X \neq \frac{Q}{E_0 - H_0 - F}.$$

Let us derive the adiabatic formulae for the operators Ω, F, and T, which are the wave operator, the energy operator, and the secular operator, respectively.

According to the adiabatic formula due to Gell–Mann and Low

$$\Psi_\gamma = \lim_{\alpha \to +0} \frac{S_\alpha(0, -\infty|g)\Phi_\gamma}{(\Phi_\gamma, S_\alpha(0, -\infty|g)\Phi_\gamma)} \tag{16}$$

and using the definition of the wave operator Ω (7), we obtain:

$$\Omega P S_\alpha(0, -\infty|g)\Phi_\gamma = S_\alpha(0, -\infty|g)\Phi_\gamma$$

Therefore,

$$\Omega P S_\alpha(0, -\infty|g) P = S_\alpha(0, -\infty|g) P \tag{17}$$

The operator $PS_\alpha(0, -\infty|g)P$ acts only in the space of the considered degenerate level E_0; so let us define for this operator an operator which is its inverse operator $(PS_\alpha\ (0,\ -\infty/g)P)^{-1}$ in the space of E_0 and zero in the complementary space, i.e.

$$(PS_\alpha(0, -\infty|g)P)^{-1} \equiv P(PS_\alpha(0, -\infty|g)P)^{-1}P$$

So we have

$$PS_\alpha(0, -\infty|g)P \,.\, (PS_\alpha(0, -\infty|g)P)^{-1} = P \tag{18}$$

Acting on both sides on the right of the Eq. (17) by $(PS_\alpha(0, -\infty/g)$ $P)^{-1}$, we obtain for the wave operator:

$$\Omega = \lim_{\alpha \to +0} S_\alpha(0, -\infty|g)P \,.\, (PS_\alpha(0, -\infty|g)P)^{-1} \tag{19}$$

From Eq. (19) together with the formula $F = gH_{\text{int}}\Omega$, we obtain the following expression for the energy operator:

$$F = \lim_{\alpha \to +0} gH_{\text{int}} S_\alpha(0, -\infty|g)P \,.\, (PS_\alpha(0, -\infty|g)P)^{-1} \tag{20}$$

and this formula (20) together with the relation $T = PF$ leads to the following expression for the secular operator:

$$T = \lim_{\alpha \to +0} PgH_{\text{int}}S_\alpha(0, -\infty|g)P \,.\, (PS_\alpha(0, -\infty|g)P)^{-1} \tag{21}$$

We note that the operator $PS_\alpha(0, -\infty/g)P$ is not unitary and therefore the construction of its inverse does not reduce to the Hermitian conjugation.

Now we use the operator identity:

$$[H_0; S_\alpha(0, -\infty|g)] = -gH_{\text{int}}S_\alpha(0, -\infty|g) + i\hbar\alpha g \frac{\partial}{\partial g} S_\alpha(0, -\infty|g) \tag{22}$$

which we shall multiply on the left and right sides by the projection operator P.

Then we obtain the following identity:

$$PgH_{\text{int}}S_\alpha(0, -\infty|g)P = i\hbar\alpha g \frac{\partial}{\partial g} PS_\alpha(0, -\infty|g)P \tag{23}$$

using the fact that $PH_0P = E_0P$. Substituting (23) into (21), we finally obtain

$$T = \lim_{\alpha \to +0} i\hbar\alpha g \frac{\partial}{\partial g} PS_\alpha(0, -\infty|g)P \,.\, (PS_\alpha(0, -\infty|g)P)^{-1} \tag{24}$$

The formulae (21) and (24) represent our main adiabatic formulae for the secular operator T. We note that in (24) we are not able to take the derivative of the logarithm of the operator $PS_\alpha(0, -\infty/g)P$ because in general, $\frac{\partial}{\partial g} \ln(1 + A) \neq \frac{\partial A}{\partial g}(1 + A)^{-1}$ when the operators $\frac{\partial A}{\partial g}$ and A do not commute with each other.

Let us take for the operator $S_\alpha(0, -\infty/g)$ its series expansion in powers of interactions:

$$S_\alpha(0, -\infty|g) = \sum_{n=0}^{+\infty} \left(-\frac{i}{\hbar}\right)^n g^n \int_{-\infty}^{0} dt_1 \int_{-\infty}^{t_1} dt_2 \ldots \int_{-\infty}^{t_{n-1}} dt_n H_{\text{int}}(t_1) \ldots H_{\text{int}}(t_n) e^{\alpha(t_1 + \cdots + t_n)} \tag{25}$$

where

$$H_{\text{int}}(t) = e^{i/\hbar H_0 t} H_{\text{int}} e^{-i/\hbar H_0 t} \tag{25a}$$

We note that $-\infty < t_n < \ldots < t_2 < t_1 < 0$ and change the integration variables in the following way:

$$t = \tau_1,\ t_2 - t_1 = \tau_2,\ \ldots,\ t_n - t_{n-1} = \tau_n \tag{26}$$

Now the new variables $\tau_1, \tau_2, \ldots, \tau_n$ will cover all the values from $-\infty$ to 0 independently. Further,

$$\begin{aligned} t_n = \tau_1 + \tau_2 + \ldots + \tau_n,\ t_1 + t_2 + \ldots + t_n \\ = n\tau_1 + (n-1)\tau_2 + \ldots + \tau_n \end{aligned}$$

After this change of variables in (25) and (25a), we obtain

$$\begin{aligned} S_\alpha(0, -\infty|g) \\ = \sum_{n=0}^{+\infty} \left(-\frac{i}{\hbar}\right)^n g^n \int_{-\infty}^{0} d\tau_1 \ldots \int_{-\infty}^{0} d\tau_n e^{i/\hbar\tau_1 H_0} H_{\text{int}} e^{i/\hbar H_0 \tau_2} \ldots \\ \times e^{i/\hbar H_0 \tau_n} H_{\text{int}} e^{-i/\hbar H_0(\tau_1 + \ldots + \tau_n)} e^{\alpha(n\tau_1 + (n-1)\tau_2 + \ldots + \tau_n)} \end{aligned}$$

and from the above relation, noting that

$$e^{-i/\hbar H_0(\tau_1 + \ldots + \tau_n)} \equiv \int_{-\infty}^{+\infty} e^{-i/\hbar E(\tau_1 + \ldots + \tau_n)} \delta(E - H_0)\, dE \tag{27}$$

we obtain finally

$$\begin{aligned} S_\alpha(0, -\infty|g) = \int_{-\infty}^{+\infty} dE \sum_{n=0}^{+\infty} \left(-\frac{i}{\hbar}\right)^n g^n \int_{-\infty}^{0} d\tau_1 e^{i/\hbar\tau_1(H_0 - E) + \alpha n\tau_1} H_{\text{int}} \ldots \\ \int_{-\infty}^{0} d\tau_n e^{i/\hbar\tau_n(H_0 - E) + \alpha\tau_n} H_{\text{int}} \delta(E - H_0) \\ = \int_{-\infty}^{+\infty} dE \sum_{n=0}^{+\infty} g^n \frac{1}{E - H_0 + i\hbar\alpha n} H_{\text{int}} \frac{1}{E - H_0 + i\hbar\alpha(n-1)} \ldots \\ \frac{1}{E - H_0 + i\hbar\alpha} H_{\text{int}} \delta(E - H_0) \end{aligned}$$

Introducing an abridged notation

$$R_\alpha(E|g) = \sum_{n=0}^{+\infty} g^n \frac{1}{E - H_0 + i\hbar\alpha n} H_{\text{int}} \frac{1}{E - H_0 + i\hbar\alpha(n-1)} H_{\text{int}} \cdots \frac{1}{E - H_0 + i\hbar\alpha} H_{\text{int}}$$

$$= 1 + \frac{1}{E - H_0 + i\hbar\alpha} gH_{\text{int}} + \frac{1}{E - H_0 + 2i\hbar\alpha} gH_{\text{int}} \frac{1}{E - H_0 + i\hbar\alpha} gH_{\text{int}} + \cdots \tag{28}$$

the result can be written in a quite laconic way:

$$S_\alpha(0, -\infty|g) = \int_{-\infty}^{+\infty} dE\, R_\alpha(E|g)\delta(E - H_0) \tag{29}$$

from which follows

$$PS_\alpha(0, -\infty|g)P = PR_\alpha(E_0|g)P \tag{30}$$

Formulae (30) and (28) will be used in Appendix A.

3. FEYNMAN DIAGRAMS FOR $PS_\alpha(0, -\infty/g)P$ AND THE ADIABATIC FORMULA FOR THE SECULAR OPERATOR T*

During the derivation of the ordinary perturbation theory formulae in the preceding Section 2 and in Appendix A we did not have to specify the detailed form of the free Hamiltonian operator H_0 and the interaction Hamiltonian H_{int}; it was sufficient to suppose them to be Hermitian. But in order to develop the field theoretic form of perturbation theory we need to point out the expressions for the free Hamiltonian H_0 and the interaction Hamiltonian H_{int} in terms of Fermi field operators.

Let us consider here for simplicity only that form of interaction Hamiltonian, which is expressed by means of a quartic form in terms of Fermi field operators. In the occupation number representation we have:

*Compare with Reference 3.

$$H = H_0 + H_{\text{int}} \tag{31}$$

$$H_0 = \sum_i \varepsilon_i \overset{+}{a_i} a_i \tag{32}$$

$$H_{\text{int}} = \tfrac{1}{2} \sum_{i'j'ji} v_{i'j',ji} \overset{+}{a_{i'}} \overset{+}{a_{j'}} a_j a_i \tag{33}$$

where the Fermi field operators $\overset{+}{a_i}$ and a_i, describing the creation and annihilation of electrons in single-electron states i, satisfy the ordinary Fermion anticommunication relations. The electronic energies ε_i are real numbers; for the interaction matrix elements we have: $v_{i'j',ji} = v_{j'i',ij}$ and $v_{i'j',ji} = v_{ij,j'i'}$.

We consider, as in the previous section, the degenerate level E_0 of the free Hamiltonian; we construct it in the following way. Let Φ_0 be some non-degenerate state of Hamiltonian H_0 with completely filled electronic shells; we shall call this state the atomic core state, or simply the core state. Let us consider then some open shell consisting of s single-electron states; we shall label these single-electron states belonging to the open shell by Greek indices β, γ and so on; all these states possess equal energy ε_0, i.e. $\varepsilon_\beta = \varepsilon_\gamma = \ldots = \varepsilon_0$. Let us apply r creation operators a_β^+, a_γ^+, . . . to the core state Φ_0 to obtain a degenerate level E_0 with the energy $E_0 + r\varepsilon_0$ and the degeneracy factor $C_s^r = \binom{r}{s}$; this is the energy level that we shall study.

As was shown in Section 2, the energy shifts ΔE for single states of degenerate level E_0 under consideration can be determined as the solution of the set of secular equations in the appropriate degenerate level space (see Eq. (4) from Section 2)

$$T\varphi_\gamma = \Delta E_\gamma \varphi_\gamma \tag{34}$$

where T is the secular operator, acting in the space of the degenerate level, φ_γ are the states belonging to this space, ΔE_γ are the unknown energy shifts.

Let us take as a starting point in this section the adiabatic formula (24) of Section 2.

$$T = \lim_{\alpha \to +0} i\hbar\alpha g \frac{\partial}{\partial g} PS_\alpha(0, -\infty|g)P \,.\, (PS_\alpha(0, -\infty|g)P)^{-1} \tag{35}$$

where the projection operator P projects any given state on the space of the degenerate level E_0 and

$$S_\alpha(0, -\infty|g) = \sum_{n=0}^{+\infty} \left(-\frac{i}{\hbar}\right)^n \frac{g^n}{n!} \int_{-\infty}^{0} dt_1 \ldots$$

$$\int_{-\infty}^{0} dt_n e^{\alpha(t_1+\ldots+t_n)} T(H_{\text{int}}(t_1) \ldots H_{\text{int}}(t_n)) \tag{36}$$

where the T product of the interaction Hamiltonians contains different field operators because

$$H_{\text{int}}(t) = \tfrac{1}{2} \sum_{i'j'ji} v_{i'j',ji} \overset{+}{a}_{i'}(t+0) \overset{+}{a}_{j'}(t+0) a_j(t) a_i(t) \tag{37}$$

and it is very important to preserve the $+0$ in the time arguments of the creation operators. According to the definition of the T product we have to arrange the field operators which stand under its sign so that each operator with larger time stands on the left of any operator with smaller time (for details see Reference 3).

For the basis of E_0 we choose the following single-determinant states

$$\overset{+}{a}_{\beta_1} \ldots \overset{+}{a}_{\beta_r} \Phi_0 \tag{38}$$

where the single-electron indices cover independently all the s permissible values in the open shell under consideration. The states (38) are orthogonal to each other, and all of them are normalized to unity, see Appendix B.

Let us write down our main formulae (34) and (35) in this basis (38). From (34) we obtain

$$\frac{1}{r!} \sum_{\gamma_1 \ldots \gamma_r} (\beta_r \ldots \beta_1 | T | \gamma_1 \ldots \gamma_r) C_{\gamma_1 \ldots \gamma_r} = \Delta E C_{\beta_1 \ldots \beta_r} \tag{39}$$

and from (35)

$$(\beta_r \ldots \beta_1 | T | \gamma_1 \ldots \gamma_r)$$

$$= \lim_{\alpha \to +0} \frac{1}{r!} \sum_{\delta_1 \ldots \delta_r} i\hbar\alpha g \frac{\partial}{\partial g} (\beta_r \ldots \beta_1 | P S_\alpha(0, -\infty|g) P | \delta_1 \ldots \delta_r)$$

$$\times (\delta_r \ldots \delta_1 | (P S_\alpha(0 - \infty| g), P)^{-1} | \gamma_1 \ldots \gamma_r) \tag{40}$$

Now single out from (40) the term accounting for the core energy shift, dividing and multiplying both factors on the right-hand side of this equation by $(\Phi_0,\ S_\alpha(0,\ -\infty|g)\Phi_0)$, and carrying out the appropriate transformations, we get the following expression

$$(\beta_r \ldots \beta_1|T|\gamma_1 \ldots \gamma_r) = \Delta E_0 I(\beta_r \ldots \beta_1; \gamma_1 \ldots \gamma_r)$$
$$+ \lim_{\alpha\to+0} \frac{1}{r!} \sum_{\delta_1\ldots\delta_r} i\hbar\alpha g \frac{\partial}{\partial g}\left(\beta_r \ldots \beta_1\left|\frac{PS_\alpha(0,\ -\infty|g)P}{(\Phi_0,\ S_\alpha(0,\ -\infty|g)\Phi_0)}\right|\delta_1 \ldots \delta_r\right)$$
$$\times\left(\delta_r \ldots \delta_1\left|\left(\frac{PS_\alpha(0,\ -\infty|g)P}{(\Phi_0,\ S_\alpha(0,\ -\infty|g)\Phi_0)}\right)^{-1}\right|\gamma_1 \ldots \gamma_r\right) \tag{41}$$

where I is the unit matrix in the space of E_0; the core energy shift ΔE_0 is given by the expression

$$\Delta E_0 = \lim_{\alpha\to+0} i\hbar\alpha g \frac{\partial}{\partial g} \ln(\Phi_0,\ S_\alpha(0,\ -\infty|g)\Phi_0) \tag{42}$$

From now on we shall call the operator appearing in (41)

$$\sigma_\alpha(g) = \frac{PS_\alpha(0,\ -\infty|g)P}{(\Phi_0,\ S_\alpha(0,\ -\infty|g)\Phi_0)} \tag{43}$$

The diagram technique for the calculations in (41) and (42) is developed in the following way. First of all, the diagrams for the quantity $(\Phi_0,\ S_\alpha(0,\ -\infty|g)\Phi_0)$ (or, correspondingly, for the quantity $(\beta_r \ldots \beta_1|PS_\alpha(0,\ -\infty|g)P/\delta_1 \ldots \delta_r))$ are constructed; then after appropriate arguments we proceed from these diagrams to the diagrams for ΔE_0 and the second term on right-hand side of (41).

Let us discuss the procedure of the diagram derivation for $(\beta_r \ldots \beta_1|PS_\alpha(0,\ -\infty|g)P|\delta_1 \ldots \delta_r)$ (the diagrams for $(\Phi_0,\ S_\alpha(0,\ -\infty|g)\Phi_0)$ are considered analogously, see Reference 3).

In the expression for the matrix element $(\beta_r \ldots \beta_1|PS_\alpha(0,\ -\infty|g)P|\delta_1 \ldots \delta_r)$ we shall use for the operator $S_\alpha(0,\ -\infty|g)$ its expansion (36), where $H_{\text{int}}(t)$ is in the form (37). Then we shall take the T product of the resulting products of $4n$ Fermi operators. In order to evaluate expectation values of T products of field operators we shall apply the field-theoretic technique of pairings, which is based on Wick's theorem, expressing the T product of field operators as the sum of N products of these operators with all possible pairings (including the cases where only some operators are paired or as a special case the case of non-pairing). For the matrix element considered above we then obtain (see Reference 3)

$$(\beta_r \ldots \beta_1|PS_\alpha(0, -\infty|g)P|\gamma_1 \ldots \gamma_r) = (\beta_r \ldots \beta_1|S_\alpha(0, -\infty|g)|\gamma_1 \ldots \gamma_r)$$

$$= \sum_{n=0}^{+\infty}\left(-\frac{i}{\hbar}\right)^n \frac{g^n}{n!}\int_{-\infty}^{0} dt_1 \ldots \int_{-\infty}^{0} dt_n e^{\alpha(t_1+\ldots+t_n)}$$

$$\times (\Phi_0, a_{\beta_r} \ldots a_{\beta_1} T(H_{\text{int}}(t_1) \ldots H_{\text{int}}(t_n))\overset{+}{a}_{\gamma_1} \ldots \overset{+}{a}_{\gamma_r}\Phi_0)$$

$$= \sum_{n=0}^{+\infty}\left(-\frac{i}{\hbar}\right)^n \frac{g^n}{2^n n!}\int_{-\infty}^{0} dt_1 \ldots \int_{-\infty}^{0} dt_n e^{\alpha(t_1+\ldots+t_n)}$$

$$\times \Sigma_1 \ldots \Sigma_r(\Phi_0, a_{\beta_r} \ldots a_{\beta_1} T(A_1(t_1) \ldots A_n(t_n))\overset{+}{a}_{\gamma_1} \ldots \overset{+}{a}_{\gamma_r}\Phi_0)$$

$$= \sum_{n=0}^{+\infty}\left(-\frac{i}{\hbar}\right)^n \frac{g^n}{2^n n!}\int_{-\infty}^{0} dt_1 \ldots \int_{-\infty}^{0} dt_n e^{\alpha(t_1+\ldots+t_n)}\Sigma_1 \ldots \Sigma_2$$

$$\times \sum_{\text{all pairings}}(\Phi_0, a_{\beta_r} \ldots a_{\beta_1} N(\overbrace{A_1(t_1) \ldots A_n(t_n)})\overset{+}{a}_{\gamma_1} \ldots \overset{+}{a}_{\gamma_r}\Phi_0) \tag{44}$$

where in order to shorten the notation the following designations are used

$$A_k(t_k) \equiv \overset{+}{a}_{i'_k}(t_k + 0)\overset{+}{a}_{j'_k}(t_k + 0)a_{j_k}(t_k)a_{i_k}(t_k) \tag{45a}$$

$$\Sigma_k \times \equiv \sum_{i_k' j_k' j_k i_k} v_{i_k' j_k', j_k j_k} \times \tag{45b}$$

where,

$$a_j(t) = e^{-i/\hbar \varepsilon_j t}a_j \qquad \overset{+}{a}_j(t) = e^{i/\hbar \varepsilon_j t}\overset{+}{a}_j \tag{46}$$

We consider the N product for the core state Φ_0 (for the definition see Appendix C). For this N product, consisting of single pairings, we have

$$\begin{aligned}\overbrace{a_i(t)\overset{+}{a}_j(s)} &= T(a_i(t)\overset{+}{a}_j(s)) - N(a_i(t)\overset{+}{a}_j(s)) \\ &= (\Phi_0, T(a_i(t)\overset{+}{a}_j(s))\Phi_0) \\ &= \delta_{ij}e^{-i/\hbar\varepsilon_j(t-s)}((1 - n_j^0)\theta(t - s) - n_j^0\theta(s - t)) \\ &= G_j^0(t - s)\delta_{ii}\end{aligned} \tag{47}$$

where the occupation numbers are supplied by index 0, noting their relation to the core state Φ_0.

From the definition of the N product (Appendix C) it follows immediately, that in (44) all unpaired field operators under the N product sign must have as their single-electron indices the ones from the open shell. In fact, if between the operators on the right or on the left there are (see (48)) field operators with indices not belonging to the open shell, then it is always possible to carry them through all the remaining operators, so as to make them act on the right or left Φ_0 in (44) correspondingly, which gives zero (these operators anticommute with all other operators).

Let us introduce an abridged notation for the averages of operators, which remained unpaired in (44),

$$(\Phi_0, a_{\beta_r} \ldots a_{\beta_1} N(a_{\lambda_1} \ldots a_{\lambda_p} \overset{+}{a}_{\lambda'_p} \ldots \overset{+}{a}_{\lambda'_1}) \overset{+}{a}_{\gamma_1} \ldots \overset{+}{a}_{\gamma_r} \Phi_0) \\ = N(\beta_r \ldots \beta_1, \lambda'_1 \ldots \lambda'_p; \lambda_p \ldots \lambda_1, \gamma_1 \ldots \gamma_r) \qquad (48)$$

where all indices belong to the open shell.

Now let us return to Eq. (44); we shall show how to represent the different terms of this formula by means of diagrams (for details see Reference 3).

It is convenient to consider first the so-called primitive time-dependent diagrams, from which we can easily pass to the structural time-dependent diagrams or, simply, time-dependent diagrams and then to the so-called time-independent diagrams.

A. Primitive Time-Dependent Diagrams

In order to draw a primitive diagram of the order n with $2p$ external lines, it is necessary to take quartic vertices, enumerated by the numbers $1, \ldots, n$. For these vertices the dashed interaction lines have definite directions. Then the ends of electron lines of n vertices (the total number of them is $4n$) are connected with each other in such a way that an outgoing line is connected with an incoming one, and p incoming and p outgoing lines are left free (unconnected). In this way we obtain a typical primitive diagram, shown in Fig. 1b; the vertex of the primitive is shown in the Fig. 1a.

Two primitive diagrams are considered as different, if they cannot be superposed after three-dimensional topological transformations in such a way that not only their dashed lines, electronic lines, and the ends of electronic lines coincide with each other, but also the directions of electronic lines, of their ends, and dashed lines as well as the numbers labelling the quartic vertices, should coincide.

The primitive diagrams are in one-to-one correspondence with the particular pairings in Eq. (44). In fact any four field operators from each $H_{\text{int}}(t_1), \ldots, H_{\text{int}}(t_n)$ in Eq. (44) correspond with appropriate vertices of the primitive diagram. Four field operators for $H_{\text{int}}(t_k)$ can be represented by the vertex number K. The appropriate ends of electron lines in the primitive diagram

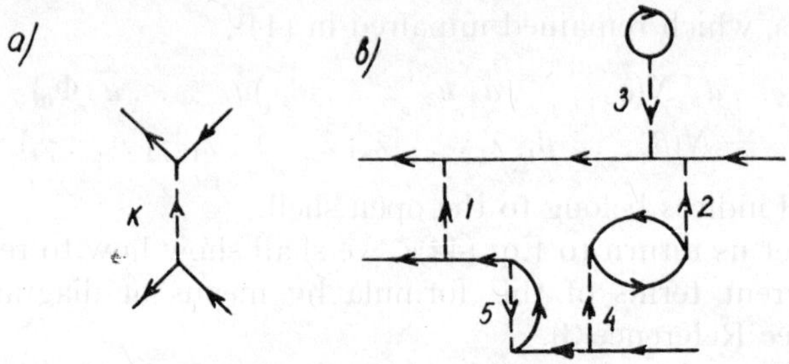

Fig. 1. (*a*) Quartic vertex of the primitive diagrams. (*b*) Typical primitive diagram.

vertex are the images of relevant operators; also the dashed interaction lines as well as the direction of the electronic lines and their ends should coincide.

The upper outgoing end of the electronic line means the first Fermi operator on the left of the two-body interaction term, the lower outgoing end of the electron line corresponds to the second Fermi operator in the vertex, the lower incoming end of the electron line the third Fermi operator, and the upper incoming end of the line corresponds to the fourth Fermi operator.

During this procedure the quartic vertex is supposed to be drawn so that its dashed line is vertical and is directed upwards. The quartic vertices of the primitive diagrams are specially drawn with directed dashed lines.

We have shown that particular Fermi operators standing under the T product sign in (44) can be represented by the ends of electron lines in primitive diagram vertices. The pairings (47) of each pair of Fermi operators are, therefore, represented as appropriate lines of primitive diagrams. The free unpaired operators (48) are represented as free ends of electron lines of primitive diagrams. However, the primitive diagrams have an inconvenient feature: many of them give equal contributions, and therefore it is necessary to proceed from the primitive to the structural diagrams, each corresponding to the group of primitive ones with equal contributions. Such contributions belong to the diagrams which are different only in the labelling of their vertices by numbers $1, \ldots n$; they differ also in the directions of the dashed lines.

B. Structural Time-Dependent Diagrams

In order to draw the structural diagram of nth order with $2p$ free external lines it is necessary to take again n vertices now

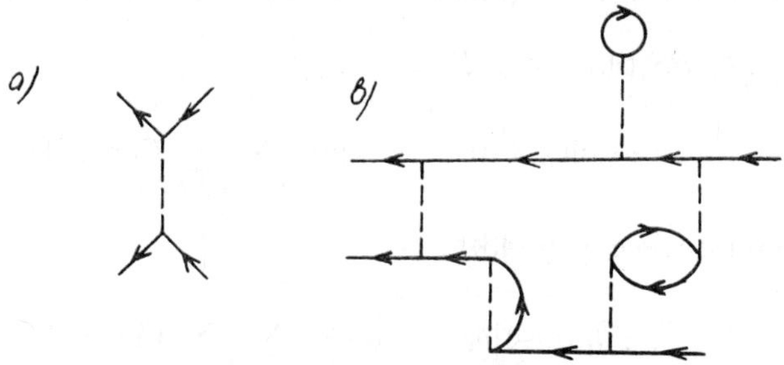

Fig. 2. (*a*) Quartic vertex of the structural diagram. (*b*) Typical structural diagram.

without any labelling on them and with undirected dashed lines.

The electron lines ending in these vertices (which are $4n$ in number) are to be connected with each other—an incoming end with an outgoing one, letting p outgoing lines and p incoming lines free (unconnected). In such a way we get structural diagrams of the type shown in Fig. 2b; the vertex of a structural diagram is presented in Fig. 2a.

Two structural diagrams are considered as different, if they cannot be superposed after three-dimensional topological transformations in such a way that not only the electron lines and their ends coincide but also the directions on its electron lines.

Denoting the primitive diagrams as D_n and structural diagrams as Γ_n, where n is the order of the diagram, we obtain finally instead of (44)

$$(\beta_r \ldots \beta_1|PS_\alpha(0, -\infty|g)P|\gamma_1 \ldots \gamma_r)$$

$$= \sum_{n=0}^{+\infty} \sum_{D_n} \text{Contr. } D_n = \sum_{n=0}^{+\infty} \sum_{\Gamma_n} \text{Contr. } \Gamma_n \qquad (49)$$

where Contr.D_n means the contribution from D_n and, correspondingly, Contr.Γ_n means the contribution from Γ_n.

The structural diagrams Γ_n are not necessarily topologically linked entitites. They may be formed from topologically unlinked parts, and in some instances topologically linked parts without free external lines—so-called vacuum parts—may be dealt with.

We can exclude the diagrams Γ_n with vacuum parts using well-known field-theoretic arguments. Then instead of (49) we obtain

$$(\beta_r \ldots \beta_1|PS_\alpha(0, -\infty|g)P|\gamma_1 \ldots \gamma_r)$$

$$= (\Phi_0, S_\alpha(0, -\infty|g)\Phi_0) \sum_{n=0}^{+\infty} \sum_{\text{link } \Gamma_n} \text{Contr. } \Gamma_n \qquad (50)$$

and, therefore (see also (43))

$$(\beta_r \ldots \beta_1|\sigma_\alpha(g)|\delta_1 \ldots \delta_r) = \sum_{n=0}^{+\infty} \sum_{\text{link } \Gamma_n} \text{Contr. } \Gamma_n \qquad (51)$$

where the index "link" means that summation includes only the linked structural diagrams, i.e. the diagrams without vacuum parts (of course, the diagram after this exclusion may be formed from topologically disconnected parts, but now every part has its own free external lines).

Both primitive and structural diagrams are time-dependent, i.e. their vertices carry the time indices t, s, and so on, over which we shall integrate; there are also the factors G_i^0 $(t - s)$ on the electron lines, and so on.

The integrations over time in the contributions of structural time-dependent diagrams may be carried through in a quite general way, and in this manner we get so-called time-independent diagrams with energy denominators.

C. Time-Independent Diagrams

In order to draw time-independent diagrams, we shall take structural time-dependent diagrams and by all possible ways order their vertices with respect to the horizontal time axis (time runs from the right to the left). Two time-independent diagrams are different if they cannot be superposed after topological transformations in the three-dimensional space so that their dashed line as well as the electron lines, their external lines and directions coincide; the order of vertices with respect to the fixed horizontal time axis should be invariable. A typical time-independent diagram shown in Figs. 3b and 3a shows its vertex (the same as for the structural time-dependent diagram).

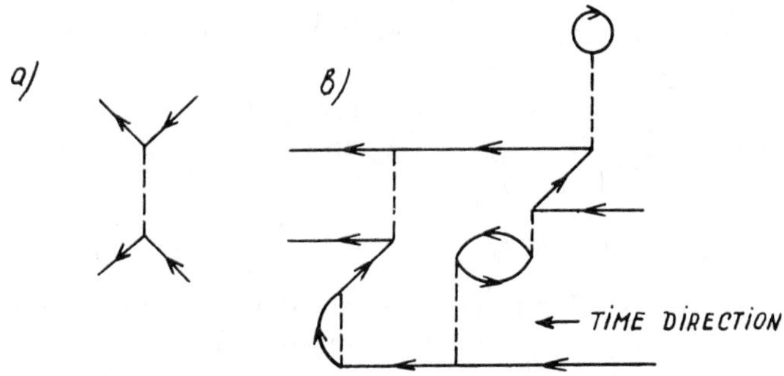

Fig. 3. (*a*) Quartic vertex of the time-independent diagram.
(*b*) Typical time-independent diagram.

Let us continue the free external electron lines of the time-independent diagram so that they come from the right infinity and go to the left infinity. These continued external electron lines would be distinguished from inner electron lines of the diagram.

We shall call the electron lines running along the time axis the particle electron lines; those running in the opposite direction are called the hole electron lines. The external lines are always the particle ones. The electron lines starting and ending at one and the same vertex should be considered the hole lines.

Let us lay down the rules for getting the contributions from time-independent diagrams of the matrix element $(\beta_r \ldots \beta_1 |\sigma_\alpha(g)|\delta_1 \ldots \delta_r)$ obtained by the foregoing procedure.

In order to write down the appropriate contribution to the matrix element given above from a time-independent diagram with $2p$ incoming and outgoing external electron lines the following procedure is proposed.

[1]. The inner electron lines of the diagram as well as its p external incoming electron lines and p outgoing ones are labelled with one-electron indices. Indices i, j, k, etc., having arbitrary values are put on the inner electron lines. Put the indices $\lambda_1, \ldots, \lambda_p$ on p external incoming lines, those indices can have the arbitrary values from the open shell, while for the external outgoing lines the indices $\lambda'_1, \ldots, \lambda'_p$ are used and these also can have the values from the open shell. The foregoing indices should be set on the external electron lines of the diagram so that their pairs λ'_1 and $\lambda_1, \ldots, \lambda'$ and λ_p label the appropriate external electron lines, being the starting and ending ones for the particular p open loops of the electron lines.

[2]. Write the product of the following factors:

(i) To external lines $\lambda_1, \ldots, \lambda_p$ and $\lambda'_1, \ldots, \lambda'_p$ of the diagram corresponds the factor

$$N(\beta_r \ldots \beta_1, \lambda'_1 \ldots \lambda'_p; \lambda_p \ldots \lambda_1, \delta_1 \ldots \delta_r) \tag{52}$$

(ii) To every two-body vertex of the diagram with the outgoing electron line ends i', j' and the appropriate incoming ends i, j corresponds the factor:

$$-gv_{i'j',ji} \tag{53}$$

(iii) To every inner particle electron line with the index i corresponds the factor:

$$1 - n_i^0 \tag{54a}$$

and to every inner hole line with the index i the factor

$$n_i^0 \tag{54b}$$

(iv) To each vertical section between the kth and $(k + 1)$th vertices (counting from the left to the right) corresponds the factor:

$$\frac{1}{\mathscr{E} + (n - k)\varepsilon - p\varepsilon_0} \tag{55}$$

where n is the order of the diagram, $\varepsilon = \frac{\hbar}{i}\alpha$, $\mathscr{E}$ is the sum of (the inner and external) energies of particle lines, crossed by the vertical section, minus the sum of (the inner and external) energies of hole lines, crossed by it. At $k = 0$ we obtain the contribution from the vertical section at the far left of the diagram (the section which lies to the left in front of the diagram).

(v) To the whole time-independent diagram belongs the factor:

$$\frac{(-1)^{l+f}}{\lambda} \tag{56}$$

where l is the number of closed loops of electron lines, f is the number of hole lines, λ is the topological factor of the time-independent diagram: λ equals the product of the topological factors for single linked parts: λ equals 2 or 1 for every linked part depending on whether a linked part is symmetrical or not under the simultaneous overturning of its every two-body vertex.

[3]. Sum over all the indices i, j, etc. labelling inner electron lines and over the indices $\lambda_1, \ldots, \lambda_p$ and $\lambda_1', \ldots, \lambda_p'$ labelling the external electron lines of the time-independent diagram.

After these preliminaries, i.e. the construction of the diagram technique for evaluation of the matrix element $(\beta_r \ldots \beta_1|\sigma_\alpha(g)|\delta_1 \ldots \delta_r)$ we return to our main adiabatic formula (41).

But before we proceed to that we modify the technique described above of time-independent diagrams in order to facilitate further discussion. Let us consider the so-called extended time-independent diagrams, i.e. the diagrams with additional external electron lines and vertical sections marked as dangerous.

Introducing these additional external electron lines we obtain the graphical interpretation of formula (C4) from Appendix C.

The difference between the extended time-independent diagrams with additional external electron lines and the time-independent diagrams without additional lines described above is that for the extended diagrams we introduce additional external electron lines coming from the infinity on the right and going to the left. For any diagram with $2p$ external electron lines $r - p$ additional lines are introduced; their one-electron indices belong to the open shell.

Now any time-independent diagram has exactly r outgoing and r incoming external electron lines.

Having introduced the vertical sections marked as dangerous on the extended time-independent diagram we obtain diagrams for which the adiabatic limit $\alpha \to +0$ is easier to carry out.

The singularities in matrix elements of the operator $\sigma_\alpha(g)$ in the $\alpha \to +0$ limit produce dangerous denominators (vanishing at $\alpha \to +0$) in some time-independent diagrams. These denominators arise just from such sections which are crossed with only particle lines and those whose indices belong to the open shell. For

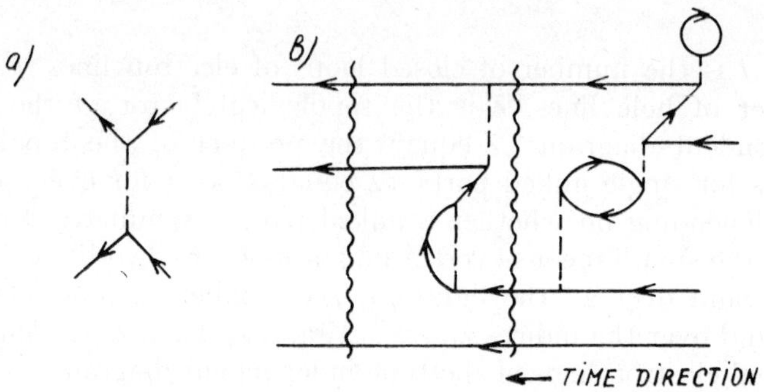

Fig. 4. (*a*) Quartic vertex of the time-independent diagram. (*b*) Typical diagram with external additional lines and with marked dangerous vertical sections.

these vertical sections $\mathscr{E} - p\varepsilon_0 = 0$ in (55) holds. The particular time-independent diagrams may have several dangerous vertical sections. The presence of at least one dangerous vertical section is obligatory for any time-independent diagram. This is the vertical section lying most to the left from all the vertices of the diagram. We indicate the dangerous vertical sections with wavy vertical lines. Only those sections of time-independent diagrams are marked which are crossed only with electron lines and are not crossed with hole lines at all. All vertical sections which cross only external electron lines must be marked as dangerous. In Fig. 4b the approximate outlook of an extended time-independent diagram is shown. Fig. 4a shows the quartic vertex of the extended time-independent diagram.

In order to write down the contribution to the matrix element $(\beta_r \ldots \beta_1|\sigma_\alpha(g)|\delta_1 \ldots \delta_r)$ from the extended time-independent diagram with additional external electron lines and vertical sections marked as dangerous the foregoing rules for the composition of contributions for the usual time-independent diagrams should be used with the following modifications.

[1]. The electron lines which are not crossed by any one of the dangerous vertical sections should be labelled with indices i, j, k, etc. taking arbitrary values. The electron lines which are crossed by dangerous vertical sections should be labelled with μ, ν, κ, etc. indices taking the values only from the open shell. The external electron lines of a diagram are labelled with the indices $\lambda_1, \ldots, \lambda_p$ and $\lambda'_1, \ldots, \lambda'_p$ and the additional external electron lines with the indices

$$\lambda'_{p+1} = \lambda_{p+1}, \ldots, \lambda'_r = \lambda_r$$

[2]. To the outgoing external electron lines $\lambda'_1, \ldots, \lambda'_r$ corresponds a factor

$$I(\beta_2 \ldots \beta_1; \lambda'_1 \ldots \lambda'_r) \tag{57a}$$

and to incoming external electron lines $\lambda_1, \ldots, \lambda_r$ corresponds a factor

$$I(\lambda_r \ldots \lambda_1; \delta_1 \ldots \delta_r) \tag{57b}$$

[3]. To the vertical sections from the energy denominators correspond factors

$$\frac{1}{\mathscr{E} + (n - k)\varepsilon - r\varepsilon_0} \tag{58}$$

where the expressions for $\mathscr{E}$ are set up taking into account the additional external electron lines.

[4]. By setting up the expression for the topological factor one should also take into consideration the factor $(r - p)!$, which arises from the additional external topologically equivalent electron lines of the diagram.

[5]. Summing over the indices labelling the electron lines one should bear in mind the restrictions excluding from the sums all the terms with vanishing denominators for those vertical sections which are not marked as dangerous in the extended time-independent diagram; such terms for which $\mathscr{E} - r\varepsilon_0 = 0$.

By means of this extended time-independent diagram we are able to define a very important operator $M_\alpha(E|g)$ whose matrix elements are $(\beta_r \ldots \beta_1|M_\alpha(E|g)|\delta_1 \ldots \delta_r)$. These are given by the sums of contributions from all kinds of extended time-independent diagrams for $(\beta_r \ldots \beta_1|\sigma_\alpha(g)|\delta_1 \ldots \delta_r)$ of non-zero order in which only the vertical sections furthest to the left are marked as dangerous (these sections are always to be marked as dangerous in any diagram).

For the evaluation of $(\beta_r \ldots \beta_1|M_\alpha(E|g)|\delta_1 \ldots \delta_r)$ both diagrams with the additional external electron lines and with the marked dangerous denominators and the usual time-independent diagrams may be examined.

The contributions from these diagrams for $(\beta_r \ldots \beta_1|M_\alpha(E|g)|\delta_1 \ldots \delta_r)$ are written down similarly to those for $(\beta_r \ldots \beta_1|\sigma_\alpha(g)|\delta_1 \ldots \delta_r)$ taking into consideration the following rules.

[1]. One should drop out the denominator from the vertical section furthest to the left which is equal to $\frac{1}{n\varepsilon}$, where n is the order of the diagram.

[2]. The external energy, E, should be added with a minus sign to all rest energy denominators, thus instead of (58) we obtain

$$\frac{1}{\mathscr{E} + (n-k)\varepsilon - r\varepsilon_0 - E} \tag{59}$$

and instead of Eq. (55) we have

$$\frac{1}{\mathscr{E} + (n-k)\varepsilon - p\varepsilon_0 - E} \tag{60}$$

[3]. The whole diagram should be multiplied by the factor

$$(-1) \tag{61}$$

[4]. In order to sum over one-electron indices of the electron lines of the diagram one should introduce special restrictions which forbid the appearance of vanishing denominators (if any appears).

Let us state the relation between the operators $M_\alpha(E|g)$ and $\sigma_\alpha(g)$. This relation is evident if we employ the extended time-independent diagrams with additional external lines and with marked dangerous denominators. To facilitate the notation we assume that

$$\frac{PS_\alpha(0, -\infty|g)P}{(\Phi_0, S_\alpha(0, -\infty|g)\Phi_0)} = \sigma_\alpha(g) \equiv \sigma = \sum_{l=0}^{+\infty} \sigma_l$$

$$M_\alpha(E|g) \equiv M(E) = \sum_{k=1}^{+\infty} M_k(E)$$

then we may assert that

$$\sigma = I - \sum_{k,l} \frac{1}{(k+l)\varepsilon} M_k(-l\varepsilon)\sigma_l \tag{62}$$

where I is a unity operator; the first factor with ε in the denominator is the contribution from the vertical section furthest to the left ($\varepsilon = \hbar/i\alpha$). In Eq. (62) we understand the product of operators as the product in the degenerate level space E_0. The relation (62) is the most important result of Section 3.

Let us discuss the meaning of this expression. First of all one can easily obtain the following relation (see Appendix D).

$$\frac{1}{r!}\sum_{\delta_1\ldots\delta_r}(\beta_r\ldots\beta_1|M(l\varepsilon;\bar{V}_k)|\delta_1\ldots\delta_r)(\delta_r\ldots\delta_1|\sigma(V_l)|\gamma_1\ldots\gamma_r)$$
$$= \sum_{V_m\in\mathscr{U}(\bar{V}_k,V_l)}(\beta_r\ldots\beta_1|\sigma(V_m)|\gamma_1\ldots\gamma_r) \tag{63}$$

where $\bar{V}_k$ and V_l are extended time-independent diagrams of kth and lth orders for the operators $M(E)$ and σ ($k + l = m$); V_m is the extended time-independent diagram of mth order for the operator σ (or rather for appropriate matrix elements of this operator). The diagrams V_m are taken from the set of diagrams $\mathscr{U}(\bar{V}_k, V_l)$. This set of diagrams is composed in the following way: draw the diagram $\bar{V}_k$ and to the right of it draw the diagram V_l; then outgoing external electron lines of the diagram V_l are linked in some way with incoming external electron lines of the diagram $\bar{V}_k$. Thus we obtain a new diagram of the order $m = k + l$. On this new diagram V_m those vertical sections which have been marked as dangerous on $\bar{V}_k$ and V_l should be marked dangerous as well, and besides the new vertical section between $\bar{V}_k$ and V_l should also be marked as dangerous.

Let us sum in the relation (63) over all possible diagrams $\bar{V}_k$ and V_l of kth and lth orders and then sum over all the $k, l (k \geqslant 1, l \geqslant 1, k + l = m)$; then we obtain

$$\sum_{\substack{k,l \\ k+l=m}} \sum_{\bar{V}_k} \sum_{V_l} \sum_{V_m \in \mathscr{U}(\bar{V}_k, V_l)} \times \cdots \equiv \sum_{V_m} \times \cdots \tag{64}$$

Hence, from (63) we obtain

$$\frac{1}{r!} \sum_{\substack{k,l \\ k+l=m}} (\beta_r \ldots \beta_1 | M_k(l\varepsilon) | \delta_1 \ldots \delta_r)(\delta_r \ldots \delta_1 | \sigma_l | \gamma_1 \ldots \gamma_r)$$
$$= (\beta_r \ldots \beta_1 | \sigma_m | \gamma_1 \ldots \gamma_r) \tag{65}$$

which written in the operator form is

$$\sum_{\substack{k,l \\ k+l=m}} M_k(l\varepsilon)\sigma_l = \sigma_m \tag{66}$$

From (62) we obtain easily, that

$$i\hbar\alpha g \frac{\partial}{\partial g} \sigma = \sum_{k,l} M_k(-l\varepsilon)\sigma_l$$

and therefore

$$T = \Delta E_0 I + T' \tag{67}$$

$$T' = \lim_{\alpha \to +0} \sum_{k,l} M_k(-l\varepsilon)\sigma_l(\sum_l \sigma_l)^{-1} \tag{68}$$

By expanding $M_k(-l\varepsilon)$ in a Taylor series we have

$$M_k(l\varepsilon) = \sum_{p=0}^{+\infty} \frac{(-1)^p}{p!} M_k^{(p)}(0) l^p \varepsilon^p$$

so that

$$T' = \lim_{\alpha \to +0} \sum_{k,l} \sum_{p=0}^{+\infty} \frac{(-1)^p}{p!} M_k^{(p)}(0)\varepsilon^p \left(g \frac{\partial}{\partial g}\right)^p \sigma \,.\, \sigma^{-1}$$

Remembering that

$$l^p \sigma_l = \left(g \frac{\partial}{\partial g}\right)^p \sigma_l$$

we get

$$T' = \lim_{\alpha \to +0} \sum_{p=0}^{+\infty} \sum_k \frac{(-1)^p}{p!} M_k^{(p)}(0)\varepsilon^p \left(g \frac{\partial}{\partial g}\right)^p \sigma \,.\, \sigma^{-1}$$

Let us now examine the adiabatic formula. Asymptotically in the limit $\alpha \rightarrow +0$ we find:

$$T' \cong -\varepsilon \left(g \frac{\partial}{\partial g}\right) \sigma \, . \, \sigma^{-1}$$

so that

$$-\frac{T'}{\varepsilon} \cong \left(g \frac{\partial}{\partial g}\right) \sigma \, . \, \sigma^{-1}$$

We have also

$$-\frac{T'}{\varepsilon} \sigma \cong \left(g \frac{\partial}{\partial g}\right) \sigma$$

and, hence

$$-g \frac{\partial}{\partial g} \frac{T'}{\varepsilon} \, . \, \sigma - \frac{T'}{\varepsilon} g \frac{\partial}{\partial g} \sigma \cong \left(g \frac{\partial}{\partial g}\right)^2 \sigma$$

Leaving only the main asymptotic term in the left-hand side (the second one) we obtain:

$$-\left(\frac{T'}{\varepsilon}\right)^2 \cong \left(g \frac{\partial}{\partial g}\right)^2 \sigma \, . \, \sigma^{-1}$$

etc. In this way we arrive at the asymptotic formula which is correct for any $p \geqslant 1$,

$$\left(-\frac{T'}{\varepsilon}\right)^p \cong \left(g \frac{\partial}{\partial g}\right)^p \sigma \, . \, \sigma^{-1}$$

Using this result we have

$$\begin{aligned} T' &= \lim_{\alpha \rightarrow +0} \sum_{p=0}^{+\infty} \sum_k \frac{(-1)^p}{p!} M_k^{(p)}(0) \varepsilon^p \left(-\frac{T'}{\varepsilon}\right)^p \\ &= \sum_{p=0}^{+\infty} \sum_k \frac{1}{p!} M_k^{(p)}(0) T'^p \end{aligned}$$

So we get finally the following equation for the secular operator

$$T' = \sum_{p=0}^{+\infty} \frac{1}{p!} M^{(p)}(0|g) T'^p \tag{69}$$

where

$$M(E|g) = \lim_{\alpha \rightarrow +0} M_\alpha(E|g) \tag{70}$$

The process of going to the limit $\alpha \to +0$ in (70) is rather straightforward and produces no additional difficulties: one should put $\alpha = 0$ in all the diagrams for $M_\alpha(E|g)$.

Let us return now to the secular equations (34). Let us take some φ_γ and ΔE_γ, which satisfy these equations, i.e.

$$T\varphi_\gamma = \Delta E_\gamma \varphi_\gamma \tag{71}$$

It follows immediately that

$$T'\varphi_\gamma = \Delta E'_\gamma \varphi_\gamma \tag{72}$$

where T' is defined by (68) and besides it is assumed that

$$\Delta E_\gamma = \Delta E_0 + \Delta E'_\gamma \tag{73}$$

According to (69) and (70) we may then write

$$\Delta E'_\gamma \varphi_\gamma = T'\varphi_\gamma = \sum_{p=0}^{+\infty} \frac{1}{p!} M^{(p)}(0|g) T'^p \varphi_\gamma$$

$$= \sum_{p=0}^{+\infty} \frac{1}{p!} M^{(p)}(0|g)(\Delta E'_\gamma)^p \varphi_\gamma = M(\Delta E\gamma|g)\varphi_\gamma$$

so that we conclude that ΔE and φ satisfy the equation

$$\Delta E'_\gamma \varphi_\gamma = M(\Delta E'_\gamma|g)\varphi_\gamma \tag{74}$$

This is the new secular equation arising naturally from the previous one when going to the limit $\alpha \to +0$.

4. THE TRANSFORMATIONS OF NEW SECULAR EQUATIONS WITH THE OPERATOR *M* AND THE FORMULATION OF PROCEDURE FOR THE EVALUATION OF ENERGY SHIFTS

We start with the short summary of final results obtained in Section 3. To evaluate the energy shifts ΔE for any degenerate level one should first write ΔE in the following form:

$$\Delta E = \Delta E_0 + \Delta E' \tag{75}$$

where ΔE_0 is the energy shift of the core found as a sum of contributions from the appropriate diagrams; the rules of setting up for these contributions are given below. The values of $\Delta E'$

should be calculated from the solution of the following equations

$$\Delta E' C_{\beta_1 \ldots \beta_r} = \frac{1}{r!} \sum_{\gamma_1 \ldots \gamma_r} (\beta_r \ldots \beta_1 | M(\Delta E') | \gamma_1 \ldots \gamma_r) C_{\gamma_1 \ldots \gamma_r} \quad (76)$$

where the matrix M is given by the sum of contributions from the particular diagrams; the rules for these are given below.

The diagrams of the first and second order for ΔE_0 are shown in Fig. 5.

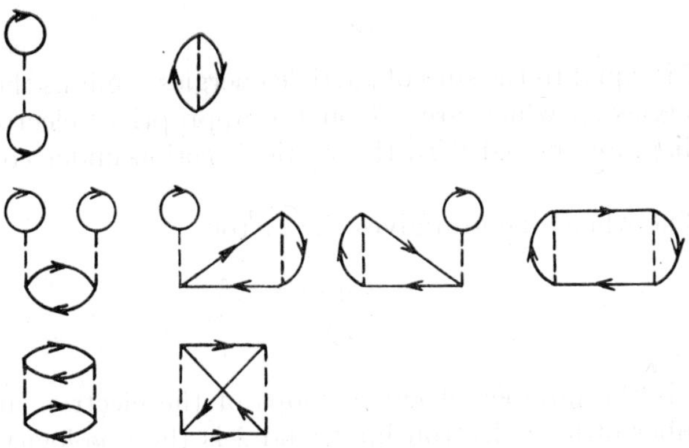

Fig. 5. Time-independent Feynman diagrams for ΔE_0 (there are two first-order diagrams and six second-order diagrams in all).

In the case of ΔE_0 the rules for time-independent Feynman diagrams are the following:

[1]. All electron lines of the diagram should be labelled with one-electron indices i, j, k, etc.: sum over all these indices is understood.

[2]. One should write down the product of factors from vertices, lines, sections and from the diagram itself.

(i) To each quartic vertex (the free end of the ith electron line comes in one of its nodes and the jth end goes out of it; the i'th end comes in the other node and j' goes out of it) corresponds the factor

$$-v_{i'j',ji}$$

with the minus sign.

(ii) The factor $1 - n_i^0$ should be put on every particle line i (going along the time axis); the factor n_i^0 is to be set on every hole line i (going in the opposite direction). The electron line starting and ending at one and the same two-body vertex should be always treated as the hole one.

(iii) Every vertical section between any two neighbouring vertices gives the factor

$$\frac{1}{\mathscr{E}}$$

where $\mathscr{E}$ is equal to the sum of particle energies ε_i minus the sum of hole energies ε_i, which are set on the appropriate electron lines of the diagram crossed with the vertical section under consideration.

(iv) The whole diagram gives the factor

$$-\frac{(-1)^{l+f}}{\lambda}$$

where l is the number of closed loops of the electron lines, f is the number of hole electron lines, and λ is the topological factor of the considered time-ordered diagram.

[3]. One should sum over all the indices i, j, k, etc. labelling the lines of the diagram.

The diagrams of the first and second orders for M are shown in Figs. 6–10. From the third order on there may appear also topologically unlinked diagrams.

[1]. The electron lines of the diagrams as well as their external lines are labelled with one-electron indices inner, lines are labelled with $i, j \ldots$, the incoming free external lines are labelled with $\lambda_1', \ldots \lambda_p'$; the outgoing external lines are labelled with $\lambda_1, \ldots, \lambda_p$. We shall sum over all these indices.

[2]. The following factor corresponds to the whole diagram:

$$\frac{1}{(r-p)!} \sum_{\lambda_{p+1}\ldots\lambda_r} I(\beta_r \ldots \beta_1; \lambda_1' \ldots \lambda_p'\lambda_{p+1} \ldots \lambda_r)$$
$$I(\lambda_r \ldots \lambda_{p+1}\lambda_p \ldots \lambda_1; \gamma_1 \ldots \gamma_r) \quad (77)$$

Here p is the number of incoming or outgoing free lines of the diagram, r is the number of electrons in the open shell. Besides, by definition

$$I(\mu_r' \ldots \mu_1'; \mu_1 \ldots \mu_r) = \sum_P (-1)^P \delta_{\mu_r', \mu_{Pr}} \ldots \delta_{\mu_1', \mu_{P1}} \qquad (78)$$

one should sum over all $r!$ permutations P.

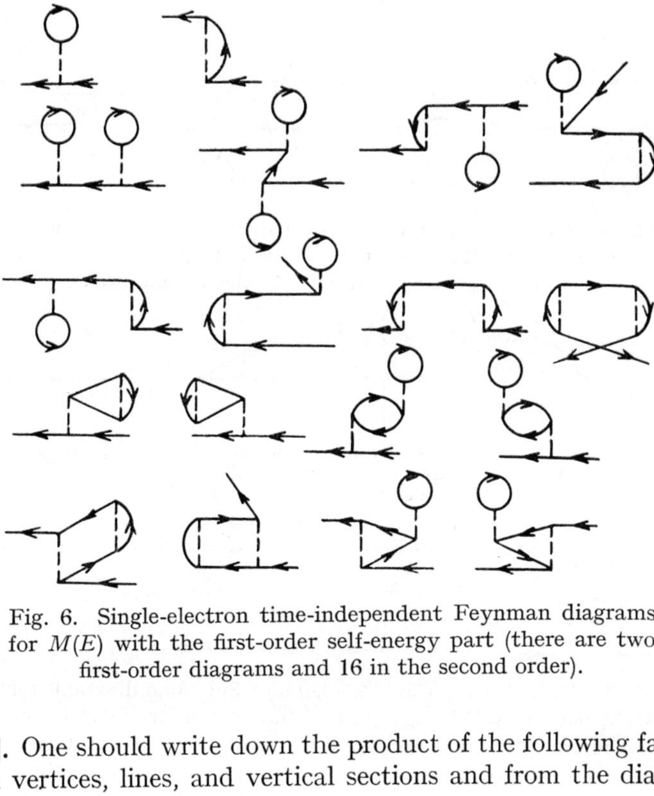

Fig. 6. Single-electron time-independent Feynman diagrams for $M(E)$ with the first-order self-energy part (there are two first-order diagrams and 16 in the second order).

[3]. One should write down the product of the following factors from vertices, lines, and vertical sections and from the diagram itself:

(i) Every quartic vertex with incoming lines i', j' and the appropriate outgoing lines i, j gives the factor

$$-v_{i'j',ji} \qquad (79)$$

(ii) The particle electron line i gives

$$1 - n_i^0 \qquad (80a)$$

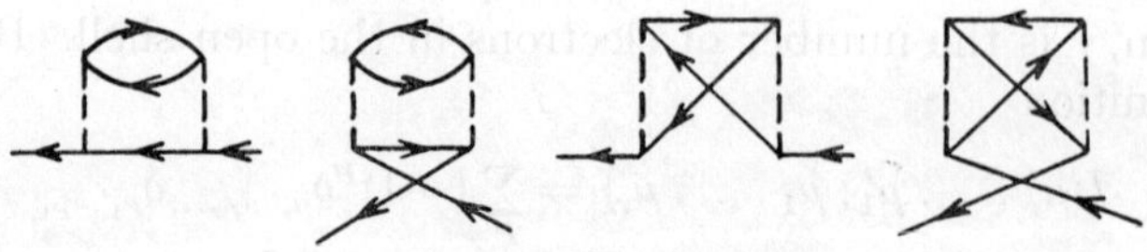

Fig. 7. Single-electron time-independent Feynman diagrams for $M(E)$ without self-energy part of the first order (there is one first-order diagram and four for the second order).

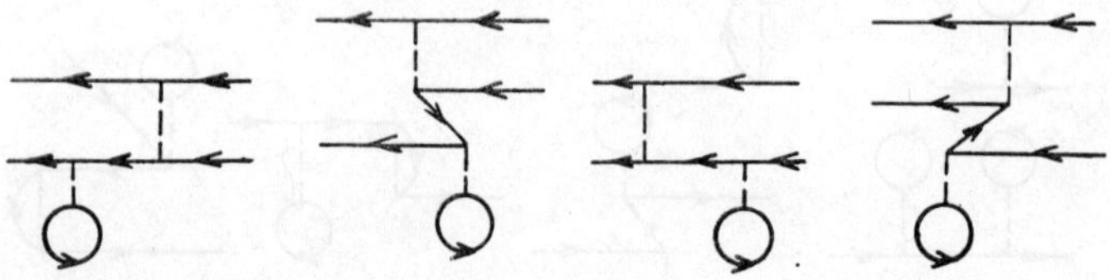

Fig. 8. Two-electron time-independent Feynman diagrams for $M(E)$ with the self-energy part of the first order (there are no first-order diagrams and eight in the second order).

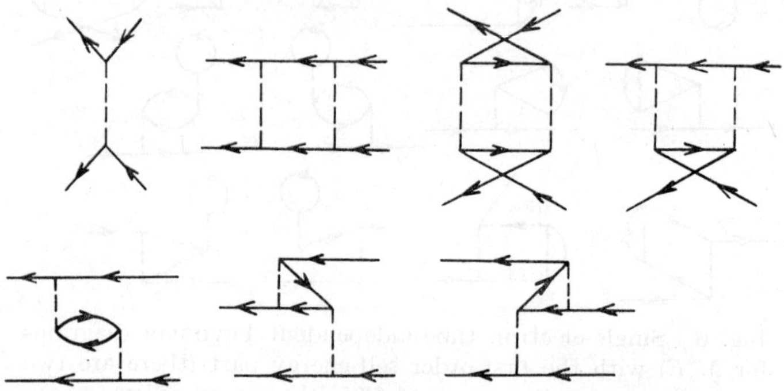

Fig. 9. Two-electron time-independent Feynman diagrams for $M(E)$ without self-energy part of the first order (there is one diagram of the first order and six second-order diagrams).

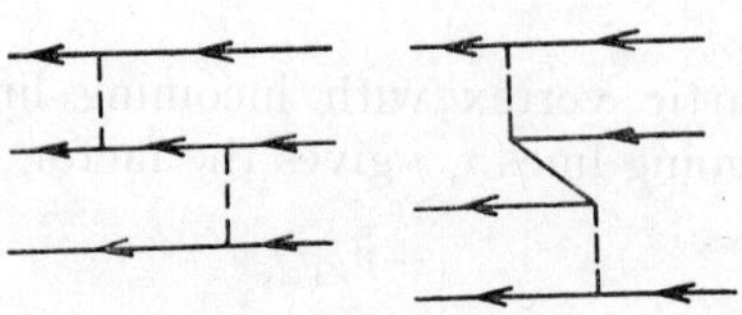

Fig. 10. Three-electron time-independent Feynman diagrams for $M(E)$ (there are no first-order diagrams and two for the second order).

and the hole electron line i

$$n_i^0 \tag{80b}$$

where n_i^0 are the occupation numbers for one-electron state i of the core state. The electron line starting and ending at one and the same vertex should be treated as the hole one.

(iii) Every internal vertical section gives the factor

$$\frac{1}{\mathscr{E} - p\varepsilon_0 - \Delta E} \tag{81}$$

where $\mathscr{E}$ is equal to the sum of particle energies, crossed with the section, minus the sum of hole energies crossed with it, p is the number of incoming (or outgoing) external electron lines of the diagram, ε_0 is the one-electron energy for electrons of the open shell.

(iv) The whole diagram gives the factor

$$-(-1)^{l+f} \tag{82}$$

where l is the number of closed loops of the electron lines on the diagram, f is the number of hole lines.

(v) The factor

$$\frac{1}{\lambda} \tag{83}$$

where λ is the number of topological automorphisms of the diagram.

[4]. One should sum over all the indices of the lines $i, j, \ldots$ and $\lambda_1, \ldots, \lambda_p; \lambda_1', \ldots \lambda_p'$ and drop all the terms with zero denominators which might appear.

In Section 3 we used in the theoretical discussion the basis of single determinantal states (see the beginning of Section 3, formulae (38) to (48)) because of its simplicity. But for practical applications it is more convenient to use another basis and therefore we must examine the secular equations for any arbitrary basis.

The space of the degenerate level is finite-dimensional. Let us consider some arbitrary (orthonormalized) basis in this space, we shall denote single states of this basis with the index K which runs over a finite number of values.

For these basis states we may write down the following formulae

$$\varphi = \frac{1}{r!} \sum_{\gamma_1 \ldots \gamma_r} C^{K}_{\gamma_1 \ldots \gamma_r} \overset{+}{a}_{\gamma_1} \ldots \overset{+}{a}_{\gamma_r} \Phi_0 \tag{84}$$

where the coefficients are in general complex and besides the restriction of antisymmetry over the indices $\gamma_1, \ldots, \gamma_r$ they also obey orthogonality and normalization conditions given below:

$$\frac{1}{r!} \sum_{\gamma_1 \ldots \gamma_r} \overset{*}{C}{}^{K'}_{\gamma_1 \ldots \gamma_r} C^{K}_{\gamma_1 \ldots \gamma_r} = \delta_{K',K} \tag{85a}$$

$$\sum_{K} \overset{*}{C}{}^{K}_{\beta_1' \ldots \beta_r'} C^{K}_{\beta_1 \ldots \beta_r} = I(\beta_r' \ldots \beta_1'; \beta_1 \ldots \beta_r) \tag{85b}$$

Our secular equations (76) in the new basis (84) may be written as

$$\Delta E' C(K') = \sum_{K} M(K', K | \Delta E') C(K') \tag{86}$$

where $C(K')$ are unknown and where

$$M(K', K | \Delta E) = \frac{1}{(r!)^2} \sum_{\substack{\beta_1 \ldots \beta_r \\ \gamma_1 \ldots \gamma_r}} \overset{*}{C}{}^{K'}_{\beta_1 \ldots \beta_r} C^{K}_{\gamma_1 \ldots \gamma_r} (\beta_r \ldots \beta_1 | M(\Delta E') | \gamma_1 \ldots \gamma_r) \tag{87}$$

$$C_{\beta_1 \ldots \beta_r} = \frac{1}{r!} \sum_{K'} C(K') C^{K'}_{\beta_1 \ldots \beta_r} \tag{88}$$

The diagrams for matrix elements $M(K, K' | \Delta E)$ are the same as those for matrix elements $(\beta_r \ldots \beta_1 | M(\Delta E') | \gamma_1 \ldots \gamma_r)$; the rules for setting up the contributions [1]–[4] differ only by the point [2]. Now [2] should be read as follows (see (87)):

[2]. To the whole graph belongs the factor

$$\frac{1}{(r-p)!} \sum_{\lambda_{p+1} \ldots \lambda_r} \overset{*}{C}{}^{K'}_{\lambda_1' \ldots \lambda_p' \lambda_{p+1} \ldots \lambda_r} C^{K}_{\lambda_1 \ldots \lambda_p \lambda_{p+1} \ldots \lambda_r} \tag{89}$$

where p is the number of incoming or outgoing external electron lines of the diagram, r is the number of electrons in the open shell; we consider only those diagrams for which $p \leqslant r$.

Let us turn now to the problem of the reduction of contributions from the diagrams. Consider any diagram D with $2p$ external electron lines, precisely such a diagram with $p \neq r$ and

$p < r$ (strictly less). The contribution from this diagram to the matrix element $M(K, K'|\Delta E)$ is denoted as

$$M_D(K'_r, K_r|\Delta E)$$

where index K is labelled with the subsidiary subscript r in order to show that K_r labels basis states arising at the presence of r electrons in the considered open shell; from now on we shall examine the basis states arising from the presence of p electrons in the open shell; these states we shall denote as K_p. Write down the contribution from the diagram described above D when there are exactly p electrons in the open shell; this contribution is called

$$M_D(K'_p, K_p|\Delta E')$$

where the index K_p is used.

Thus one may state that the contribution $M_D(K_r, K'_r|\Delta E')$ is simply reduced to the contribution $M_D(K_p, K'_p|\Delta E')$. In fact, the contribution $M_D(K_r, K'_r|\Delta E')$ is accompanied with the factor (89). The summation is carried over the indices $\lambda'_1, \ldots, \lambda'_p$ and $\lambda_1, \ldots, \lambda_p$ on which this factor depends and which label incoming and outgoing external electron lines of the diagram D.

Therefore, we may write

$$M_D(K'_r, K_r|\Delta E') = \sum_{\substack{\lambda_1' \ldots \lambda_p' \\ \lambda_1 \ldots \lambda_p}} C(\lambda'_p \ldots \lambda'_1; \lambda_1 \ldots \lambda_p) \times A(\lambda'_p \ldots \lambda'_1; \lambda_1 \ldots \lambda_p)$$

where C is the factor (89) which is an antisymmetric function of both $\lambda'_1, \ldots, \lambda'_p$ and $\lambda_1, \ldots, \lambda_p$; A is a function which in general is not antisymmetric. We can now write

$$\begin{aligned} M_D(K'_r, K_r|\Delta E') &= \sum_{\substack{\mu_1' \ldots \mu_p' \\ \mu_1 \ldots \mu_p}} \sum_{\substack{\lambda_1' \ldots \lambda_p' \\ \lambda_1 \ldots \lambda_p}} C(\lambda'_p \ldots \lambda'_1; \lambda_1 \ldots \lambda_p) A(\mu'_p \ldots \mu'_1; \mu_1 \ldots \mu_p) \\ &\quad \times \delta_{\mu_p', \lambda_p'} \ldots \delta_{\mu_1', \lambda_1'} \delta_{\mu_1, \lambda_1} \ldots \delta_{\mu_p, \lambda_p} \\ &= \frac{1}{(p!)^2} \sum_{\substack{\mu_1' \ldots \mu_p' \\ \mu_1 \ldots \mu_p}} \sum_{\substack{\lambda_1' \ldots \lambda_p' \\ \lambda_1 \ldots \lambda_p}} C(\lambda'_p \ldots \lambda'_1; \lambda_1 \ldots \lambda_p) A(\mu'_p \ldots \mu'_1; \mu_1 \ldots \mu_p) \\ &\quad I(\mu'_p \ldots \mu'_1; \lambda'_1 \ldots \lambda'_p) I(\mu_p \ldots \mu_1; \lambda_1 \ldots \lambda_p) \end{aligned}$$

where we have used the symmetry properties of the function C. Using the orthogonality condition (85b) twice, we obtain

$$M_D(K_r', K_r|\Delta E') = \frac{1}{(p!)^2} \sum_{\substack{\lambda_1'\ldots\lambda_p' \\ \lambda_1\ldots\lambda_p}} C(\lambda_p' \ldots \lambda_1'; \lambda_1 \ldots \lambda_p)$$

$$\times \sum_{K_p', K_p} C^{K_p'}_{\lambda_1'\ldots\lambda_p'} \overset{*}{C}{}^{K_p'}_{\mu_1'\ldots\mu_p'} C^{K_p}_{\mu_1\ldots\mu_p} \overset{*}{C}{}^{K_p}_{\lambda_1\ldots\lambda_p} A(\mu_p' \ldots \mu_1'; \mu_1 \ldots \mu_p)$$

from this expression the contribution $M_D(K_p', K_p|\Delta E')$ can be easily singled out (see (89)) and we finally get a very useful reduction formula

$$M_D(K_r', K_r|\Delta E') = \sum_{K_p', K_p} g(K_r', K_p'; K_p, K_r) M_D(K_p', K_p|\Delta E') \tag{90}$$

where

$$g(K_r', K_p'; K_p, K_r) = \frac{1}{(p!)^2(r-p)!} \sum_{\substack{\lambda_1'\ldots\lambda_p' \\ \lambda_1\ldots\lambda_p}} \sum_{\lambda_{p+1}\ldots\lambda_r} \overset{*}{C}{}^{K_r'}_{\lambda_1'\ldots\lambda_p'\lambda_{p+1}\ldots\lambda_r}$$

$$\times C^{K_p'}_{\lambda_1'\ldots\lambda_p'} \overset{*}{C}{}^{K_p}_{\lambda_1\ldots\lambda_p} C^{K_p}_{\lambda_1\ldots\lambda_p\lambda_{p+1}\ldots\lambda_z} \tag{91}$$

The quantities $g(K_r', K_p'; K_p, K_n)$ appearing in the reduction formula do not depend upon the interaction and have no connection with the perturbation theory; they may be evaluated once and for all. The reduction formula shows that actually we have to consider only those diagrams where $p = r$ (the number of external electron lines equals the number of electrons in the open shell).

In conclusion we emphasize a very important property of the quantities $g(K_r', K_p'; K_p, K_r)$ introduced above. This property enables us to reduce the problem of their evaluation according to (91) to the problem of the calculation with the same formula of only the terms of the type $g(K_{p+1}', K_p'; K_p, K_{p+1})$ corresponding to the case of one added electron.

This property is given by the following simple formula, which is proved below.

$$g(K_r', K_p'; K_p, K_r) = \frac{(r-q)!(q-p)!}{(r-p)!} \sum_{K_q', K_q} g(K_r', K_q'; K_q, K_r) g(K_q', K_p'; K_p, K_q) \tag{92}$$

The proof is rather easy. In fact according to (91) we get for the right-hand side of the formula (92)

$$\frac{1}{(q!)^2(p!)^2(r-p)!} \sum_{K_q' K_q} \sum_{\substack{\lambda_1' \ldots \lambda_p' \lambda_p'+1 \ldots \lambda_q' \\ \lambda_1 \ldots \lambda_p \lambda_{p+1} \ldots \lambda_q}} \sum_{\lambda_{q+1} \ldots \lambda_r} \sum_{\substack{\mu_1' \ldots \mu_p' \\ \mu_1 \ldots \mu_p}} \sum_{\mu_{p+1} \ldots \mu_q}$$

$$\times \overset{*}{C}{}^{K_r'}_{\lambda_1' \ldots \lambda'_p \lambda'_{p+1} \ldots \lambda_q' \lambda_{p+1} \ldots \lambda_z} C^{K_q'}_{\lambda_1' \ldots \lambda_p' \lambda_{p+1}' \ldots \lambda_q'} \overset{*}{C}{}^{K_q'}_{\mu_1' \ldots \mu_p' \mu_{p+1} \ldots \mu_q} C^{K_p'}_{\mu_1' \ldots \mu_p}$$

$$\times \overset{*}{C}{}^{K_p}_{\mu_1 \ldots \mu_p} C^{K_q}_{\mu_1 \ldots \mu_p \mu_{p+1} \ldots \mu_q} \overset{*}{C}{}^{K_q}_{\lambda_1 \ldots \lambda_p \lambda_{p+1} \ldots \lambda_q} C^{K_r}_{\lambda_1 \ldots \lambda_p \lambda_{p+1} \ldots \lambda_q \lambda_{q+1} \ldots \lambda_r}$$

using (85b) twice, we may explicitly sum over K_q', K_q, because

$$\sum_{K_q'} C^{K_q'}_{\lambda_1' \ldots \lambda_p' \lambda_p'+1 \ldots \lambda_q'} \overset{*}{C}{}^{K_q'}_{\mu_1' \ldots \mu_p' \mu_{p+1} \ldots \mu_q}$$
$$= I(\lambda_q' \ . \ . \ . \ \lambda'_{p+1} \lambda_p' \ . \ . \ . \ \lambda_1'; \mu_1' \ . \ . \ . \ \mu_p' \mu_{p+1} \ . \ . \ . \ \mu_q)$$

$$\sum_{K_q} C^{K_q}_{\mu_1 \ldots \mu_p \mu_{p+1} \ldots \mu_q} \overset{*}{C}{}^{K_q}_{\lambda_1 \ldots \lambda_p \lambda_{p+1} \ldots \lambda}$$
$$= I(\mu_q \ . \ . \ . \ \mu_{p+1} \mu_p \ . \ . \ . \ \mu_1; \lambda_1 \ . \ . \ . \ \lambda_p \lambda_{p+1} \ . \ . \ . \ \lambda_q)$$

Using then the antisymmetry properties of the former factor C in the former line of the formula written as well as of the latter factor C in the latter line of it, we may substitute the functions I appropriately by the following products of δ symbols (simultaneously dropping out the factor $1/(q!)^2$)

$$\delta_{\lambda'_q, \mu_q} \ . \ . \ . \ \delta_{\lambda'_{p+1}, \mu_{p+1}} \delta_{\lambda'_p, \mu_p'} \ . \ . \ . \ \delta_{\lambda'_1, \mu'_1}$$

$$\delta_{\mu_q, \lambda_q} \ . \ . \ . \ \delta_{\mu_{p+1}, \lambda_{p+1}} \delta_{\mu_p, \lambda_p} \ . \ . \ . \ \delta_{\mu_1, \lambda_1}$$

These products of δ symbols enable us to remove the summation over $\mu_1', \ldots, \mu_p'$, over $\mu_1, \ldots, \mu_p$ and over $\mu_{p+1}, \ldots, \mu_q$; this yields the following result for the right-hand side of the formula (92)

$$\frac{1}{(p!)(r-p)!} \sum_{\substack{\lambda_1' \ldots \lambda_p' \lambda_p'+1 \ldots \lambda_q' \\ \lambda_1 \ldots \lambda_p \lambda_{p+1} \ldots \lambda_q}} \sum_{\lambda_{q+1} \ldots \lambda_r} \delta_{\lambda'_q, \lambda_q} \ . \ . \ . \ \delta_{\lambda'_{p+1}, \lambda_{p+1}}$$

$$\times \overset{*}{C}{}^{K_r'}_{\lambda_1' \ldots \lambda_p' \lambda_p'+1 \ldots \lambda_q' \lambda_{q+1} \ldots \lambda_r} C^{K_p'}_{\mu_1' \ldots \mu_p'} \overset{*}{C}{}^{K_p}_{\mu_1 \ldots \mu_p} C^{K_r}_{\lambda_1 \ldots \lambda_p \lambda_{p+1} \ldots \lambda_q \lambda_{q+1} \ldots \lambda_r}$$

Comparing this expression with (91) we find that it is identical to the left-hand side of the formula (92). Thus the validity of (92) is proved.

5. CONCLUSIONS

The theory developed in this work may be applied for the evaluation of excitation energies of states in many-electron atoms and molecules. There are reasons to believe that rather satisfactory results may be obtained even in the first and second orders of the perturbation theory at least for systems of not very many electrons.

Particular emphasis should be placed upon the very important property of compensation of large contributions in the foregoing formulae. We have always dealt not with the energy shifts ΔE_γ, but with the energy shifts $\Delta E'_\gamma$ reckoned from the core energy shift ΔE_0.

$$\Delta E_\gamma = \Delta E_0 + \Delta E'_\gamma$$

Evidently, in many-electron molecules and atoms the main contribution is given by the electron core $\Delta E_0 \gg \Delta E_\gamma$ (if there are one, two, three, or four electrons in the open shell, and the total number of electrons in the molecule or the atom is several tens). The more complicated the molecule, or the atom, the more exactly the inequality $\Delta E_0 \gg \Delta E'_\gamma$ holds.

If we are concerned not with the absolute values of energies of excited states but with more useful excitation energies (differences of these absolute values) the large contribution from ΔE_0 is completely dropped out

$$\Delta E_\gamma - \Delta E_{\gamma 0} \equiv \Delta E'_\gamma - \Delta E'_{\gamma 0}$$

Having calculated the values of $\Delta E'_\gamma$ (not ΔE_γ) we can immediately evaluate the energy of excitation without a doubtful procedure of calculation of absolute energies and then take the differences. Practically always this is the method used in other calculations of excitation energies for molecules and atoms.

APPENDIX A

The Derivation of the Formulae of Ordinary Perturbation Theory from the Adiabatic Formula for the Secular Operator

We shall derive the formulae for energy shifts ΔE_γ in the third order of ordinary perturbation theory from the adiabatic formula (24) and the secular equations (4).

We expand $PS_\alpha(0, -\infty|g)P$ in series:

$$PS_\alpha(0, -\infty|g)P = \sum_{n=0}^{+\infty} g^n S^{(n)} = 1 + gS^{(1)} + g^2S^{(2)} + g^3S^{(3)} + \dots \tag{A1}$$

According to (30) and (28) the terms of this expansion are:

$$\begin{aligned} S^{(1)} &= P \frac{1}{E_0 - H_0 + i\hbar\alpha} H_{\text{int}} P \\ S^{(2)} &= P \frac{1}{E_0 - H_0 + 2i\hbar\alpha} H_{\text{int}} \frac{1}{E_0 - H_0 + i\hbar\alpha} H_{\text{int}} P \\ S^{(3)} &= P \frac{1}{E_0 - H_0 + 3i\hbar\alpha} H_{\text{int}} \frac{1}{E_0 - H_0 + 2i\hbar\alpha} \\ &\qquad \times H_{\text{int}} \frac{1}{E_0 - H_0 + i\hbar\alpha} H_{\text{int}} P \end{aligned} \tag{A2}$$

from which we get for matrix elements

$$\begin{aligned} S^{(1)}_{\beta\gamma} &= \frac{1}{i\hbar\alpha} V_{\beta\gamma} \\ S^{(2)}_{\beta\gamma} &= \frac{1}{2i\hbar\alpha} \sum_l \frac{V_{\beta l} V_{l\gamma}}{E_0 - E_l + i\hbar\alpha} \\ S^{(3)}_{\beta\gamma} &= \frac{1}{3i\hbar\alpha} \sum_{l,s} \frac{V_{\beta l} V_{ls} V_{s\gamma}}{(E_0 - E_l + 2i\hbar\alpha)(E_0 - E_s + i\hbar\alpha)} \end{aligned} \tag{A3}$$

The subscripts l,s in these formulae can take on also the values from the space of the degenerate level E_0, the subscripts β, γ take on the values only from this space.

We shall write the secular operator T according to the adiabatic formula (24) to the third order of perturbation theory:

$$T = \sum_{n=1}^{+\infty} g^n T^{(n)} = gT^{(1)} + g^2T^{(2)} + g^3T^{(3)} + \dots \tag{A4}$$

from which we get

$$T^{(1)}_{\beta\gamma} = \lim_{\alpha\to+0} i\hbar\alpha S^{(1)}_{\beta\gamma}$$

$$T^{(2)}_{\beta\gamma} = \lim_{\alpha\to+0} i\hbar\alpha\{2S^{(2)}_{\beta\gamma} - \sum_{\lambda} S^{(1)}_{\beta\lambda}S^{(1)}_{\lambda\gamma}\}$$

$$T^{(3)}_{\beta\gamma} = \lim_{\alpha\to+0} i\hbar\alpha\{3S^{(3)}_{\beta\gamma} - 2\sum_{\sigma} S^{(2)}_{\beta\sigma}S^{(1)}_{\sigma\gamma} - \sum_{\lambda} S^{(1)}_{\beta\lambda}S^{(2)}_{\lambda\gamma} + \sum_{\lambda,\sigma} S^{(1)}_{\beta\lambda}S^{(1)}_{\lambda\sigma}S^{(1)}_{\sigma\gamma}\} \tag{A5}$$

where the Greek subscripts β, γ, λ, σ, τ label states of the degenerate level E_0.

In the first order, according to (A5) and (A3) we have

$$T^{(1)}_{\beta\lambda} = V_{\beta\gamma} \tag{A6}$$

In the second order, according to (A3) we obtain from (A5) for the expression in braces

$$\{\ldots\} = \frac{1}{(i\hbar\alpha)^2}\sum_{\lambda} V_{\beta\lambda}V_{\lambda\gamma} + \frac{1}{i\hbar\alpha}\sum_{l,E_l\neq E_0}\frac{V_{\beta l}V_{l\gamma}}{E_0 - E_l} + \ldots$$

$$- \frac{1}{(i\hbar\alpha)^2}\sum_{\lambda} V_{\beta\lambda}V_{\lambda\gamma} = \frac{1}{i\hbar\alpha}\sum_{l,E_l\neq E_0}\frac{V_{\beta l}V_{l\gamma}}{E_0 - E_l}$$

The virtual terms, proportional to $1/\alpha^2$, are cancelled out. Hence,

$$T^{(2)}_{\beta\gamma} = \sum_{l,E_l\neq E_0}\frac{V_{\beta l}V_{l\gamma}}{E_0 - E_l} \tag{A7}$$

In the third order using (A3) in order to modify the expression in braces in (A5) we get

$$\{\ldots\} = \frac{1}{i\hbar\alpha}\sum_{l,s}\frac{V_{\beta l}V_{ls}V_{s\gamma}}{(E_0 - E_l + 2i\hbar\alpha)(E_0 - E_s + i\hbar\alpha)}$$

$$-\frac{1}{(i\hbar\alpha)^2}\sum_{\sigma,l}\frac{V_{\beta l}V_{l\sigma}V_{\sigma\gamma}}{E_0 - E_l + i\hbar\alpha} - \frac{1}{2(i\hbar\alpha)^2}\sum_{\lambda,s}\frac{V_{\beta\lambda}V_{\lambda s}V_{s\gamma}}{E_0 - E_s + i\hbar\alpha}$$

$$+ \frac{1}{(i\hbar\alpha)^3}\sum_{\lambda,\sigma} V_{\beta\lambda}V_{\lambda\sigma}V_{\sigma\gamma}$$

For small α there are virtual terms in this expression which are proportional to $1/\alpha^3$, $1/\alpha^2$ and which are cancelled out (the sum of terms proportional to $1/\alpha^3$ as well as the sum of terms proportional to $1/\alpha^2$ is equal to zero). Thus only the terms proportional to $1/\alpha$ are left and we obtain

$$T^{(3)}_{\beta\lambda} = \sum_{\substack{l,s \\ E_l \neq E_0, E_s \neq E_0}} \frac{V_{\beta l} V_{ls} V_{s\gamma}}{(E_0 - E_l)(E_0 - E_s)} - \sum_{\substack{l,\sigma \\ E_l \neq E_0}} \frac{V_{\beta l} V_{l\sigma} V_{\sigma\gamma}}{(E_0 - E_l)^2} \tag{A8}$$

It is important to emphasize that there are no terms with V in the denominator in the expressions for matrix elements of the secular operator as there are in the ΔE_γ expansion and therefore we have not applied the perturbation theory for the energy shifts ΔE_γ.

Let us now solve the secular equations to the third order and derive the formulae of the ordinary perturbation theory for ΔE^{1-3} in the case *when even the first order of the perturbation theory completely removes the generacy.*

We shall choose the right linear combinations as the basic states β, γ etc. . . ., then

$$V_{\beta\gamma} = V_{\beta\beta}\delta_{\beta\gamma} \tag{A9}$$

Using (A9) we rewrite the formulae for $T_{\beta\gamma}{}^{(1-3)}$ in the following form

$$T^{(1)}_{\beta\gamma} = V_{\beta\beta}\delta_{\beta\gamma}$$

$$T^{(2)}_{\beta\gamma} = \sum_{l, E_l \neq E_0} \frac{V_{\beta l} V_{l\gamma}}{E_0 - E_l}$$

$$T^{(3)}_{\beta\gamma} = \sum_{\substack{l,s \\ E_l \neq E_0, E_s \neq E_0}} \frac{V_{\beta l} V_{ls} V_{s\gamma}}{(E_0 - E_l)(E_0 - E_s)} - \sum_{E_l \neq E_0} \frac{V_{\beta l} V_{l\gamma} V_{\gamma\gamma}}{(E_0 - E_l)^2} \tag{A10}$$

The secular equations with T operator have the form

$$\sum_{\gamma} T_{\beta\gamma} C_{\gamma\lambda} = \Delta E_\lambda C_{\beta\lambda} \tag{A11}$$

or approximately

$$V_{\beta\beta} C_{\beta\lambda} + \sum_{\gamma} (T^{(2)}_{\beta\gamma} + T^{(3)}_{\beta\gamma} + \ldots) C_{\gamma\lambda} = \Delta E_\lambda C_{\beta\lambda} \tag{A12}$$

We find the solution of secular equations (A11) in the form of expansion in powers of the small parameters of weak interaction.

$$C_{\beta\lambda} = C^{(0)}_{\beta\lambda} + C^{(1)}_{\beta\lambda} + C^{(2)}_{\beta\lambda} + \ldots \tag{A13}$$

$$\Delta E_\lambda = \Delta E^{(1)}_\lambda + \Delta E^{(2)}_\lambda + \ldots \tag{A14}$$

Substituting (A13), (A14) into (A12) and putting separately equal to zero the terms of zeroth, first, and second order we obtain a set of equations for the zeroth, first, and second approximation, respectively.

The set of equations for the zeroth approximation has a very simple form

$$V_{\beta\beta}C^{(0)}_{\beta\lambda} = \Delta E^{(1)}_{\lambda}C^{(0)}_{\beta\lambda} \tag{A15}$$

from which follows immediately:

$$\Delta E^{(1)}_{\lambda} = V_{\lambda\lambda}$$

$$C^{(0)}_{\beta\lambda} = C_0\delta_{\beta\lambda} \tag{A16}$$

The set for the first order is given by

$$V_{\beta\beta}C^{(1)}_{\beta\lambda} + \sum_{\gamma} T^{(2)}_{\beta\gamma}C^{(0)}_{\gamma\lambda} = \Delta E^{(2)}_{\lambda}C^{(0)}_{\beta\lambda} + \Delta E^{(1)}_{\lambda}C^{(1)}_{\beta\lambda}$$

or

$$(V_{\beta\beta} - V_{\lambda\lambda})C^{(1)}_{\beta\lambda} = \Delta E^{(2)}_{\lambda}C_0\delta_{\beta\lambda} - C_0T^{(2)}_{\beta\lambda}$$

From which at $\beta = \gamma$ we have

$$0 = \Delta E^{(2)}_{\lambda}C_0 - C_0T^{(2)}_{\lambda\lambda}$$

$$\Delta E^{(2)}_{\lambda} = T^{(2)}_{\lambda\lambda} \tag{A17}$$

and at $\beta \neq \gamma$

$$(V_{\beta\beta} - V_{\lambda\lambda})C^{(1)}_{\beta\lambda} = -C_0T^{(2)}_{\beta\lambda}$$

$$C^{(1)}_{\beta\lambda} = C_1\delta_{\beta\lambda} - C_0\frac{T^{(2)}_{\beta\lambda}}{V_{\beta\beta} - V_{\lambda\lambda}}(1 - \delta_{\beta\lambda}) \tag{A18}$$

Substituting the expression (A10) for $T^{(2)}_{\beta\gamma}$ into (A17) we finally get the following expression for the second-order correction

$$\Delta E^{(2)}_{\lambda} = \sum_{l,E_l\neq E_0}\frac{V_{\lambda l}V_{l\lambda}}{E_0 - E_l} \tag{A19}$$

The set of equations for the second order is given by

$$V_{\beta\beta}C^{(2)}_{\beta\lambda} + \sum_{\gamma} T^{(3)}_{\beta\gamma}C^{(0)}_{\gamma\lambda} + \sum_{\gamma} T^{(2)}_{\beta\gamma}C^{(1)}_{\gamma\lambda}$$
$$= \Delta E^{(3)}_{\lambda}C^{(0)}_{\beta\lambda} + \Delta E^{(2)}_{\lambda}C^{(1)}_{\beta\lambda} + \Delta E^{(1)}_{\lambda}C^{(2)}_{\beta\lambda}$$

or

$$(V_{\beta\beta} - V_{\lambda\lambda})C_{\beta\lambda}^{(2)} = \Delta E_{\lambda}^{(3)} C_0 \delta_{\beta\lambda} + C_1 \Delta E_{\lambda}^{(2)} \delta_{\beta\lambda}$$

$$-C_0 \Delta E_{\lambda}^{(2)} \frac{T_{\beta\lambda}^{(2)}}{V_{\beta\beta} - V_{\lambda\lambda}} (1 - \delta_{\beta\lambda}) - T_{\beta\lambda}^{(3)} C_0 - T_{\beta\lambda}^{(2)} C_1$$

$$+ C_0 \sum_{\gamma \neq \lambda} \frac{T_{\beta\gamma}^{(2)} T_{\gamma\lambda}^{(2)}}{V_{\gamma\gamma} - V_{\lambda\lambda}}$$

From which at $\beta = \gamma$ we have

$$0 = \Delta E_{\lambda}^{(3)} C_0 + C_1 \Delta E_{\lambda}^{(2)} - T_{\lambda\lambda}^{(3)} C_0 - C_1 T_{\lambda\lambda}^{(2)} + C_0 \sum_{\gamma \neq \lambda} \frac{T_{\lambda\gamma}^{(2)} T_{\gamma\lambda}^{(2)}}{V_{\gamma\gamma} - V^{\lambda\lambda}}$$

$$E\Delta_{\lambda}^{(3)} = T_{\lambda\lambda}^{(3)} + \sum_{\gamma \neq \lambda} \frac{T_{\lambda\gamma}^{(2)} T_{\gamma\lambda}^{(2)}}{V_{\lambda\lambda} - V_{\gamma\gamma}} \tag{A20}$$

and at $\beta \neq \gamma$

$$C_{\beta\lambda}^{(2)} = \left(C_0 \frac{T_{\beta\lambda}^{(3)}}{V_{\lambda\lambda} - V_{\beta\beta}} + C_1 \frac{T_{\beta\lambda}^{(2)}}{V_{\lambda\lambda} - V_{\beta\beta}} + C_0 \frac{1}{V_{\lambda\lambda} - V_{\beta\beta}} \right.$$

$$\left. \sum_{\gamma \neq \lambda} \frac{T_{\beta\gamma}^{(2)} T_{\gamma\lambda}^{(2)}}{V_{\lambda\lambda} - V_{\gamma\gamma}} - C_0 \frac{T_{\lambda\lambda}^{(2)} T_{\beta\lambda}^{(2)}}{(V_{\lambda\lambda} - V_{\beta\beta})^{-2}} \right) (1 - \delta_{\beta\lambda}) + C_2 \delta_{\beta\lambda} \tag{A21}$$

Using the expressions (A10) for $T^{(3)}$ and $T^{(2)}$ we get from (A20):

$$\Delta E_{\lambda}^{(3)} = \sum_{\substack{l,s \\ E_l \neq E_0, E_s \neq E_0}} \frac{V_{\lambda l} V_{ls} V_{s\lambda}}{(E_0 - E_l)(E_0 - E_s)} - V_{\lambda\lambda} \sum_{\substack{l \\ E_l \neq E_0}} \frac{V_{\lambda l} V_{l\lambda}}{(E_0 - E_l)^2}$$

$$+ \sum_{\substack{\gamma \neq \lambda, l, t \\ E_l \neq E_0, E_t \neq E_0}} \frac{V_{\lambda l} V_{l\gamma} V_{\gamma t} V_{t\lambda}}{(E_0 - E_l)(E_0 - E_t)(V_{\lambda\lambda} - V_{\gamma\gamma})}$$

APPENDIX B

The Basic System of Single-Determinantal States of the Degenerated Level E_0

Using the basis (38) one should bear in mind that every single state of this system occurs $r!$ times in the notation (38) and besides, there are many states equal to zero (for those indices $\beta_1, \ldots, \beta_r$ which are equal to each other).

In spite of these facts the basis (38) is very convenient. Any state φ belonging to the space of the degenerate level E_0 can be uniquely represented in the following form:

$$\varphi = \frac{1}{r!} \sum_{\gamma_1 \ldots \gamma_z} C_{\gamma_1 \ldots \gamma_r} \overset{+}{a}_{\gamma_1} \ldots \overset{+}{a}_{\gamma_r} \Phi_0 \tag{B1}$$

where the coefficients $C_{\gamma_1 \ldots \gamma_r}$ are antisymmetric in permutations of their indices. Consider the matrix elements of an arbitrary operator A in the basis (38):

$$(\beta_r \ldots \beta_1 | A | \gamma_1 \ldots \gamma_r) = (\Phi_0, a_{\beta_r} \ldots a_{\beta_1} A \overset{+}{a}_{\gamma_1} \ldots a_{\gamma_r} \Phi_0) \tag{B2}$$

These matrix elements are antisymmetric functions both in the indices $\beta_r, \ldots, \beta_1$, and in the indices $\gamma_1, \ldots, \gamma_2$.

Matrix elements of the unity operator I in the space of the degenerated level E_0 in our basis are given by

$$\begin{aligned}(\beta_r \ldots \beta_1 | I | \gamma_1 \ldots \gamma_r) &= \sum_P (-1)^P \delta_{\beta_1, \gamma_{P1}} \ldots \delta_{\beta_r, \gamma_{Pr}} \\ &= I(\beta_r \ldots \beta_1, \gamma_1 \ldots \gamma_r)\end{aligned} \tag{B3}$$

For the inverse operator A^{-1} in the space of the degenerate level E_0 we have

$$\begin{aligned}&\frac{1}{r!} \sum_{\delta_1 \ldots \delta_r} (\beta_r \ldots \beta_1 | A | \delta_1 \ldots \delta_r)(\delta_r \ldots \delta_1 | A^{-1} | \gamma_1 \ldots \gamma_r) \\ &\qquad = I(\beta_r \ldots \beta_1, \gamma_1 \ldots \gamma_z)\end{aligned} \tag{B4}$$

The matrix elements of the product of any two operators A and B in the space of the degenerate level can be written down as:

$$\begin{aligned}&(\beta_r \ldots \beta_1 | AB | \gamma_1 \ldots \gamma_r) \\ &\qquad = \frac{1}{r!} \sum_{\delta_1 \ldots \delta_r} (\beta_r \ldots \beta_1 | A | \delta_1 \ldots \delta_r)(\delta_r \ldots \delta_1 | B | \gamma_1 \ldots \gamma_r)\end{aligned} \tag{B5}$$

and the result of applying on operator A to the state φ is given by

$$A\varphi = \frac{1}{r!} \sum_{\beta_1 \ldots \beta_r} \left[\frac{1}{z!} \sum_{\gamma_1 \ldots \gamma_r} (\beta_r \ldots \beta_1 | A | \gamma_1 \ldots \gamma_r) C\gamma_1 \ldots \gamma_r \right] \overset{+}{a}_{\beta_1} \ldots \overset{+}{a}_{\beta_r} \Phi_0 \tag{B6}$$

APPENDIX C

The Definition of the N Product Operation and Derivation of the Formula for $N(\beta_r \ldots \beta_1, \lambda_1' \ldots \lambda_p'; \lambda_p \ldots \lambda_1, \gamma_1 \ldots \gamma_r)$

We define the N product for the state Φ_0, in the following way: the operators under the N product sign are arranged so that the operators a_i with $i \in G_0$ and a_i^+ with $i \in F_0$ stand to the right, and the operators a_i with $i \in F_0 a_i^+$ with $i \in G_0$ stand to the left and all sign changes on anticommuting are included:

$$\begin{array}{l|l} a_i \text{ for } i \in F_0 & a_i \text{ for } i \in G_0 \\ \overset{+}{a}_i \text{ for } i \in G_0 & \overset{+}{a}_i \text{ for } i \in F_0 \end{array} \tag{C1}$$

Here $i \in F_0$ means that i refers to occupied one-electron states in the core state Φ_0; $i \in G_0$ denotes that i belongs to free (non-occupied) one-electron states in Φ_0.

To calculate the factor $N(\beta_r \ldots \beta_1, \lambda_1' \ldots \lambda_p'; \lambda_p \ldots \lambda_1, \gamma_1 \ldots \gamma_r)$ we can write at once (according to the definition of the N product just mentioned)

$$\begin{aligned} N(\beta_r \ldots \beta_1, \lambda_1' \ldots \lambda_p', \lambda_p \ldots \lambda_1, \gamma_1 \ldots \gamma_r) \\ = (\Phi_0, a_{\beta_r} \ldots a_{\beta_1} N(a_{\lambda_1} \ldots a_{\lambda_p} \overset{+}{a}_{\lambda_p'} \ldots \overset{+}{a}_{\lambda_1'}) \overset{+}{a}_{\gamma 1} \ldots \overset{+}{a}_{\gamma_r} \Phi_0) \\ = (\Phi_0, a_{\beta_r} \ldots a_{\beta_1} \overset{+}{a}_{\lambda_1'} \ldots \overset{+}{a}_{\lambda_p'} a_{\lambda_p} \ldots a_{\lambda_1} \overset{+}{a}_{\gamma_1} \ldots \overset{+}{a}_{\gamma_r} \Phi_0) \end{aligned} \tag{C2}$$

and the expectation values (C2) may be further expanded by means of ordinary pairing technique.

In order to obtain a non-zero result, we have to drop out all those systems of pairings, in which anyone of the operators from the group $a_{\lambda_1'}^+, \ldots, a_{\lambda_p'}^+$ is paired with any operator in the group $a_{\lambda_p}, \ldots, a_{\lambda_1}$.

In fact, we get zero for this case:

$$\overset{+}{a}_{\lambda'} a_\lambda = (\Phi_0, \overset{+}{a}_{\lambda'} a_\lambda \Phi_0) = 0$$

We have to consider only those pairings where operators from the group $a_{\lambda_1'}^+, \ldots, a_{\lambda_p'}^+$ are paired with operators from the group $a_{\beta_r}, \ldots, a_{\beta_1}$ and operators form the group $a_{\lambda_p}, \ldots a_{\lambda_1}$ are paired with operators from the group $a^+_{\gamma_1}, \ldots, a^+_{\gamma_r}$ (it is necessary that $r \geqslant p$). Taking into account the pairing systems above, we obtain:

$$(\Phi_0, \ldots \Phi_0) = \frac{1}{(r-p!)} \sum_P (-1)^P P_\beta \sum_Q (-1)^Q Q_\gamma \times$$

$$\times (\Phi_0, a_{\beta_r}, \ldots a_{\beta_{p+1}} a_{\beta_p} \ldots a_{\beta_1} a^+_{\lambda_1}, \ldots a^+_{\lambda_p}, a_{\lambda_p} \ldots a_{\gamma_1} a^+_{\gamma_1} \ldots a^+_{\gamma_p} a^+_{\gamma_{p+1}} \ldots a^+_{\gamma_r} \Phi_0)$$

where P_β is an operator which permutes the indices $\beta_1, \ldots, \beta_r$ and Q_γ is an operator of permutation of the indices $\gamma_1, \ldots, \gamma_r$. Any (possible) system of pairing is obtained from the permutation shown above by means of permutations of the field operators (which anticommute with each other). However, each system of pairing arises $(r-p)!$ times, in these permutations therefore the result should be divided by $(r-p)!$. As a matter of fact, by pairing an operator with an asterisk and an operator without an asterisk, which belong to two different groups, only the operators with an asterisk or the operators without an asterisk should be interchanged; but we interchange both groups of operators.

Thus, we have

$$N(\beta_r \ldots \beta_1, \lambda_1' \ldots \lambda_p'; \lambda_p \ldots \lambda_1, \gamma_1 \ldots \gamma_r)$$
$$= \frac{1}{(r-p)!} \sum_P (-1)^P \sum_Q (-1)^Q \delta_{\beta_{Pr}, \gamma_{Qr}} \ldots \delta_{\beta_{P(p+1)}, \gamma_{Q(p+1)}}$$
$$\times \delta_{\beta_{Pp}, \lambda_p'} \ldots \delta_{\beta_{P1}, \lambda_1'} \delta_{\lambda_p, \gamma_{Qp}} \ldots \delta_{\lambda_1, \gamma_{Q1}} \tag{C3}$$

The formula (C3) seems rather complicated; we shall try to simplify it. To this end we shall first represent it in another form (introducing the additional summation over $\lambda_r', \ldots, \lambda_{p+1}'$ and $\lambda_r, \ldots, \lambda_{p+1}$). We have

$$N(\beta_r \ldots \beta_1, \lambda_1' \ldots \lambda_p'; \lambda_p \ldots \lambda_1, \gamma_1 \ldots \gamma_r)$$
$$= \frac{1}{(r-p)!} \sum_P (-1)^P \sum_Q (-1)^Q$$
$$\times \sum_{\lambda'_{p+1} \ldots \lambda_r'} \sum_{\lambda_{p+1} \ldots \lambda_r} \delta_{\beta_{Pp}, \lambda'_p} \ldots \delta_{\beta_{P1}, \lambda_1'} \delta_{\lambda_p, \gamma_{Qp}} \ldots \delta_{\lambda_1, \gamma_{Q1}}$$
$$\times \delta_{\beta_{Pr}, \lambda_r'} \delta_{\lambda_r', \lambda_r} \delta_{\lambda_r, \gamma_{Qr}} \ldots \delta_{\beta_{P(p+1)}, \lambda'_{p+1}} \delta_{\lambda'_{p+1}, \lambda_{p+1}} \delta_{\lambda_{p+1}, \gamma_{Q(p+1)}}$$

from which we get finally:

$$N(\beta_z \ldots \beta_1, \lambda'_1 \ldots \lambda'_p; \lambda_p \ldots \lambda_1, \gamma_1 \ldots \gamma_p)$$

$$= \frac{1}{(r-p)!} \sum_{\lambda'_{p+1} \ldots \lambda_r'} \sum_{\lambda_{p+1} \ldots \lambda_r} \delta_{\lambda'p+1,\, \lambda p+1} \ldots \delta_{\lambda'_r,\, \lambda_r}$$

$$\times I(\beta_r \ldots \beta_1; \lambda'_1 \ldots \lambda'_r) I(\lambda_r \ldots \lambda_1; \gamma_1 \ldots \gamma_r) \qquad \text{(C4)}$$

where I is the unit matrix (B3).

APPENDIX D

The Proof of the Formula Connecting the Operators M and σ

In order to prove the validity of (63) we have to consider the rules of setting up time-independent diagrams with additional external electron lines and vertical sections marked as dangerous. By "cutting" the diagram with some vertical section in two parts all the factors prescribed by these rules factorize. The only exceptions are the factors (57a), (57b), and the topological factor λ; we have also to take care of the additional term $(n-k)\varepsilon$ in energy denominators (see (58)). In the left-hand side of (63) we have:

$$\frac{1}{r!} \sum_{\delta_1 \ldots \delta_r} I(\mu'_r \ldots \mu'_1; \delta_1 \ldots \delta_r) I(\delta_r \ldots \delta_1; \mu_1 \ldots \mu_r) \qquad \text{(D1)}$$

which is equal to

$$I(\mu'_r \ldots \mu'_1; \mu_1 \ldots \mu_r) = \sum_P (-1)^P \delta_{\mu_r', \mu_{Pr}} \ldots \delta_{\mu_1', \mu_{P1}} \qquad \text{(D2)}$$

Using these δ symbols we may carry out the summation over $\mu'_1, \ldots, \mu'_r$. It reduces to the equalizing of $\mu'_1, \ldots, \mu'_r$ to some interchanged $\mu_1, \ldots, \mu_r$, i.e. it integrates the diagrams $\bar{V}_k$ and V_l into the diagrams $V_m \in \mathscr{U}(\bar{V}_k, V_l)$; using this procedure $r!$ terms arise. Some of them lead, however, to the same diagram V_m. It can be shown that each diagram V_m is met $\lambda^{(1)}\lambda^{(2)}/\lambda$ times among these $r!$ diagrams, where $\lambda^{(1)}$ is the topological factor of $\bar{V}_k$, $\lambda^{(2)}$ is the topological factor of V_l, λ is the topological factor of V_m. The factors $\lambda^{(1)}$ and $\lambda^{(2)}$ cancel out with the factors $1/\lambda^{(1)}$

and $1/\lambda^{(2)}$ in the contributions from the diagram $\bar{V}_k$ and V_l. The factor $1/\lambda$ enters into the contribution from the diagram V_m. Now one must clarify the origin of the argument $l\varepsilon$ in (63); but it hardly requires any further comment, if we examine attentively Eq. (58).

References

1. Tolmachev, V. V., *Vestnik of the Leningrad University LGU*, **4**, 11 (1962).
2. Tolmachev, V. V., *Lit. Fiz. Sb.*, **3**, 47 (1963).
3. Tolmachev, V. V., *The Field-Theoretic Form of the Perturbation Theory for Many-Electron Problem in Atoms and Molecules*, Rotaprint, in Russian, Tartu, 1963.
4. Safronova, U. I., and Tolmachev, V. V., *Lit. Fiz. Sb.*, **4**, 249 (1964).
5. Bolotin, A. B., Levinson, I. B., and Tolmachev, V. V., *Lit. Fiz. Sb.*, **4**, 25 (1964).
6. Safronova, U. I., Vilnus, Candidate thesis (1964).
7. Matulis, A. Yu., Vilnus, Candidate thesis (1965).
8. Kelly, H. P., *Phys. Rev.*, **131**, 684 (1963).
9. Kelly, H. P., *Phys. Rev.*, **136**, B 896 (1964).
10. Kelly, H. P., *Phys. Rev.*, **144**, 39 (1966).
11. Judd, B. R., *Second Quantization and Atomic Spectroscopy*, Baltimore, 1966.
12. Goldstone, J., *Proc. Roy. Soc. (London), Ser. A*, **239**, 267 (1957).
13. Hubbard, J., *Proc. Roy. Soc. (London), Ser. A*, **240**, 539 (1957).
14. Hugenholtz, N. M., *Physica*, **23**, 481 (1957).
15. Brueckner, K. A., *Phys. Rev.*, **97**, 1353 (1955); **100**, 36 (1955); *The Many-Body Problem*, John Wiley & Sons Inc., New York, 1959.
16. Bloch, C., and Harowitz, J., *Nucl. Phys.*, **8**, 91 (1958).
17. Morita, T., *Progr. Theoret. Phys.*, **29**, 351 (1963).
18. Sandars, P. G. H., Clarendon Laboratory, University of Oxford. Preprint.
19. Bogolubov, N. N., The Lectures on Quantum Statistics, Radyanska Shkola, Kiev, § 5, 145–147 (1947).
20. Heitler, V., *Quantum Theory of Radiation*, Clarendon Press, Oxford, § 16 (1954).
21. Primas, H., *Rev. Mod. Phys.*, **35**, 710–712 (1963).
22. Löwdin, P.-O., *Rev. Mod. Phys.*, **35**, 702 (1963).

THE FIELD-THEORETIC FORM OF THE PERTURBATION THEORY FOR MANY-ELECTRON ATOMS

II. ATOMIC SYSTEMS

V. V. TOLMACHEV, *Mathematics–Mechanics Faculty, University of Moscow, Moscow, U.S.S.R. and Theoretical Department, Institute of Electrochemistry, Academy of Sciences, Moscow, U.S.S.R.*

CONTENTS

1. INTRODUCTION

In the first part of the present work an abstract theory was developed which may be applied to the general system of interacting Fermions (the interaction term contains four Fermi field operators) for the case of perturbation theory for a degenerate level. In the present (second) part we shall deal with atomic systems.[1–11]

16 12pp.

The Schrödinger equation for the many-electron atomic system has the following form:

$$\left\{\sum_{(1\leq i\leq N)}\left(-\frac{1}{2}\Delta\rho_i-\frac{Z}{\rho_i}\right)+\sum_{(1\leq i\leq j\leq N)}\frac{1}{\rho_{ij}}\right\}\Phi(\boldsymbol{\rho}_1,\ldots,\boldsymbol{\rho}_N)=\mathscr{E}\Phi(\boldsymbol{\rho}_1\ldots,\boldsymbol{\rho}_N) \tag{1}$$

where $\boldsymbol{\rho}_1\ldots,\boldsymbol{\rho}_N$ are the coordinates of N electrons, ρ_i is the distance of the ith electron from the nucleus, ρ_{ij} is the mutual distance between ith and jth electrons, and Z is the nuclear charge.

Let us transform the variables in the Eq. (1) $\boldsymbol{\rho}_1=\dfrac{\mathbf{r}_1}{Z}\ldots,$ $\boldsymbol{\rho}_N=\dfrac{\mathbf{r}_N}{Z}$ introducing the notations $\Psi(\mathbf{r}_1,\ldots,\mathbf{r}_N)\equiv\Phi\left(\dfrac{\mathbf{r}_1}{Z},\ldots,\dfrac{\mathbf{r}_N}{Z}\right)$ and $E=\dfrac{\mathscr{E}}{Z^2}$; then we get immediately the Schrödinger equation

$$\left\{\sum_{(1\leq i\leq N)}\left(-\frac{1}{2}\Delta r_i-\frac{1}{r_i}\right)+\frac{1}{Z}\sum_{(1\leq i<j\leq N)}\frac{1}{r_{ij}}\right\}\Psi(\mathbf{r}_1,\ldots,\mathbf{r}_N)=E\Psi(\mathbf{r}_1,\ldots,\mathbf{r}_N) \tag{2}$$

Thus, instead of considering a many-electron atom with the nuclear charge Z we may study the fictitious problem of an atom with nuclear charge $Z=1$; for this atom the interelectron interaction is reduced by the factor $1/Z$ and in order to obtain the energy eigenvalues for the problem (1) we should multiply the eigenvalues of the problem (2) by the factor Z^2. The perturbation theory considered below is a small interaction parameter perturbation theory; the relevant parameter is the strength of the interaction between the electrons, i.e. due to Eq. (2) the parameter is equal to $1/Z$.

As the zero-order problem we have the following: N non-interacting electrons move in the field created by a nucleus with the charge $Z=1$. Therefore we have to choose as a system $\varphi_i(x)$ of single-electron functions in the second quantization representation (these functions formed the basis of the abstract theory presented in the first part of this work) the set of hydrogenic functions. Due to spherical symmetry of the problem they have the following form:

$$\varphi_i(x) = R_{nl}(r) Y_{lm}(\Omega) \delta_{s,\sigma} \tag{3}$$

$$\varepsilon_i = -\frac{1}{2n^2} \tag{4}$$

Here $x \equiv (\mathbf{r}, \sigma)$, $\mathbf{r} \equiv (r, \Omega)$, $\Omega \equiv (\theta, \rho)$ where $\mathbf{r}$ is the radius vector, r is the radial coordinate, Ω is the solid angle, θ is the polar angle, φ is the azimuthal angle, σ is the spin coordinate of electron $i = (n, l, m, s)$, where n is the principal quantum number, l is the orbital angular momentum of the electron for the fixed value of quantum number n, m is the projection of the orbital angular momentum l on the z axis, s is the projection of the spin on the z axis. For the discrete spectrum r is equal to 1, 2, and so on; for the continuous spectrum $n = 1/ik$, where k is the wave number; therefore $\varepsilon_i = k^2/2$ for the continuous spectrum. The functions Y_{lm} which enter into (3) are the ordinary spherical functions (see Appendix A).

We shall not quote here the detailed expressions for the radial functions $R_{nl}(r)$, because we do not need to use them in the present work; we shall only assume these radial functions $R_{nl}(r)$ to be real; in fact, we need only the formulae (3) and (4).

As the basis states of the zero-order problem (2) for the open shell containing p electrons and characterized by the principal quantum number n_0, we take the simultaneous eigenstates of the set of operators L^2, L_z, S^2, S_z. Then for the coefficients $C^{K_p}_{\gamma_1 \ldots \gamma_p}$ of these states we have $K_p \equiv (l_a^{n_a} \ldots l_c^{n_c} \alpha S L M_S M_L)$ where $l_a^{n_a} \ldots l_e^{n_c}$ are the indices of a possible electron configuration of the open shell considered ($n_a + \ldots + n_c = p$; $l_a, \ldots, l_c$ are the electronic orbital angular momenta for the principal quantum number n_0), $\alpha S L$ is the index of the electronic configuration term (α is the frequency of the term, S is the total spin, L is the total orbital angular momentum), $M_S M_L$ is the index of function for the term and configuration (M_S is the projection of total spin angular momentum), M_L is the projection of the total orbital angular momentum. The indices of single-electron states from the open shell are to be interpreted as the composite ones

$\gamma \equiv (lms)$, where l is the electronic orbital angular momentum for the principal quantum number n_0, m is the projection of orbital angular momentum, and s is the projection of spin.

The energy ε_i of any electronic line with index i is determined only by the principal quantum number n of this index and does not depend upon the remaining quantum numbers l, m, s of the line.

Analogously, the occupation numbers n_i^0 for the core state Φ_0 depend only upon the principal quantum number n of the index i; the state Φ_0 is the state with all shells completely filled, and in the case of a hydrogenic atom they are characterized by principal quantum numbers.

Thus, let us define

$$\sum_i n_i^0 \times \equiv \sum_{nl \in f_0} \sum_m \sum_s \times \quad \sum_i (1 - n_i^0) \times \equiv \sum_{nl \in g_0} \sum_m \sum_s \times$$

where $nl \in f_0$ is assumed when $n_i = 1$, and $nl \in g_0$ when $1 - n_i^0 = 1$. The sums over nl include also the integrals over the continuous spectrum states.

$$\sum_{nl} \equiv \sum_{n=1}^{+\infty} \sum_{l=0}^{n-1} + \sum_{l=0}^{+\infty} \int dk$$

In the Sections 2 and 3 we formulate the rules of construction for Feynman atomic diagrams; the diagrams of the abstract theory developed in the first part of the present work are taken as the starting point. We use the formula for the interaction matrix elements which express these matrix elements in terms of the radial integrals $R_l(n_1'l_1',\ n_2'l_2';\ n_2l_2,\ n_1l_1)$ and Wigner's $3j$ coefficients (see formula (8)) as well as the recurrence formula for the coefficients $C^{l_a^{n_a} \ldots l_c^{n_c}\, \alpha SLM_SM_L}_{l_1m_1s_1 \ldots l_pm_ps_p}$ (see formula (16)); using this formula the above coefficients may be expressed in terms of Wigner's $3j$ coefficients and fractional parentage coefficients. These formulae enable us to proceed easily from the abstract theory Feynman diagrams to atomic Feynman diagrams.

In the course of this procedure we have to use a new definition for the fractional parentage coefficients; we are working in the space of antisymmetric functions only, and therefore we have to get a definition of fractional parentage coefficients which would use only the concepts relevant for this space. This new definition of fractional parentage coefficients is given in Appendix D; it allows us to consider the fractional parentage coefficients not only for the one-momentum configurations $l_a^{n_a}$ but also for the arbitrary many-momentum configurations $l_a^{n_a} \ldots l_c^{n_c}$.

The summations over the projections of the orbital and spin angular momenta on the atomic diagram lines are carried out below; for this purpose we have to use the graphical technique for the angular momentum diagrams, which is presented in detail in Appendix C.

2. ATOMIC DIAGRAMS FOR ΔE_0

Let us derive first the important formula for the interaction matrix element; we have by definition:

$$v_{i_1'i_2',i_2i_1} = \sum_{\sigma_1,\sigma_2} \int\int d\mathbf{r}_1 d\mathbf{r}_2 \varphi_{i_1'}^*(x_1)\varphi_{i_2'}^*(x_2) \frac{1}{|\mathbf{r}_1 - \mathbf{r}_2|} \varphi_{i_2}(x_2)\varphi_{i_1}(x_1) \tag{5}$$

where the summation is carried over the spin and the integration over the space coordinates. Let us substitute (3) into (5) and use the formulae (A4) and (A4a) from the Appendix A: we obtain

$$\begin{aligned} v_{i_1'i_2',i_2i_1} = \delta_{s_1',s_1}\delta_{s_2',s_2} \sum_{l=0}^{+\infty} R_l(n_1'l_1', n_2'l_2'; n_2l_2, n_1l_1) \\ \times \frac{4\pi}{2l+1} \sum_{m=-l}^{+l} \int d\Omega_1 Y_{l_1'm_1'}^*(\Omega_1) Y_{lm}(\Omega_1) Y_{l_1m_1}(\Omega_1) \\ \times \int d\Omega_2 Y_{l'_2m_2'}^*(\Omega_2) Y_{lm}^*(\Omega_2) Y_{l_2m_2}(\Omega_2) \end{aligned} \tag{6}$$

where a special notation for the radial integral is introduced:

$$\begin{aligned} R_l(n_1'l_1', n_2'l_2'; n_2l_2, n_1l_1) = \int_0^{+\infty} r_1^2 dr_1 \int_0^{+\infty} r_2^2 dr_2 R_{n_1'l_1'}(r_1) R_{n_2'l_2'}(r_2) \\ \times u_l(r_1, r_2) R_{n_2l_2}(r_2) R_{n_1l_1}(r_1) \end{aligned} \tag{7}$$

We point out the following important properties of the radial integral. According to the definition (7), we have

$$R_l(n_1'l_1', n_2'l_2'; n_2l_2, n_1l_1) \equiv R_l(n_2'l_2', n_1'l_1'; n_1'l_1', n_2l_2)$$

$$R_l(n_1'l_1', n_2'l_2'; n_2l_2) \equiv R_l(n_1l_1, n_2l_2; n_2'l_2', n_1'l_1')$$

it is possible to interchange simultaneously the components in those pairs of arguments of the radial integral, which are separated by the semicolon. Furthermore, the radial integral is invariant under the operation of arranging its arguments in the inverse order.

Let us express the integrals of three spherical harmonics, which enter into Eq. (6) in terms of $3j$ coefficients with the aid of the formulae (A2) and (A3) from Appendix A. After a simple transformation of one of the two arising $3j$ coefficients using the formulae (B9) we obtain finally for the matrix element of interaction the formula

$$\begin{aligned} v_{i_1'i_2',i_2i_1} = {} & \delta_{s_1',s_1}\delta_{s_2',s_2}\sum_{l=0}^{+\infty} R_l(n_1'l_1', n_2'l_2'; n_2l_2, n_1l_1) \\ & \times \sqrt{(2l_1'+1)(2l_2'+1)(2l_2+1)(2l_1+1)}\,[l_1'l_1l][l_2'l_2l] \\ & \times \sum_{m=-l}^{+1} (-1)^{l-m} \begin{pmatrix} l_1' & l_1 & l \\ m_1' & -m_1 & -m \end{pmatrix} \\ & \begin{pmatrix} l_2' & l_2 & l \\ m_2' & -m_2 & m \end{pmatrix} (-1)^{\frac{1}{2}(l_1'-m_1')+\frac{1}{2}(l_2'-m_2')+\frac{1}{2}(l_2-m_2)+\frac{1}{2}(l_1-m_1)} \end{aligned} \tag{8}$$

(we have used the definition (B12)). The formula (8) for the interaction matrix element completely characterizes the interaction between electrons. First the δ symbols guarantee not only the conservation of the projection of the spin angular momentum during the elementary interaction (i.e. the condition $s_1' + s_2' = s_1 + s_2$), but also the fulfilment of conversation laws $s_1' = s_2$, and $s_2' = s$, which express the independence of the electronic interaction upon the spins. Further, the $3j$ coefficients are non-zero only if $m_1' - m_1 = -m_2' + m_2$, and, therefore, $m_1' + m_2' = m_1' + m_2$ must hold—this is the law of conservation of the projection of the orbital angular momentum in the elementary interaction. Finally, the moduli of the $3j$ coefficients with zero projections of angular momenta are non-zero only if $l_1' + l_1 + l$ equals an even number, and so is $l_2' + l_2 + l$; therefore $(-1)^{l_1+l_2} = (-1)^{l_1+l_2}$, and this is the law of conservation of the parity for the elementary interaction.

Now let us represent the different factors, entering the formula (8) for the matrix element $v_{i'_1i'_2,i_2i_1}$ with the aid of the appropriate diagram for a quartic vertex.

There are two outgoing and two ingoing electronic lines in the quartic vertex; both types of lines are quite symmetric at the vertex, and therefore they are indistinguishable (Fig. 1a).

It is not clear, for example, which of two outgoing electronic lines corresponds to the index i'_1, and which one to the index i'_2 (the same is true for the ingoing ends of electronic lines and indices i_1, i_2).

In order to make the ends of electronic lines in the vertex distinguishable, we shall consider the dashed interaction line in the

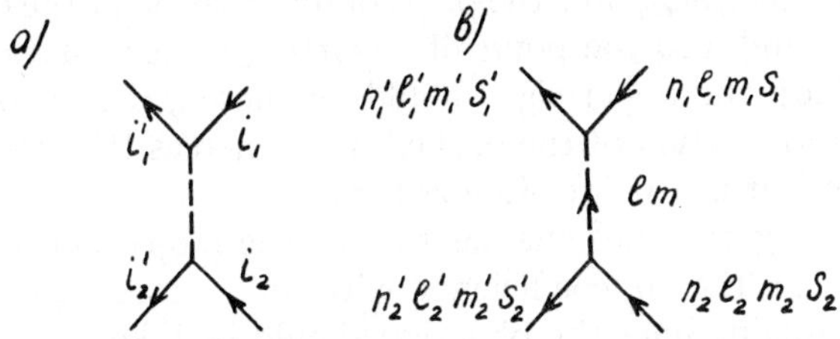

Fig. 1. Illustrations for the formula for the matrix element $v_{i'_1 j'_1, j_1 i_1}$.

vertex as a directed one, and we suggest the following way of labeling the outgoing and ingoing electronic lines of the vertex. The index i'_1 labels the outgoing electronic line belonging to that node of the vertex which is entered by the dashed line; index i'_2 labels the outgoing electronic line for the second node of the vertex, which is left by the dashed line (Fig. 1b).

According to the formula (8) we shall label the electronic lines of the vertex with the indices n'_1, l'_1, m'_1, s'_1; n'_2, l'_2, m'_2, s'_2; n_1, l_1, m_1, s_1; n_2, l_2, m_2, s_2 (Fig. 1b), and the directed dashed line with a pair of indices lm (Fig. 1b).

According to the formula (8) we now put on each of the four electronic lines, attached to the vertex, the appropriate factor

$$(-1)^{\frac{1}{2}(l_a - m_a)} \sqrt{2l_a + 1}$$

where $l_a m_a$ is the pair of indices, labeling the electronic line under consideration; on the dashed vertex line we set the factor

$$(-1)^{l-m}$$

According to formula (8), we put in each of two vertex nodes the appropriate $3j$ coefficients as factors; these $3j$ coefficients should be multiplied by their moduli with zero projections of angular momenta.

The factor

$$[l'_a l_a l] \begin{pmatrix} l'_a & l_a & l \\ m'_a & -m_a & \pm m \end{pmatrix} \delta_{s_a', s_a}$$

corresponds to the node, which is the point of departure of the electronic line $l'_a m'_a s'_a$ and the point of termination of the electronic line $l_a m_a s_a$ and also the point of departure or termination of the dashed line; the upper sign should be chosen, if the node is the point of termination of the dashed line, whereas the lower should be chosen if it is a point of departure.

For the ingoing lines one has to take the projections of angular momenta in the $3j$ coefficients with the minus sign, whereas for the outgoing lines the plus sign should be taken.

The order of writing appropriate angular momenta in the $3j$ coefficient is unimportant, because the sum $l'_a + l_a + l$ is always equal to an even number. However, it is convenient for the sake of definiteness to consider the lines in every node according to the following order: the outgoing electronic line, the ingoing electronic line, the dashed line.

Finally according to formula (8) the total two body vertex is supplied by the radial integral

$$R_l(n'_1 l'_1, n'_2 l'_2; n_2 l_2, n_1 l_1)$$

as a factor; its arguments are written in the following order: the outgoing electronic line of the node, which is the point of termination of the dashed line, another outgoing electronic line for another node, then the ingoing electronic line and finally the remaining electronic line.

We consider time-independent time-ordered Feynman diagrams for ΔE_0 and this means, that all dashed lines of vertices are represented by vertical lines, and also all of these vertices are placed in a definite way relative to each other and with respect to the horizontal time axis, which goes from right to left. Electronic lines, which go from right to left are particle lines, the electronic lines going from left to right or starting and ending in the same vertex are hole lines.

The time-independent diagrams do not carry any indices on their lines; in order to get the contributions from them, we have to put indices on their lines.

It appears during this procedure, that for a given time-independent diagram one should consider two corresponding diagrams with the indices put upon their lines: the radial diagram and the orbital angular momentum diagram.

The radial diagram is a time-independent diagram where electronic lines are labeled by double indices nl, and vertex by a simple index l. The lines of a radial diagram do not carry any projections of orbital angular momenta m or spin s.

The orbital angular momentum diagram is obtained from the radial one in the following way. In the diagram all directions on each of its electronic lines are removed. Then all dashed lines are replaced by full lines.

We attach to the lines of an orbital angular momentum diagram those orbital angular momentum indices, which label corresponding lines of radial diagrams. Those lines, which were previously dashed, are labeled by the orbital angular momenta l, which labeled the same lines of the radial diagram. The lines of the orbital angular momentum diagram which were previously the electronic lines, are supplied by the indices l, entering into the double indices nl, which labeled the corresponding electronic lines of the radial diagram.

The contribution for ΔE_0 from the given atomic Feynman diagram is set up according to its radial diagram with the aid of the orbital angular momentum diagram. The contribution is equal to the sum over all indices nl, labeling the electronic lines of the radial diagram, and also over all indices l, labeling its dashed lines, of the product, which is set up from the radial and angular momentum factors, i.e.

Contribution of the diagram = Contribution of the radial diagram

$$= \sum_{nl} \sum_{l} \text{Radial factor} \times \text{Orbital angular momentum factor} \tag{9}$$

Summing over nl one has to take into account that the summation for the hole lines is to be carried for $nl \in f_0$, and for the particle lines for $nl \in g_0$; the summations over l are carried from zero to infinity. The radial factor in (9) is set up according to the radial diagram, and is composed of the following single factors:

[1]. Every quartic vertex of the radial diagram with the dashed line l, and outgoing electronic lines $n_1'l_1'$, $n_2'l_2'$, and incoming electronic lines n_1l_1, n_2l_2, gives the following factor:

$$R_l(n_1'l_1', n_2'l_2'; n_2l_2, n_1l_1)[l_1'l_1l][l_2'l_2l] \tag{10}$$

The ordering of arguments in the radial integral corresponds to the following ordering of the electronic line in the vertex: the outgoing electronic line, another outgoing electronic line, the incoming electronic line, corresponding to the latter, and another incoming electronic line.

[2]. To every electronic line nl of the radial diagram corresponds the factor

$$2l + 1 \tag{11}$$

[3]. To every vertical section between two neighbouring vertices of the radial diagram corresponds the factor

$$\frac{1}{\mathscr{E}} \tag{12}$$

where $\mathscr{E}$ equals the sum of energies of particle lines minus the sum of all the electronic energies of the hole lines.

[4]. To all the radial diagram corresponds the factor

$$2^l \frac{1}{\lambda}(-1)^{n+l+f+1} \tag{13}$$

where n is the number of quartic vertices in the diagram, l is the number of closed loops of electronic lines of the diagram, f is the number of hole lines of the diagram, λ is the topological factor of the time-independent diagram for ΔE_0 considered.

The orbital angular momentum factor in (9) is set up by means of the orbital angular momentum diagram. This must be transformed beforehand arbitrarily into the ordinary angular momentum diagram, putting directions on its lines and plus and minus signs on its nodes. Then using the rules formulated in Appendix

C we have to put the angular momentum diagram in correspondence with the appropriate angular momentum sum, where the summation is carried over the projections of the angular momentum from the relevant product of the $3j$ coefficients.

Now let us clarify these rules by an example of the first-order diagram.

We have only two diagrams 1A and 1B in the first order; they are represented in Fig. 2, where the time-independent diagrams without indices are shown as well as the appropriate radial and orbital angular momentum diagrams.

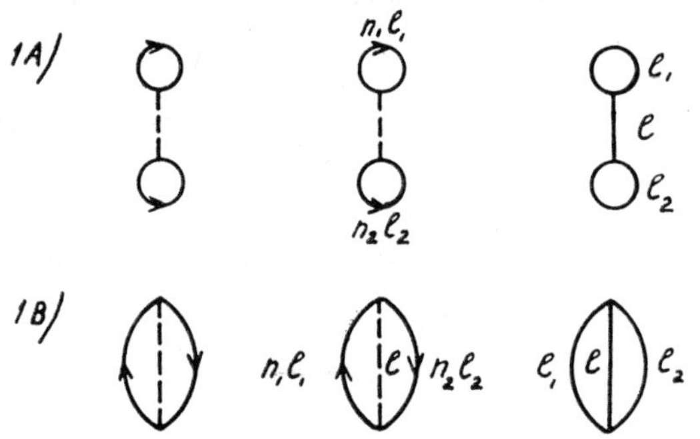

Fig. 2. First-order vacuum diagrams for ΔE_0. Diagrams of abstract theory, radial diagrams, and orbital angular momentum diagrams. (There are two diagrams 1A and 1B.)

Let us first take the diagram 1A. Considering the corresponding radial diagram (see Fig. 2), we obtain

$$\Delta E^0_{1\mathrm{A}} = 2 \sum_{\substack{n_1 l_1 \in f_0 \\ n_2 l_2 \in f_0}} \sum_{l} R_l(n_1 l_1, n_2 l_2; n_2 l_2, n_1 l_1)[l_1 l_1 l][l_2 l_2 l]$$

$$\times (2l_1 + 1)(2l_2 + 1)$$

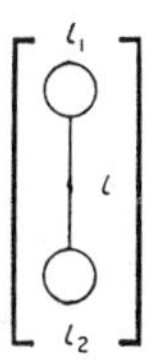

where the symbol in brackets denotes the appropriate orbital angular momentum factor (see Fig. 2).

Using the formulae (C6), (C2) of Appendix C, we can write

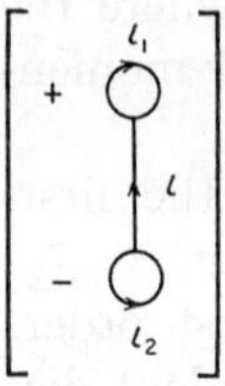
$$= $$
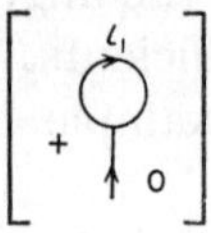
$$\times$$
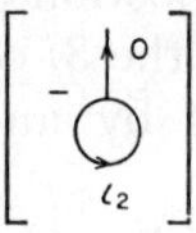
$$\times \delta_{l,0} = \sqrt{2l_1+1}\,\sqrt{2l_2+1}\,\delta_{l,0}$$

Then using the result and also noting that $[ll0] = \dfrac{1}{\sqrt{2r+1}}$ (see (B13a)), we get immediately the value of ΔE^0_{1A}, which is listed in Table 1.

TABLE 1. Contributions of the first-order vacuum diagrams.

1A		$2\sum_{n_1l_1\in f_0, n_2l_2\in f_0} (2l_1+1)(2l_2+1)R_0(n_1l_1, n_2l_2; n_2l_2, n_1l_1)$
1B		$\sum_{n_1l_1\in f_0, n_2l_2\in f_0} (2l_1+1)(2l_2+1)\sum_l [l_1l_2l]^2 R_l(n_1l_1, n_2l_2; n_1l_1, n_2l_2)$

Let us then take the diagram 1B. Considering the corresponding radial diagram (see Fig. 2), we obtain:

$$\Delta E^0_{1B} = -\sum_{\substack{n_1l_1\in f_0\\ n_2l_2\in f_0}} \sum_l [l_1l_2l]^2 R_l(n_1l_1, n_2l_2; n_1l_1, n_2l_2)$$

$$\times (2l_1+1)(2l_2+1)$$
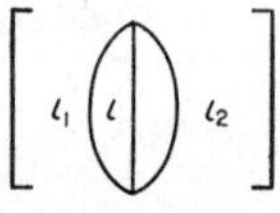

where the orbital angular momentum factor, according to (C3), is

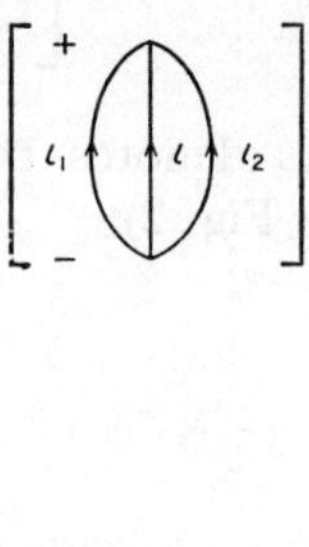
$$= \delta_{l_1,l_2,l}$$

Using this value in the formula for ΔE^0_{1B} written above, we obtain for ΔE^0_{1B} the value listed in Table 1.

3. ATOMIC DIAGRAMS FOR $M_p(K_p', K_p|E')$

The time-independent diagrams for the matrix elements of the secular matrix $M(K_r', K_r|\Delta E')$ are diagrams having external electronic lines. If we have r electrons in the open shell, we must consider all contributions from the time-independent diagram which have all possible $2p$ electronic lines, where $1 \leqslant p \leqslant r$.

Accordingly,

$$M(K_r', K_r|\Delta E') = \sum_{p=1}^{r} M_p(K_r', K_r|\Delta E') \tag{14}$$

By virtue of the reduction formula

$$M(K_r', K_r|\Delta E') = \sum_{p=1}^{r} \sum_{K_p'K_p} g(K_r', K_p'; K_p, K_r) M_p(K_p', K_p|\Delta E') \tag{15}$$

Thus, it is necessary to consider only those diagrams, for which the number of external (incoming or outgoing) electronic lines is equal to the number of electrons in the open shell.

Let us start the interpretation of the formula (D1) of Appendix D. This formula expresses the coefficients $C^{K_p}_{\gamma_1 \ldots \gamma_p}$ for p electrons in terms of the coefficients $C^{K_{p-1}}_{\gamma_1 \ldots \gamma_{p-1}}$ for $p - 1$ electrons and relevant fractional parentage coefficients. We rewrite the formula (D1), expressing the Clebsch–Gordan coefficients in terms of Wigner $3j$ coefficients. Then (D1) yields

$$\begin{aligned}
C^{l_a^{n_a} \ldots l_c^{n_c} \alpha S L M_s M_L}_{l_1 m_1 s_1, \ldots, l_p m_p s_p} &= \sqrt{p} \sum_{b=a}^{c} \sum_{\alpha' S' L'} C^{l_a^{n_a} \ldots l_b^{n_b - 1} \ldots l_c^{n_c} \alpha' S' L' M'_L M'_b}_{l_1 m_1 s_1, \ldots, l_{p-1} m_{p-1} s_{p-1}} \delta_{l_p, l_b} \\
&\times (l_a^{n_a} \ldots l_c^{n_c} \alpha S L \{| l_a^{n_a} \ldots l_b^{n_b - 1} \ldots l_c^{n_c} (\alpha' S' L') l_b S L) \\
&\times (-1)^{-\frac{3S'}{2} + \frac{1}{4} - \frac{S}{2}} (-1)^{\frac{1}{2}(S - M_S)} (-1)^{\frac{1}{2}(S' - M_S')} (-1)^{\frac{1}{2}(\frac{1}{2} - s_p)} \\
&\times (-1)^{-\frac{3L'}{2} + \frac{l_p}{2} - \frac{L}{2}} (-1)^{\frac{1}{2}(L - M_L)} (-1)^{\frac{1}{2}(L' - M_L')} (-1)^{\frac{1}{2}(l_p - m_p)} \\
&\times \sqrt{2S+1} \sqrt{2L+1} \begin{pmatrix} S & S' & \frac{1}{2} \\ -M_S & M_S' & s_p \end{pmatrix} \begin{pmatrix} L & L' & l_p \\ -M_L & M_L' & m_p \end{pmatrix}
\end{aligned} \tag{16}$$

Let us note the possibility of making the following substitution in the formula (16)

$$
\begin{aligned}
&(l_a^{n_a} \ldots l_c^{n_c} \alpha SL\{|l_a^{n_a} \ldots l_b^{n_b-1} \ldots l_c^{n_c}(\alpha' S' L') l_b SL) \\
&\quad = (l_a^{n_a} \ldots l_b^{n_b-1} \ldots l_c^{n_c}(\alpha' S' L') l_b SL|\} l_a^{n_a} \ldots l_c^{n_c} \alpha SL) \\
&\begin{pmatrix} S & S' & \frac{1}{2} \\ -M_S & M'_S & s_p \end{pmatrix} = \begin{pmatrix} S' & S & \frac{1}{2} \\ -M'_S & M_S & -s_p \end{pmatrix} \\
&\begin{pmatrix} L & L' & l_p \\ -M_L & M'_L & m_p \end{pmatrix} = \begin{pmatrix} L' & L & l_p \\ -M'_L & M_L & -m_p \end{pmatrix}
\end{aligned} \tag{17}
$$

The formula (16) in the unchanged form may be employed for the left coefficient function in matrix element. Formula (16) with substitution (17) for the remaining coefficient in the matrix element.

The formula (16) and the same formula with the substitution of (17) is represented by the diagrams of Figs. 3a and 3b. The different factors in the formulae are set in correspondence with different graphical elements of Figs. 3a and 3b.

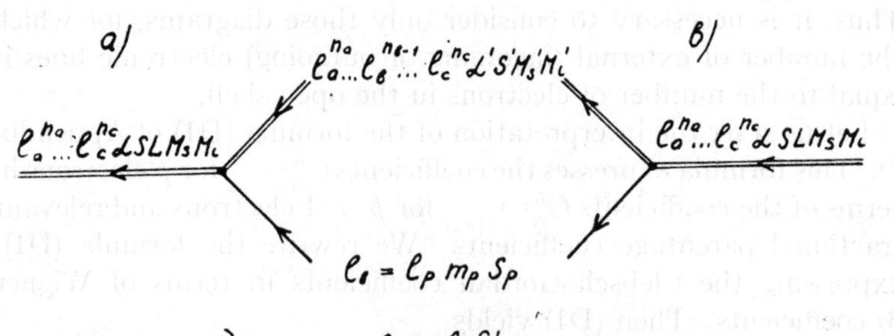

Fig. 3. Illustrations for the formula expressing $C^{K_p}_{\gamma_1 \ldots \gamma_p}$ in terms of $C^{K_{p-1}}_{\gamma_1 \ldots \gamma_{p-1}}$ and also formula for $C^{K_1}_{\gamma_1}$.

The $3j$ coefficients are made to correspond to the nodes of diagrams. Accordingly, from the node of the diagram of Fig. 4a we get the factor

$$
\begin{pmatrix} S & S' & \frac{1}{2} \\ -M_S & M'_S & s_p \end{pmatrix} \begin{pmatrix} L & L' & l_p \\ -M_L & M'_L & m_p \end{pmatrix}
$$

and from the node of diagram Fig. 3b we get

$$\begin{pmatrix} S' & S & \frac{1}{2} \\ -M'_S & M_S & -s_p \end{pmatrix} \begin{pmatrix} L' & L & l_p \\ -M'_L & M_L & -m_p \end{pmatrix}$$

The angular momenta in the $3j$ coefficients are written in such an order that the lines connected with the diagram nodes are counted as follows: the outgoing double line, the incoming double line, and the incoming or outgoing electron line. The projections of the angular momenta for the outgoing lines are taken with the plus sign, the projections of the angular momenta for the ingoing lines are taken with the minus sign.

The fractional parentage coefficients of the formulae are also put in correspondence with the nodes of the diagram. Accordingly, we get from the diagram Fig. 3a the factor

$$\sqrt{p}(l_a^{n_a} \ldots l_b^{n_b-1} \ldots l_c^{n_c}(\alpha' S' L') l_b \alpha SL|\} l_a^{n_a} \ldots l_c^{n_c} \alpha SL)$$

from the diagram Fig. 3b the factor

$$\sqrt{p}(l_a^{n_a} \ldots l_c^{n_c} \alpha SL\{| l_a^{n_a} \ldots l_b^{n_b-1} \ldots l_c^{n_c}(\alpha' S' L') l_b SL)$$

The factor

$$\sqrt{2S+1}\sqrt{2L+1}$$

is attached to the outgoing double line for the diagram Fig. 3a and the incoming double line for the diagram Fig. 3b.

From the nodes of every diagram on Figs. 3a and 3b we get also sign factors

$$(-1)^{-\frac{3S'}{2}+\frac{1}{4}-\frac{S}{2}} (-1)^{-\frac{3L'}{2}+\frac{l_p}{2}-\frac{L}{2}}$$

The sign factors

$$(-1)^{\frac{1}{2}(S-M_S)} (-1)^{\frac{1}{2}(L-M_L)}, (-1)^{\frac{1}{2}(S'-M'_S)} (-1)^{(L'-M'_L)}$$

and

$$(-1)^{\frac{1}{2}(\frac{1}{2}-s_p)} (-1)^{\frac{1}{2}(l_p-m_p)}$$

are obtained from the ends of appropriate lines of each diagram Figs. 3a and 3b.

The formulae (16) and (17) are recurrent formulae, and in

order to find coefficients we have to know the coefficients for the one-electron case $p = 1$. In this case we have

$$C_{l_1 m_1 s_1}^{l_a SLM_S M_L} = \delta_{l_a, l_1} \delta_{S, \frac{1}{2}} \delta_{L, l_1} \delta_{M_S, s_1} \delta_{M_L, m_1} \tag{18}$$

which is represented by the diagram of Fig. 3c.

Let us formulate now the rules for the construction of atomic Feynman diagrams and for getting contributions for $M_p(K_p', K_p|\Delta E')$.

Atomic Feynman diagrams for $M_p(K_p', K_p|\Delta E')$ are extended time-independent diagrams.

In order to construct an extended time-independent diagram we add to a given ordinary time-independent diagram (Fig. 4a) special diagrams with double lines (Fig. 4b) to the left-hand as well as to the right-hand end; during this procedure one left and one right external electronic line become double lines. In general, if the starting time-independent diagram contains one incoming and outgoing external electronic lines, then to get the extended time-independent diagram it is sufficient to transform only these external electronic lines into double lines (Fig. 4c).

The vertices of extended time-independent diagrams are quartic vertices (which are drawn vertically) or left and right ternary vertices. The lines of the extended diagram are double lines, electronic lines, and dashed lines.

Each of the quartic vertices has a vertical dashed line as well as two outgoing and two incoming electronic lines. Every left ternary vertex has one outgoing double line, one entering double line, and one outgoing electronic line. Every right ternary vertex has one outgoing double line, one outgoing electronic line, and one incoming double line.

The electronic lines may be internal or external. The external electronic lines are attached to quartic vertices only by one end having their second end attached to right or left ternary vertices. The external electronic lines may be left or right. In any diagram the number of outgoing left external electronic lines is equal to the number of right incoming external electronic lines.

The electronic lines may be particle or hole lines. The particle lines are directed from right to left. The hole lines are directed from left to right, they may also start and finish in the same quartic vertex.

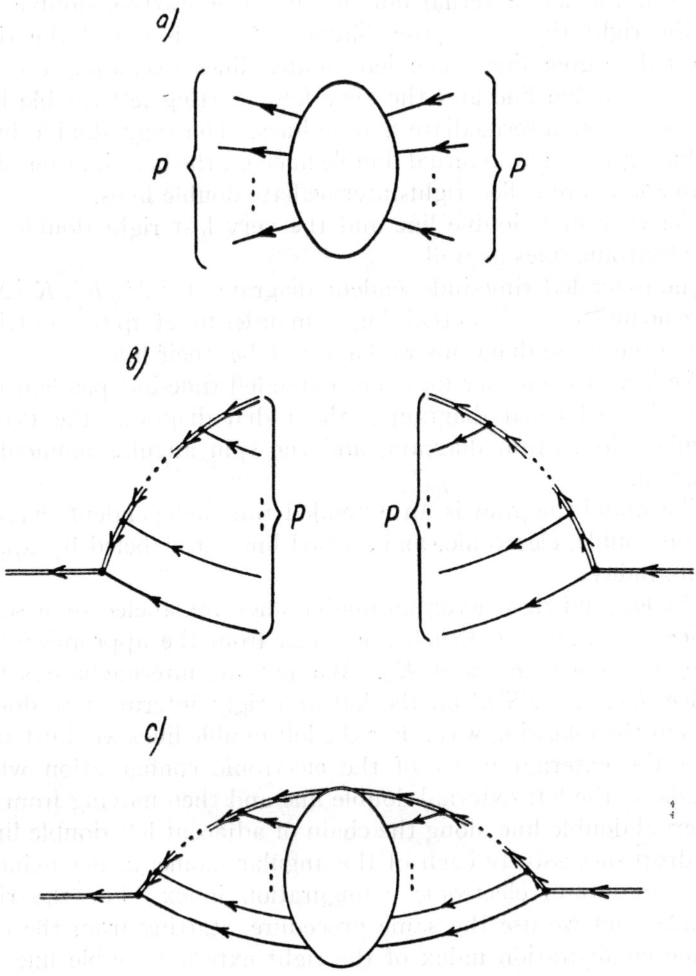

Fig. 4. Illustration of the procedure of the construction of the extended time-independent diagram for $M_p(K'_p, K_p|\Delta E)$.

The double lines may be also of right and left types. The terminal double line on the left (by which the diagram terminates) is called the left external double line. The starting double line on the right (by which the diagram starts) is called the right external double line. The left double lines excluding the left external double line and the very first starting left double line, are called left intermediate double lines. The right double lines, excluding the right external double line and the last right terminal double line, are called right intermediate double lines.

The very first double line and the very last right double line are electronic lines as well.

The extended time-independent diagrams for $M_p(K_p', K_p|\Delta E')$ have no indices at all on their lines; in order to set up the contributions from these diagrams we have to label their lines.

We have to consider for every extended time-independent diagram three labeled diagrams: the radial diagram, the orbital angular momentum diagram, and the spin angular momentum diagram.

The radial diagram is an extended time-independent diagram whose double, electronic, and dashed lines are labeled by appropriate indices.

The left and right external double lines are labeled by a set of indices $l_a^{n_a} \ldots l_c^{n_c}\alpha SL$ which are taken from the appropriate full external indices K_p' and K_p. We put an intermediate set of indices $l_a^{n_a} \ldots l_c^{n_c}\alpha' S' L'$ on the left and right intermediate double lines in the following way. For the left double lines we must start from the external index of the electronic configuration which stands on the left external double line and then moving from the external double line along the chain of adjacent left double lines, we drop successively each of the angular momenta belonging to the appropriate electronic configuration index. For the right double lines we use the same procedure, starting from the electronic configuration index of the right external double line and then moving from it along the chain of the right intermediate double lines. On the first left double line and the last right double line (which are also electronic lines) we put the orbital angular momenta which are left after the procedure of dropping out orbital angular momenta described above.

The right and left external electronic lines are labeled by

those orbital angular momenta which were dropped out by passing through the appropriate ternary vertex to which these external electronic lines were attached.

The internal electron lines are labeled by double indices $nl \in g_0$ for particle lines and $nl \in f_0$ for hole lines. The dashed lines of the diagram are labeled by arbitrary indices of orbital angular momenta l.

The orbital angular momentum diagram is constructed from the radial diagram as follows. We transform double and dashed lines as well as electron lines to usual single lines without directions and join the two external double lines to each other.

All nodes of the right and left ternary vertices of the orbital angular momentum diagram have a plus or minus sign. The plus sign is associated with a clockwise passing of the node counting the lines in this node in the following order: the outgoing double line, the incoming double line, the in- or outgoing electronic line; if we pass the node counter clockwise we must associate to it the minus sign.

The lines of the orbital diagram are labeled by indices of those angular momenta which labeled corresponding lines of the radial diagram.

The spin diagram is also constructed from the radial diagram. We transform double lines of the radial diagram into ordinary single lines by joining the external lines of the radial diagrams to each other. The dashed lines in quartic vertices as well as the closed loops formed by the internal electronic lines of the radial diagram must be dropped out.

All nodes of the spin diagram have a plus or minus sign. The plus sign in the quartic vertex node corresponds to a clockwise passing of the node counting the lines in this node in the following order: the outgoing electronic line, the incoming electronic line. the dashed interaction line; the minus sign corresponds to a counter clockwise counting. The plus sign of the node of the left or right ternary vertices means that we pass the node clockwise by counting the lines in this node in the following order: the outgoing line, the incoming double line, the in- or outgoing electronic line; the minus sign corresponds to the case of counter clockwise passing in the process of the counting.

The lines of the spin diagram are labeled by indices of those spin angular momenta which labeled the corresponding lines of the radial diagrams; this means that electronic lines are labeled by angular momentum $\frac{1}{2}$.

The contribution of an extended time-independent diagram for $M_p(K'_p, K_p|\Delta E')$ corresponds to the radial diagram plus the orbital angular momentum diagram plus the spin diagram.

The contribution from the diagram is equal to the product of the radial factor, orbital angular momentum factor and spin factor, summed over the indices, belonging to the lines of the radial diagram. This summation is carried over: (*a*) all possible arrangements of electronic configuration indices $l_a^{n_a}, \ldots, l_c^{n_c}$ on the intermediate right and left double lines of the diagram as well as over all possible arrangements of orbital angular momentum indices $l_a \ldots, l_c$ on the right and left external electronic lines; (*b*) all indices of terms $\alpha'S'L'$ labeling intermediate right and left double lines; (*c*) all double indices nl, belonging to the internal electronic lines; (*d*) all indices l, labeling dashed lines. Thus

$$\text{Contribution of the diagram} = \text{Contribution of the radial diagram} = \sum_{\alpha'L'S'} \sum_{nl} \sum_{l} \text{The radial factor} \times \text{Orbital factor} \times \text{Spin factor} \tag{19}$$

When carrying out the summation, we have to take into account for double indices of hole lines $nl \in f_0$, and for double indices of the particle lines $nl \in g_0$. The summation over every l labeling a dashed line is from 0 to $+\infty$. Besides, we have to take into account possible vanishing denominators in some terms and neglect these terms in the sum.

The radial factor associated with the radial diagram is composed of the following factors:

[1]. To every quartic vertex of the radial diagram with the dashed line l; outgoing electronic lines n_1l_1, n_2l_2, and appropriate incoming electronic lines n_1l_1, n_2l_2 corresponds the factor

$$R_l(n_1'l_1', n_2'l_2'; n_2l_2, n_1l_1)[l_1'l_1l][l_2'l_2l] \tag{20}$$

To obtain this factor it is necessary to take into account that the ordering of arguments in the radial integral corresponds to the following order of counting the electronic ends of the vertex: the outgoing end, another outgoing end, incoming end, corresponding to the latter, another incoming end.

[2]. The factor (21a) corresponds to every ternary vertex of the radial diagram with outgoing double line $l_a^{n_a} \dots l_c^{n_c}\alpha SL$, incoming double line $l_a^{n_a} \dots l_b^{n_b-1} \dots l_c^{n_c}\alpha' S' L'$ and incoming electronic line l_b

$$\sqrt{p}(l_a^{n_a} \dots l_c^{n_c}\alpha SL\{|l_a^{n_a} \dots l_b^{n_b-1} \dots l_c^{n_c}(\alpha' S' L')l_b SL) \quad \text{(21a)}$$

Every right ternary vertex of the radial diagram with incoming double line $l_a^{n_a} \dots l_c^{n_c}\alpha SL$, outgoing double line $l_a^{n_a} \dots l_b^{n_b-1} \dots l_c^{n_c}\alpha' S' L'$, and outgoing electronic line l_b gives the factor

$$\sqrt{p}(l_a^{n_a} \dots l_b^{n_b-1} \dots l_c^{n_c}(\alpha' S' L')l_b SL|\{l_a^{n_a} \dots l_c^{n_c}\alpha SL) \quad \text{(21b)}$$

In the formulae above the number p is equal to the number of electrons in the configuration $l_a^{n_a} \dots l_c^{n_c}$, i.e. we have to take

$$p = n_a + \dots + n_c$$

[3]. Every internal electronic line nl gives the factor

$$(2l + 1) \quad \text{(22a)}$$

and every external electronic line $n_0 l$ the factor

$$\sqrt{2l + 1} \quad \text{(22b)}$$

Every left and right intermediate double line $l_a^{n_a} \dots l_c^{n_c}\alpha SL$ gives the factor

$$\sqrt{2S + 1}\sqrt{2L + 1} \quad \text{(22c)}$$

External double lines do not give any factors.

[4]. Every vertical section between two adjacent quartic vertices gives the factor

$$\frac{1}{\mathscr{E} - p\varepsilon_0 - \Delta E} \quad \text{(23)}$$

where $\mathscr{E}$ is equal to the sum of single-electron energies of particle lines of the diagram minus the sum of single-electron energies of the hole lines.

[5]. The whole diagram gives the factor

$$2^l \frac{1}{\lambda} (-1)^{n+p+l+f+1}(-1)^{S+\frac{p}{2}+L+\frac{1}{2}\Sigma l+\frac{1}{2}\Sigma l} \quad \text{(24)}$$

where n is the number of quartic vertices in the diagram, p is the number of the external electronic lines (outgoing or incoming), l is the number of closed loops of electronic lines in the diagram, f is the number of hole lines of the diagram, S and L are the total spin and the total orbital angular momentum, labeling the external double lines of the diagram, the sums Σl are formed from the single electronic orbital angular momenta included in both electronic configuration indices, labeling the external double diagram lines; λ is the topological factor of the unextended initial time-independent diagram.

Let us calculate the contributions from the first-order diagrams for $M_p(K_p', K_p|\Delta E')$ using previously formulated rules.

There are three first-order diagrams: 1C, 1D, and 1E. They are shown in Fig. 5, which also depicts the corresponding abstract time-independent diagrams, radial diagrams, orbital angular momentum diagrams, and spin diagrams.

TABLE 2. Contribution of the first-order single-electron diagrams.

1C		$4 \sum_{n_1 l_1 \in f_0} (2l_1 + 1) R_0(n_1l_1, n_0l_a; n_0l_a, n_1l_1)$
1D	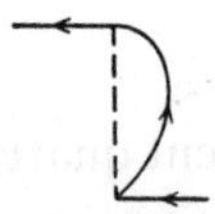	$-2 \sum_{n_1 l_1 \in f_0} (2l_1 + 1) \sum_l [l_a l_1 l]^2 R_l(n_1l_1, n_0l_a; n_1l_1, n_0l_a)$

Table 2 gives the expressions for $M(n_0l_a|\Delta E')$ for the diagrams 1C and 1D; these quantities determine the contributions from these diagrams

$$M(n_0l_a'S'L', n_0l_aSL|\Delta E') = \delta_{l_a', l_a}\delta_{L', l_a'}\delta_{L, l_a}\delta_{S', \frac{1}{2}}\delta_{S, \frac{1}{2}} \cdot M(n_0l_a|\Delta E')$$

Tables 3, 4, 5, and 6 give the expressions for the contributions of the diagram 1E to quantities $M_{1E}(l_a^2SL;\ l_B^2SL|\Delta E')$, $M_{1E}(l_al_bSL;\ l_c^2SL|\Delta E')$, $M_{1E}(l_a^2SL;\ l_cl_dSL|\Delta E')$, $M_{1E}(l_al_bSL;\ l_cl_dSL|\Delta E')$ respectively.

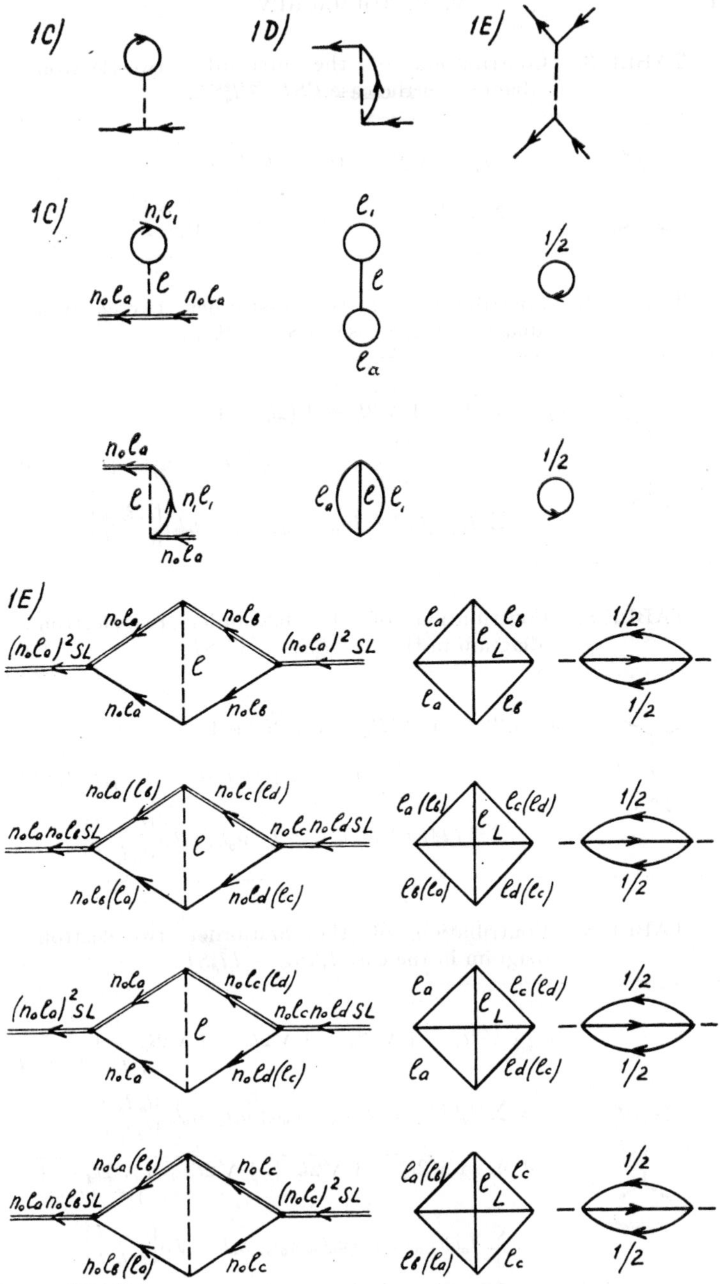

Fig. 5. The single-electron and two-electron first-order diagrams for $M_p(K_p', K_p|\Delta E)$. Diagrams of abstract theory, radial diagrams, orbital angular momentum diagrams, and spin diagrams. (There are two single-electron diagrams 1C and 1D and one two-electron diagram 1C.)

TABLE 3. Contribution of the first-order two-electron diagram in the case $l_a^2SL - l_b^2SL$.

1E

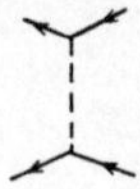

$$\delta_{S,\frac{1}{2},\frac{1}{2}}(2l_a+1)(2l_b+1)(-1)^{L+l_a+l_b} \times \sum_l [l_al_bl]^2 R_l(n_0l_a, n_0l_a; n_0l_b, n_0l_b)\begin{Bmatrix} l_a & l_b & l \\ l_b & l_a & L \end{Bmatrix}$$

TABLE 4. Contribution of the first-order two-electron diagram in the case $l_al_bSL - l_c^2SL$.

1E

$$\delta_{S,\frac{1}{2},\frac{1}{2}}\sqrt{2l_a+1}\,\sqrt{2l_b+1}\,(2l_c+1) \frac{1}{\sqrt{2}}(1+(-1)^{S+L+l_a+l_b})(-1)^{L+l_c+\frac{1}{2}(l_a+l_b)} \times \sum_l [l_al_cl][l_bl_cl] R_l(n_0l_a, n_0l_b; n_0l_c, n_0l_c)\begin{Bmatrix} l_a & l_c & l \\ l_c & l_d & L \end{Bmatrix}$$

TABLE 5. Contribution of the first-order two-electron diagram in the case $l_a^2SL - l_cl_dSL$.

1E

$$\delta_{S,\frac{1}{2},\frac{1}{2}}(2l_a+1)\sqrt{2l_c+1}\,\sqrt{2l_d+1} \frac{1}{\sqrt{2}}(1+(-1)^{S+L+l_c+l_d})(-1)^{L+l_a+\frac{1}{2}(l_c+l_d)} \times \sum_l [l_al_cl][l_al_dl] R_l(n_0l_a, n_0l_a; n_0l_c, n_0l_d)\begin{Bmatrix} l_a & l_c & l \\ l_d & l_a & L \end{Bmatrix}$$

TABLE 6. Contribution of the first-order two-electron diagram in the case $l_al_bSL - l_cl_dSL$.

1E

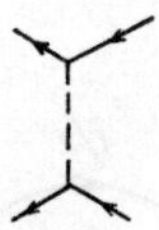

$$\delta_{S,\frac{1}{2},\frac{1}{2}}\sqrt{2l_a+1}\,\sqrt{2l_b+1}\,\sqrt{2l_c+1}\,\sqrt{2l_d+1}\,(-1)^{L+\frac{1}{2}(l_a+l_b+l_c+l_d)} \times \sum_l [l_al_cl][l_bl_dl] R_l(n_0l_a, n_0l_b; n_0l_d, n_0l_c)\begin{Bmatrix} l_a & l_b & l \\ l_d & l_c & L \end{Bmatrix}$$
$$+\,\delta_{S,\frac{1}{2},\frac{1}{2}}\sqrt{2l_a+1}\,\sqrt{2l_b+1}\,\sqrt{2l_c+1}\,\sqrt{2l_d+1}\,(-1)^{S+\frac{1}{2}(l_a+l_b-l_c-l_d)} \times \sum_l [l_al_dl][l_bl_cl] R_l(n_0l_a, n_0l_b; n_0l_c, n_0l_d)\begin{Bmatrix} l_a & l_d & l \\ l_c & l_b & L \end{Bmatrix}$$

4. THE EVALUATION OF $g(K_r', K_p'; K_p, K_r)$

As was shown previously (Section 4, Part 1) the problem of evaluation of general quantities $g(K_r', K_p';\ K_p, K_r)$ can be reduced to the evaluation of partial quantities $g(K_p', K_{p-1}';\ K_{p-1}, K_p)$. For these we have

$$g(K_p', K_{p-1}'; K_{p-1}, K_p) = \sum_{\mu} f^*(K_p', K_{p-1}'|\mu) f(K_p, K_{p-1}|\mu) \quad (25)$$

where

$$f(K_p, K_{p-1}|\mu) = \frac{1}{(p-1)!} \sum_{\lambda_1 \ldots \lambda_{p-1}} C^{K_p}_{\lambda_1 \ldots \lambda_{p-1}\mu} \overset{*}{C}{}^{K_{p-1}}_{\lambda_1 \ldots \lambda_{p-1}} \quad (26)$$

Assume that

$$K_p' \equiv (y_u^{n_u} \ldots y_w^{n_w} \beta SLM_S M_L)$$
$$K_{p-1}' \equiv (y_u^{n_u} \ldots y_v^{n_v - 1} \ldots y_w^{n_w} \beta' S' L' M_S' M_L')$$
$$K_{p-1} \equiv (x_a^{n_a} \ldots x_b^{n_b - 1} \ldots x_c^{n_c} \alpha' S' L' M_S' M_L')$$
$$K_p \equiv (x_a^{n_a} \ldots x_c^{n_c} \alpha SLM_S M_L),\ \mu \equiv lms$$

Then according to Eqs. (26) and (D1) f is given by:

$$\begin{aligned} f(K_p, K_{p-1}|\mu) &= \sqrt{p}\,\delta_{l,x_b} \\ &\times (x_a^{n_a} \ldots x_b^{n_b-1} \ldots x_c^{n_c}(\alpha' S' L') x_b SL| \} x_a^{n_a} \ldots x_c^{n_c} \alpha SL) \\ &\qquad \times (S' \tfrac{1}{2} M_S' s | S' \tfrac{1}{2} S M_S)(L' l M_L' m | L' l L M_L) \end{aligned}$$

$$\begin{aligned} f^*(K_p', K_{p-1}'|\mu) &= \sqrt{p}\,\delta_{l,y_v} \\ &\times (y_u^{n_u} \ldots y_w^{n_w} \beta SL \{| y_u^{n_u} \ldots y_v^{n_v-1} \ldots y_w^{n_w}(\beta' S' L') y_v SL) \\ &\qquad \times (S' \tfrac{1}{2} M_S' s | S' \tfrac{1}{2} S M_S)(L' l M_L' m | L' l L M_L) \end{aligned}$$

Substituting these expressions into Eq. (25) and using orthogonality properties of Clebsch–Gordan coefficients we obtain

$$\begin{aligned} g(K_p', K_{p-1}'; K_{p-1}, K_p) &= p\,\delta_{x_b, y_v} \\ &\times (x_a^{n_a} \ldots x_b^{n_b-1} \ldots x_c^{n_c}(\alpha' S' L') x_b SL| \} x_a^{n_a} \ldots x_c^{n_c} \alpha SL) \\ &\times (y_u^{n_u} \ldots y_w^{n_w} \beta SL \{| y_u^{n_u} \ldots y_v^{n_v-1} \ldots y_w^{n_w}(\beta' S' L') y_v SL) \end{aligned}$$

APPENDIX A

Collected Formulae for Spherical Harmonics[12,13]

The spherical harmonics $Y_{lm}(\Omega)$ where $l = 0, 1, 2,$ etc. and where for each l the projection m runs over $2l + 1$ values, $m = -l, -(l-1), \ldots l-1, l$, form the complete orthonormalized system for functions defined on the sphere.

$$\int d\Omega Y^*_{l'm'}(\Omega) Y_{lm}(\Omega) = \delta_{ll'}\delta_{mm'} \tag{A1}$$

where

$$\int d\Omega \equiv \int_0^{2\pi} d\varphi \int_0^{\pi} \sin\theta d\theta \tag{A1a}$$

The phases of the spherical harmonics are chosen so that

$$Y^*_{lm}(\Omega) = (-1)^m Y_{lm}(\Omega) \tag{A2}$$

For the spherical harmonics the following formula is valid

$$\int d\Omega Y_{l_1m_1}(\Omega) Y_{l_2m_2}(\Omega) Y_{l_3m_3}(\Omega)$$
$$= \left[\frac{(2l_1+1)(2l_2+1)(2l_3+1)}{4\pi}\right]^{\frac{1}{2}} \begin{pmatrix} l_1 & l_2 & l_3 \\ 0 & 0 & 0 \end{pmatrix} \begin{pmatrix} l_1 & l_2 & l_3 \\ m_1 & m_2 & m_3 \end{pmatrix} \tag{A3}$$

where Wigner's $3j$ coefficients are on the right-hand side; as well as the formula

$$\frac{1}{r_{12}} = \sum_{l=0}^{+\infty} u_l(r_1, r_2) \frac{4\pi}{2l+1} \sum_{m=-l}^{+l} Y_{lm}(\Omega_1) Y^*_{lm}(\Omega_2) \tag{A4}$$

where

$$u_l(r_1, r_2) = \begin{cases} \dfrac{r_1^l}{r_2^{l+1}} \text{ for } r_2 > r_1 \\[2ex] \dfrac{r_2^l}{r_1^{l+1}} \text{ for } r_2 < r_1 \end{cases} \tag{A4a}$$

In (A4) and (A4a) we have $r_{12} = |\mathbf{r}_1 - \mathbf{r}_2|$, $\mathbf{r}_1 \equiv (r_1, \Omega_1)$, $\mathbf{r}_2 = (r_2, \Omega_2)$.

APPENDIX B

Properties of Wigner's 3j Coefficients and their Relation to Clebsch–Gordan Coefficients[12–14]

The Clebsch–Gordan coefficients are needed for the coupling of two angular momenta:

$$\psi_{j_1j_2jm} = \sum_{m_1m_2} (j_1j_2m_1m_2|j_1j_2jm)\psi_{j_1j_2m_1m_2} \tag{B1}$$

where $\psi_{j_1j_2jm}$ are simultaneous eigenfunctions of the operators j_1^2, j_2^2, j^2, j_z and $\psi_{j_1j_2m_1m_2}$ are simultaneous eigenfunctions of the operators j_1^2, j_2^2, j_{1z}, j_{2z}.

The inverse relation has the following form:

$$\psi_{j_1j_2m_1m_2} = \sum_{jm} (j_1j_2jm|j_1j_2m_1m_2)\psi_{j_1j_2jm} \tag{B2}$$

and

$$(j_1j_2jm|j_1j_2m_1m_2) = (j_1j_2m_1m_2|j_1j_2jm) \tag{B3}$$

so that Clebsch–Gordan coefficients are real.

For fixed j_1, j_2, Clebsch–Gordan coefficients form an orthogonal matrix and we have two orthogonality relations

$$\sum_{m_1m_2} (m_1m_2|j'm')(m_1m_2|jm) = \delta_{j'j}\delta_{m'm} \tag{B4a}$$

$$\sum_{jm} (m_1'm_2'|jm)(m_1m_2|jm) = \delta_{m_1'm_1}\delta_{m_2'm_2} \tag{B4b}$$

by virtue of these relations the formulae (B1) and (B2) really describe the transition from one orthonormalized system of functions to another one. The Wigner 3j symbols are more convenient than Clebsch–Gordan coefficients; according to the definition

$$(j_1j_2m_1m_2|j_1j_2jm) = (-1)^{j_1-j_2+m}\sqrt{2j+1}\begin{pmatrix} j_1 & j_2 & j \\ m_1 & m_2 & -m \end{pmatrix} \tag{B5}$$

We collect some important properties of 3j coefficients:

$$\begin{pmatrix} j_1 & j_2 & j_3 \\ m_1 & m_2 & m_3 \end{pmatrix} \quad \text{is a real number not equal to zero only if } m_1 + m_2 + m_3 = 0 \text{ and if the vectors } j_1, j_2, j_3 \text{ form a triangle} \tag{B6}$$

$$\begin{pmatrix} j_1 & j_2 & j_3 \\ m_1 & m_2 & m_3 \end{pmatrix} \quad \text{is invariant in cyclic permutations of 123} \tag{B7}$$

$$\begin{pmatrix} j_1 & j_2 & j_3 \\ m_1 & m_2 & m_3 \end{pmatrix} = (-1)^{j_1+j_2+j_3} \begin{pmatrix} j_2 & j_1 & j_3 \\ m_2 & m_1 & m_3 \end{pmatrix} \tag{B8}$$

$$\begin{pmatrix} j_1 & j_2 & j_3 \\ -m_1 & -m_2 & -m_3 \end{pmatrix} = (-1)^{j_1+j_2+j_3} \begin{pmatrix} j_1 & j_2 & j_3 \\ m_1 & m_2 & m_3 \end{pmatrix} \tag{B9}$$

Here, the sum $j_1 + j_2 + j_3$ is always integer (either all three moments are integers, or two of them are half-integers).

An important $3j$ coefficient is the one with three projections equal to zero $m_1 = m_2 = m_3 = 0$.

For this $3j$ coefficient we have

$$\begin{pmatrix} j_1 & j_2 & j_3 \\ 0 & 0 & 0 \end{pmatrix} = 0 \text{ if the sum } j_1 + j_2 + j_3 \text{ is odd} \tag{B10}$$

$$\begin{pmatrix} j_1 & j_2 & j_3 \\ 0 & 0 & 0 \end{pmatrix} = (-1)^{\frac{1}{2}(j_1+j_2+j_3)} \left| \begin{pmatrix} j_1 & j_2 & j_3 \\ 0 & 0 & 0 \end{pmatrix} \right| \tag{B11}$$

Where $\|$ signify that the absolute value is taken. We shall use an abridged notation:

$$[j_1 j_2 j_3] = \left| \begin{pmatrix} j_1 & j_2 & j_3 \\ 0 & 0 & 0 \end{pmatrix} \right| \tag{B12}$$

According to (B10) $[j_1 j_2 j_3]$ is non-vanishing only when the sum $j_1 + j_2 + j_3$ equals an even integer.

For the special case of $j_3 = 0$ the $3j$ coefficient has a very simple form:

$$\begin{pmatrix} j & j & 0 \\ m & -m & 0 \end{pmatrix} = \frac{(-1)^{j-m}}{\sqrt{2j+1}} \tag{B13}$$

from which, in particular, it follows that

$$[jj0] = \frac{1}{\sqrt{2j+1}} \tag{B13a}$$

Using (B5) the orthogonality relations can be written in the form:

$$\sum_{m_1 m_2} \begin{pmatrix} j_1 & j_2 & j' \\ m_1 & m_2 & m' \end{pmatrix} \begin{pmatrix} j_1 & j_2 & j \\ m_1 & m_2 & m \end{pmatrix} = \frac{1}{2j+1} \delta_{j',j} \delta_{j_1,j_2,j} \delta_{m',m} \tag{B14}$$

$$\sum_{jm} (2j+1) \begin{pmatrix} j_1 & j_2 & j \\ m_1' & m_2' & m \end{pmatrix} \begin{pmatrix} j_1 & j_2 & j \\ m_1 & m_2 & m \end{pmatrix} = \delta_{m_1' m_1} \delta_{m_2' m_2} \tag{B15}$$

where $\delta_{j_1,j_2,j_3} = 1$ when the vectors j_1, j_2, j form a triangle and $\delta_{j_1,j_2,j_3} = 0$ when this is not the case. Because the sum $m_1 + m_2 + m_3 = 0$, we get for (B14) a more convenient form

$$\sum_{m_1 m_2} (-1)^{j-m}(-1)^{j_1-m_1}(-1)^{j_2-m_2} \begin{pmatrix} j_1 & j_2 & j' \\ m_1 & m_2 & m' \end{pmatrix} \begin{pmatrix} j_1 & j_2 & j \\ -m_1 & -m_2 & -m \end{pmatrix} = \frac{1}{2j+1} \delta_{j',j}\delta_{j_1,j_2,j}\delta_{m',m} \tag{B16}$$

Combining (B14) and (B13) we obtain one more formula

$$\sum_{m_1} (-1)^{j_1-m_1} \begin{pmatrix} j_1 & j_1 & j \\ m_1 & -m_1 & m \end{pmatrix} = \sqrt{2j_1+1}\, \delta_{j,0}\delta_{m,0} \tag{B17}$$

Let us consider now Racah's W coefficient, which is defined as a sum of four Clebsch–Gordan coefficients; thus according to the definition

$$\begin{aligned} &[(2J'+1)(2J''+1)]^{\frac{1}{2}} W(j_1 j_2 J j_3; J'J'') \\ &\quad = \sum_{m_1 m_2 m_3 M' M''} (J' j_3 J M | J' j_3 M' m_3)(j_1 j_2 J' M' | j_1 j_2 m_1 m_2) \\ &\qquad \times (j_2 j_3 m_2 m_3 | j_2 j_3 J'' M'')(j_1 J'' m_1 M'' | j_1 J'' J M) \end{aligned} \tag{B18}$$

Just as instead of Clebsch–Gordan coefficients it is more convenient to consider more symmetrical Wigner $3j$ coefficients, so instead of the Racah W coefficient we consider for convenience the so-called $6j$ coefficient. According to the definition we have for the $6j$ coefficient:

$$\begin{Bmatrix} j_1 & j_2 & j_3 \\ l_1 & l_2 & l_3 \end{Bmatrix} = (-1)^{j_1+j_2+l_1+l_2} W(j_1 j_2 l_2 l_1; j_3 l_3) \tag{B19}$$

We now point out the most important properties of the $6j$ coefficients

$\begin{Bmatrix} j_1 & j_2 & j_3 \\ l_1 & l_2 & l_3 \end{Bmatrix}$ is a real number (B20)

$\begin{Bmatrix} j_1 & j_2 & j_3 \\ l_1 & l_2 & l_3 \end{Bmatrix}$ is invariant in any permutation of columns (B21)

$\begin{Bmatrix} j_1 & j_2 & j_3 \\ l_1 & l_2 & l_3 \end{Bmatrix}$ is invariant against interchange of the upper and lower arguments in each of any two columns (B22)

Writing the corresponding formula to (B18) using $3j$ and $6j$ coefficients we obtain the important formula

$$\begin{Bmatrix} j_1 & j_2 & j_3 \\ l_1 & l_2 & l_3 \end{Bmatrix} = \sum_{\substack{m_1m_2m_3 \\ n_1n_2n_3}} (-1)^{j_1-m_1}(-1)^{j_2-m_2}(-1)^{j_3-m_3}(-1)^{l_1-n_1}(-1)^{l_2-n_2}(-1)^{l_3-n_3}$$

$$\times \begin{pmatrix} j_1 & j_2 & j_3 \\ m_1 & m_2 & -m_3 \end{pmatrix}\begin{pmatrix} j_1 & l_2 & l_3 \\ -m_1 & n_2 & n_3 \end{pmatrix}\begin{pmatrix} l_1 & l_2 & j_3 \\ n_1 & -n_2 & m_3 \end{pmatrix}\begin{pmatrix} l_1 & j_2 & l_3 \\ -n_1 & -m_2 & -n_3 \end{pmatrix} \quad \text{(B23)}$$

Using Eq. (B18) we now derive the expressions for the product of two Clebsch–Gordan coefficients in terms of the sum of two other Clebsch–Gordan coefficients.

First let us introduce the functions f and g, defined below in which these are functions of m_1, m_2, m_3;

$$f_{J'JM,m_1m_2m_3} = (j_1j_2m_1m_2|j_1j_2J'M')(J'j_3M'm_3|J'j_3JM) \quad \text{(B24)}$$

$$g_{J''JM,m_1m_2m_3} = (j_2j_3m_2m_3|j_2j_3J''M'')(j_1J''m_1M''|j_1J''JM) \quad \text{(B25)}$$

It is easy to see that these functions satisfy orthonormalization conditions (the functions are real)

$$\sum_{m_1m_2m_3} f_{J'JM,m_1m_2m_3} f_{\bar{J}'\bar{J}\bar{M},m_1m_2m_3} = \delta_{J',\bar{J}'}\delta_{J,\bar{J}}\delta_{M,\bar{M}} \quad \text{(B26a)}$$

$$\sum_{J'JM} f_{J'JM,m_1m_2m_3} f_{J'JM,\bar{m}_1\bar{m}_2\bar{m}_3} = \delta_{m_1\bar{m}_1}\delta_{m_2,\bar{m}_2}\delta_{m_3,\bar{m}_3} \quad \text{(B26b)}$$

$$\sum_{m_1m_2m_3} g_{J''JM,m_1m_2m_3} g_{\bar{J}''\bar{J}\bar{M},m_1m_2m_3} = \delta_{J''\bar{J}''}\delta_{J,\bar{J}}\delta_{M,\bar{M}} \quad \text{(B27a)}$$

$$\sum_{J''JM} g_{J''JM,m_1m_2m_3} g_{J''JM,\bar{m}_1\bar{m}_2\bar{m}_3} = \delta_{m_1,\bar{m}_1}\delta_{m_2,\bar{m}_2}\delta_{m_3\bar{m}_3} \quad \text{(B27b)}$$

We prove, for instance, the first relation (B26a); substituting the expressions for f functions (B24) into the Eq. (B26a) we obtain

$$\begin{aligned}
&\sum_{m_1m_2m_3} f_{J'JM,m_1m_2m_3} f_{\bar{J}'\bar{J}\bar{M},m_1m_2m_3} \\
&= \sum_{m_1m_2m_3} (j_1j_2m_1m_2|j_1j_2J'M')(J'j_3M'm_3|J'j_3JM) \\
&\qquad\qquad \times (j_1j_2m_1m_2|j_1j_2\bar{J}'\bar{M}')(\bar{J}'j_3\bar{M}'m_3|\bar{J}'j_3\bar{J}\bar{M}) \\
&= \sum_{m_3} \delta_{J',\bar{J}'}\delta_{M',\bar{M}'}(J'j_3M'm_3|J'j_3JM)(\bar{J}'j_3\bar{M}'m_3|\bar{J}'j_3\bar{J}\bar{M}) \\
&= \delta_{J,\bar{J}'} \sum_{m_3M'} (J'j_2M'm_3|J'j_3JM)(J'j_3M'm_3|J'j_3\bar{J}\bar{M}) \\
&= \delta_{J',\bar{J}'}\delta_{J,\bar{J}}\delta_{M,\bar{M}}
\end{aligned}$$

We have introduced an additional summation over m' and used twice the orthogonality condition (B4a).

Expanding the orthonormalized system of functions $g_{J'JM,m_1m_2m_3}$ in terms of the orthonormalized system of functions $f_{J'J\bar{M},m_1m_2m_3}$; we have

$$g_{J'JM,m_1m_2m_3} = \sum_{J'J\bar{M}} c_{J'J\bar{M},J'JM} f_{J'J\bar{M},m_1m_2m_3} \tag{B28}$$

Multiplying this relation from the left and right by the function $f_{J'J\bar{M},m_1m_2m_3}$, summing over all m_1, m_2, m_3 and using (B26a) we obtain

$$c_{J'J\bar{M},J'JM} = \sum_{m_1m_2m_3} f_{J'J\bar{M},m_1m_2m_3} g_{J'JM,m_1m_2m_3} \tag{B29}$$

It is evident that

$$c_{J'J\bar{M},J'JM} = c^J_{J',J'} \delta_{J',J} \delta_{\bar{M},M} \tag{B30}$$

from which according to Eq. (B29)

$$c^J_{J',J'} = \sum_{m_1m_2m_3} f_{J'JM,m_1m_2m_3} g_{J'JM,m_1m_2m_3} \tag{B29a}$$

Substituting the explicit expressions for f and g into Eqs. (B24) and (B25) and introducing an additional summation over M, M' we get

$$c^J_{J',J'} = \sum_{m_1m_2m_3} \sum_{\bar{M}'M'} (j_1j_2m_1m_2|j_1j_2\bar{J}'\bar{M}')(\bar{J}'j_3\bar{M}'m_3|\bar{J}'j_3JM) \\ \times (j_2j_3m_2m_3|j_2j_3J'M')(j_1J'm_1M'|j_1J'JM)$$

or by virtue of Eq. (B18)

$$c^J_{J',J''} = [(2J'+1)(2J''+1)]^{\frac{1}{2}} W(j_1j_2Jj_3; J'J'') \tag{B31}$$

Let us now return to the Eq. (B28), substituting (B24), (B25), and (B30), (B31) into it we obtain the following useful formula

$$(j_2j_3m_2m_3|j_2j_3J'M')(j_1J'm_1M'|j_1J'JM) \\ = \sum_{J''} [(2J'+1)(2J''+1)]^{\frac{1}{2}} W(j_1j_2Jj_3; J''J') \\ \times (j_1j_2m_1m_2|j_1j_2J''M'')(J''j_3M''m_3|J''j_3JM) \tag{B32}$$

which can be written

$$
\begin{aligned}
&(-1)^{j_2+j_3-J'}(j_2j_3m_2m_3|j_2j_3J'M')(J'j_1M'm_1|J'j_1JM) \\
&\quad = (j_3j_2m_3m_2|j_3j_2J'M')(J'j_1M'm_1|J'j_1JM) \\
&\quad = \sum_{J''} [(2J'+1)(2J''+1)]^{\frac{1}{2}} W(j_1j_2Jj_3; J''J')(-1)^{j_1+j_2+j_3-J} \\
&\qquad\qquad \times (j_1j_2m_1m_2|j_1j_2J''M'')(J''j_3M''m_3|J''j_3JM) \qquad \text{(B32a)}
\end{aligned}
$$

$$
\begin{aligned}
&(-1)^{j_3+j_1-J}(j_1j_3m_1m_3|j_1j_3J'M')(J'j_2M'm_2|J'j_2M'm_2|J'j_2JM) \\
&\quad = (j_3j_1m_3m_1|j_3j_1J'M')(J'j_2M'm_2|J'j_2JM) \\
&\quad = \sum_{J''} [(2J'+1)(2J''+1)]^{\frac{1}{2}} W(j_2j_1Jj_3; J''J')(-1)^{J-j_3-J''} \\
&\qquad\qquad \times (j_1j_2m_1m_2|j_1j_2J''M'')(J''j_3M''m_3|J''j_3JM) \qquad \text{(B32b)}
\end{aligned}
$$

Note also that

$$(j_2j_1m_2m_1|j_2j_1JM) = (-1)^{j_1+j_2-J}(j_1j_2m_1m_2|j_1j_2JM) \qquad \text{(B33)}$$

APPENDIX C

Angular Momentum Diagrams[12]

The sums over projections of angular momenta in products of $3j$ coefficients as well as transformations of these sums are conveniently interpreted by means of angular momentum diagrams.

The angular momentum diagram is defined as an assembly of directed lines with fixed angular momenta j, which begin and end at ternary nodes of the diagram. The ternary nodes are the meeting points of three lines, every node is assigned with a plus or minus sign.

In order to set up the appropriate angular momentum sum from the given angular momentum diagram, one must also label all lines of the angular momentum diagram projections of angular momenta.

Every angular momentum sum corresponding to a given angular momentum diagram is equal to the sum over projections of angular momenta m (which label the lines of the diagram) of the product of the following factors:

[1]. Every line jm of the angular momentum diagram gives the sign factor:

$$(-1)^{j-m} \tag{C1a}$$

[2]. The appropriate $3j$ coefficient

$$\begin{pmatrix} j_1 & j_2 & j_3 \\ \pm m_1 & \pm m_2 & \pm m_3 \end{pmatrix} \tag{C1b}$$

is attached to every node of the angular momentum diagram where lines j_1m_1, j_2m_2, j_3m_3, meet. These lines are counted clockwise if the appropriate node has a plus sign and counted counter clockwise when the appropriate node has a minus sign. In the expression for the $3j$ coefficient the projections of outgoing lines of the node are taken with a plus sign, the incoming lines are taken with a minus sign.

As an example we consider some simple diagrams. The most simple angular momentum sum (B17) is conventionally represented by a special diagram (this diagram has a free end with zero angular momentum).

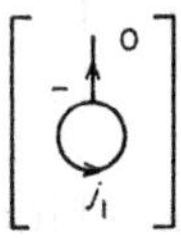

$$= \sum_{m_1} (-1)^{j_1-m_1} \begin{pmatrix} j_1 & j_1 & 0 \\ m_1 & -m_1 & 0 \end{pmatrix} = \sqrt{2j_1+1} \tag{C2}$$

The angular momentum sum (B16) is represented by the following diagram

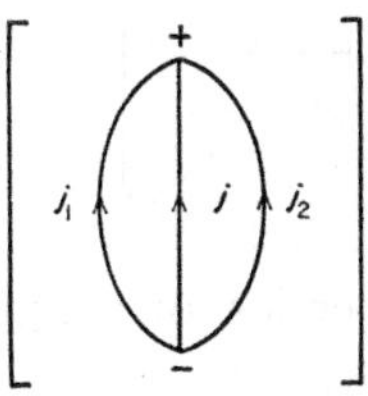

$$= \sum_{m_1 m m_2} (-1)^{j_1-m_1}(-1)^{j_2-m_2}(-1)^{j-m} \begin{pmatrix} j_1 & j_2 & j \\ m_1 & m_2 & m \end{pmatrix} \begin{pmatrix} j_1 & j_2 & j \\ -m_1 & -m_2 & -m \end{pmatrix} = \delta_{j_1,j_2,j} \tag{C3}$$

And the sum (B23) can be represented by the diagram

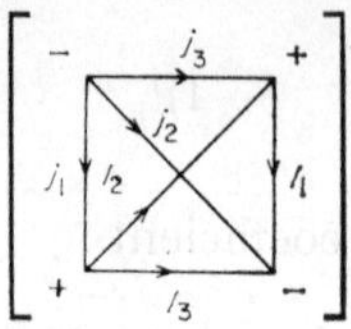

$$= \sum_{\substack{m_1 m_2 m_3 \\ n_1 n_2 n_3}} (-1)^{j_1-m_1}(-1)^{j_2-m_2}(-1)^{j_3-m_3}(-1)^{l_1-n_1}(-1)^{l_2-n_2}(-1)^{l_3-n_3}$$

$$\times \begin{pmatrix} j_1 & j_2 & j_3 \\ m_1 & m_2 & -m_3 \end{pmatrix} \begin{pmatrix} j_1 & l_2 & l_3 \\ -m_1 & n_2 & n_3 \end{pmatrix} \begin{pmatrix} l_1 & l_2 & j_3 \\ -n_1 & -m_2 & -n_3 \end{pmatrix} \begin{pmatrix} l_1 & j_2 & l_3 \\ -n_1 & -m_2 & -n_3 \end{pmatrix}$$

$$= \begin{Bmatrix} j_1 & j_2 & j_3 \\ l_1 & l_2 & l_3 \end{Bmatrix} \tag{C4}$$

From previously formulated rules it follows immediately that:

[1]. The sum of the angular momentum diagram where the direction of one of its lines is changed equals the sum of the original diagram multiplied by a sign factor (using the fact that $(-1)^{j+m} = (-1)^{2j}(-1)^{j-m}$):

$$\left[\text{diagram}\right] = (-1)^{2j}\left[\text{diagram}\right] \tag{C5}$$

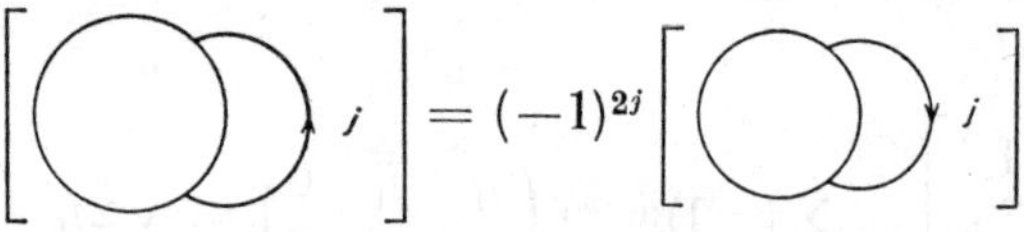

[2]. The sum of the diagram where the sign of one of its nodes is changed is also reduced to the sum of the original diagram multiplied by a sign factor; thus we have

$$\left[\text{diagram}\right] = (-1)^{j_1+j_2+j_3}\left[\text{diagram}\right] \tag{C6}$$

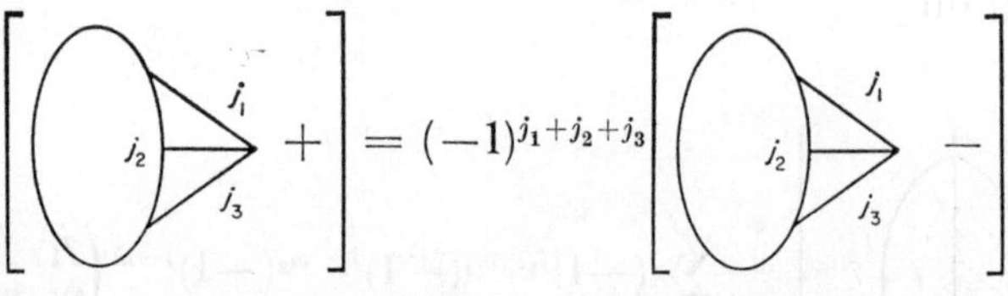

Let us now discuss some less trivial properties of angular momentum diagrams and the appropriate sums.

[A]. In any sum where the summation is carried out over (all) projections of angular momenta, which label the lines of the diagram, it turns out that the last summation does not depend on the projection of angular momentum over which one is summing. So this last summation is reduced merely to a multiplication of the expression considered by the number of terms of the sum, i.e. by $(2l + 1)$, where l is the angular momentum corresponding to the last projection of angular momentum. Certainly, it does not matter which projection one starts with while summing over the projections of angular momenta. The graphical representation of this property is that the sum of the diagram with a disconnected line is merely reduced to the sum of an ordinary diagram obtained by joining the free ends to each other.

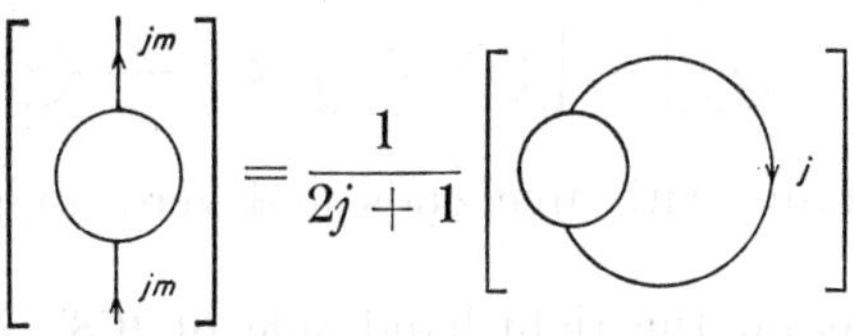

(C7)

[B]. If we imagine that any angular momentum diagram is cut by an arbitrary vertical section, so that the two parts thus formed are connected to each other,

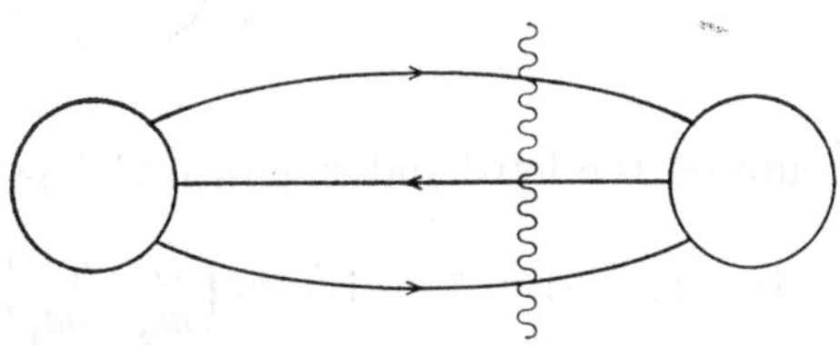

then the appropriate sum of the diagram considered is non-vanishing only if the angular momenta which label the lines cut by the section form a polygon (a triangle in the case of three angular momenta).

Besides, the algebraic sum of magnetic quantum numbers, which label lines cut by the vertical section must be equal to zero (the sign of the projection depends on the direction of the line).

The following useful corollaries of the statements previously formulated deal with the sums of diagrams which can be divided into two parts joined either by one or two lines.

Consider the diagram which can be divided into two parts joined by only *one* line

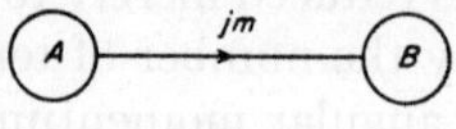

The sum of this diagram is non-vanishing only if the angular momentum of the joining line is equal to zero, $j = 0$, and therefore the projection of angular momentum of this line equals zero, $m = 0$.

If we may assume that $j = m = 0$, then the sum of the diagram considered is factorizable into two factors.

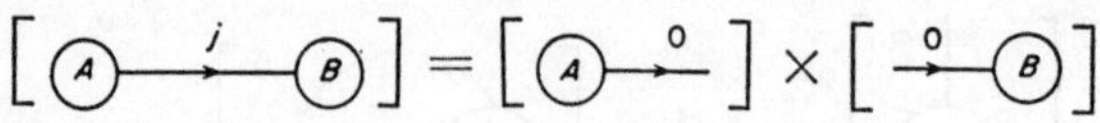

$\times\ \delta_{j,0}$ (C8)

in which diagrams with free ends and zero angular momenta appear.

The diagrams on the right-hand side of (C8) are reduced to ordinary diagrams; by a straightforward calculation we get

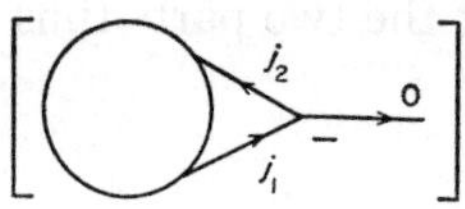

$$= \frac{\delta_{j_1,j_2}}{\sqrt{2_{j_1} + 1}}$$

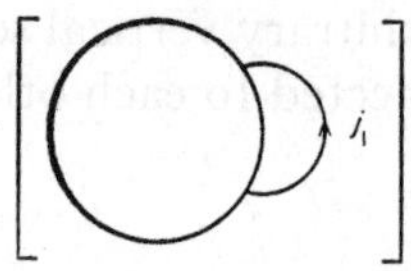

(C9)

In fact the sum on the left-hand side of (C9) has the form

$$\sum_{m_1 m_2} A(m_1 m_2)(-1)^{j_1 - m_1}(-1)^{j_2 - m_2} \begin{pmatrix} j_2 & j_1 & 0 \\ m_2 & -m_1 & 0 \end{pmatrix}$$

where $A(m_1 m_2)$ is some quantity. Using the formula (B13) for the $3j$ coefficient with a zero in the last column, we obtain

$$\sum_{m_1 m_2} A(m_1 m_2)(-1)^{j_1 - m_1}(-1)^{j_2 - m_2} \frac{(-1)^{j_2 - m_2}}{\sqrt{2j_1 + 1}} \delta_{m_2, m_1} \delta_{j_2, j_1}$$

$$= \frac{\delta_{j_1, j_2}}{\sqrt{2j_1 + 1}} \sum_{m_1} (-1)^{j_1 - m_1} A(m_1, m_1)$$

This sum is exactly the sum of the right-hand side of (C9); so the validity of (C9) is proved. There is only one exception when

(C9) cannot be used; this is the most simple diagram with only one electron line. But in this case we can use Eq. (C2).

Let us now discuss the diagram which can be divided into two parts connected by *two* lines.

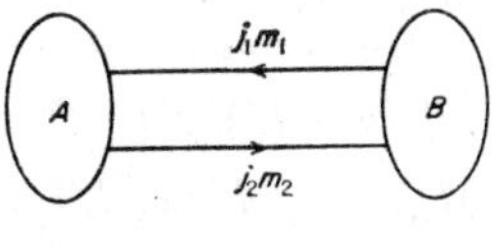

The sum of this diagram is non-vanishing only if $j_1 = j_2$, and moreover in this sum $m_1 = m_2$.

If $j_2 = j_1$, $m_2 = m_1$, we obtain

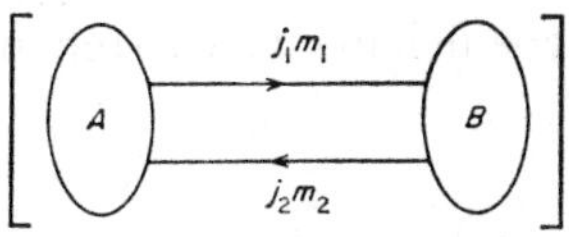

$= \delta_{j_1j_2}$

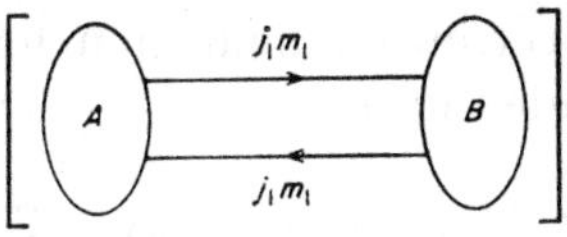

$= \dfrac{\delta_{j_1j_2}}{2j_1+1} \times$

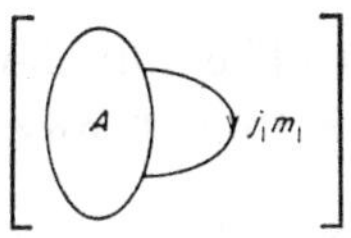

$\times$

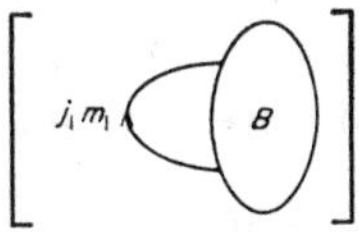

(C10)

There is only one sum over m_1 on the left-hand side of (C10) but there are two such sums on the right-hand side; that is why the factor $\dfrac{1}{2j_1+1}$ appears on the right-hand side of (C10).

APPENDIX D

Fractional Parentage Coefficients and their Evaluation[14]

The definition of fractional parentage coefficients is given here by the formula expressing orthonormalized antisymmetric functions $C^{l_a^{n_a}\ldots l_c^{n_c}\alpha SLM_SM_L}_{l_1m_1s_1\ldots l_pm_ps_p}$ for p electrons ($n_a + \ldots + n_c = p$) in terms of orthonormalized antisymmetric functions of the same kind $C^{l_a^{n_a}\ldots l_b^{n_b-1}\ldots l_c^{n_c}\alpha SLM_SM_L}_{l_1m_1s_1\ldots l_{p-1}m_{p-1}s_{p-1}}$ for $p - 1$ electrons (the configuration

$l_a^{n_a} \ldots l_c^{n_c}$ has one more electron l_b compared to the configuration $l_a^{n_a} \ldots l_b^{n_b-1} \ldots l_c^{n_c}$). This definition is given by:

$$
\begin{aligned}
& C^{l_a^{n_a} \ldots l_c^{n_c} \alpha SLM_S M_L}_{l_1 m_1 s_1, \ldots, l_p m_p s_p} \\
&\quad = \sqrt{p} \sum_{b=a}^{c} \sum_{\alpha' S' L'} \sum_{M_S' M_L'} C^{l_a^{n_a} \ldots l_b^{n_b-1} \ldots l_c^{n_c} \alpha' S' L' M_S' M_L'}_{l_1 m_1 s_1, \ldots, l_{p-1} m_{p-1} s_{p-1}} \delta_{l_p, l_b} \\
&\quad \times (l_a^{n_a} \ldots l_b^{n_b-1} \ldots l_c^{n_c}(\alpha' S' L') l_b SL | \} l_a^{n_a} \ldots l_c^{n_c} \alpha SL) \\
&\quad \times (S' \tfrac{1}{2} M_S' s_p | S' \tfrac{1}{2} SM_S)(L' l_p M_L' m_p | L' l_p LM_L) \qquad \text{(D1)}
\end{aligned}
$$

Besides the appropriate fractional parentage coefficients there are Clebsch–Gordan coefficients in this formula as well. Both are real.

Besides these true fractional parentage coefficients (as we call them) we shall consider for convenience the auxiliary fractional parentage coefficients defined below. The index $l_a^{n_a} \ldots l_c^{n_c}$ of the electron configuration for these coefficients is written in an expanded form

$$
l_1^0 \ldots l_p^0 \equiv \overbrace{l_a \ldots l_a}^{n_a} \ldots \overbrace{l_c \ldots l_c}^{n_c} \qquad \text{(D2)}
$$

where $n_a + \ldots + n_c = p$. The auxiliary fractional parentage coefficients are defined in a similar way to the true coefficients

$$
\begin{aligned}
C^{l_1^0 \ldots l_p^0 \alpha SLM_S M_L}_{l_1 m_1 s_1, \ldots, l_p m_p s_p} &= \sqrt{p} \sum_{i=1}^{p} \sum_{\alpha' S' L'} \sum_{M_S' M_L'} C^{l_1^0 \ldots l_{i-1}^0 l_{i+1}^0 \ldots l_p^0 \alpha' S' L' M_S' M_L'}_{l_1 m_1 s_1, \ldots, l_{p-1} m_{p-1} s_{p-1}} \delta_{l_p, l_i^0} \\
&\times (l_1^0 \ldots l_{i-1}^0 l_{i+1}^0 \ldots l_p^0 (\alpha' S' L') l_i^0 SL | \} l_1^0 \ldots l_p^0 \alpha SL) \\
&\times (S' \tfrac{1}{2} M_S' s_p | S' \tfrac{1}{2} SM_S)(L' l_p M_L' m_p | L' l_p LM_L) \qquad \text{(D3)}
\end{aligned}
$$

where the configuration $l_1^0 \ldots l_{i-1}^0 l_{i+1}^0 \ldots l_p^0$ is obtained from the initial configuration $l_1^0 \ldots l_p^0$ by the cancellation of the ith angular momentum l_i^0. The auxiliary fractional parentage coefficients also are real.

The true fractional parentage coefficients can be easily expressed in terms of the auxiliary coefficients; as a straightforward corollary of (D1), (D2) we have

$$
\begin{aligned}
& (l_a^{n_a} \ldots l_b^{n_b-1} \ldots l_c^{n_c}(\alpha' S' L') l_b SL | \} l_a^{n_a} \ldots l_c^{n_c} \alpha SL) \\
&\quad = \sum_{i \in \mathscr{A}_b} (\overbrace{l_a \ldots l_a}^{n_a} \ldots \overbrace{l_b \ldots l_b}^{n_{b-1}} \ldots \overbrace{l_c \ldots l_c}^{n_c} \\
&\quad (\alpha' S' L') l_b SL | \} l_a \ldots l_a \ldots l_c \ldots l_c \alpha SL)_i \qquad \text{(D4)}
\end{aligned}
$$

where the sum $i \in \mathscr{A}_b$ goes over those indices i which label the place occupied by angular momenta l_b in the configuration $l_a \ldots l_a \ldots l_b \ldots l_b \ldots l_c \ldots l_c$; compare with (D2).

The coefficients C are orthonormalized; using this property we get

$$\delta_{\alpha_a,\alpha_b} = \frac{1}{p!}\sum_{l_1m_1s_1,\ldots,l_pm_ps_p} C^{l_1^0\ldots l_p^0\alpha_aSLM_SM_L}_{l_1m_1s_1,\ldots,l_pm_ps_p} C^{l_1^0\ldots l_p^0\alpha_bSLM_SM_L}_{l_1m_1s_1,\ldots,l_pm_ps_p}$$

$$= \sum_{l_pm_ps_p}\sum_{i=1}^{p}\sum_{j=1}^{p}\sum_{\alpha_a'S_a'L_a'}\sum_{M_{Sa}'M_{La}'}\sum_{\alpha_b'S_b'L_b'}\sum_{M_{Sb}'M_{Lb}'} \delta_{l_p,l_i^0}\delta_{l_p,l_j^0}$$

$$\times\, \delta(l_1^0 \ldots l_{i-1}^0 l_{i+1}^0 \ldots l_p^0;\ l_1^0 \ldots l_{j-1}^0 l_{j+1}^0 \ldots l_p^0)$$

$$\delta_{\alpha_a',\alpha_b'}\delta_{S_a',S_b'}\delta_{L_a',L_b'}\delta_{M_{Sa}',M_{Sb}'}\delta_{M_{La}',M_{Lb}'}$$

$$\times\, (l_1^0 \ldots l_{i-1}^0 l_{i+1}^0 \ldots l_p^0(\alpha_a'S_a'L_a')l_i^0SL|\}l_1^0 \ldots l_p^0\alpha_aSL)_i$$

$$\times\, (l_1^0 \ldots l_{j-1}^0 l_{j+1}^0 \ldots l_p^0(\alpha_b'S_b'L_b')l_j^0SL|\}l_1^0 \ldots l_p^0\alpha_bSL)_j$$

$$\times\, (S_a'\tfrac{1}{2}M_{Sa}'s_p|S_a'\tfrac{1}{2}M_{Sa})(L_a'l_pM_{La}'m_p|L_a'l_pL_aM_{La})$$

$$\times\, (S_b'\tfrac{1}{2}M_{Sb}'s_p|S_b'\tfrac{1}{2}S_bM_{Sb})(L_b'l_pM_{Lb}'m_p|L_b'l_pL_bM_{Lb})$$

The δ symbols are used to remove the extra summations (the appropriate prime indices for a and b are identified). Using the orthonormalization conditions of Clebsch–Gordan coefficients we obtain:

$$\delta_{\alpha_a,\alpha_b} = \sum_{i=1}^{p}\sum_{j=1}^{p}\sum_{\alpha'S'L'} \delta_{l_i^0,l_j^0}$$

$$\times\, (l_1^0 \ldots l_{i-1}^0 l_{i+1}^0 \ldots l_p^0(\alpha S'L')l_i^0SL|\}l_1^0 \ldots l_p^0\alpha_aSL)_i$$

$$\times\, (l_1^0 \ldots l_{j-1}^0 l_{j+1}^0 \ldots l_p^0(\alpha S'L')l_j^0SL|\}l_1^0\alpha_bSL)_j$$

$$\times \sum_{s_pM_S'} (S'\tfrac{1}{2}M_S's_p|S'\tfrac{1}{2}SM_S)(S'\tfrac{1}{2}M_S's_p|S'\tfrac{1}{2}SM_S)$$

$$\times \sum_{m_pM_L'} (L'l_pM_L'm_p|L'l_pLM_L)(L'l_pM_L'm_p|L'l_pLM_L)$$

In this way we obtain the following useful condition for the auxiliary fractional parentage coefficients

$$\sum_{i=1}^{p}\sum_{j=1}^{p}\sum_{\alpha'S'L'} (l_1^0 \ldots l_{i-1}^0 l_{i+1}^0 \ldots l_p^0(\alpha'S'L')l_i^0SL|\}l_1^0 \ldots l_p^0\alpha_aSL)_i$$

$$\times\, (l_1^0 \ldots l_{j-1}^0 l_{j+1}^0 \ldots l_p^0(\alpha'S'L')l_j^0SL|\}l_1^0 \ldots l_p^0\alpha_bSL)_j\, \delta_{l_i^0,l_j^0} = \delta_{\alpha_a,\alpha_b} \tag{D5}$$

from which we immediately obtain the relation for true fractional parentage coefficients

$$\sum_{b=a}^{c}\sum_{\alpha'S'L'}(l_a^{n_a}\ldots l_b^{n_b-1}\ldots l_c^{n_c}(\alpha'S'L')l_bSL|\}l_a^{n_a}\ldots l_c^{n_c}\alpha_aSL)$$
$$\times(l_a^{n_a}\ldots l_b^{n_b-1}\ldots l_c^{n_c}(\alpha'S'L')l_bSL|\}l_a^{n_a}\ldots l_c^{n_c}\alpha_bSL)=\delta_{\alpha_a,\alpha_b} \tag{D6}$$

Coupling of Two Electrons

Except for an unimportant normalization factor the correct basis states φ for the case of two electrons can be formed merely by the coupling of second quantization operators according to the rules of coupling angular momenta (the frequency index of terms becomes redundant for the case of two electrons).

$$\tilde{\varphi}_{l_1{}^0l_2{}^0SLM_SM_L}=\sum_{l_1m_1s_1,l_2m_2s_2}\overset{+}{a}_{l_1m_1s_1}\overset{+}{a}_{l_2m_2s_2}\Phi_0$$
$$\times(\tfrac{1}{2}\tfrac{1}{2}s_1s_2|\tfrac{1}{2}\tfrac{1}{2}SM_S)(l_1l_2m_1m_2LM_L)\delta_{l_1,l_1{}^0}\delta_{l_2,l_2{}^0} \tag{D7}$$

But the coefficient function thus obtained

$$\tilde{C}^{l_1{}^0l_2{}^0SLM_SM_L}_{l_1m_1s_1,l_2m_2s_2}=2(\tfrac{1}{2}\tfrac{1}{2}s_1s_2|\tfrac{1}{2}\tfrac{1}{2}SM_S)(l_1l_2m_1m_2|l_1l_2LM_L)\delta_{l_1,l_1{}^0}\delta_{l_2,l_2{}^0} \tag{D8}$$

is neither antisymmetric, nor normalized. It must be antisymmetrized and normalized, so we get

$$C^{l_1{}^0l_2{}^0SLM_SM_L}_{l_1m_1s_1,l_2m_2s_2}=N(l_1^0l_2^0SL)(\tfrac{1}{2}\tfrac{1}{2}s_1s_2|\tfrac{1}{2}\tfrac{1}{2}SM_S)(l_1l_2m_1m_2|l_1l_2LM_L)$$
$$\times(\delta_{l_1,l_1{}^0}\delta_{l_2,l_2{}^0}+(-1)^{S+L-l_1{}^0-l_2{}^0}\delta_{l_1,l_2{}^0}\delta_{l_2,l_1{}^0}) \tag{D9}$$

where $N(l_1^0l_2^0SL)$ is the normalization factor.

Let us evaluate this factor. Using the orthogonality condition (B4a) we find

$$1=\tfrac{1}{2}\sum_{l_1m_1s_1,l_2m_2s_2}(C^{l_1{}^0l_2{}^0SLM_SM_L}_{l_1m_1s_1,l_2m_2s_2})^2=\tfrac{1}{2}N^2(l_1^0l_2^0SL)$$
$$\times\sum_{l_1m_1s_1,l_2m_2s_2}(\tfrac{1}{2}\tfrac{1}{2}s_1s_2|\tfrac{1}{2}\tfrac{1}{2}SM_S)(l_1l_2m_1m_2|l_1l_2LM_L)$$
$$\times(\tfrac{1}{2}\tfrac{1}{2}s_1s_2|\tfrac{1}{2}\tfrac{1}{2}SM_S)(l_1l_2m_1m_2|l_1l_2LM_L)$$
$$\times(\delta_{l_1,l_1{}^0}\delta_{l_2,l_2{}^0}+(-1)^{S+L+l_1{}^0+l_2{}^0}\delta_{l_1,l_2{}^0}\delta_{l_2,l_1{}^0})^2$$
$$=\tfrac{1}{2}N^2(l_1^0l_2^0SL)\sum_{l_1,l_2}(\delta_{l_1,l_1{}^0}\delta_{l_2,l_2{}^0}+\delta_{l_1,l_2{}^0}\delta_{l_2,l_1{}^0}+$$
$$\times 2(-1)^{S+L+l_1{}^0+l_2{}^0}\delta_{l_1,l_1{}^0}\delta_{l_2,l_2{}^0}\delta_{l_1,l_2{}^0}\delta_{l_2,l_1{}^0})$$
$$=N^2(l_1^0l_2^0SL)(1+(-1)^{S+L}\delta_{l_1{}^0,l_2{}^0}) \tag{D10}$$

from which it follows that for two identical electrons $l_1^0 = l_2^0$ only those terms are possible for which $S + L$ equals an even number. Finally, according to (D10) we get for the normalization factor:

$$N^{-2}(l_1^0 l_2^0 SL) = 1 + (-1)^{S+L}\delta_{l_1^0, l_2^0} \tag{D11}$$

Let us now calculate the expressions for the appropriate fractional parentage coefficients evaluating first the auxiliary fractional parentage coefficients. By virtue of (D3) we have

$$\begin{aligned} C^{l_1^0 l_2^0 SLM_SM_L}_{l_1m_1s_1, l_2m_2s_2} = \sqrt{2}\sum_{S'L'}\sum_{M_S'M_L'} &(S'\tfrac{1}{2}M_S's_2|S'\tfrac{1}{2}SM_S)(L'l_2M_L'm_2|L'l_2LM_L) \\ &\times (C^{l_2^0 S'L'M_S'M_L'}_{l_1m_1s_1}\delta_{l_1^0, l_2}(l_2^0(S'L')l_1^0SL|\}l_1^0l_2^0SL)_1 \\ &+ C^{l_1^0 S'L'M_S'M_L'}_{l_1m_1s_1}\delta_{l_2^0, l_2}(l_1^0(S'L')l_2^0SL|\}l_1^0l_2^0SL)_2) \end{aligned} \tag{D12}$$

Using

$$C^{l_1^0 SLM_SM_L}_{l_1m_1s_1} = \delta_{l_1^0, l_1}\delta_{L, l_1}\delta_{S, \frac{1}{2}}\delta_{M_S, s_1}\delta_{M_L, m_1} \tag{D13}$$

we obtain

$$\begin{aligned} C^{l_1^0 l_2^0 SLM_SM_L}_{l_1m_1s_1, l_2m_2s_2} = &\sqrt{2}(\tfrac{1}{2}\tfrac{1}{2}s_1s_2|\tfrac{1}{2}\tfrac{1}{2}SM_S)(l_1l_2m_1m_2|l_1l_2LM) \\ &\times [(l_2^0(S'L')l_1^0SL|\}l_1^0l_2^0SL)_1\delta_{l_1, l_2^0}\delta_{l_2, l_1^0} \\ &+ (l_1^0(S'L')l_2^0SL|\}l_1^0l_2^0SL)_2\delta_{l_1, l_1^0}\delta_{l_2, l_2^0}] \end{aligned} \tag{D12a}$$

Comparing (D12a) with (D9) we obtain the following expressions for auxiliary fractional parentage coefficients.

$$(l_2^0(S'L')l_1^0SL|\}l_1^0l_2^0SL)_1 = \frac{(-1)^{S+L-l_1^0-l_2^0}}{\sqrt{2}} N(l_1^0l_2^0SL)\delta_{S', \frac{1}{2}}\delta_{L', l_2^0}$$

$$(l_1^0(S'L')l_2^0SL|\}l_1^0l_2^0SL)_2 = \frac{1}{\sqrt{2}} N(l_1^0l_2^0SL)\delta_{S', \frac{1}{2}}\delta_{L', l_1^0} \tag{D14}$$

By a straightforward calculation using (D4) we get for the true fractional parentage coefficients:

$$(l_a(S'L')l_bSL|\}l_al_bSL) = \frac{(-1)^{S+L-l_a-l_b}}{\sqrt{2}}\delta_{S', \frac{1}{2}}\delta_{L', l_a} \tag{D15}$$

$$(l_b(S'L')l_aSL|\}l_al_bSL) = \frac{1}{\sqrt{2}}\delta_{S', \frac{1}{2}}\delta_{l_b, L'} \tag{D16}$$

$$(l_a(S'L')l_aSL|\}l_a^2SL) = \delta_{S', \frac{1}{2}}\delta_{L', l_a} \tag{D17}$$

Tables D1 to D3 list some representative values of these fractional parentage coefficients.

TABLE D1. $(s(S'L')pSL|\}spSL)(p(S'L')sSL|\}spSL)$.

sp	N	s ^2S	p ^2P
^1P	$2^{-\frac{1}{2}}$	1	1
^3P	$2^{-\frac{1}{2}}$	1	1

TABLE D2. $(s(S'L')sSL|\}s^2SL)$.

s^2	N	s ^2S
^1S	1	1

TABLE D3. $(p(S'L')pSL|\}p^2SL)$.

p^2	N	p ^2P
^1S	1	1
^3P	1	1
^1D	1	1

Coupling of Three Electrons

The construction of basis states φ for the case of three electrons is similar to the case of two electrons. First we couple the second quantization operators according to the coupling scheme of angular momenta

$$\tilde{\varphi}_{l_1{}^0 l_2{}^0 l_3{}^0 \alpha S L M_S M_L} = \sum_{l_1 m_1 s_1, l_2 m_2 s_2, l_3 m_3 s_3} \overset{+}{a}_{l_1 m_1 s_1} \overset{+}{a}_{l_2 m_2 s_2} \overset{+}{a}_{l_3 m_3 s_3} \Phi_0$$
$$\times (\tfrac{1}{2}\tfrac{1}{2}s_1 s_2 | \tfrac{1}{2}\tfrac{1}{2}S' M_S')(S'\tfrac{1}{2}M_S' s_3 | S'\tfrac{1}{2}S M_S)$$
$$\times (l_1 l_2 m_1 m_2 | l_1 l_2 L' M_L')(L' l_3 M_L' m_3 | L' l_3 L M_L) \delta_{l_1, l_1{}^0} \delta_{l_2, l_2{}^0} \delta_{l_3, l_3{}^0} \tag{D18}$$

The subscript $\alpha \equiv S'L'$. The coefficient C thus obtained is neither antisymmetric, nor normalized; it is given by:

$$\tilde{C}^{l_1{}^0 l_2{}^0 l_3{}^0 \alpha S L M_S M_L}_{l_1 m_1 s_1, l_2 m_2 s_2, l_3 m_3 s_3} = 6 \delta_{l_1, l_1{}^0} \delta_{l_2, l_2{}^0} \delta_{l_3, l_3{}^0}$$
$$\times (\tfrac{1}{2}\tfrac{1}{2}s_1 s_2 | \tfrac{1}{2}\tfrac{1}{2}S' M_S')(S'\tfrac{1}{2}M_S' s_3 | S'\tfrac{1}{2}S M_S)$$
$$\times (l_1 l_2 m_1 m_2 | l_1 l_2 L' M_L')(L' l_3 M_L' m_3 | L' l_3 L M_L) \tag{D19}$$

Antisymmetrizing (D19) we then obtain

$$C^{l_1{}^0 l_2{}^0 l_3{}^0 \alpha S L M_S M_L}_{l_1 m_1 s_1, l_2 m_2 s_2, l_3 m_3 s_3} = N(l_1^0 l_2^0 l_3^0 \alpha S L) \{ \delta_{l_3, l_1{}^0} \delta_{l_1, l_2{}^0} \delta_{l_2, l_3{}^0}$$
$$\times (\tfrac{1}{2}\tfrac{1}{2}s_3 s_1 | \tfrac{1}{2}\tfrac{1}{2}S' M_S')(S'\tfrac{1}{2}M_S' s_2 | S'\tfrac{1}{2}S M_S)$$
$$\times (l_3 l_1 m_3 m_1 | l_3 l_1 L' M_L')(L' l_2 M_L' m_2 | L' l_2 L M_L)$$
$$- \delta_{l_3, l_1{}^0} \delta_{l_2, l_2{}^0} \delta_{l_1, l_3{}^0}$$
$$\times (\tfrac{1}{2}\tfrac{1}{2}s_3 s_2 | \tfrac{1}{2}\tfrac{1}{2}S' M_S')(S'\tfrac{1}{2}M_S' s_1 | S'\tfrac{1}{2}S M_S)$$
$$\times (l_3 l_2 m_3 m_2 | l_3 l_2 L' M_L')(L' l_1 M_L' m_1 | L' l_1 L M_L)$$
$$+ \delta_{l_2, l_1{}^0} \delta_{l_3, l_2{}^0} \delta_{l_1, l_3{}^0}$$
$$\times (\tfrac{1}{2}\tfrac{1}{2}s_2 s_3 | \tfrac{1}{2}\tfrac{1}{2}S' M_S')(S'\tfrac{1}{2}M_S' s_1 | S'\tfrac{1}{2}S M_S)$$
$$\times (l_2 l_3 m_2 m_3 | l_2 l_3 L' M_L')(L' l_1 M_L' m_1 | L' l_1 L M_L)$$
$$- \delta_{l_1, l_1{}^0} \delta_{l_3, l_2{}^0} \delta_{l_2, l_3{}^0}$$
$$\times (\tfrac{1}{2}\tfrac{1}{2}s_1 s_3 | \tfrac{1}{2}\tfrac{1}{2}S' M_S')(S'\tfrac{1}{2}M_S' s_2 | S'\tfrac{1}{2}S M_S)$$
$$\times (l_1 l_3 m_1 m_3 | l_1 l_3 L' M_L')(L' l_2 M_L' m_2 | L' l_2 L M_L)$$
$$+ \delta_{l_1, l_1{}^0} \delta_{l_2, l_2{}^0} \delta_{l_3, l_3{}^0}$$
$$\times (\tfrac{1}{2}\tfrac{1}{2}s_1 s_2 | \tfrac{1}{2}\tfrac{1}{2}S' M_S')(S'\tfrac{1}{2}M_S' s_3 | S'\tfrac{1}{2}S M_S)$$
$$\times (l_1 l_2 m_1 m_2 | l_1 l_2 L' M_L')(L' l_3 M_L' m_3 | L' l_3 L M_L)$$
$$- \delta_{l_2, l_1{}^0} \delta_{l_1, l_2{}^0} \delta_{l_3, l_3{}^0}$$
$$\times (\tfrac{1}{2}\tfrac{1}{2}s_2 s_1 | \tfrac{1}{2}\tfrac{1}{2}S' M_S')(S'\tfrac{1}{2}M_S' s_3 | S'\tfrac{1}{2}S M_S)$$
$$\times (l_2 l_1 m_2 m_1 | l_2 l_1 L' M_L')(L' l_3 M_L' m_3 | L' l_3 L M_L) \}$$

where $N(l_1^0 l_2^0 l_3^0 \alpha S L)$ is a normalization factor. (D20)

Using (B32a), (B32b), and (B32) for the convertion of particular terms in (D20) we obtain

$$
\begin{aligned}
C^{l_1{}^0 l_2{}^0 l_3{}^0 \alpha SLM_S M_L}_{l_1 m_1 s_1, l_2 m_2 s_2, l_3 m_3 s_3} &= N(l_1^0 l_2^0 l_3^0 \alpha SL) \\
&\times \sum_{S''L''} (\tfrac{1}{2}\tfrac{1}{2}s_1 s_2|\tfrac{1}{2}\tfrac{1}{2}S''M_S'')(S''\tfrac{1}{2}M_S'' s_3|S''\tfrac{1}{2}SM_S) \\
&\times (l_1 l_2 m_1 m_2|l_1 l_2 L''M_L'')(L''l_3 M_L'' m_3|L''l_3 LM_L) \\
&\times \{\delta_{l_3,l_1{}^0}\delta_{l_1,l_2{}^0}\delta_{l_2,l_3{}^0} \\
&\times [(2S'+1)(2S''+1)]^{\frac{1}{2}} W(\tfrac{1}{2}\tfrac{1}{2}S\tfrac{1}{2}; S''S')(-1)^{S-\frac{1}{2}-S''} \\
&\times [(2L'+1)(2L''+1)]^{\frac{1}{2}} W(l_3^0 l_2^0 L l_1^0; L''L')(-1)^{L-l_1{}^0-L''} \\
&- \delta_{l_3,l_1{}^0}\delta_{l_2,l_2{}^0}\delta_{l_1,l_3{}^0} \\
&\times [(2S'+1)(2S''+1)]^{\frac{1}{2}} W(\tfrac{1}{2}\tfrac{1}{2}S\tfrac{1}{2}; S''S')(-1)^{\frac{3}{2}-S} \\
&\times [(2L'+1)(2L''+1)]^{\frac{1}{2}} W(l_3^0 l_2^0 L l_1^0; L''L')(-1)^{l_1{}^0+l_2{}^0+l_3{}^0-L} \\
&+ \delta_{l_2,l_1{}^0}\delta_{l_3,l_2{}^0}\delta_{l_1,l_3{}^0} \\
&\times [(2S'+1)(2S''+1)]^{\frac{1}{2}} W(\tfrac{1}{2}\tfrac{1}{2}S\tfrac{1}{2}; S''S')(-1)^{\frac{1}{2}+S'-S} \\
&\times [(2L'+1)(2L''+1)]^{\frac{1}{2}} W(l_3^0 l_1^0 L l_2^0; L''L')(-1)^{l_3{}^0+L'-L} \\
&- \delta_{l_1,l_1{}^0}\delta_{l_3,l_2{}^0}\delta_{l_2,l_3{}^0} \\
&\times [(2S'+1)(2S''+1)]^{\frac{1}{2}} W(\tfrac{1}{2}\tfrac{1}{2}S\tfrac{1}{2}; S''S')(-1)^{S+\frac{1}{2}-S'-S''} \\
&\times [(2L'+1)(2L''+1)]^{\frac{1}{2}} W(l_3^0 l_1^0 L l_2^0; L''L')(-1)^{L+l_1{}^0-L'-L''} \\
&+ \delta_{l_3,l_3{}^0}\delta_{S',S''}\delta_{L',L''} \\
&\times (\delta_{l_1,l_1{}^0}\delta_{l_2,l_2{}^0} + (-1)^{S'+L'-l_1{}^0-l_2{}^0}\delta_{l_2,l_1{}^0}\delta_{l_1,l_2{}^0})\} = N(l_1^0 l_2^0 l_3^0 \alpha SL) \\
&\times \sum_{S''L''} (\tfrac{1}{2}\tfrac{1}{2}s_1 s_2|\tfrac{1}{2}\tfrac{1}{2}S''M_S'')(S''\tfrac{1}{2}M_S'' s_3|S''\tfrac{1}{2}SM_S) \\
&\times (l_1 l_2 m_1 m_2|l_1 l_2 L''M_L'')(L''l_3 M_L'' m_3|L''l_3 LM_L) \\
&\times \{\delta_{l_3,l_1{}^0}(-1)^{S+L-l_1{}^0-l_2{}^0-l_3{}^0-\frac{1}{2}}[(2S'+1)(2S''+1)]^{\frac{1}{2}}[(2L'+1)(2L''+1)]^{\frac{1}{2}} \\
&\times W(\tfrac{1}{2}\tfrac{1}{2}S\tfrac{1}{2}; S''S') W(l_3^0 l_2^0 L l_1^0; L''L') \\
&\times ((-1)^{S''+L''-l_2{}^0-l_3{}^0}\delta_{l_1,l_2{}^0}\delta_{l_2,l_3{}^0} + \delta_{l_2,l_2{}^0}\delta_{l_1,l_3{}^0}) \\
&+ \delta_{l_3,l_2{}^0}(-1)^{S+L-\frac{1}{2}-l_2{}^0-S'-L'}[(2S'+1)(2S''+1)]^{\frac{1}{2}}[(2L'+1)(2L''+1)]^{\frac{1}{2}} \\
&\times W(\tfrac{1}{2}\tfrac{1}{2}S\tfrac{1}{2}; S''S') W(l_3^0 l_1^0 L l_2^0; L''L') \\
&\times (\delta_{l_2,l_1{}^0}\delta_{l_1,l_3{}^0} + (-1)^{S''+L''-l_1{}^0-l_3{}^0}\delta_{l_1,l_1{}^0}\delta_{l_2,l_3{}^0}) \\
&+ \delta_{l_3,l_3{}^0}\delta_{S',S''}\delta_{L',L''} \\
&\times (\delta_{l_1,l_1{}^0}\delta_{l_2,l_2{}^0} + (-1)^{S''+L''-l_1{}^0-l_2{}^0}\delta_{l_2,l_1{}^0}\delta_{l_1,l_2{}^0})\} \qquad \text{(D21)}
\end{aligned}
$$

By virtue of Eq. (D3) for auxiliary fractional parentage coefficients we write

$$
\begin{aligned}
&C^{l_1{}^0 l_2{}^0 l_3{}^0 \alpha SLM_SM_L}_{l_1m_1s_1,l_2m_2s_2,l_3m_3s_3} \\
&\quad = \sqrt{2} \sum_{S''L''} (S''\tfrac{1}{2}M''_S s_3|S''\tfrac{1}{2}SM_S)(L''l_3M''_L m_3|L''l_3LM_L) \\
&\qquad \times \{(l_2^0l_3^0(S''L'')l_1^0SL|\}l_1^0l_2^0l_3^0\alpha SL)_1 C^{l_2{}^0 l_3{}^0 S''L''M_S{}''M_L{}''}_{l_1m_1s_1,l_2m_2s_2}\delta_{l_3,l_1{}^0} \\
&\qquad + (l_1^0l_3^0(S''L'')l_2^0SL|\}l_1^0l_2^0l_3^0\alpha SL)_2 C^{l_1{}^0 l_3{}^0 S''L''M_S{}''M_L{}''}_{l_1m_1s_1,l_2m_2s_2}\delta_{l_3,l_2{}^0} \\
&\qquad + (l_1^0l_2^0(S''L'')l_3^0SL|\}l_1^0l_2^0l_3^0\alpha SL)_3 C^{l_1{}^0 l_2{}^0 S''L''M_S{}''M_L{}''}_{l_1m_1s_1,l_2m_2s_2}\delta_{l_3,l_3{}^0}\}
\end{aligned} \tag{D22}
$$

where $\alpha = l_1^0\, l_2^0 L'S'$.

Comparing Eq. (D21) with Eq. (D22) we get the following formulae for auxiliary fractional parentage coefficients

$$
\begin{aligned}
&(l_2^0l_3^0(S''L'')l_1^0SL|\}l_1^0l_2^0l_3^0SL(l_1^0l_2^0S'L'))_1 \\
&\quad = \frac{N(l_1^0l_2^0l_3^0SL(l_1^0l_2^0S'L'))}{\sqrt{3}N(l_3^0l_2^0S''L'')}(-1)^{S+L-l_1{}^0-S''-L''} \\
&\qquad \times [(2S'+1)(2S''+1)]^{\frac{1}{2}}[(2L'+1)(2L''+1)]^{\frac{1}{2}} \\
&\qquad \times W(\tfrac{1}{2}\tfrac{1}{2}S\tfrac{1}{2};S''S')W(l_3^0l_2^0Ll_1^0;L''L') \\
&\quad = \frac{N(l_1^0l_2^0l_3^0SL(l_1^0l_2^0S'L'))}{\sqrt{3}N(l_3^0l_2^0S''L'')}(-1)^{S''+L''-l_2{}^0-l_3} \\
&\qquad \times [(2S'+1)(2S''+1)]^{\frac{1}{2}}\begin{Bmatrix} \frac{1}{2} & \frac{1}{2} & S'' \\ \frac{1}{2} & S & S' \end{Bmatrix} \\
&\qquad \times [(2L'+1)(2L''+1)]^{\frac{1}{2}}\begin{Bmatrix} l_3^0 & l_2^0 & L'' \\ l_1^0 & L & L' \end{Bmatrix}
\end{aligned} \tag{D23a}
$$

$$
\begin{aligned}
&(l_1^0l_3^0(S''L'')l_2^0SL|\}l_1^0l_2^0l_3^0SL(l_1^0l_2^0S'L'))_2 \\
&\quad = \frac{N(l_1^0l_2^0l_3^0SL(l_1^0l_2^0S'L'))}{\sqrt{3}N(l_3^0l_1^0S''L'')}(-1)^{S+L-\frac{1}{2}+l_1{}^0-S'-L'-S''-L''} \\
&\qquad \times [(2S'+1)(2S''+1)]^{\frac{1}{2}}[(2L'+1)(2L''+1)]^{\frac{1}{2}} \\
&\qquad \times W(\tfrac{1}{2}\tfrac{1}{2}S\tfrac{1}{2};S''S')W(l_3^0l_1^0Ll_2^0;L''L') \\
&\quad = \frac{N(l_1^0l_2^0l_3^0SL(l_1^0l_2^0S'L'))}{\sqrt{3}N(l_3^0l_1^0S''L'')}(-1)^{l_3{}^0-l_2{}^0+S'+L'-S''-L''} \\
&\qquad \times [(2S'+1)(2S''+1)]^{\frac{1}{2}}\begin{Bmatrix} \frac{1}{2} & \frac{1}{2} & S'' \\ \frac{1}{2} & S & S' \end{Bmatrix} \\
&\qquad \times [(2L'+1)(2L''+1)]^{\frac{1}{2}}\begin{Bmatrix} l_3^0 & l_1^0 & L'' \\ l_2^0 & L & L' \end{Bmatrix}
\end{aligned} \tag{D23b}
$$

$$
\begin{aligned}
&(l_1^0 l_2^0 (S''L'') l_3^0 SL | \{ l_1^0 l_2^0 l_3^0 SL (l_1^0 l_2^0 S'L'))_3 \\
&\quad = \frac{N(l_1^0 l_2^0 l_3^0 SL(l_1^0 l_2^0 S'L'))}{\sqrt{3} N(l_1^0 l_2^0 S''L'')} \delta_{S',S''} \delta_{L',L''} \delta_{\frac{1}{2},\frac{1}{2},S'} \delta_{l_1{}^0,l_2{}^0,L,} \delta_{\frac{1}{2}'S',S} \delta_{L',l_3{}^0,L}
\end{aligned} \tag{D23c}
$$

From Eqs. (D23a, b, c) together with Eq. (D4) the formulae for true fractional parentage coefficients are easily derived; in particular, we have

$$
\begin{aligned}
&C^{l_a{}^3 \alpha SLM_S M_L}_{l_1 m_1 s_1, l_2 m_2 s_2, l_3 m_3 s_3} \\
&\quad = \sqrt{3} \sum_{S''L''} \sum_{M_S{}'' M_L{}''} (S'' \tfrac{1}{2} M''_S s_3 | S'' \tfrac{1}{2} S M_S)(L'' l M''_L m_3 | L'' l_a L M_L) \\
&\quad \times (l_a^2 (S''L'') l_a SL |\} l_a^3 \alpha SL) C^{l_a{}^2 S'' L'' M_S{}'' M_L{}''}_{l_1 m_1 s_1, l_2 m_2 s_2} \delta_{l_3, l_a}
\end{aligned} \tag{D24}
$$

where

$$
\begin{aligned}
&(l_a^2 (S''L'') l_a SL |\} l_a^2 SL (l_a^2 S'L')) = \frac{N(l_a^3 SL(l_a^2 S'L'))}{\sqrt{3} N(l_a^2 S''L'')} \\
&\quad \times (\delta_{S'',S'} \delta_{L'',L} \delta_{\frac{1}{2},\frac{1}{2},S'} \delta_{\frac{1}{2},S',S} \delta_{l_a,l_a,L'} \delta_{l_a,L',L}) \\
&\quad + 2[(2S'+1)(2S''+1)]^{\frac{1}{2}} \begin{Bmatrix} \frac{1}{2} & \frac{1}{2} & S'' \\ \frac{1}{2} & S & S' \end{Bmatrix} [(2L'+1)(2L''+1)]^{\frac{1}{2}} \begin{Bmatrix} l_a & l_a & L'' \\ l_a & L & L' \end{Bmatrix} \\
&\quad N^{-2}(l_a^3 SL(l_a^2 S'L')) = 2\delta_{S',S''} \delta_{\frac{1}{2},\frac{1}{2},S'} \delta_{\frac{1}{2},S',S} \delta_{L',L''} \delta_{l_a,l_a,L'} \delta_{l_a,L',L} \\
&\quad + 4(2S'+1)(2L'+1) \begin{Bmatrix} \frac{1}{2} & \frac{1}{2} & S' \\ \frac{1}{2} & \frac{1}{2} & S' \end{Bmatrix} \begin{Bmatrix} l_a & l_a & L' \\ l_a & L & L' \end{Bmatrix} \\
&\quad N^{-2}(l_a^2 S''L'') = 2
\end{aligned} \tag{D25}
$$

Moreover we have

$$
\begin{aligned}
&C^{l_a{}^2 l_b \alpha SLM_S M_L}_{l_1 m_1 s_1, l_2 m_2 s_2, l_3 m_3 s_3} = \sqrt{3} \sum_{S''L''} (S'' \tfrac{1}{2} M''_S s_3 | S'' \tfrac{1}{2} S M_S)(L'' l M''_L m_3 | L'' l L M_L) \\
&\quad \times (l_a^2 (S''L'') l_b SL |\} l_a^2 l_b \alpha SL) C^{l_a{}^2 S'' L'' M_S{}'' M_L{}''}_{l_1 m_1 s_1, l_2 m_2 s_2} \delta_{l_3, l_b} \\
&\quad + \sqrt{3} \sum_{S''L''} (S'' \tfrac{1}{2} M''_S s_3 | S'' \tfrac{1}{2} S M_S)(L'' l M''_L m_3 | L'' l L M_L) \\
&\quad \times (l_a l_b (S''L'') l_a SL |\} l_a^2 l_b \alpha SL) C^{l_a l_b S'' L'' M_S{}'' M_L{}''}_{l_1 m_1 s_1, l_2 m_2 s_2} \delta_{l_3, l_a}
\end{aligned} \tag{D26}
$$

where

$$(l_a^2(S''L'')l_bSL|\}l_a^2l_bSL(l_a^2S'L')) = \frac{N(l_a^2l_bSL(l_a^2S'L'))}{\sqrt{3}N(l_a^2S''L'')}$$

$$\times\ \delta_{S'',S'}\delta_{L'',L'}\delta_{\frac{1}{2},\frac{1}{2},S'}\delta_{S',\frac{1}{2},S}\delta_{l_a,l_a,L'}\delta_{l_b,L',L}$$

$$\times\ (l_al_b(S''L'')l_aSL|\}l_a^2l_bSL(l_a^2S'L')) = \frac{N(l_a^2l_bSL(l_a^2S'L'))}{\sqrt{3}N(l_a^2S''L'')}$$

$$\times\ 2[(2S'+1)(2S''+1)]^{\frac{1}{2}}\begin{Bmatrix}\frac{1}{2} & \frac{1}{2} & S''\\ \frac{1}{2} & S & S'\end{Bmatrix}[(2L+1)(2L''+1)]^{\frac{1}{2}}\begin{Bmatrix}l_b & l_a & L''\\ l_a & L & L'\end{Bmatrix}$$

$$N^{-2}(l_a^2l_bSL(l_a^2S'L')) = 2\delta_{\frac{1}{2},\frac{1}{2},S'}\delta_{\frac{1}{2},S',S}\delta_{l_a,l_a,L}\delta_{l_b,L',L} \tag{D27}$$

$$(l_a^2(S''L'')l_bSL|\}l_a^2l_bSL(l_al_bS'L')) = \frac{N(l_a^2l_bSL(l_al_bS'L'))}{\sqrt{3}N(l_al_bS''L'')}$$

$$\times\ [(2S'+1)(2S''+1)]^{\frac{1}{2}}\begin{Bmatrix}\frac{1}{2} & \frac{1}{2} & S''\\ \frac{1}{2} & S & S'\end{Bmatrix}$$

$$[(2L'+1)(2L''+1)]^{\frac{1}{2}}\begin{Bmatrix}l_a & l_a & L''\\ l_b & L & L'\end{Bmatrix}(-1)^{S'+L'}$$

$$(l_al_b(S''L'')l_aSL|\}l_a^2l_bSL(l_al_bS'L')) = \frac{N(l_a^2l_bSL(l_al_bS'L'))}{\sqrt{3}N(l_al_bS''L'')}$$

$$\times\ (\delta_{S'',S'}\delta_{L'',L'}\delta_{\frac{1}{2},\frac{1}{2},S'}\delta_{\frac{1}{2},S',S}\delta_{l_a,l_b,L'}\delta_{L',l_a,L}$$

$$+\ [(2S'+1)(2S''+1)]^{\frac{1}{2}}\begin{Bmatrix}\frac{1}{2} & \frac{1}{2} & S''\\ \frac{1}{2} & S & S'\end{Bmatrix}[(2L'+1)(2L''+1)]^{\frac{1}{2}}\begin{Bmatrix}l_a & l_b & L''\\ l_b & L & L'\end{Bmatrix}$$

$$N^{-2}(l_a^2l_bSL(l_al_bS'L')) = \delta_{\frac{1}{2},\frac{1}{2},S'}\delta_{S',\frac{1}{2},S}\delta_{l_a,l_b,L'}\delta_{L',l_a,L}$$

$$+\ (2S'+1)\begin{Bmatrix}\frac{1}{2} & \frac{1}{2} & S'\\ \frac{1}{2} & S & S'\end{Bmatrix}(2L'+1)\begin{Bmatrix}l_a & l_b & L'\\ l_b & L & L'\end{Bmatrix}$$

$$N^{-2}(l_al_bS''L'') = 1 \tag{D28}$$

Tables D4a, b, D5a, b, and D6 list some representative fractional parentage coefficients evaluated by these formulae.

TABLE D4a. $(p^2(S'L')sSL|\}sp^2SL(p^2S''L''))$
$(sp(S'L')pSL|\}sp^2SL(p^2S''L''))$.

sp^2	N	p^2			sp	
		^1S	^3P	^1D	^1P	^3P
$^4P(p^2\ {}^3P)$	$3^{-\frac{1}{2}}$	0	1	0	0	$\sqrt{2}$
$^2D(p^2\ {}^1D)$	$6^{-\frac{1}{2}}$	0	0	$\sqrt{2}$	-1	$\sqrt{3}$
$^2P(p^2\ {}^3P)$	$6^{-\frac{1}{2}}$	0	$\sqrt{2}$	0	$-\sqrt{3}$	1
$^2S(p^2\ {}^1S)$	$6^{-\frac{1}{2}}$	$\sqrt{2}$	0	0	-1	$\sqrt{3}$

TABLE D4b. $(p^2(S'L')sSL|\}sp^2SL(spS''L''))$
$(sp(S'L')pSL|\}sp^2SL(spS''L''))$.

sp^2	N	p^2			sp	
		^1S	^3P	^1D	^1P	^3P
$^4P(sp\ {}^3P)$	$3^{-\frac{1}{2}}$	0	1	0	0	$\sqrt{2}$
$^2D(sp\ {}^3P)$	$6^{-\frac{1}{2}}$	0	0	2	1	$\sqrt{3}$
$^2D(sp\ {}^1P)$	$6^{-\frac{1}{2}}$	0	0	2	1	$\sqrt{3}$
$^2P(sp\ {}^3P)$	$6^{-\frac{1}{2}}$	0	$-\sqrt{2}$	0	$-\sqrt{3}$	1
$^2P(sp\ {}^1P)$	$6^{-\frac{1}{2}}$	0	$\sqrt{2}$	0	$\sqrt{3}$	-1
$^2S(sp\ {}^3P)$	$6^{-\frac{1}{2}}$	$\sqrt{2}$	0	0	1	$\sqrt{3}$
$^2S(sp\ {}^1P)$	$6^{-\frac{1}{2}}$	$\sqrt{2}$	0	0	1	$\sqrt{3}$

TABLE D5a. $(s^2(S'L')pSL|\}s^2pSL(s^2S''L''))$
$(sp(S'L')sSL|\}s^2pSL(s^2S''L''))$.

s^2p	N	s^2	sp	
		^1S	^1P	^3P
$^2P(s^2\,{}^1S)$	$6^{-\frac{1}{2}}$	$\sqrt{2}$	-1	$\sqrt{3}$

TABLE D5b. $(s^2(S'L')pSL|\}s^2pSL(spS''L''))$
$(sp(S'L')sSL|\}s^2pSL(spS''L''))$.

s^2p	N	s^2	sp	
		^1S	^1P	^3P
$^2P(sp\ ^3P)$	$6^{-\frac{1}{2}}$	$\sqrt{2}$	1	$\sqrt{3}$
$^2P(sp\ ^1P)$	$6^{-\frac{1}{2}}$	$\sqrt{2}$	1	$\sqrt{3}$

TABLE D6. $(p^2(S'L')pSL|\}p^3SL(p^2S''L''))$.

p^3	N	p^2		
		^1S	^3P	^1D
$^4S(p^2\ ^3P)$	1	0	-1	0
$^2P(p^2\ ^1S)$	$18^{-\frac{1}{2}}$	2	-3	$\sqrt{5}$
$^2P(p^2\ ^3P)$	$18^{-\frac{1}{2}}$	-2	3	$\sqrt{5}$
$^2P(p^2\ ^1D)$	$18^{-\frac{1}{2}}$	-2	3	$\sqrt{5}$
$^2D(p^2\ ^3P)$	$2^{-\frac{1}{2}}$	0	1	-1
$^2D(p^2\ ^1D)$	$2^{-\frac{1}{2}}$	0	-1	1

References

1. Bolotin, A. B., Levinson, I. B., and Tolmachev, V. V., *Lit. Fiz. Sb.*, **4,** 25 (1964).
2. Safronova, U. I., and Tolmachev, V. V., *Lit. Fiz. Sb.*, **4,** 13 (1964).
3. Matulis, A. Yu., Safronova, U. I., and Tolmachev, V. V., *Lit. Fiz. Sb.*, **4,** 331 (1964).
4. Safronova, U. I., and Tolmachev, V. V., *Lit. Fiz. Sb.*, **7,** 53 (1967).
5. Ivanova, A. I., Safronova, U. I., and Tolmachev, V. V., *Lit. Fiz. Sb.*, **7,** 35 (1967).
6. Ivanova, A. N., Ivanov, L. N., Safronova, U. I., and Tolmachev, V. V., *Lit. Fiz. Sb.*, **7,** 285 (1967).
7. Ivanova, A. I., Safronova, U. I., and Tolmachev, V. V., *Lit. Fiz. Sb.*, **7,** 303 (1967).

8. Safronova, U. I., and Tolmachev, V. V. *Theoret. Expl Chem., Kiev,* **3,** 579 (1967).
9. Ivanova, A. N., Ivanov, L. N., Safronova, U. I., and Tolmachev, V. V., *Theoret. Exp. Chem., Kiev,* **3,** 577 (1967).
10. Matulis, A. Yu., Naizles, E. P., and Bandzaytis, A. A., *Lit. Fiz. Sb.,* **5,** 289 (1965).
11. Matulis, A. Yu., and Bandzaytis, A. A., *Lit. Fiz. Sb.,* **5,** 453 (1965).
12. Yutsis, A. P., Levinson, I. B., and Vanagas, V. P., *The Mathematical Apparatus in the Theory of Angular Momentum,* Vilnus, in Russian, 1960.
13. Wigner, E. P., *Group Theory,* Academic Press, New York, 1959.
14. Racah, G., *Phys. Rev.,* **63,** 367 (1943).

AUTHOR INDEX

Entries in square brackets denote reference numbers (if any) cited in the preceding page numbers. Entries in bold type refer to authors of articles.

SUBJECT INDEX FOR VOLUME XIV

CUMULATIVE INDEX TO VOLUMES I-XIV

Authors of Articles

Numerals in bold type are volume numbers.

Titles of Articles